Statistical Models for the Social and Behavioral Sciences

Statistical Models for the Social and Behavioral Sciences

JAMES H. DWYER
State University of New York at Stony Brook

New York Oxford
OXFORD UNIVERSITY PRESS
1983

Library of Congress Cataloging in Publication Data

Dwyer, James H., 1946–
Statistical models for the social and behavioral sciences.

Bibliography: p.
Includes index.
1. Social sciences—Mathematical models.
2. Social sciences—Statistical methods. I. Title.
H61.25.D85 300′.724 81-22341
ISBN 0-19-503145-8 AACR2

Printing (last digit): 9 8 7 6 5 4 3 2 1

Printed in the United States of America

For Kathleen, Alicia, Jesse, Michael,
and my parents

Preface

Structural equation models are the subject of this book, and they are an elaboration of two fundamental insights. The first of these insights is usually associated with Charles Spearman: that sets of linear equations can be used to construct testable models of stochastic measurement errors. The second is associated with Sewall Wright: that sets of linear equations can be used to model multivariate stochastic causal processes. With only a modicum of exaggeration, in fact, much of the work in psychometrics, econometrics, and biometry in the last 50 years may be characterized as elaborations of those two insights. And, if we carry the pursuit for predecessors even further, the logical foundation on which these insights were built may be found in the seminal writings of Josiah Willard Gibbs (1839–1903). In his 1901 book, entitled *Elementary principles in statistical mechanics,* we find the *randomness* interpretation of *entropy*. This idea not only signals the transition from deterministic to statistical theorizing in physics and chemistry, it also paves the way for statistical modeling in the social and behavioral sciences. Some remnants of Gibbs' terminology may be found in the final chapter of this book.

Two recent developments have brought work based on these original insights to a kind of plateau that calls for some consolidation and the reorganization of graduate training in the social and behavioral sciences. The first of these developments was the careful work of methodologists concerning rules of causal attribution and criteria for the validity of indirect measurement. The writings of Donald Campbell and his coauthers are probably the most notable in this regard. The second development was Karl Joreskog's creative combination of measurement and causal models into the LISREL model and computer program. He and his colleagues solved the estimation problems presented by a very general linear model, and they provided many of the available guideposts for potential applications of the model. This general linear model then serves as a basis for mathematical expression of the ideas developed under the rubric of "methodology." For example, the multitrait–multimethod approach to measurement can be translated into a mathematical model that makes very specific predictions about the intercorrelation of multiple indicators. A set of assumptions necessary to estimate a model parameter and interpret it as a causal effect can be unambiguously specified. Threats to the internal validity of a particular research design can be depicted in terms of equation models and path diagrams. Thus, structural equation models serve a dual purpose: the medium for precise statement of a hypothesis or theoretical proposal and the medium of methodological discourse.

In addition to these more technical bases for the approach to a quantitative human science presented here, the approach has also been shaped by *meta*methodological debates within my own specialty of social psychology. Since social psychology is situated at the crossroads of the more social and the more behavioral/individualistic disciplines in the human sciences, it suffers from multiple identity crises. These identity crises revolve around concerns about how to do a science of ourselves: What are the essentials of the scientific method and can they be applied to the study of human action in a rapidly developing social system that may be primarily extrabiological in dynamics? Can we circumvent our personal biases and pursue a science of individual and collective human history with the same methods used to build a science of particle, chemical, or biological evolution? The response of many in social psychology to these concerns was to seek security in the internal validity of the randomized experiment. However, restricting the scientific study of social psychological processes to experiments necessarily restricts our quest for determinants of human action to those factors we, as scientists, are currently able to manipulate. Furthermore, the internal validity of randomized field experiments is often less than expected: uncertainty about unmeasured "third variables" in nonrandomized studies is exchanged for uncertainty about just what was manipulated in a field experiment (the problem of irrelevant mediation).

The resolution, it seems to me, of this apparent conflict between internal validity and the capacity to study what is considered important by an individual scientist or a society is rather straightforward: develop the best possible bases for causal inference from the entire range of available research designs. Then these methods can be combined with randomized experiments in a total research program. This book is intended as a step in the direction of a unified quantitative methodology that is applicable to randomized experiments, quasi-experiments, and passive observational designs.

To some extent, then, this book is an attempt to reorient my own discipline's view of "statistics"—to encourage a progression from rough estimation of associations to the use of linear equations as multivariate causal models. The short-term goal of this reorientation is to make explicit (mathematize) what otherwise remains implicit—or, perhaps, obscure—in the research work of psychologists and sociologists. The longer-term goal is proliferation of a statistical language that facilitates a shift away from bivariate "cause and effect" theorizing and toward models of human action that incorporate temporally accumulating and reciprocal causality.

But the approach to statistical modeling presented here is also intended as a bridge between the various divisions of the human sciences: history, economics, medicine, psychology, sociology, etc. Artificial barriers, in the form of statistical conventions, restrict communication among these branches of what may become a unified science of social history. At present, psychologists speak the language of analysis of variance and factor analysis; economists the language of structural equations; sociologists the language of log linear models; epidemiologists the language of logistic regression; and geneticists the language of path analysis. In each instance, however, some form of a linear equation with one or more of Gibbs' "randomness" terms is being used to model a probabilistic causal system; and the distinctions among these various applications may be less important than their commonalities. Thus, an attempt is made here to focus on the logical and mathematical commonalities of all these statistical procedures while emphasizing more important distinctions: (1) the differential capacity of various research designs to provide evidence of a causal effect, and (2) the difference between the use of a linear equation as a causal model (structural equation model) and the use of a linear equation as a representation of associations (regression equation).

The crossing of disciplinary lines is not always received with applause. However, the trail has already been blazed by such scholars as Arthur Goldberger and Otis Dudley Duncan. The first is an eminent econometrician who has devoted many years and many pages to bridging the gaps between his own field and others. The latter is the dean of quantitative sociology who has imported from genetics and econometrics and then re-exported to psychology and other fields. The approach to equation modeling presented in the following pages is addressed primarily to graduate students and researchers in psychology, but it draws heavily from, and is I hope an extension of, these previous efforts to cross disciplinary boundaries and to develop a universal language of quantitative social science.

The synthesis of work in various fields that is presented here has several elements: (1) The use of stochastic linear equations as the general and unifying tool of model specification and model testing. (2) Emphasis upon the logic of causal inference that flows from the randomized experiment and the application of this logic to interpretation of data generated by other designs. (3) An emphasis upon the serious distortions of associations that can arise from measurement error and upon the use of multitrait–multimethod measurement models to circumvent the measurement error problem. (4) The use of Joreskog's LISREL model and computer program for the estimation and testing of causal models that incorporate measurement models. (5) An emphasis upon the use of longitudinal designs and the estimation of reciprocal effects when randomization is not feasible. These elements constitute the plateau in mathematics for the social sciences that I mentioned above. And it is my judgment that this plateau provides social scientists, perhaps for the first time, with quantitative tools that approach adequacy for the task of investigating complex social psychological systems. These systems may involve numerous reciprocally related variables, variables that cannot be measured precisely or even directly, and variables that are extremely difficult to manipulate randomly.

Chapters 1 through 7 recast the material covered in basic statistics courses in the language and logic of structural equation models. These chapters have been used at Stony Brook as the text for a semester course required of all doctoral students in psychology. The chapter on analysis of variance (Chapter 7), for example, demonstrates how the usual F ratios for main effects and interactions may be derived from the fit of various linear structural models. But the chapter recommends a focus upon estimation of structural coefficients and their standard errors rather than upon isolated significance tests. Chapters 5 and 6 develop the simple linear regression model and contrast it with a causal model.

Chapters 8 through 11 have been used in a course for advanced graduate students from psychology, sociology, political science, and urban policy. Chapter 8 generalizes regression models to the multivariate case, and Chapter 9 develops linear equation models of flawed measurement (multiple indicator models). These two components are then combined and extended to the general LISREL model in Chapter 10. Finally,

Chapter 11 demonstrates the application of structural equation modeling to causal inference from longitudinal designs. I would suggest to instructors planning to use the text in a course that one of two prerequisites is appropriate: (1) a good background in undergraduate statistics for the social and behavioral sciences (probability distributions, sampling distributions, analysis of variance, and regression) or (2) mathematics training up to calculus for those who have no background in statistics. The book is intended for university graduate students in quantitatively oriented research programs. But it may also be useful in seminars for advanced undergraduates at colleges where a number of students receive their quantitative training in their earlier years.

An elementary review of matrix algebra is presented in Appendix A. Appendix B outlines the use of the logit transformation and specification of a multiplicative model for regression with a dichotomous dependent variable. Although the logic of these models is quite contiguous with that of the linear models in the previous chapters, two unresolved problems led to the placement of this material in an appendix. The first problem is the modeling of measurement errors, the second the absence of a stochastic term for individual cases. The two problems are linked, and the general use of these equations as causal models awaits a practical solution.

For the specialist in structural equation models, Chapter 11 will be of most interest. In many ways, this chapter is the *raison d'être* of the book. Some new models for the analysis of nonequivalent control group designs are proposed, and a general approach to causal inference from longitudinal passive observation is introduced. Many social scientists, and social psychologists, in particular, were simultaneously attracted to and suspicious of causal statements derived from nonrandomized designs. The attraction stemmed from the frequent impracticality of randomization and the intuition that observation of change was somehow superior to inferring change from cross-sectional differences between cases. The suspicion stemmed from the nagging feeling that only randomization could sort out the problem of unmeasured third variables and reverse causation. My conclusions in Chapter 11 support the validity of both the suspicion and the attraction. As is argued, in some instances much can be gleaned from longitudinal data concerning causality that is not available from cross-sectional data. In particular, reciprocal effects can be estimated and *restricted* null hypotheses about unmeasured variables can be tested. However, these procedures always involve assumptions that are much less plausible than the assumption that a coin toss produces equivalent groups. Ultimately, there is no rival to randomization when causal inference is concerned.

But, from a more practical perspective, most of us are not as concerned with ultimate knowledge as we are with the best knowledge we can obtain in a particular substantive area of research. The absence of human experiments on the determinants of psychological distress and physical disease, for example, will not deter us from investigating possible causal relations. The use of the scientific method where it is most easily applied is certainly worthwhile; but, in Einstein's words, scientific work aimed at "the great unsolved problems of the organization of labour and the distribution of goods" is, for many of us, a higher priority.

Stony Brook, New York — J. H. D.
October 1982

Acknowledgments

I have spent many years developing and writing this book. Almost necessarily, many people have helped me in many different ways. One of my greatest debts is to Professor Barry McLaughlin. Through our early collaboration and his unwavering intellectual and emotional support, this work was given birth.

Most recently, David Kenny and Rafael Ramirez provided crucial criticisms of Chapter 11, and Tom Phelan provided invaluable assistance by reviewing the entire manuscript. In between these contributions, Donald Campbell wrote an encouraging letter that came at "just the right time." Others who have helped include Dane Archer, Richard Balkin, Dana Bramel, David Cross, Ron Friend, William Ibarra, Fred Levine, Samuel Modoono, Wilem Saris, William A. Scott, and numerous other colleagues, students, and anonymous reviewers. I must also acknowledge an unexpected debt to an exceptional group at Oxford. Marcus Boggs has mastered the editor's art of simultaneously serving as an author's friend and superego; and Carol Miller, Victoria Wong, and Ellen Leventhal were a pleasure to work with and most competent in their contributions to the production of the book. Any errors that remain, of course, are my responsibility.

Finally, I must express my debt to Kathleen, Alicia, and Jesse. Although it is doubtful that A and J hastened the completion of this work, it is certainly true that they helped make it worth doing. And to K, I only hope that I can repay you in kind for the seemingly endless hours of editing and general support that has made it all possible.

Contents

List of Symbols

$E(Y) = \text{Exp}(Y)$ = expected value of Y with random sampling from a population (or a sample treated as a population)

$E(Y \mid X)$ = expected value of Y given that X is held constant

$E(Y \mid X = a)$ = expected value of Y given that X is equal to a

if b^* is a population parameter, then b = sample estimate of b^*

if c is a population parameter, then $\text{est}(c) = \hat{c}$ = sample estimate of c

$\text{Var}(X) = \text{Cov}(X,X) = \sigma_{XX} = E\{[X - E(X)]^2\}$ = population variance of X

$S(X,X) = S_{XX}$ = sample variance of X [and maximum likelihood estimate of $\text{Var}(X)$]

$N\,S(X,X)/(N - 1) = \hat{\sigma}_{XX}$ = unbiased estimate of $\text{Var}(X)$ from a random sample of N cases

$\text{Cov}(Y,X) = \sigma_{YX} = E\{[Y - E(Y)][X - E(X)]\}$ = population covariance of Y and X

$S(Y,X) = S_{YX}$ = sample covariance of Y and X [and maximum likelihood estimate of $\text{Cov}(Y,X)$]

$N\,S(Y,X)/(N - 1) = \hat{\sigma}_{YX}$ = unbiased estimate of $\text{Cov}(Y,X)$ from a random sample of N cases

$\text{SD}(X) = \sigma_X = [\text{Var}(X)]^{.5}$ = population standard deviation of X

$S(X) = S_X = S(X,X)^{.5}$ = sample standard deviation of X

$S(X)\{[N/(N - 1)]^{.5}\}$ = estimate of $\text{SD}(X)$ from a random sample of N cases

$\text{Var}(Y \mid X)$ = conditional population variance; variance of Y given that X is held constant

$\text{Var}(Y \mid X = a)$ = conditional population variance; variance of Y given that X is equal to a

if b = sample estimate of b^*, then $\text{SE}(b) = S_b$ = estimated standard error of sample estimate of b^*

Corr $(Y,X) = \rho_{YX}$ = population correlation of Y and X

$r(Y,X) = r_{YX} = S(Y,X)/[S(X)\ S(Y)]$ = sample correlation of Y and X

$r(Y,X \mid Z = a) = r_{YX\mid Z=a}$ = sample conditional correlation between Y and X given that $Z = a$

$c(Y,X) = c_{YX} = S(Y,X)/S(X,X)$ = OLS slope of Y regressed on X

$R(t) = R^2_{Y.X_1,X_2,\ldots,X_p}$ = sample R-square between Y and all variables X_i ($i = 1, \ldots, p$)

$R(r)$ = sample R-square between Y and X_i given some linear constraints on the linear combination of X_i

$\underset{(n \times p)}{\underline{Y}}$ = a matrix with n rows and p columns

$\det(\underline{Y})$ = determinant of the matrix $\underline{Y}$

$\underline{Y}'$ = transpose of the matrix $\underline{Y}$

$\log(Y)$ = natural (Naperian) logarithm of Y

Structural equation model:

$\underline{B}\,\underline{E} = \underline{G}\,\underline{F} + \underline{z}$ (causal model)

$\underline{Y} = \underline{Ly}\,\underline{E} + \underline{e}$
$\underline{X} = \underline{Lx}\,\underline{F} + \underline{d}$ (measurement model)

where

$\underline{E}$ = eta = vector of endogenous latent variables

$\underline{F}$ = xi = vector of exogenous latent variables

$\underline{z}$ = zeta = vector of structural disturbances

$\underline{B}$ = beta = matrix of structural coefficients (paths) defining causal relations among $\underline{E}$

$\underline{G}$ = gamma = matrix of structural coefficients (paths) defining causal effects from $\underline{F}$ to $\underline{E}$

$\underline{Y}$ = vector of observed indicators of $\underline{E}$

$\underline{e}$ = epsilon = vector of measurement errors in $\underline{Y}$

$\underline{Ly}$ = lambda Y = matrix of factor loadings defining measurement relations between observed indicators $\underline{Y}$ and latent variables $\underline{E}$

$\underline{X}$ = vector of observed indicators of $\underline{F}$

$\underline{d}$ = delta = vector of measurement errors in $\underline{X}$

$\underline{Lx}$ = lambda X = matrix of factor loadings defining measurement relations between observed indicators $\underline{X}$ and latent variables $\underline{F}$

Statistical Models for the Social and Behavioral Sciences

Concern for man himself and his fate must always form the chief interest for all technical endeavours, concern for the great unsolved problems of the organization of labour and the distribution of goods—in order that the creations of our mind shall be a blessing and not a curse for mankind. Never forget this in the midst of your diagrams and equations.

Albert Einstein, in a speech delivered at the California Institute of Technology, 1937

1

Introduction

1.1 ABOUT THIS BOOK

The approach to data analysis that is presented in this book has been derived, of course, from my practice as a researcher and teacher in the social sciences. Several comments concerning my approach to social science may be helpful to the potential reader—aiding you in your decision about whether or not to invest the considerable amount of time that will be necessary to develop a working knowledge of linear models.

My sympathies and commitments are almost antique, finding more comfort in the context of the nineteenth century than in the fragmented trivialities of value-free professionalism. St. Simon's vision of social science for political decision-making and Marx's and Weber's promise of a science that unearths dynamics of historical transformation speak to my hopes for my work with others. My motivation for writing this book stems from my desire to understand how and why we do what we do—as individuals and as social groups—so that we can democratically control our own lives. It is my sense that in some small way the unification of statistical methodologies will combat some of the pressures toward fragmentation that divide the human sciences. This unification comes from a focus on linear models and the rules of causal inference that have become the defining characteristic of an ideal science. More importantly, however, it is my intuition that greater use of historical designs—time series and longitudinal or panel studies—provide social scientists with one of the best opportunities to understand social and individual functioning via models of historical change. In fact, most of the more difficult mathematical developments presented in the following chapters are of the greatest use to those seeking to untangle the complex processes of individual or group development.

The fallout of the above orientation is the following list of emphases that characterize this book:

1. An emphasis on the heuristic meaning of the parameters estimated in multivariate models. This emphasis can be contrasted with a focus on the derivation of estima-

tion procedures. The latter is fascinating and often represents a *tour de force* of human reasoning. However, like the working of a modern computer, a good understanding of the output is possible without detailed knowledge of machine language or silicon electronics. Furthermore, I have found that few users of multivariate techniques, whether mathematically sophisticated or not, actually understand the relationship between theoretically relevant statements about their data and the output of multivariate computer programs. Exploratory factor analysis is probably the most misused and most misunderstood, but even regression coefficients are misinterpreted with regularity. The maxim urged upon readers of this text is that prior to the use of any multivariate technique the user should concoct data that reflect the kind of relationship that she hopes to find in her real data. She should then analyze the concocted data to see if the relationships that were embedded in the data can be rediscovered with the multivariate technique.

2. The logical and mathematical treatment of measurement is integrated into the entire development of statistical model testing. When only bivariate relationships are of interest, and the primary goal is to determine whether or not a direct or inverse association can be inferred from the data, then measurement considerations can be limited to "eyeballing" a multimethod–multitrait correlation matrix. However, virtually all *multi*variate techniques make explicit use of the magnitude of covariances or correlations. Thus, the imprecise attitude toward measurement that was not disastrous in the bivariate era has produced great mounds of confusion—in the form of computer printouts and research reports—in the multivariate era. It is my intention that readers of this book will not add to this confusion because they will have the knowledge necessary to test measurement models and estimate the relationships among errorless hypothetical constructs.
3. An effort is made to present almost all statistical methods so that computations can be performed with one computer program—or, at most, two. A general regression program will perform most computations, and programs analogous to Jöreskog's Linear Structural Relations program (LISREL) will perform the most complex procedures.
4. Non-randomized designs are given special emphasis, since most politically important social science questions cannot be addressed practically via randomized designs. Furthermore, an emphasis on experimentation in human sciences has accompanied an ahistorical emphasis that is in need of redress.
5. The importance of significance tests for hypothesis testing is downplayed. The main function of significance tests has been to make trivial relationships appear important. The view advocated here is that standard errors are rough *under*estimates of what to expect from a new sample. And significance tests are very poor substitutes for replication.
6. As often as possible, path diagrams are used to portray the statistical procedure being described. I have found that once these diagrams are understood, the learning of new topics is greatly enhanced.

1.2 CENTRAL RESEARCH EXAMPLE

I find it useful to use the same research example throughout the development of linear models. The advantages of this strategy should become apparent, but there are some *dis*advantages. The first of these disadvantages becomes apparent early on: when intuitively continuous variables are treated as dichotomies. However, the continuity that the same example achieves is probably worth the problems.

The data presented for examples have been concocted with the inspiration of Sales. Before his untimely death, Sales was investigating the relationship between a hypothetical personality dimension (known to social scientists as "authoritarianism") and characteristics of social contexts labeled "threat." Sales proposed that various kinds of social threat, especially economically oppressive conditions, tended to shape people's personalities such that they became more authoritarian. The opposite pole of this personality dimension (or cluster of dimensions) is termed "democratic." The theoretical and methodological issues raised by such an hypothesis are numerous (cf. Christie and Jahoda, 1954), but they are not unrepresentative of the problems faced in most social science projects that seek to measure and explore the interrelations of elusive human realities.

1.3 SCIENCE AND MODELS

The modeling approach presented in the following sections—especially the inclusion of measurement hypotheses—should help the developing researcher to understand the nature of constructs such as authoritarianism and threat in the practice of science. Such questions as "Do authoritarian personalities really exist?" or "Is threat an observable entity?" should recede from your thinking as you learn to treat hypothetical constructs as ways of talking about the world and generating predictions about the interrelations of publicly reproducible observations. Even the constructs used by physicists, such as gravity or electrons, are simply metaphors about which the question of existence is irrelevant. Whether or not electrons exist is a philosophical or theological question, what is of interest to the scientist is whether or not theorizing in terms of electrons gives organization to otherwise disparate observations and gives rise to publicly replicable observational predictions.

There are no rules, other than the requirement of logic, governing the generation of basic constructs in scientific models. Rather, there are rules governing how one goes about testing a model once it has been proposed. At an abstract level, there are two sets of rules governing model testing: those pertinent to construct validity or measurement (can public procedures for observing more or less authoritarianism be specified such that consistent rankings result?); and those pertinent to causal relations between constructs (does a change in threat cause a change in authoritarianism?). Most scientific models consist of these two kinds of statements—measurement and causation, and we will develop the statistical methods for testing the extent to which objective observations of the world are consistent with a given model.

1.4 MODEL TESTING AND IDEOLOGY

It is often argued that ideological biases are embedded in quantitative social science methodology. Furthermore, it is quite obvious that the compulsion for experimentation in American social psychology has served the ideological function of diverting attention away from nonmanipulatable independent variables such as race, class, family structure, and so on. But the appeal of the argument that an ideological bias is actually embedded in the logic of the *method itself* has faded for me. It has faded because the diversity of political ends to which linear models have been devoted is too great. Ideologues certainly may foster particular methodological/statistical fads that suit their purposes in a particular historical context; but when that context changes, the method may take on a new meaning. For example, in one context social experimentation may mean the pursuit of more effective means of control and oppression; in another context emphasis on experimentation may be part of a genuine search for greater justice and freedom. But the logic of causal inference based on randomization remains the same in either context.

Similarly, the models and model testing procedures presented here do not favor individuals over aggregates (although they are most often applied to data from individuals), linear relationships over nonlinear, unidirectional over reciprocal models, ahistorical over historical, etc. In the examples concocted, in fact, I have attempted to demonstrate the versatility of these models.

It seems to me that the model testing techniques presented in this text are simply formalized procedures for responding to the Missouri dictum of "show me." That is, the essence of an ideal scientific methodology is a set of rules specifying how one may demonstrate the consistency of a hypothesis with observations that are public. The logic of science includes a democratization of truth testing. A scientific statement is *supposed* to gain acceptance only if all interested parties can make observations of the world that support the statement. This means that the ability to make these observations must not depend upon social position (authority), extraordinary states of consciousness (mysticism), or unusual powers of reason (abstract philosophy); rather, to paraphrase Walter Wallace, scientific evidence must be observable by the body politic.

The linear model, therefore, is a way of organizing observations: the root of the linear model is the two-dimensional graph. In a very real sense then, linear models summarize patterns of points that could be displayed on such graphs. (This is an important point to keep in mind, since the newcomer to these models is tempted to believe that something non-concrete is reflected by the coefficients in these models). These graphs describe or portray both the distribution of each variable and, most importantly, the *relationship* between two or more variables. Portrayal of such relationships with linear equations (which are not limited to linear relationships) is the essence of this book, and the portrayal of relationships does not *necessarily* serve any ideological function.

1.5 SCIENCE AND IDEOLOGY

Discussion of the "logic of science" focuses attention on an ideal type or prototype. The real world activities that make up the social project known as Science, however, may deviate from that ideal in an irregular or a regular manner. When the work of scientists is primarily ideological, we would expect patterned (regular) deviations from scientific canons. History is replete with good examples of such deviations: Stalin's influence on evolutionary biology in the Soviet Union, Cyril Burt's fraudulent promotion of a genetic theory of intelligence in capitalist societies, and so on. Instances such as these often lead people to the belief that the democratic ideals of science serve as little more than a cover up for the political purposes of practicing scientists or those who fund scientific research.

It is undeniable that much of what we normally call science constitutes a systematic deviation from an ideal science. Most scientists do what they are funded to do, and funding agencies are dominated by the elite in any society. This means that scientific achievements tend to serve the purposes of those elites. Does it then follow that the rules of ideal science are merely a cover up for politically motivated research? I think not, for several reasons. Firstly, when science works for those elites (e.g., in producing an atomic bomb or reliable survey sampling techniques), it does not succeed because political pressure is the crucial ingredient. Political pressure on theologians or philosophers would not have produced the bomb—even though such groups may have wanted to do so. Although some may argue that this particular inability is to their credit, it seems clear that the method of science provides knowledge about the world that is unlikely to be produced by other means. Whether or not such knowledge is good or justly applied is another matter. Secondly, the canons of ideal science are more than a cover up because their application can also lead to findings that are not in the interest of dominant groups in a society. For example, the heliocentric model of the solar system did not please the clergy-aristocracy that was dominant during the life of Copernicus (16th century). Rather, destruction of the geocentric model was most congenial with the anticlerical goals of the French Revolution (18th century). Thirdly, radical critiques of quantitative methodology either offer no alternative or retreat from materialism and resort to some form of rationalism that loses touch with the real world. In my view, the emphasis in science upon observational testing of hypotheses is consistent with a materialist philosophical position. The ultimate test of an idea is to be found in the material world. Without this corrective, social theory becomes idealistic, and the ultimate test of an idea then becomes the relation between that idea and other ideas.

1.6 IDEAS AND THE WORLD

At this point, the root assumption of scientific logic becomes apparent. That assumption is the distinction between ideas and the world. Those who argue that human beings construct the "outside" world—as in a dream—do not make this distinction. For those idealists, it makes little sense to speak of observations (or facts) as independent tests of hypotheses, since both are ideas constructed either individually or collectively. The alternative assumption implicit in most scientific methodology is that people can overcome their subjectivity to some extent and systematically observe the material world in a way that is partially independent of their beliefs. This is achieved collectively; that is, the findings of an investigator and her coworkers are treated skeptically. Only upon replication by peers should a set of observations become "solid" evidence. (It is interesting to note here that Sir Cyril Burt's apparently fictitious data on the IQ of monozygotic twins reared in disparate environments *should* not have been influential among psychologists, since they were not replicable by ideological opponents—since Burt's sample was the only sizable sample of such twins. Nevertheless, for political and other reasons, most psychologists—especially in Britain and the Commonwealth—accepted Burt's "observations" as definitive. Only the application of strict science by Leon Kamin was successful, after Burt's death, in dislodging Burt's fabrications from the annals of "solid" findings.)

Although it cannot be proven in any logical sense that the world intrudes upon us to correct our misconceptions, the assumption of this possibility is the essential assumption of scientific method. Furthermore, most of us operate as if this principle applies in our daily lives. My own experience has convinced me that my observations of the world can change my mind about important issues.

1.7 SCIENCE AND POLITICS

If the scientific method enables a collective to transcend individual subjectivity, can applied social science resolve political questions? For example, can social science tell us whether or not to enact full employment legislation? In my view, social science cannot tell us whether or not to enact such legislation. Rather, scientific method could be applied to estimate the effects of such legislation before enactment and to monitor the actual effects in the event of enactment. (M. Harvey Brenner, in a report to the Joint Economic Committee, estimated that the increase in unemployment in the United States in 1970 caused as many American deaths—due to stress-related disease, crime, and so on—as the Vietnam War.) Then values must be applied to those effects and a decision made.

The extent to which such a process can actually operate in the United States is an empirical question. It is not clear, for example, that employers who believe that some unemployment serves to discipline the labor force could ever be convinced by social scientific evidence that full employment would, on balance, be an overall good for the society. Furthermore, these employers may have sufficient power in Washington to block such legislation—even in the face of compelling evidence concerning its positive effects. Only through such St. Simonian attempts to influence legislation, however, will social scientists learn whether or not their efforts yield positive results. The methods in this book are presented with the express purpose of maximizing the policy relevance of social science research. The focus upon non-randomized designs, multivariate causal modeling, and a multiple indicator measurement strategy all serve this purpose.

1.8 THE CONCEPT OF CAUSE

Scientists and non-scientists use the word cause to refer to a cluster of ideas. Since the notion of a causal relationship is central to this book, it is important that the reader is informed about the meaning of cause that is used.

Cook and Campbell's important book, *Quasi-Experimentation,* includes a chapter on the concept of cause. Although this chapter is presented as only a brief introduction to a burgeoning literature, it provides important background for any reader of this book. The goal of this philosophy of science is the production of general statements or models concerning the development and practice of science as a social project. A crucial element in any philosphy of science is an explication of both abstract (or ideal) rules for causal inference and the implicit rules governing causal attribution in practice. Cook and Campbell present their set of statements concerning causality under the rubric of an *evolutionary critical realist perspective.*

Cook and Campbell use this label to indicate their position on several basic issues in the philosophy of science concerned with causality. These issues have been raised by arguments between various factions in philosophy: positivism from Hume to Russell, essentialism, linguistic philosophy from Wittgenstein, falsification from Popper, and postpositivism from Thomas Kuhn. The positivists argued that causality could not be observed and therefore that it was a superfluous construct indistinguishable from concomitance or correlation. Bertrand Russell, for example, argued that physicists and astronomers deal with functional relationships in which near perfect correlations, rather than causal relations, are sought. Cook and Campbell reject this position in favor of a realist perspective in which the existence of causal relations "in the world" is assumed. But they qualify this assumption with the *critical*-realist corollary that human perceptions of causality are always flawed and, therefore, "constructed by the mind." The practical implication of this critical realism is that researchers must develop methods that sort out causal relations from spurious correlations if they seek to change people or their environments—or both.

In keeping with this critical-realist perspective on causation, Cook and Campbell propose that the human penchant to develop causal assertions probably stems from evolutionarily adaptive mind-brain structures. In other words, the ability to develop valid causal hypotheses was naturally selected to produce a species that adapts via environmental manipulation. The essence of causality, then, is the observation that a change in the presumed cause is (usually) followed by a change in the presumed effect. This definition of causation matches with our everyday usage of the word and follows from the method of linguistic philosophy. This match enhances the plausibility of the evolutionary argument.

This hypothesis of a connection between causal inference and adaptive mind-brain mechanisms calls attention to some intriguing parallels between visual perception and the rules of causal inference—connections "pointed out" to me by my colleague Alan Gilchrist. It appears, for example, that the visual system partitions the motion of two dots of light into common (shared) motion and unique motion—in the same way that a correlation coefficient partitions the variance (analogous to motion) of two variables into shared and unique components. Furthermore, the visual system can be tricked into making invalid causal inferences (illusions) through the presentation of concomitancies that usually allow valid inferences. While positivists would argue that these parallels between our methods of causal inference and the workings of the visual system are simply imposed rather than suggestive, it seems plausible to me that the parallels may suggest a more substantive connection between rules of "higher level" thought and cognitive structures of visual perception. Future research in that area will, of course, be necessary to resolve these contradictory speculations. If my intuitions are correct, then further developments in models of visual perception may even inform us in our attempts to develop formal methodologies for causal inference.

The evolutionary critical-realist perspective on causation, then, is rooted in a very materialist world view, a world view in which our ideas (causal assertions) are shaped by a real world. Those ideas are not perfect reflections of the world, but they are not independent of it either. There is room for false consciousness, but there is also enough connection with the world so that the world can intrude to impress us with the invalidity of incorrect causal models. In the following paragraphs I will discuss some of the implications of this perspective.

The first implication of the Cook and Campbell perspective is a rejection of the postpositivist argument that competing theories cannot be subjected to "critical tests."

The postpositivists have criticized Popper's method of falsification with the assertion that so-called "facts" are actually "theory-laden" constructs. This means that a newly accepted theory will derive its support from observations that were irrelevant to the previously established theory. From the postpositivist perspective, then, Popper was deceived into the belief that science proceeds through attempts to falsify competing theories—in which the victorious theory is not yet contradicted rather than true. Cook and Campbell do not accept this critique of Popper for two basic reasons: (1) It seems to them, and to me, that many "facts" remain intact after a shift from one theory to another. (2) The postpositivists rely upon major paradigm shifts that overemphasize the theory ladenness of observations relative to the competing hypotheses of most scientific research (for example, do certain programs of early childhood education increase the reading skills of children from poverty families by a specified amount or not?).

The second implication of the evolutionary critical-realist perspective is a rejection of the essentialist insistance on restricting causal attributions to the most immediately preceding cause. Cook and Campbell's acceptance of the everyday language meaning of cause leads them to accept as causal any link in a causal chain, the manipulation of which *usually* produces the predicted effect. In my view, this rejection of essentialism is crucial for the avoidance of reductionistic explanations in the social and behavioral sciences. For example, the restriction in behavior modification to causal explanations in terms of immediate reinforcers tends to suppress explanations that attribute causality to structures of social control over valued resources (reinforcers). In the case of psychology and the social sciences, then, an insistence on essentialism may serve political, as well as philosophical or scientific, purposes.

Finally, Cook and Campbell conclude that their everyday language approach to model testing through falsification leads to an emphasis upon experimentation rather than causal inference from passive observation. They state that

> If we define the meaningfulness of causes in terms of their ability to create testable, dependable, and planned changes, then the most meaningful causes are those which can be deliberately manipulated. Such a concept of cause mirrors the unique feature of experimentation—the manipulation of putative causes (p. 36).

They support this position by citing the dangers of incorrect causal inference in passive observational studies and their impression that the products of nonexperimental research have been disappointing over the last decade.

Although I agree that the experiment is the best procedure for testing causal assertions, I must part ways with Cook and Campbell as they define the *meaningfulness* of causes in terms of our ability to deliberately manipulate them. Such a position is tantamount to a restriction of research efforts to those causal variables that may be manipulated within existing political constraints. In other words, a science that does not challenge the status quo is indirectly advocated via methodological recommendations. Are we to avoid study of the effects of social inequalities because the inequalities are not now manipulatable? Are social inequalities of wealth and power less *meaningful* causes of mental illness or heart disease because the current political system resists change in those inequalities? To whom are these causes less meaningful?

Now, Cook and Campbell might argue that I am confusing meaningfulness with importance or other value considerations. However, I fail to see how their notion of meaningfulness can be construed in any way that does not overlap some kind of prescriptive about what social and behavioral scientists should and should not study. My own view is that the causal relations studied by a researcher should be chosen on the basis of the researcher's values—in the broadest sense. A researcher should then strive to test causal assertions in the "least dangerous" manner available. Furthermore, those who choose to study currently nonmanipulatable causes must make every effort to reduce the dangers to validity as much as possible (as with the measurement models advocated in this book) and work toward a situation in which those causes may become manipulatable.

(In fact, a careful reading of Cook and Campbell's chapter, "Inferring cause from passive observation," reveals a position more similar to mine than the one gleaned from their first chapter. Nevertheless, their first chapter is, in my opinion, in need of the above rebuttal.)

1.9 CONCLUDING REMARKS

With the conclusion of the discussion of causality we will leave most of the metamethodological issues behind us. Although I have attempted to interweave methodological and metamethodological considerations with the presentation of statistical procedures, what follows Chapter 2 is clearly more statistics than method.

2

Review of Research Designs

The goal of this chapter is to present a general overview of research designs and the linear equations used to model the processes explored by those designs. It is assumed that the reader is already familiar with this elementary material although probably in a different format.

2.1 PATH DIAGRAMS AND LINEAR EQUATIONS

A linear model can be expressed in three ways: (1) in words; (2) with equations; (3) with a path diagram. The first and last expressions are the easiest to understand, but the equations make actual computations possible. We will begin with a model expressed all three ways.

Verbal Model If the amount of threat in a person's social environment is increased, that person will become more authoritarian. There are, however, other factors that influence the degree of authoritarianism and these factors are unrelated to the degree of threat.

Path Diagram Figure 2.1 depicts a model in which a change in X (threat) will produce a change in Y (authoritarianism). Other unspecified factors (S) also influence Y and they are uncorrelated with X. The letter b_1 on the arrow from X to Y is called a regression coefficient and it indicates the extent to which a change in X is translated into a change in Y. The larger b is, the larger the change in Y that results from a given change in X. We could also put a coefficient on the arrow from S to Y, but the arrows from unspecified factors are usually assumed to have a coefficient of one—so none is included. (This is not always the case, as we shall see.) The requirement that S not be correlated with X is shown in the diagram by the absence of a line connecting the two variables.

Equation Model

$$Y = b_0 + b_1X + S \tag{2.1}$$

where b_0 and b_1 are constant numbers and the correlation between X and S is zero. This equation is assumed to apply to any person in a given population.

This simple model includes no statements about measurement; it only deals with causal relations. Equation 2.1 is thus referred to as a *structural* equation because it describes a causal structure in which a change in X produces a change in Y. Were we to actually gather data in an attempt to test the model and estimate b_1, then it would be necessary to develop *indicators* of X and Y. In other words, *operational definitions* of authoritarianism and threat would be required. These operational definitions should specify how authoritarianism (and threat) scores would be assigned to individuals—with the assumption that the indicator scores were flawed estimates of some "true" level of authoritarianism. The word "true" is not used in the existential sense; rather, the true score for a person is a hypothetical (i.e., unobservable) position of that person along a hypothetical dimension. Temperature, for example, is a useful hypothetical dimension; and it can be useful to imagine that measurements of a temperature are flawed estimates of the true temperature being measured.

Thus, we can expand the structural model in Figure 2.1 to that depicted in Figure 2.2. The structural portion of the model remains the same: a change in X produces a change in Y to the extent b_1. But this new model specifies that X_1 and X_2 are observed indicators of the hypothetical, or *latent,* variable X. Similarly, Y_1 and Y_2 are observed indicators of Y.

To be more concrete, suppose that Y_1 is a score on a paper and pencil personality inventory (call it the F scale) and Y_2 is a score assigned to individuals by an interviewer who is familiar with the authoritarianism construct. The model specifies that different individuals receive different scores on Y_1 (and Y_2) because people differ along the hypothetical dimension "authoritarianism" (this relationship is depicted by the arrow leading from Y to Y_1 in Figure 2.2) *and* because of other unidentified influences on Y_1 (depicted by the arrow leading from e_3 to Y_1) that are uncorrelated with Y. In other words, people's scores on the F scale described in Chapter 1 (Y_1) are partly determined by their true position on the authoritarianism dimension (Y) and partly determined by factors other than Y (e_3)—such as beliefs about what is expected of participants in the study, their physical health, random responses, etc.

One of the most important aspects of this model is that Y_1 and Y_2 are assumed to coincide (i.e., give similar authoritarianism rankings of people) because each is a measure of the same true scores, while the other influences on each of these indicators (e_3 and e_4) are *not* connected in the diagram. In other words, Y_1 and Y_2 rank people similarly *only* because they each reflect—although imperfectly—the latent scores Y. If e_3 and e_4 had zero influence on Y_1 and Y_2, respectively, then and only then would Y_1 and Y_2 agree exactly.

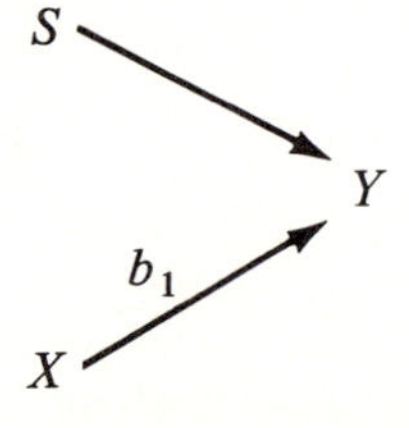

Figure 2.1 A simple path diagram of a model in which a change in social threat (X) leads to a change in authoritarianism (Y).

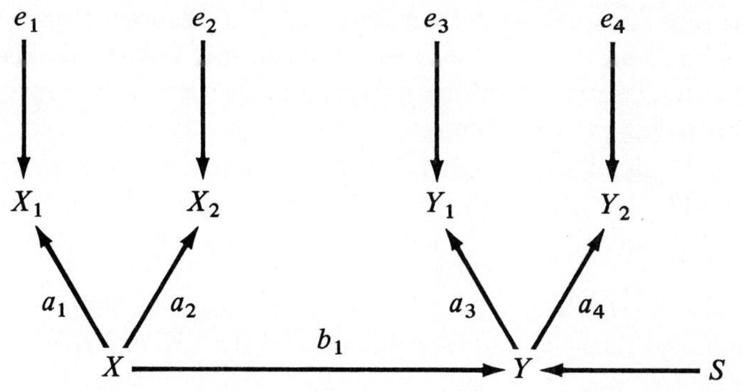

Figure 2.2 A path diagram of a model that includes both causal structure and measurement specifications.

The coefficients a_1, a_2, a_3, and a_4 in Figure 2.2 reflect the closeness or accuracy with which the indicators measure the latent variables. The degree of relationship between an indicator and a latent variable is termed the *validity* of the indicator. Suppose that X_1 measures threat by assigning a score to each individual that is based on the violent crimes in the person's neighborhood; X_2, on the other hand, measures threat by asking the person's employer about the proportion of workers in jobs similar to that held by the individual that had been laid off in the last two years. The validities of X_1 and X_2 are then defined as the correlations between X and X_1 (validity of X_1) and between X_2 and X (validity of X_2).

The model in Figure 2.2 can be written in equation form as follows:

$$Y = b_0 + b_1X + S \tag{2.2}$$

where S and X are uncorrelated.

$$\begin{aligned} X_1 &= a_{01} + a_1X + e_1 \\ X_2 &= a_{02} + a_2X + e_2 \\ Y_1 &= a_{03} + a_3Y + e_3 \\ Y_2 &= a_{04} + a_4Y + e_4 \end{aligned}$$

where the e's are uncorrelated with X, Y, and S. The full meaning of these equations is the subject of this text, but the reader should be able to make some sense of them if the translation from the diagram to the equations is closely inspected.

2.2 MODEL TESTING

The essence of model testing is summarized as follows:

1. The data are gathered according to the operational definitions of the indicators;
2. The parameters (coefficients of arrows in the path diagram, usually) are estimated by some technique;
3. The "fit" of the model is assessed by some statistical criterion. That is, do the observations interrelate in the manner predicted by the model?

4. If the fit is poor, then the model is rejected and alternatives may be concocted. If the fit is good, then the model has survived the test and awaits disconfirmation. (Notice that, unlike logical proofs, empirical models are never proved; rather, the "good" ones have yet to be disproved.)

The structural and measurement portions of a model can usually be tested separately. (Some problems arise when the tests of the two portions are interdependent—"confounding of unobservables." But in some instances this interdependence is unavoidable.)

The fundamental logic of measurement model testing is quite simple: indicators of the same latent variable(s) should relate more strongly with one another than they do with indicators of separate—although related—latent variables. If a model specifies the exact nature of the relationships between hypothetical latent variables and indicators, then much more stringent constraints on the interrelations of the indicators must be met in order to achieve a good fit. But, in the final analysis, most all of these constraints are refinements of the fundamental logic stated above.

Tests of the causal portion of a structural model derive from rules of causal inference. And the most accepted rules of causal inference are embodied in the research design termed the true, or randomized, experiment.

2.3 CAUSAL INFERENCE AND THE EXPERIMENT

For the purpose of organizing this discussion of causal inference, we will define three classes of model testing strategy: (1) cross sectional, (2) longitudinal, and (3) experimental. Each of these strategies consists of a method for gathering observations (data) and of organizing them into a test of a structural (causal) model. And, in general, the three strategies are ordered above in terms of ascending internal validity. That is, a cross-sectional design provides the least definite demonstration of a causal link, whereas the experiment is the most convincing. Further elaboration of these three strategies for model testing will be presented in the context of the authoritarianism/threat model.

Suppose that previous work has satisfied us that neighborhood crime rate is a good indicator of social threat and that the F-scale is a good indicator of authoritarianism. In other words, we have already tested the measurement portion of the model in Figure 2.2, and we are now proceeding to test the structural portion. In general, the better tests of causal models are those involving multiple indicators of each latent variable. For the sake of simplicity in this discussion, however, we will assume that threat (X) and authoritarianism (Y) are validly measured by crime rate (X_1) and the F-scale (Y_1). In fact, we will assume that X_1 is a perfect measure of X and Y_1 is a perfect measure of Y.

We will now develop a series of data-gathering projects that escalate in terms of the strength with which they test our causal model.

The Model An increase in social threat produces an increase in an individual's authoritarianism *even if* all other factors in a person's life remain constant or unchanged. In equation form

$$Y = b_0 + b_1X + S \tag{2.3}$$

where S is a summary term for all those "other" factors that influence an individual's level of authoritarianism and b_1 is the amount that Y changes if X is increased by one unit. The crucial concept for understanding this model is that b_1 is the effect of X on Y that is independent of *all* other influences (S) on Y.

There are at least two ways to conceptualize an effect of X on Y that is independent of all other influences on Y. First, b_1 is the change in Y that would result from a change in X if all other influences on Y were held constant (restrained from changing). Second, b_1 is the *average* change in Y resulting from a change in X if changes in X occur such that they are uncorrelated with all changes in other factors that influence Y. This latter conception is rather subtle and should become clearer in the context of the ensuing concrete examples, but the reader can try to imagine the following. Suppose that the crime rate in a neighborhood increases by a certain amount and that the authoritarianism of each adult in that neighborhood increases by an amount specified in our model. The problem presented to the potential observer of the increased authoritarianism is that many of the other influences on authoritarianism also changed during this time, and F-scale scores would be pushed up and down such that the change in an *individual's* F-scale score attributable to the increased crime rate would be impossible to identify. However, suppose that we observed many people (a sample of the neighborhood population) and *could assume* that the changes in the other factors (S) influencing the authoritarianism (Y) of individuals in the neighborhood occurred in such a manner that, *when averaging* across individuals, the pressures S to increase and decrease authoritarianism tend to cancel one another. For example, if being in good physical health decreases—and poor health increases—authoritarianism, then the effects of health would cancel one another when averaging F scale increases *if* the improvement/deterioration of health among individuals was unrelated to the change in crime rate. That is, although everyone in the neighborhood was under the pressure of increased crime to increase in authoritarianism, there is no average change due to health changes because the proportion of people with deteriorating health (experiencing an increase in authoritarianism) has not changed along with the change in crime rate. In other words, changes in health were uncorrelated with changes in crime rate. If we could make this assumption about all influences on Y (other than X), then the *average* change in Y associated with differences would be in X the effect of the change in X.

Of course, it is probably *not* true that changes in crime rates are uncorrelated with changes in average health levels. But the point is that if they were, then the average change in authoritarianism would not be affected by the effects of health on authoritarianism. Thus it is the task of the empirical social scientist to sort out causal sequences from complex individual/social systems. This is especially true for the social scientist whose goal is specification of the changes in conditions that are usually sufficient to produce desired outcomes. Several strategies for such "sorting out" will now be outlined: cross sectional, longitudinal, experimental.

Non-randomized Cross-Sectional Designs

Non-randomized cross-sectional designs are often labeled "correlational"; however, I do not want to encourage this use of the term correlation since the term refers to a statistic that summarizes relationships and, as such, is logically independent of the research design adopted.

Our goal is to test the model that states that increases in social threat (X, operationalized as neighborhood crime rate) cause changes in individual's authoritarianism (Y, F-scale scores):

$$Y = b_0 + b_1X + S \qquad (b_1 > 0) \tag{2.4}$$

The two characteristics of a cross-sectional test of this model are: (1) the observations of X and Y are made at one point in time for each person observed and all persons are observed at roughly the same point in time; (2) the observations are made without any attempt to influence or change people along the dimensions X, Y, or any other variable relevant to X and Y. The following design exemplifies a cross-sectional approach to testing the model in (2.4) with passive observation.

Since a cross-sectional design involves only one point in time, assessment of change is generally ruled out. (Exceptions to this generalization arise when, for example, respondents to surveys are asked historical questions that are assumed to be indicators of change: "What was your father's occupation when you were 15?"). So, instead of looking at change, suppose that we sampled a population of persons that included many neighborhoods with a wide range of crime rates. These people are then given the F-scale. Then, if the model in equation 2.4 is true (a hypothetical statement), and especially if b_1 is fairly large, we would expect to find that the people in higher crime neighborhoods would have higher F scale scores than people in lower crime neighborhoods (i.e., a positive correlation between X and Y).

Now, suppose that this was our finding, what does the finding tell us about the model (equation 2.4)? The result is *consistent* with the model. But the crucial consideration in testing a causal model (i.e., causal inference) is the extent to which alternative models, or competing causal explanations, can be contradicted by the observations. And, in this respect, our cross-sectional evidence is rather weak. To be more specific, there are three classes of alternative models that are just as plausible in light of the finding that X and Y are positively correlated in a cross-sectional study.

Alternative Model 1: Spuriousness

This model is depicted in Figure 2.3A. Notice that no arrow leads from X to Y, indicating that a change in X will not influence Y. The curved line connecting X and S, however, indicates that X is correlated with factors S that do influence Y. This means that X and Y are correlated, but for spurious reasons. In equation form, the spurious alternative might be:

$$Y = c_0 + c_1X + S \tag{2.5a}$$

$$X = c_2 + c_3S + W \tag{2.5b}$$

where c_1 is zero or so small that it is inconsequential. Clearly, when $c_1 = 0$, equation 2.5a specifies no causal influence between X and Y; increases in X have zero impact on Y.

What are some factors that might be included in S that could give rise to a spurious

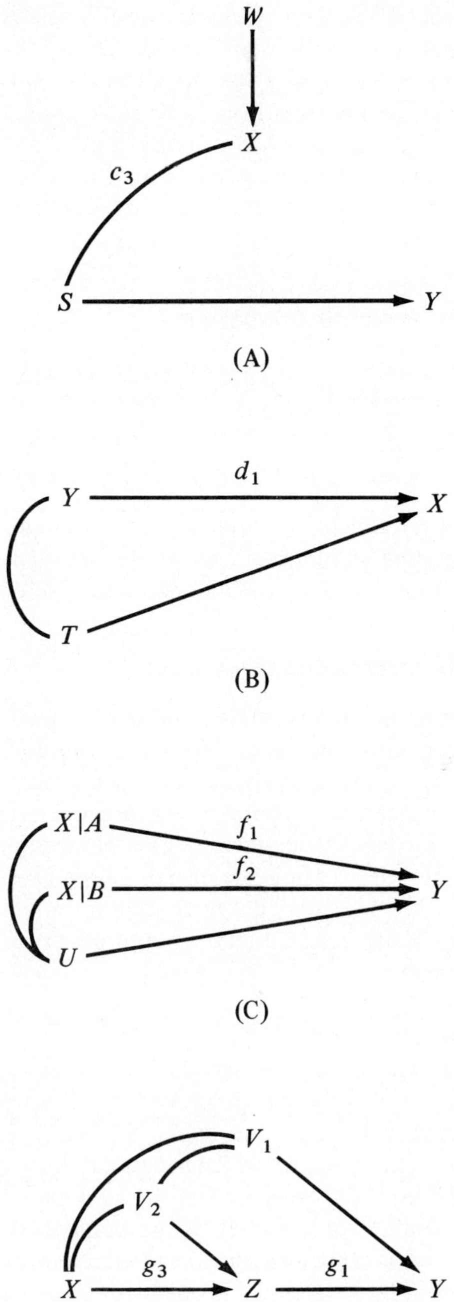

Figure 2.3 Four alternative causal models that are consistent with the observation of a positive correlation between an indicator of X and an indicator of Y in a cross-sectional study.

correlation between X and Y? A likely candidate would be educational level. Suppose that people with less education are, by some unspecified process, pushed to seek abode in high crime areas *and* that having less education leads to higher F-scale scores. Thus, a relationship between crime rate and F-scale score across a sample of individuals is equally consistent with either model (2.4) or (2.5), and in the latter case a change in crime rate would not affect authoritarianism, since the actual cause of the observed correlation was a set of factors (such as level of education) that were not measured in the study.

Alternative Model 2: Reverse Causation

The third model that is consistent with a positive correlation between crime rate and F-scale scores is diagrammed in Figure 2.3B. Notice in this instance that an increase in Y produces a change in X and not vice versa. In equation form:

$$X = d_0 + d_1 Y + T \qquad (d_1 > 0) \tag{2.6a}$$

where the magnitude of d_1 reflects the extent to which changes in Y influence changes in X. A verbal interpretation of equation 2.6a would be that high authoritarianism in an individual produces more crime in the individual's neighborhood.

Alternative Model 3: Interactive Causation

The distinguishing feature of the interactive alternative model is that the magnitude of the effect of X on Y is different under different conditions or for different subsets of people. Suppose, for example, that the effect of social threat on a person depends upon whether the person interprets the increased threat (i.e., increased crime) as primarily a result of inadequacies in the social system or as primarily a result of inadequacies in the individuals who commit the crimes. If we label the "system blamers" group A and the "individual blamers" group B, then we might find that increasing X has almost no effect on Y for group A and a large effect for group B. This can be written in equation form as:

$$Y = f_0 + f_1 X|\mathrm{A} + f_2 X|\mathrm{B} + U \tag{2.6b}$$

where f_1 is close to zero and f_2 is large and positive. (The notation $X|\mathrm{A}$ is read "X given A." Thus, f_1 is the effect of a change in X given group A, and f_2 is the effect of a change in X given group B.)

This model is an alternative to equation 2.4 because it does not specify that a change in X produces a change in Y for every person in the population. Rather, it specifies that the causal effect applies for a subset of the population and that the observed correlation between X and Y occurs because a sufficient number of people from group B are included in the sample. This model is depicted in Figure 2.3c.

From a policy or intervention point of view, the interactive alternative is not as serious a challenge to the usefulness of the original model (equation 2.4) as are the spurious and reverse causation alternatives. This is the case because changes in X *would* produce change in Y for at least a subset of the population if the interactive model were true. Reducing the crime rate would reduce authoritarianism, but only among people that attribute the reduction to individual processes (which might be the bulk of the population).

Alternative Model 4: Mediated Causation

The fourth challenger to model (2.4) is depicted in Figure 2.3D. In this instance, changes in X produce changes in Z; then changes in Z produce changes in Y. In equation form:

$$Y = g_0 + g_1 Z + V_1 \tag{2.7a}$$

$$Z = g_2 + g_3 X + V_2 \tag{2.7b}$$

This model states that changes in X influence Y only as those changes are transmitted (or mediated) by the intervening variable Z. The model (2.4) is then said to be the *reduced* form of the more complete model.

Suppose, for example, that an increase in social threat induces an increase in the level of fear experienced by a person, and the increased fear, in turn, leads to increased authoritarianism. In other words, if other factors influencing fear (V_2) operate on a particular individual such that the increase in fear due to increasing social threat is counteracted by these other factors, then no increase in authoritarianism would occur. On the average, however, an increase in X would lead to an increase in Y, thus explaining the positive correlation between X and Y.

Finally, a fifth category of alternatives to equation 2.4, consistent with an observed positive correlation between X and Y in a cross-sectional study, might be termed "any combination of the above." The four alternatives described above are simplistic logical types that can be combined to construct models that may be appropriate to the complexities of our lives. Furthermore, the above models have been restricted to specification in terms of linear equations.

Statistical Control The reader may now ask, "Is there anything that can be done to improve the inferential power of cross-sectional studies?" My answer is that *something* can be done, and that something is statistical control. But statistical control is useful only in the following contexts: (1) to reject *specific* alternative explanations (as opposed to a *class* of alternative models) of any type except those involving reverse causation; (2) when all the variables relevant to a causal model are theoretically known and measured. And the latter is still only a hope for most social scientists.

Much of our discussion of linear models will be relevant to the first of these contexts: rejecting specific alternative models. For example, recall the spurious alternative to equation 2.4 involving educational level, where crime rate and F-scale scores correlate because each is influenced by the amount of schooling completed. We will develop methods for testing this particular alternative explanation. However, even if that model were rejected, we could not be assured that some other variable(s) did not render our cross-sectional correlation as spurious.

The logic of statistical control is quite simple. In order to test the "spurious due to education" model we could simply break down our sample into subgroups of people such that all members of each subgroup had completed the same number of years of schooling. Then, we could look at the relationship between X and Y within all these subgroups. If the relationships were about the same as in the total sample, then we could reject this particular alternative explanation of our cross-sectional finding. On the other hand, if the correlations within the subgroups were near zero (or much smaller than in the full sample), we could not reject this spurious alternative, and we may be motivated to reject the original model (2.4). In an analogous manner, specific interactive and mediating alternatives could be tested. But the reverse causation class of alternative explanations remains impervious to testing with cross-sectional data, and the rejection of any specific alternative in one of the other classes leaves the myriad of remaining models in those classes as plausible alternative explanations.

An additional qualification concerning the potential of statistical control for model testing should also be mentioned at this point. The numerical results of statistical controlling procedures depend heavily on how well the chosen indicators measure the hypothetical constructs. In most instances where such controls have been used, however, little is known about the validity of indicators—resulting in almost meaningless statistics.

Concluding Remarks on Non-randomized Cross-Sectional Designs It may seem at this point that the range of alternative causal models consistent with findings from a cross-sectional study render such studies as not very useful. This is not the point of the previous discussion. Rather, the point is to communicate the limitations placed upon causal inferences from cross-sectional studies and to set the stage for description of other designs that enable the researcher to reject more of the alternative causal explanations than is possible with cross-sectional data. But the importance of the cross-sectional design is its flexibility for exploring the patterns of covariation among a wide array of variables. This exploration can lead to important theoretical refinements that can then be tested with the more costly designs involving observation over time, or experimentation.

The complexity and flexibility of psychological-social systems are apparently so great that changes in many, many variables *can* produce changes in a particular outcome variable. Thus, the essential contribution of cross-sectional studies to the human sciences is in pointing us toward what are the *major* causes of an important outcome in our current (or past) historical context.

Non-randomized Longitudinal Designs

The most glaring weaknesses of the non-randomized cross-sectional design for causal inference were: (1) we are unable to test reverse causation alternative models; (2) every spurious alternative must be tested by measurement of suspected third variables. Longitudinal designs, involving measurement of change, offer *some* advantages in this respect.

Types of Longitudinal Designs There are two basic longitudinal designs used by social science researchers: *panel* designs and *time series* designs.

The panel design is essentially a *repeated* cross-sectional design in which the same sample of people is observed (or serve as respondents) at two or more points in time. The most extensive panel study conducted in the United States is the National Longitudinal Survey (NLS) sponsored by the Department of Labor. In this major study of work and economic status, roughly 10,000 men and 10,000 women were selected in 1966, and most of these people have been interviewed or telephoned every other year since 1966.

Time series designs usually involve observations of a single person or aggregate of persons over an extended period of time. For example, M. Harvey Brenner has collected data on mental hospital admission in the state of New York for each year from the late 1800's to the 1960's. He reports that the ups and downs in admission rates over these several decades correlate strongly, and negatively (about −.80), with ups and downs in indices of economic production.

Clues about Causal Direction If two variables evidence a cross-sectional correlation, a longitudinal design can be used to obtain clues as to the direction of the causal effects—if any. The simple principle that enables us to specify those clues is that if a change in X causes a change in Y, then the causally linked events must occur in the obvious sequence: change in X, some time lag, change in Y. This temporal sequencing, which is at least implicit in most all conceptions of causation, can be seized upon to test hypotheses concerning causal direction.

A nice example from the social psychological literature on alienation demonstrates the approach. Numerous investigators found cross-sectional correlation for women between questionnaire measures of alienation and the number of children born. That is, women scoring higher on the measures of alienation tended to have more children. This is a classic chicken and egg problem: Does having a lot of childrem make a woman more alienated or are more alienated women induced by some process to have more children? A panel study of this relationship gave some clues. A sample of women was given the alienation scales before they bore any children. These women were then contacted some years later and the number of children born was obtained. After attempting to control for other variables known to be relevant to birth rates (religion, class, etc.), it was found that (to a small, but reliable) degree, the women who were more alienated before child bearing had given birth to more children than the less alienated women. Furthermore, having a large number of children did not result in an assessable increase in alienation relative to those women who had fewer children. This is fairly convincing evidence that alienation, or some factor *correlated with alienation* (which was not adequately measured and controlled), has a causal influence on the number of children a woman bears. And these data are an even stronger demonstration that the number of children that women bear *does not* influence their level of alienation.

It is important to reiterate here that this panel study does not solve the spuriousness problem. That is, the factor actually influencing the number of births could be distinct from, although correlated with, alienation. So we can at least say that alienation or a related factor influences child bearing, whereas a causal effect in the other direction is inconsistent with the observations gathered in the panel studies.

Returning to our social threat/authoritarianism research project, suppose that a panel study was conducted in which a sample of people were observed every six months

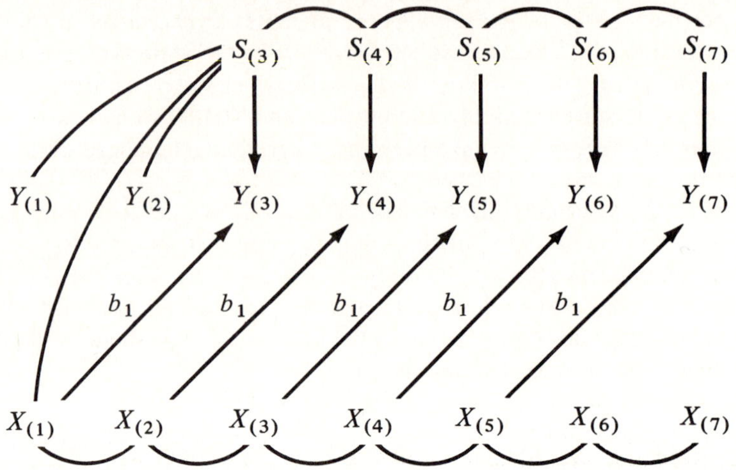

Figure 2.4 A structural model of lagged causation that could be tested with data from a seven-wave panel study.

for three years. The resulting data would consist of seven measures of X and seven of Y (see Figure 2.4). Now suppose that an increase in X causes an increase in Y with a 12-month lag. In equation form:

$$Y_{(j)} = b_0 + b_1 X_{(j-2)} + S_{(j)} \qquad (j = 3, \ldots, 7) \tag{2.8}$$

where subscripts in parentheses indicate the *wave* of the panel study. Model (2.8) portrays Y at wave j as a function of X at the time wave $(j - 2)$ (i.e., 12 months earlier) and other factors $S_{(j)}$ that may be correlated with X or Y at any point. This model involving exact time points might be considered a simplification of a more realistic model in which Y is a cumulative function of X over the previous 8 to 16 months.

The problem now is to determine how we could use the observed data to test the model in equation 2.8. This is a complex issue with no one answer, and the general prediction about the pattern of relationships between X and Y cannot be stated in simple terms of the correlations or covariances among these variables. Rather, specific causal models generate predictions about the relative magnitudes of the various correlations or covariances and variances. These predictions constitute "overidentifying restrictions" upon the observations, and procedures outlined in Chapter 10 can be used to compare the model's predictions with observations.

Unfortunately, we seldom have the opportunity to conduct large panel studies with more than two or three waves, and only a few theoretical models in developmental psychology make predictions about the length of the lag in lagged effects. Furthermore, realistic models of psychological or social processes would seldom be as simplistic as the model in Figure 2.4. Thus, statistical methods will be presented for extracting as many clues about causal direction as seems reasonable.

Suppose now that we designed a time series study of the relationship between social threat and authoritarianism and that the *unit of analysis* is a particular neighborhood. Thus, we will attempt to measure the average level of authoritarianism and social threat for the neighborhood as a whole—and do so over a long period of time. Although it is difficult to imagine, imagine that average F-scale scores were available for the neighborhood for each of the last 30 years. If we also had crime rates for each of those 30 years, then we could see if the annual ups and downs (relative to the long-term trend) in these two indicators evidence a positive correlation (i.e., moved up and down in phase with one another). However, the testing of causal models concerning the directionality of causal influence between these two variables is a complex problem that is not developed in this book. The exclusion of models for time series was motivated by the relative infrequency of adequate time series data outside the fields of economics and demography. Introductions to the testing of structural models with time series data are available in basic econometrics texts (e.g., Kmenta, 1971; Johnston, 1972), and the use of Box-Jenkins autoregressive integrated moving average (ARIMA) models is clearly outlined in McCleary and Hay (1980).

Longitudinal Designs and Change In progressing from cross-sectional to longitudinal designs, an important shift, which may not have been noticed, has occurred in the congruence between the observations used and the language of the causal model. The language of the models always includes the word "change": that a change in X causes a change in Y. (In terms of our standard example: an increase in social threat leads to an increase in authoritarianism—even if all other relevant variables are held constant.) But the cross-sectional design only includes observations of X and Y at one point in time; there is no attempt to measure a change in X or Y for any individual. Rather, differences *between* individuals in X and Y are substituted for observations of change, and a kind of logical leap is made. In a cross-sectional design, we must extract information about what would happen to Y if we changed X for a *single* individual by observing *different* people at different levels of X. For example, differences in the authoritarianism of one group of persons in a low threat situation and another group in a high threat situation must be substituted for observations of what happens to a group of people when they themselves change from one threat condition to another. The longitudinal designs are not so constrained. They can actually be used to track changes in X and Y over time—and for the same individuals (or cities, countries, etc.).

There are, of course, special problems that arise in longitudinal models—especially when indicators with systematic errors are used repeatedly. As mentioned previously, the most sophisticated mathematics presented in this book are primarily necessary to deal with the knotty problems that arise in longitudinal studies.

The True Randomized Experiment

The true experiment is the ideal demonstration of a causal relation because, *ideally,* the spuriousness and reverse causation alternative explanations of a correlation can be rejected. That is, these entire *classes* of competing models can be rejected without resorting to statistical control of other variables. (The interactive and mediating alternatives are still challengers, but these challengers do not totally contradict the basic notion of a causal relation between the two variables of interest.)

Now, the true experiment achieves this superior status because the model

$$Y = b_0 + b_1X + S \tag{2.9}$$

can be tested in a manner that *almost* assures the following: (1) the change in X occurs before the change in Y; (2) all other factors S (that could be responsible for the change in Y in a cross-sectional or longitudinal design) are correlated with changes in X only by *chance*—meaning that observing many people in the experiment will almost guarantee that no third variable is actually responsible for the observed change in Y. These wonders of causal inference are made possible through the careful use of *randomization:* the manipulation of X by the experimenter such that changes in X are determined by a *random* process. For example, if changes in X are determined by the toss of a fair coin, then it is logically *impossible* for changes in Y to have influenced changes in X (reverse causation alternative), *and* it is logically *improbable* that any of the other factors that influence Y (i.e., S) would, for a large number of people, change in unison with the changes in X (since, by assumption, all factors are uncorrelated with outcomes of a coin toss—in the long run). Thus, since no third variable is likely to correlate with randomly determined changes in X, any subsequent changes in Y that correlate with changes in X are not subject to explanation by spurious alternatives.

As we discuss a concrete example of a true experiment it will become apparent that the "wonders" described above are overstated. In practice, an experimenter can never be certain that he has not inadvertently changed some factor other than that which was intentionally manipulated. However, one can be strongly convinced that a reverse causation model is implausible and that the true causal factor(s) is closely associated with manipulations of X. Through a strategy involving multiple methods of manipulation and multiple measures of hypothetical constructs, spurious alternatives can be rejected with greater confidence.

Types of True Experiments It is useful to distinguish between two types of true experiments: *cross sectional* and *longitudinal.* In a manner analogous to nonexperimental design, cross-sectional experiments operationalize changes in X in terms of differences between groups of people, whereas longitudinal experiments focus on changes in Y for individuals. However, the advantages of longitudinality in nonexperimental designs are not necessarily reproduced in experiments. The reason for this is that randomization gives the cross-sectional experiment an *initial* comparability of groups at various levels of X that can only be rivaled with longitudinal observations in non-randomized studies.

Examples of Experimental Designs We can now return to the social threat/authoritarianism model and design true experiments to test the model, which states that decreases and increases in threat cause decreases and increases, respectively, in the authoritarianism of individuals. Accepting the F-scale as a good measure of Y, the first task in designing an experiment is to invent a procedure for manipulating the level of social threat (X) for individuals such that the amount and direction of change in social threat is determined by a random process. The second task is to manipulate X such that change in other relevant variables is minimized.

Now, if we are to continue thinking of neighborhood crime rate as a good indicator of social threat, the experiment presents us with a problem: How are we to change the crime rate of a neighborhood? And when we add the requirement that the manipulation of the crime level must be accomplished without concomitant changes in other relevant variables, the impracticality of the task is obvious. Even if a social scientist had enough power to arbitrarily move people from one neighborhood to another—and wanted to do so—the resulting changes in crime rate would be accompanied by changes in all the other variables that correlate with crime rate.

The constraints of an experimental design are, in this instance, so great that a nonexperimental longitudinal design would probably be the best strategy for testing the causal model. Upon some reflection, it should be apparent to the reader that certain types of causal models in the human sciences are especially uncongenial with experimentation. For example, models in which the causal variables are social structural are usually not testable by experimentation. This means that the effects of social class, ethnic and racial discrimination, economic downturns, family characteristics, etc., must be studied nonexperimentally. An exhaltation of experimentation as a route to knowledge, then, can have the effect of distracting attention away from these crucial areas of concern.

But, for the sake of discussion, let us suppose that financial income security (or insecurity) is a good indicator of social threat and that a social scientist has an opportunity to change the level of income security for a sample of people by providing them with a government-sponsored "income insurance." This insurance could be in the form of a guarantee that, if the person were laid off during the next three years, the government would reward them for their past contributions to society by maintaining their income from wages or salary.

The aspect of such a study that is crucial to making it a true experiment is the random manipulation of the income security variable. Let us consider a longitudinal design first. Here the researcher could first measure the level of authoritarianism in each person in the sample by administration of the F-scale. Then the sample could be *randomly* divided into two groups—perhaps by flipping a fair coin once for each participant in the study. If the coin came up heads, the participant would have the income insurance, thereby decreasing the level of social threat. If the coin came up tails, the income insurance would not be provided. In this way the change in X has been determined by a random process: who increases in income security and who remains the same has been determined by a coin toss and any other variable influencing authoritarianism could not have caused the changes in income security. Furthermore, as the sample size increases, it becomes less and less likely that any other variable influencing Y will even be correlated with changes in X. This means that pressures from changing social and personal factors to increase authoritarianism are likely to be the same for both the manipulated (or experimental) group and the non-manipulated (or control) group. Similarly, pressures to decrease authoritarianism would only favor one group over the other by chance, and the probability of such is very small when more than about a hundred people are included in each group. Thus, when F-scale scores are gathered a second time, say three months after informing members of the experimental group about their insurance, a decrease in the average F-score for the experimental group—relative to the control group—could be safely attributed to the causal effect of a change in income security. The "safety" of this attribution arises from the temporal sequence of events (X changes before Y) and from the assumption that no third variable is likely to be correlated with the manipulated changes in X.

This experiment would not, however, allow the rejection of interactive or mediative alternatives. A very complex chain of changes may be necessary to actually achieve the change in authoritarianism. Furthermore, if we relax the unrealistic assumption that the income insurance necessarily manipulated social threat, then we have the additional problem of not having any evidence that a change in social threat resulted from the manipulation and that the change in F-scale scores resulted from the change in threat. Evidence for this part of the model can be gleaned from such a design by measuring changes in social threat and using a more complex linear model to test the hypothesis that changes in Y resulting from changes in income security were indeed *mediated* by changes in social threat. A path diagram of such a model is presented in Figure 2.5. In equation form, this model would be specified as follows:

$$Y_2 = a_0 + a_1X_1 + a_2X_2 + a_3Y_1 + a_4X^* + S_1 \qquad (\text{where } a_4 = 0)$$
$$X_2 = b_0 + b_1X_1 + b_2X^* + S_2 \qquad (\text{where } b_2 > 0)$$

Note that the randomized variable is unconnected (uncorrelated) with the other time 1 variables and that X^* only affects Y through change in X. This is a complex model that will be explained in a later chapter, but the logic of it is quite simple. Furthermore, if observed data are consistent with the predictions of such a model, then those observations constitute strong support for the model.

The cross-sectional version of this experiment would be similar except that the pretests (measurement of X, and Y, before the manipulation) would not have been conducted. It would then be necessary to assume that randomization had achieved equivalence of the experimental and control groups in the average level of authoritarianism, and changes for individuals would be inferred from group differences. The advantage of the pretreatment measures is that the equivalence of the treatment and control groups can be tested. That is, the observations can be used to test the null hypothesis that population correlations between X^* and Y_1 are zero. If, by chance, the procedure of random assignment produced groups that differed on X or Y (i.e., X^* evidenced a substantial correlation with X_1 or Y_1), then the pretreatment measures could be used

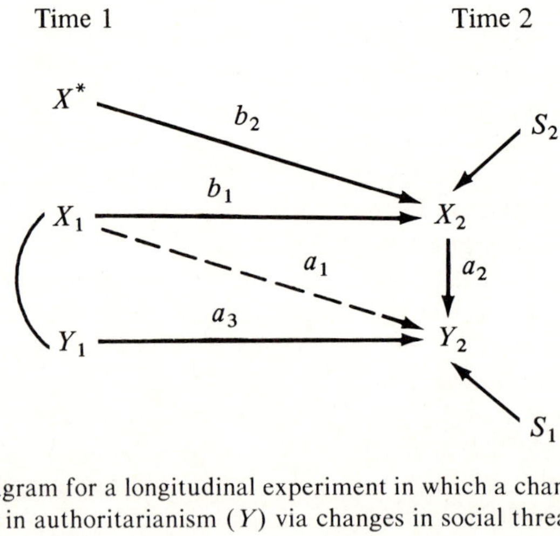

Figure 2.5 Path diagram for a longitudinal experiment in which a change in income security (X^*) causes changes in authoritarianism (Y) via changes in social threat (X).

to adjust for these *nonequivalencies*. However, it is crucial for the reader to understand that adjustment for such nonequivalencies is possible *only* when indicators are extremely valid—with only small amounts of measurement error—or when multiple indicators have been used in the manner prescribed earlier. The use of multiple indicators allows the researcher to estimate the validities of indicators and to adjust for measurement error prior to adjustment for nonequivalencies. (See Chapter 11 for a detailed discussion of these issues.)

This discussion of accidental nonequivalencies in randomized experiments raises the issues associated with research designs involving manipulations of presumed causal variables, but *not* involving randomization. Such designs are often labeled "quasi-experimental." This type of design is the one most used in the physical sciences in which laboratory conditions allow enough control of extraneous variables to satisfy the researcher that "other" variables are not spuriously producing the observed correlation between the manipulation and the dependent variable. For example, when a chemist increases the size of the flame under a beaker of solution, she is not concerned that some change in the weather, in her personal life, or in the state of the economy may covary with the increased flame and actually produce the observed change in the solution. Her assurance derives from the control of pressure and temperature in the laboratory and the immediacy of the effect relative to the manipulation. However, when we leave the laboratory, the threats to valid causal inference become apparent, and randomization is an important tool. Randomization enables a researcher to methodologically control, through research design, what he cannot control physically or statistically. Randomization, for example, was used on an extensive scale when the scientific method was applied to agribusiness. Field studies of cultivation techniques, fertilizers, and so on, raised the problem of how to account for differences in soil, amount of light, and so on, when different plots of ground were used to compare different plant strains or methods of growing. The idea of numerous replications and random assignment of treatments essentially solved this problem.

But numerous research situations with human beings do not allow the random assignment of treatments. Special education programs are a good example of such situations. For political or moral reasons, it is often necessary to include all children whose parents have certain characteristics. This means that children for a control group must be obtained from some other population, and the problem of nonequivalencies arises. Such designs are termed *nonequivalent control group designs* and will be the subject of considerable discussion in Chapter 11. Models for analysis of data from such studies are analogous to the one depicted in Figure 2.5 with the exception that we would *expect* the manipulation (X^*) to be correlated with the pretreatment measures. As noted, statistical control of pretreatment differences may be achieved satisfactorily only when a tested measurement model has been incorporated into the overall design. For this reason, we will not deal with the analysis of nonequivalent control group designs until specification of measurement models has been discussed.

2.4 INFERENTIAL VALIDITY

Virtually all the discussion thus far has focused on *internal* validity. That is, how confident are we that an observed association of variables results from a causal relation?

However, another major consideration in evaluating a design is the *external* validity of the causal inference: How generalizable are the conclusions of the study beyond the actual participants? External validity depends heavily on representative sampling from relevant populations and, in the case of experiments, manipulations that either are or can be reproduced in the real world.

There tends to be a "trade off" between internal and external validity when nonexperimental and experimental strategies are contrasted. The randomized experiment maximizes internal validity, but the practicalities of randomization and manipulation of causal variables usually dictate very inadequate sampling from any interesting population. Thus, researchers conducting experiments dealing with nonbiological human phenomena usually find that their causal effects are difficult to replicate. Conversely, large sample surveys uncover very replicable relationships—such as the income differentials of males with differing amounts of education. But the *causal* explanation of why more men with more education receive more income is problematic. Perhaps, people from advantaged backgrounds achieve more education, and they earn more because of those advantages—rather than because they developed valued skills (termed "human capital" by many economists) during schooling. And it is virtually impossible to imagine an experiment in which the number of years completed could be randomly manipulated. Nevertheless, the determinants of income inequality are much too important a topic to discourage investigation in the face of these difficulties. Instead, social scientists must apply their ingenuity to develop methods of causal inference from nonexperimental data. (A good example of such ingenuity is the study of education-earnings relationships among brothers; this provides a control for background that complements statistical control of measured background.) Several of the subsequent chapters will be devoted to such methods.

Finally, I would urge the reader not to despair over the difficulties of nonexperimental strategies for yet another reason. Namely, social psychological systems are probably of sufficient complexity that an almost infinite variety of conditions *can* be manipulated to produce a change in a particular outcome variable. Practically oriented social scientists, however, are usually concerned with discovery of which factors actually *do* have large effects on important outcomes. In short, experiments point us toward what *can* happen, nonexperiments point us toward what *is* happening. Put differently, nonexperiments can *only* discover large and currently operating causal processes, since strong spurious relationships are indications of even stronger causal processes involving "third" variables.

3

Review of Statistical Concepts

It is assumed that the reader has completed some kind of work in which the basic ideas of measurement scales, central tendency, variation, and covariation have been developed. They are briefly reviewed here to clarify nomenclature for the discussion following.

3.1 MEASUREMENT SCALES

It is customary to delineate four types of measurement scales. Each can be construed as an abstract map with two or more "locations." The process of measurement then consists of assigning each person (or aggregate or other unit of analysis) to a location on the imagined map.

Nominal Scales A nominal scale is a map with discrete locations and no distances between the locations. Ethnicity, for example, is a set of categories to which most people can be mapped (i.e., assigned to a category); but no ethnic category (map location) is any higher or lower in "ethnicity" than any other ethnic category.

Ordinal Scales Ordinal scales are abstract maps in which exact distances between locations are not known, but a rank order of distances is specified. For example, an ordinal map of occupational status might locate physicians above high school teachers and high school teachers above used car salesmen, but it would not specify whether the distance between physician and teacher is greater or lesser than the distance between teacher and salesman. Measurement utilizing ordinal scales leads to a rank ordering of people in which the distances between ranks (map locations) are unknown—although they may be theoretically measurable. Locations along an ordinal scale are labeled with positive integers and referred to as "ranks."

Interval Scales and Ratio Scales An interval scale is a straight line map. Locations on this map consist of two or more points along a straight line, and distances between locations have the metric properties usually associated with physical distances: a unit increase is equivalent at all points along the scale. A ratio scale is an interval scale with a zero point such that ratios of location distances (from the zero point) reflect the relative magnitudes of the two locations. For example, ten degrees on the Kelvin scale represents twice the amount of "heat" as five degrees. However, the ratio of centigrade degrees, say ten and five, has no clear meaning. Thus, degrees centigrade is an interval scale whereas degrees Kelvin is a ratio scale. Interval scales usually arise when the absolute zero point of a scale is not known or when ordered dichotomies are used. [It is often stated that variables included in regression models must be measured on an interval or ratio scale. A less misleading statement, however, is that variables included in regression models must increase (or decrease) linearly as other variables in the model increase. This means that transformations of interval scales that do not preserve the original interval equivalencies may still be included in multiple regression models.]

The notion of an interval scale can be clarified by distinguishing between observed variables (indicators) and the latent (unobserved) dimensions they measure. For example, suppose that reported annual income is used as an indicator of purchasing power. In one sense, the indicator is a ratio scale: an increase of one thousand dollars is an equivalent increase *in dollars* whether it is added to an income of five thousand or fifty thousand. However, as a measure of purchasing power, an increase of one thousand dollars in annual income may reflect differing increases in purchasing power for people at different positions in the income distribution. For example, each additional dollar may buy more if one has already received sufficient income to secure reliable transportation, purchase consumer information, etc. Thus, some transformation of dollar income may be necessary to achieve an interval scale of purchasing power.

3.2 MEASURES OF CENTRAL TENDENCY

Once a scale has been defined and a population of people has been mapped onto the scale, the resulting set of scores is said to define a *distribution.* In the case of a nominal scale, the distribution of scores consists of a histogram or table giving the number of people mapped into each category. The notion of central tendency has no meaning in this context.

In the case of ordinal scales the *median* is the most popular measure of central tendency. The median is the location on the scale that divides the distribution into the top and bottom halves. The median of an ordinal scale of occupational status would be the occupation that one-half of the people in the distribution exceed in status. As another example, suppose that scores on the F-scale are used as indicators of authoritarianism, but there is no evidence that these scores represent an interval scale. Then, the average F-scale score would best be represented by the median: the score one-half the distribution exceeded.

When variables are measured on an interval or ratio scale, the distance between scores can be used to compute the measure of central tendency known as the arithmetic or additive *mean.* If X is a variable and $X(i)$ is the value of that variable assigned to person i, then the mean of X for persons 1, 2, . . . , i, . . . , N is given by:

$$\overline{X} = \sum_{i=1}^{n} X(i)/N = \sum_{i=1}^{n} X_i/N \tag{3.1}$$

The mean of X is also referred to as the *expected value* of X. Expected value is written as:

$$E(X) = \overline{X} \tag{3.2}$$

The mean of a distribution of scores is a good representative of the "center" of the distribution if that distribution is symmetric. The center of skewed distributions is better represented by the median, and the center*s* of multimodal distributions are better described by local modes (most frequent scores).

3.3 MEASURES OF VARIABILITY

Measures of variability reflect the spread of a distribution about its center. The variability measure associated with the median is the *interquantile range.* The lower limit of this range is defined by the score exceeded by three-fourths of the distribution, and the upper limit is defined by the score exceeded by the upper one-fourth of persons in the distribution. This means that one-half of the people—in the middle half—are assigned scores in the interquantile range.

Variability about the mean of a distribution is usually measured by the *variance* (or the square root of the variance, the *standard deviation*). Variance is defined as the average (expected) squared distance of a score from the mean:

$$\text{variance } (X) = S_{XX} = \sum_{i=1}^{N} [X(i) - \overline{X}]^2/N \tag{3.3}$$

Similarly, in terms of expected value,

$$E(X - \overline{X})^2 = S_{XX} \tag{3.4}$$

An alternative definition of variance that is intuitively more appealing to some is the following: Randomly select two scores [$X(i)$ and $X(j)$] from the distribution, with replacement, and square the distance between them $[X(i) - X(j)]^2$. The expected value of this squared difference is twice the variance about the mean.

The standard deviation of a distribution is simply the square root of the variance, and it is labeled S_X. The standard deviation of a distribution is thus the root mean square (RMS) of deviations about the mean.

Problem Generate a few distributions of scores with a small number of cases and compute the variance. Can you concoct an alternative measure of variability? What is the problem with using the average deviation about the mean? Can you calculate the expectation of the squared distance between two scores?

The equation 3.3 is appropriate for computation of variance for a population. When a random sample has been selected from a population and the goal is to estimate the population variance from the sample, use of equation 3.3 results in an underestimate. If the population variance is σ_{XX}, then the expected value of the sample variance for samples of size N is:

$$E(S_{XX}) = \frac{N-1}{N}\sigma_{XX} \tag{3.5}$$

Thus, in order to obtain an unbiased estimate of σ_{XX}, it is necessary to multiply the sample variance by $N/(N-1)$:

$$\hat{\sigma}_{XX} = NS_{XX}/(N-1)$$

where $\hat{\sigma}_{XX}$ signifies an unbiased sample estimate of the population variance σ_{XX}.

Problem If the population mean, $\overline{X}^*$, was known, then squared deviations of sample scores would provide an unbiased estimate of σ_{XX} without adjustment. That is,

$$E(\Sigma(X(i) - \overline{X}^*)^2/N) = \sigma_{XX}$$

Can you speculate why it is that computation of the sample variance about the sample mean produces an underestimate of σ_{XX}?

3.4 MEASURES OF COVARIATION

The two basic measures of covariation used in this book are *covariance* and standardized covariance (or *correlation*). Each is a measure of the extent to which two variables are linearly related to one another. If X and Y are two variables that have been observed in a sample of N persons, then the covariance of X and Y is given by:

$$S_{XY} = \sum_{i=1}^{N} [(X(i) - \overline{X})[Y(i) - \overline{Y})]/N \tag{3.6}$$

Notice that S_{XY} can be positive or negative, whereas S_{XX} is always positive or zero.

The correlation coefficient is a standardized version of the covariance in that the product of the deviations that defines the covariance is divided by the standard deviation of each variable:

$$r_{XY} = S_{XY}/S_XS_Y \tag{3.7}$$

The correlation coefficient, r_{XY}, ranges from -1 to $+1$, thus providing a standardized index of the degree of linear relationship between X and Y.

The correlation coefficient is given several special names, depending upon the type of measurement scale involved. If X and Y are dichotomous, then r is called the *phi coefficient*; if X is dichotomous and Y is continuous, r is the *point biserial* correlation; if X and Y are ordinal ranks, r is the *Spearman's rho*; and if X and Y are continuous variables, r is *Pearson's* product moment correlation, or simply a correlation coefficient. In all these instances formula 3.7 is appropriate for computation. The main reason for differentiating these various applications of r is that the sampling distributions are slighly different for each, meaning that different significance tests are appropriate for each.

An alternative definition of the correlation coefficient can be developed by first defining standardized scores Z_X and Z_Y as:

$$Z_X = (X - \overline{X})/S_X \qquad Z_Y = (Y - \overline{Y})/S_Y$$

Then the correlation between X and Y is merely the mean of the product of the standardized scores:

$$r_{XY} = \sum_{i=1}^{N} Z_X(i)Z_Y(i)/N$$

3.5 POPULATION PARAMETERS AND SAMPLE STATISTICS

Much attention is often paid to the different computational formulas for population parameters and sample statistics. For example, the discovery that the use of $(N - 1)$ as the denominator of a sample variance makes it an unbiased estimator of the population variance is an intriguing finding. However, these details are of minimal practical concern. When sample size is large, the maximum likelihood estimate (i.e., the sample variance) is only trivially different from the unbiased estimate (when $N - 1$ is used in the denominator). (The meaning of "maximum likelihood estimation" is explained in Chapters 9 and 10.) Thus, it is in the small sample case that the bias can be substantial (say $N < 40$), and such small samples are usually not of interest in the testing of multivariable models. The more important issue is the conceptual distinction between a parameter and a statistic.

A population parameter is a fixed number. For example, the mean height of all Americans aged 21 at midnight is a theoretically knowable, although practically only estimable, number of centimeters. We might label this parameter m^* (after the Greek letter mu). Now, suppose that we select a random sample of Americans aged 21 and measure their heights at midnight tonight. We label this mean $m(X)$. The observed value of $m(X)$ in our sample is now a particular value between one and three meters. As a random variable, $m(X)$ has a certain distribution or probability density that describes how likely different values of $m(X)$ are to occur during random sampling. This density function is referred to as the *sampling distribution* of a sample statistic such as the mean. Sampling distributions exist for any sample statistic, and assumptions about the form of sampling distributions are at the heart of standard procedures for hypothesis testing or "significance" testing.

When a particular value of $m(X)$ is obtained from a sample, it may be treated as an *estimate* of the population parameter m^*. For most any distribution of X, the *expected value* of $m(X)$ is m^*. Expected value in this context means the average upon repeated sampling:

$$\text{expected value } [m(X)] = E[m(X)] = \text{m}^*$$

It is important at this point to distinguish between the two ways in which the notion of expectation can be applied. First, there is the above use: the expected value of $m(X)$

is the average value of $m(X)$ when $m(X)$ is computed for many samples. In this sense, then, we are referring to the "mean of the sample means." However, we could also treat a particular sample as a population and take the expected value of X for that finite population. Thus, we would have the expected value of X within the sample equal to the sample mean:

$$E(X) = m(X)$$

whereas the expected value of X *across samples* from the population is m^*. Thus, we might write

$$\text{expected value (within sample) of } [X - m(X)]^2 = S(X,X)$$
$$\text{expected value (across samples) of } [X - m(X)]^2 = [(N - 1)/N] \operatorname{Var}(X,X)$$

where $S(X,X)$ is the variance of the sample and $\operatorname{Var}(X,X)$ is the variance of the population from which the sample was randomly drawn.

The distinction between a sample statistic and a population parameter is crucial to the logic of hypothesis testing. All hypothesis testing procedures described in this book are a ramification of the following question: If a population parameter, say $\operatorname{Var}(X,X)$, has the value c^*, and this random sample has yielded a value of c, say for $S(X,X)$, as an estimate of c^*, then how likely is it that the observed value c would have occurred? In other words, the value assigned to c^* is a hypothetical "state of nature," and the task is to assess how likely our sample result c would be if the hypothesized state of nature were true.

4

Introduction to Regression Models

4.1 INTRODUCTION

In order to demonstrate the use of regression equations most simply, we will begin with the case in which all variables are dichotomous. There are certain circumstances in which problems arise from the application of regression techniques to dichotomous data. However, these problems are often trivial and, in practice, compensated for by the ease with which regression models can be applied. From the perspective of learning about data analysis, there is the additional advantage of having to learn only one statistical technique (multiple regression) for analysis of both dichotomous and continuous variables.

It is important to remember that causal attributions are not based on statistical procedures used to analyze the data. Causal inference rests on the *design* of the study. The focus of this chapter is the use of linear regression models to organize observations and paint a picture of the interrelationships that are discovered. It turns out that the coefficients of a linear regression model are easily obtained when dichotomous variables are involved. Thus, we can delay coverage of estimation procedures (used by computer programs) until a clear understanding of the meaning of the coefficients has been developed in this chapter.

4.2 SIMPLE EXPERIMENT

Let us begin with the results of a simple experiment in which a single independent variable (X) is manipulated to assess the effect on a single dependent variable (Y). Suppose that 200 people participate in the study and that one-half of the participants are randomly selected to undergo a "threat-reducing" experience—this is the experimental group. The other one-half of the participants are treated as similarly as possible, except that they are not subjected to the threat-reducing manipulation. At a point in time assumed to be sufficient for the manipulation to influence the authoritarianism of the experimental group members, the authoritarianism of both groups is assessed. The hypothetical results are presented in Table 4.1.

Table 4.1 Results of a hypothetical experiment testing the effect of reduced threat on authoritarianism.

	Democratic personality $Y = 0$	Authoritarian personality $Y = 1$
Low threat conditions: $X = 0$	70	30
Normal threat conditions: $X = 1$	50	50

Notice that members of the low threat group are assigned a score of zero on X, whereas members of the control group are assigned a score of one on X. Similarly, participants classified as authoritarian personalities are assigned a one for variable Y, and democratic personalities are assigned a zero on Y. These score assignments are written more succinctly as follows:

$$X_i = 0 \quad \text{if person } i \text{ in low threat condition}$$
$$X_i = 1 \quad \text{if person } i \text{ in normal threat condition}$$

where X_i is the score on X for the ith person in the sample ($i = 1, 2, \ldots, N$; $N = 200$), and Y_i is similarly defined.

The results of the experiment (Table 4.1) clearly suggest that reducing the level of threat has reduced the likelihood that a person will be authoritarian in personality. Without the threat-reducing manipulation ($X = 1$) the probability that a person would be assessed as authoritarian is .5 ($^{50}/_{100}$). With the manipulation ($X = 0$) this probability is reduced to .3 ($^{30}/_{100}$). This result may be written as:

$$P(Y = 1 | X = 1) = .50 \tag{4.1a}$$
$$P(Y = 1 | X = 0) = .30 \tag{4.1b}$$

Equation 4.1a is read as "the probability that Y equals one—*given* that X equals one—is one-half." Similarly, equation 4.1b states that "the probability that Y equals one—given that X equals zero—is three-tenths." In a sense, then, the *effect* of the manipulation is reflected in the difference between the two probabilities in equation 4.1. The effect of the treatment was to reduce the probability of authoritarianism by two-tenths (.50 − .30); or, conversely, the greater threat condition increases the probability of authoritarianism by .20. This effect is written as:

$$\Delta p_{YX} = P(Y = 1 | X = 1) - P(Y = 1 | X = 0) \tag{4.2}$$

where the Y precedes the X in the subscript of Δp to indicate that Y is the dependent variable and X is the independent variable.

A Linear Model The next task is to construct a linear model of the form:

$$Y_i = b_0 + b_1 X_i + T_i \tag{4.3}$$

such that the coefficients (b_0 and b_1) describe the experimental results given in Table 4.1. This could be accomplished by entering the X and Y scores for each participant into a regression program that estimates b_0 and b_1 by a procedure known as *ordinary least squares*. But there is a simpler procedure when two dichotomous variables are involved.

Equation 4.3 states that whether Y_i is zero (democratic) or one (authoritarian) for the ith person *depends* on whether X_i is zero (low threat) or one (normal threat) and on other factors (T_i) that determine whether or not the ith participant conforms to a democratic or authoritarian personality type—although the factors T_i have not been specified or measured.

The definitions of b_0 and b_1 in this instance are quite straightforward:

$$b_0 = P(Y = 1 | X = 0) \tag{4.4a}$$

$$b_1 = \Delta p_{YX} \tag{4.4b}$$

Thus, b_1 is intuitively defined as the magnitude of change in Y that results from a change in X. In the example presented in Table 4.1, a one-unit change in X (from zero to one) produces a .20 change in the probability that Y is one. Since the mean of Y is also the probability that Y equals one, the effect b_1 can also be interpreted as the extent to which a one-unit change in X increases the mean of Y. This latter interpretation will be the one used when we apply regression models to continuous variables.

The reason for defining b_0 and b_1 as in equation 4.4 becomes more apparent when the expectation (or mean) of equation 4.3 is written for separate values of X:

$$E(Y) = E(b_0 + T) \quad \text{when } X = 0 \tag{4.5a}$$

$$E(Y) = E(b_0 + b_1 + T) \quad \text{when } X = 1 \tag{4.5b}$$

or

$$\overline{Y}_{X=0} = b_0 + \overline{T}_{X=0} \tag{4.5c}$$

$$\overline{Y}_{X=1} = b_0 + b_1 + \overline{T}_{X=1} \tag{4.5d}$$

where $\overline{Y}_{X=0}$ is read as the mean of Y when X equals zero. Equations 4.5c and 4.5d can be further simplified since change in X was determined by a random process (true experimental randomization), and, presumably, *only* X was changed by the manipulation. The simplifications consist of the assumptions that

$$\overline{T}_{X=0} = \overline{T}_{X=1} = 0$$

The equality of the two means follows from the fact that randomization makes it very unlikely that the two groups differ on any factors other than X, and the assumption of zero means is simply for convenience. (Since the factors represented by T are not actually measured anyway, the mean of such factors is arbitrary and zero is a convenient choice.)

Equations 4.5 now can be written:

$$P(Y = 1 | X = 0) = \overline{Y}_{X=0} = b_0 \tag{4.6a}$$

$$P(Y = 1 | X = 1) = \overline{Y}_{X=1} = b_0 + b_1 \tag{4.6b}$$

For the results of our hypothetical experiment, equation 4.6 is given by:

$$Y = .3 + .2X + T \tag{4.7a}$$

$$P(Y = 1 | X) = .3 + .2X \tag{4.7b}$$

where $P(Y = 1 | X)$ is .3 when X equals zero and .5 when X equals one.

The coefficient b_0 is called the *intercept* (mean value of Y when X is zero) and the coefficient b_1 is a *regression coefficient.*

4.3 FACTORIAL EXPERIMENT WITHOUT INTERACTION

This next example will demonstrate the application of a regression model to data generated by a design involving two independent variables (X_1 and X_2) and their joint effects on one dependent variable (Y). Suppose that the experiment described in Section 4.2 was expanded to include a second experimental manipulation (X_2). The original sample of 200 people would be divided randomly into four groups of 50 people. Each of the four groups would then be subjected to a combination of two experimental conditions.

It might be argued, for example, that rational explanations of "how authoritarian personality characteristics can be a reaction to social threat" could lead to reduced authoritarianism. This could be tested by providing one-half of the participants in the reduced threat ($X_1 = 0$) and normal threat ($X_1 = 1$) conditions with such explanations. Who received the explanations and who did not would, of course, also be determined randomly. Table 4.2 presents hypothetical results of such a study.

A careful inspection of the frequencies in Table 4.2 will generate the following conclusion: (1) it appears that the effect of threat reduction ($X_1 = 0$) on the likelihood of authoritarianism ($Y = 1$) is comparable to the effect found in the previous experiment when no explanation of threat and authoritarianism is provided ($X_2 = 0$); (2) furthermore, the effect of giving the explanation ($X_2 = 1$) seems to have been *added on to* the effect of the threat-reducing manipulation. This interpretation can be expressed in terms of the probability of authoritarianism in each experimental condition:

$$P(Y = 1 | X_1 = 0 \text{ and } X_2 = 0) = 15/50 = .3 \tag{4.8a}$$
$$P(Y = 1 | X_1 = 1 \text{ and } X_2 = 0) = 25/50 = .5 \tag{4.8b}$$
$$P(Y = 1 | X_1 = 0 \text{ and } X_2 = 1) = 10/50 = .2 \tag{4.8c}$$
$$P(Y = 1 | X_1 = 1 \text{ and } X_2 = 1) = 20/50 = .4 \tag{4.8d}$$

Notice that when X_1 increases from zero to one, the probability of authoritarianism (or, equivalently, the mean of Y) increases by .2 (from .3 to .5 when $X_2 = 0$ and from .2 to .4 when $X_2 = 1$). Similarly, the effect of increasing X_2 from zero to one is a .1 decrease in the probability of authoritarianism (from .3 to .2 when $X_1 = 0$ and from

Table 4.2 Hypothetical results of a factorial experiment with no interaction.

		Democratic personality $Y = 0$	Authoritarian personality $Y = 1$
Explanation not given: $X_2 = 0$	Low threat: $X_1 = 0$	35	15
	Normal threat: $X_1 = 1$	25	25
Explanation given: $X_2 = 1$	Low threat: $X_1 = 0$	40	10
	Normal threat: $X_1 = 1$	30	20

.5 to .4 when $X_1 = 1$). These differences between the probabilities in equation 4.8 can be rewritten as:

$$\Delta p_{YX_1|X_2=0} = .5 - .3 = .2 \tag{4.9a}$$

$$\Delta p_{YX_1|X_2=1} = .4 - .2 = .2 \tag{4.9b}$$

$$\Delta p_{YX_2|X_1=0} = .2 - .3 = -.1 \tag{4.9c}$$

$$\Delta p_{YX_2|X_1=1} = .4 - .5 = -.1 \tag{4.9d}$$

Now the data in Table 4.2 were concocted so that equation 4.9a equals equation 4.9b and equation 4.9c equals 4.9d. That is, the effect of a change in X_1 on Y is *not* dependent upon which level of X_2 we are considering. Similarly, the effect of X_2 is not dependent upon X_1. This situation is described as additivity or *no interaction* between the effects of X_1 and X_2 on Y.

These results can easily be fit to a linear model of the following form:

$$Y = b_0 + b_1X_1 + b_2X_2 + T \tag{4.10}$$

where we say that Y has been regressed on X_1 and X_2. The goal of entering the scores on Y, X_1, and X_2 for each participant into a regression program would be to compute b_0, b_1, and b_2. However, we can obtain the ordinary least squares (OLS) solution of those coefficients by adopting the following:

$$b_0 = P(Y = 1 | X_1 = 0 \text{ and } X_2 = 0) = .3$$

$$b_1 = \Delta p_{YX_1|X_2=0} = .2$$

$$b_2 = \Delta p_{YX_2|X_1=0} = -.1$$

Thus, we can rewrite equation 4.10 as:

$$Y = .3 + .2X_1 - .1X_2 + T \tag{4.11}$$

Then, taking the mean of both sides of equation 4.11 gives:

$$P(Y = 1 | X_1 \text{ and } X_2) = \overline{Y}_{X_1X_2} = b_0 + b_1X_1 + b_2X_2 \tag{4.12a}$$

$$= .3 + .2X_1 - .1X_2 \tag{4.12b}$$

if the mean of T is zero for each combination of values for X_1 and X_2. The reader should play with equation 4.12b, substituting different values for X_1 and X_2, to convince herself that the model reproduces the probabilities of authoritarianism for each row of Table 4.2 (and equations 4.8). Furthermore, the coefficients b_1 and b_2 reflect the magnitudes of effects of changes in X_1 and X_2, respectively, on Y.

The causal model in equation 4.11 is depicted as a path diagram in Figure 4.1. How do we know that X_1, X_2, and T are uncorrelated (most likely)?

4.4 FACTORIAL EXPERIMENT WITH INTERACTION

The next task is to develop a linear model that fits and describes (and can reproduce) the results of causal models that involve interactive effects.

Suppose that the factorial experiment described in Section 4.2 gave the results displayed in Table 4.3. There are two ways to talk about these results. One, the effect of

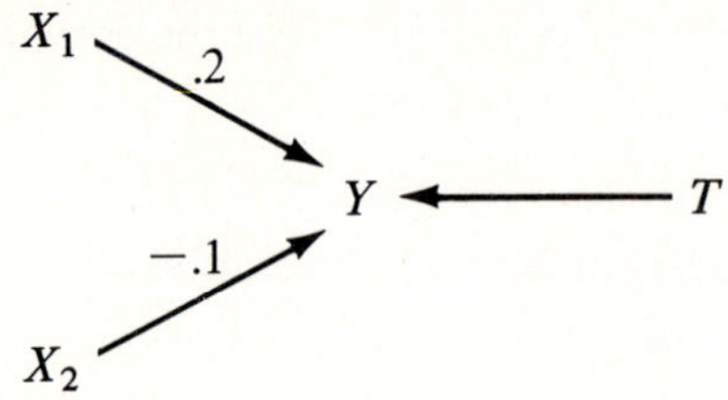

Figure 4.1 Path diagram for linear model with no interaction between causal variables. See Table 4.2 for basic data.

the threat-reducing treatment appears to be greater in the explanation condition than in the no-explanation condition. Or, two, there seems to be an effect of the rational explanation *only* in the lower threat condition. The message here is that the effect of *either* independent variable is dependent upon the level of the other independent variable. Our goal, then, is to construct a linear model that describes these results.

One strategy for constructing such a model is to add a multiplicative term (X_1X_2) to the model used in the additive case (equation 4.12). Clearly, this term has a value of one only when both X_1 and X_2 are one, and is zero otherwise. The model is then:

$$Y = b_0 + b_1X_1 + b_2X_2 + b_3X_1X_2 + T \tag{4.13}$$

It is probably not immediately apparent just how the addition of this interaction term will accommodate the nonadditivity of Table 4.3. It is easier to understand if we first write out the probabilities of authoritarianism in the following manner:

$$\begin{aligned}\Delta p_{YX_1|X_2=0} &= P(Y = 1|X_1 = 1 \text{ and } X_2 = 0) - P(Y = 1|X_1 = 0 \text{ and } X_2 = 0)\\ &= .60 - .30 = .30\\ \Delta p_{YX_1|X_2=1} &= P(Y = 1|X_1 = 1 \text{ and } X_2 = 1) - P(Y = 1|X_1 = 0 \text{ and } X_2 = 1)\\ &= .60 - .10 = .50\end{aligned}$$

Clearly, the "effect" of threat (X_1) is greater under the condition of rational explanation ($X_2 = 1$) than under the condition of no explanation ($X_2 = 0$). This is what is meant by interaction or nonadditivity. Thus, no constant coefficient for X_1 can model the effect of X_1 on Y. Rather, the effect itself must be a function of X_2. If we rewrite equation 4.13 in a slightly different manner, it is possible to see how the multiplicative term accomplishes this goal:

$$P(Y = 1|X_1X_2) = b_0 + (b_1 + b_3X_2)X_1 + b_2X_2 \tag{4.14}$$

Notice that the coefficient for X_1 is ($b_1 + b_3X_2$); that is, the coefficient is b_1 when X_2 is zero and ($b_1 + b_3$) when X_2 is one. Thus the model incorporates the observation in Table 4.3 that the effect of change in X_1 on Y is .3 when X_2 is zero and .5 when X_2 is one.

The model can then be fitted to the data by assigning the following values to the coefficients:

$$b_0 = P(Y = 1|X_1 = 0, X_2 = 0) = \overline{Y}_{X_1=0,X_2=0} = .3 \tag{4.15a}$$

$$b_1 = \Delta p_{YX_1|X_2=0} = \overline{Y}_{X_1=1,X_2=0} - b_0 = .6 - .3 = .3 \tag{4.15b}$$

$$b_2 = \Delta p_{YX_2|X_1=0} = \bar{Y}_{X_1=0,X_2=1} - b_0 = .1 - .3 = -.2 \tag{4.15c}$$

$$b_3 = \Delta p_{YX_1|X_2=1} - \Delta p_{YX_1|X_2=0} = \bar{Y}_{X_1=1,X_2=1} - b_0 - b_1 - b_2 \tag{4.15d}$$
$$= .6 - .3 - .3 - (-.2) = .2$$

Thus, the model in equations 4.13 and 4.14 becomes

$$P(Y = 1|X_1X_2) = \bar{Y}_{X_1X_2} = .3 + .3X_1 - .2X_2 + .2(X_1X_2) \tag{4.16a}$$
$$= .3 + (.3 + .2X_2)X_1 - .2X_2 \tag{4.16b}$$

Problem Substitute values of X_1 and X_2 into equations 4.16 to verify that the results given in Table 4.3 are reproduced by the model. Note how the value of the coefficient for X_1 in (4.16b) changes as X_2 changes.

Now, equations 4.15d and 4.16b depict the model in terms of how the effect of X_1 is dependent on X_2. However, as was noted earlier, the interaction between the effects of threat and rational explanation on authoritarianism can be described alternatively as the dependence of the effect of X_2 on X_1. Thus, equation 4.15d can be written as:

$$b_3 = \Delta p_{YX_2|X_1=1} - \Delta p_{YX_2|X_1=0} = (.6 - .6) - (.1 - .3) = .2$$

and the basic model (4.16a) can be rewritten as:

$$P(Y = 1|X_1X_2) = \bar{Y}_{X_1X_2} = b_0 + b_1X_1 + (b_2 + b_3X_1)X_2 \tag{4.17}$$

Problem Verify that equation 4.17 reproduces the appropriate probabilities of authoritarianism found in Table 4.3.

The result of fitting an interactive linear model to the frequencies in Table 4.3 can be depicted in a single path diagram, as in Figure 4.2, or with multiple path diagrams for different levels of one of the independent variables. Notice that the latter results when values for X_1 and X_2 are given for the diagram in Figure 4.2.

Problem How would you interpret the *magnitude* of b_3 in equation 4.16a? Demonstrate with the data in Table 4.3 that X_1 and X_2 are uncorrelated. How can the effects of X_1 and X_2 on Y "interact" if X_1 and X_2 are uncorrelated?

4.5 NONEXPERIMENTAL MODELS

Except under very special circumstances, fitting linear models to data from studies based on nonexperimental designs results in model coefficients that must be inter-

Table 4.3 Hypothetical results of a factorial experiment in which the effects of threat (X_1) and rational explanation (X_2) on authoritarianism (Y) are interactive (not additive).

		$Y = 0$	$Y = 1$	$p(Y = 1)$
No explanation: $X_2 = 0$	Low threat: $X_1 = 0$	35	15	.30
	Normal threat: $X_1 = 1$	20	30	.60
Explanation: $X_2 = 1$	Low threat: $X_1 = 0$	45	5	.10
	Normal threat: $X_1 = 1$	20	30	.60

preted differently from those in experiments. If a coefficient is treated as a causal effect, then we expect a change in X to actually produce the indicated change in Y (for most persons) when all other variables are held constant. But in a model fit to cross-sectional data, it is usually more appropriate to think of the coefficient as the difference between the average value of Y for people at one level of X and the average value for people at one unit higher on X. Whether or not an actual manipulation of X would produce a comparable change in Y is usually not known with any degree of confidence. Thus, regression coefficients for nonexperimental designs are usually *biased* estimates of structural (i.e., causal effect) coefficients. The bias arises because of uncontrolled third variables, reverse causation, etc. (Bias also occurs because of measurement error and sampling error. These sources of bias will be treated in subsequent chapters.)

The bias in nonexperimental regression coefficients that arises because of uncontrolled third variables, reverse causation, or nonlinearity (a problem that does not appear until we consider nondichotomous variables) is referred to as *specification error*. That is, the linear model to which sample data have been fit specifies the causal relations between variables in an inappropriate way: important sources of spuriousness are not controlled, direction of causality is specified incorrectly, etc. For this reason, we shall refer to noncausal regression coefficients as measures of *association* and give them the label *slope*. That is, given a causal system in approximate equilibrium, the slope measures the extent to which *observed* change in a variable Y (treated as dependent) is associated with *observed* (as opposed to manipulated) change in a variable X_1. If other variables (say $X_2, X_3, \ldots$) have been controlled, then the slope for X_1 is a *partial* slope.

There are some instances in which slopes may be interpreted as structural coefficients—primarily when a researcher has some reason to assume that all potential sources of spuriousness have been included in a model. However, as noted in Chapter 2, the potential for causal inference from nonexperimental data is usually very limited. Now that we have distinguished between effects and slopes, it is possible to be more specific about those limitations.

The inferential logic of the experiment can be summarized as follows:

1. Spurious and reverse causation alternative explanations are rejected *by design* (randomization and controlled manipulation);
2. Therefore, nonzero slopes can be interpreted as causal effects (structural coefficients).

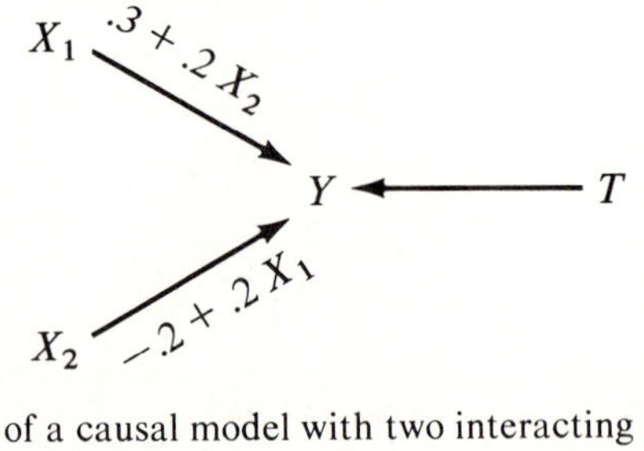

Figure 4.2 Path diagram of a causal model with two interacting (but uncorrelated) independent variables. See Table 4.3 for data.

Table 4.4 Hypothetical results of a cross-sectional study of democratic/authoritarian personality (Y), low/high social threat (X_1), and low/high education (X_2). Although each of these variables is best construed as continuous, they are treated as dichotomous for the sake of illustration.

	Y	
X_1	0	1
0	80	50
1	30	80

(A)

	X_1	
X_2	0	1
0	30	90
1	100	20

(B)

		Y		
X_2	X_1	0	1	$\bar{Y} = P(Y = 1)$
0	0	5	25	.833
0	1	15	75	.833
1	0	75	25	.250
1	1	15	5	.250

(C)

In contrast, only specific alternative models can be tested with cross-sectional data, and all relevant variables must actually be measured. And, in general, virtually no competing explanations can be rejected by design; all rejections rely upon inferences based on the magnitudes of slopes. And this reliance on magnitudes leads to further problems, since the magnitudes of slopes are meaningful only when variables are errorless. The situation is somewhat improved for longitudinal designs (since some such designs allow categorical rejection of reverse causation alternative models), but the magnitudes of slopes are still crucial—necessitating tedious and time-consuming attention to measurement models.

The following examples will demonstrate the use of linear models to *reject* causal models. As in the previous discussion, perfect measurement of all variables is assumed.

Spuriousness Suppose that a sample survey of 240 people from the adult population of New York State is conducted. Each respondent is classified according to three dichotomous variables: low/high social threat (X_1), democratic/authoritarian personality (Y_1), and low/high educational attainment (X_2). Hypothetical results of such a study are presented in Table 4.4. Notice (in part B of Table 4.4) that, unlike our experimental examples, the two variables to be treated as independent (X_1 and X_2) are correlated with one another. This means that the association between Y and X_1 overlaps with the association between Y and X_2. Therefore, we must develop a method for separating the *direct* association of Y with X_1 or (X_2) from the association that arises from the fact that both Y and X_1 are associated with X_2. This direct association is measured by the *partial slope,* and the partial slope for X_1 results from regressing Y on X_1 while holding X_2 constant. That is, computing the slope of Y regressed on X_1 separately for people with $X_2 = 0$ and people with $X_2 = 1$: computing the association between authoritarianism and social threat separately for people with a low amount of education and for people with a high amount of education.

The difference between a simple slope and a partial slope can easily be seen with the data in Table 4.4. Part A of that table presents the association between Y and X_1. Regressing Y on X_1 means that those data are fit to a simple linear model:

$$Y = b_0 + b_1X_1 + U \tag{4.18}$$

where the coefficients of the model are defined as before

$$b_0 = p(Y = 1|X = 0) = \overline{Y}_{X=0} = 50/130 = .385$$
$$b_1 = \Delta p_{YX} = p(Y = 1|X = 1) - b_0 = \overline{Y}_{X=1} - \overline{Y}_{X=0}$$
$$= 80/110 - .385 = .342$$

Notice that the slope ($b_1 = .342$) reflects the extent to which changing from the low threat to the high threat situation changes the probability of authoritarianism.

This result is clearly *consistent* with a hypothesis that actually increasing threat would increase authoritarianism. However, these results are equally consistent with models depicting the observed association as due to a third variable, reverse causation, and so on. The only strategy available for increasing confidence in a particular causal explanation is to demonstrate that other competing explanations are implausible. Or, perhaps, to find that an alternative model is preferable to the one currently favored. The strategy consists in generating predictions based on competing models and then checking the data to see which models' predictions are contradicted or upheld. The model for which the predictions are contradicted then diminishes in stature while the uncontradicted model is retained for additional testing.

Such a contest between competing models is portrayed in Figure 4.3. Part A of that figure presents a model in which social threat *does* have a causal effect on authoritarianism and that effect is in addition to whatever spurious correlation may be attributed to education. In equation form:

$$Y = c_0 + c_1X_1 + c_2X_2 + V \tag{4.19}$$

where c_1 is positive and sizable.

Part B of Figure 4.3 depicts a competing model: that the association between Y and X_1 (measured by the slope b_1 in equation 4.18) is spurious due to educational attainment influencing authoritarianism—and educational attainment covarying with social threat (Part B of Table 4.4). In equation form, this model corresponds to equation 4.19, except that c_1 is predicted to be small (close to zero). We now turn to the data in Table 4.4 to compute the slopes that provide the basis for contrasting the "fit" of

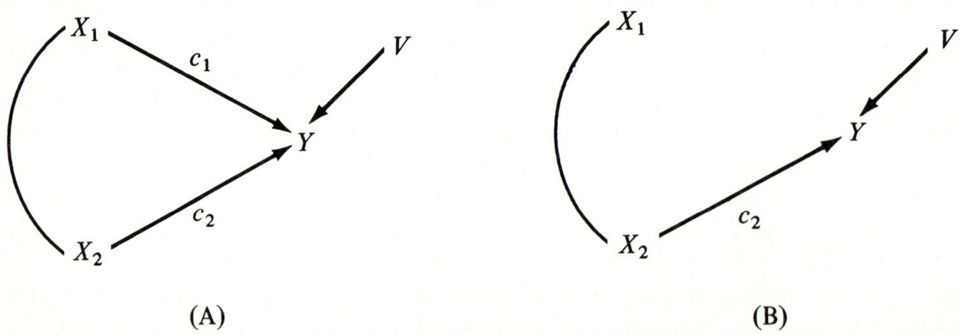

(A) (B)

Figure 4.3 Two competing causal models for the hypothetical survey data presented in Table 4.4.

the two models. This is accomplished, for the moment, by assuming that equation 4.19 is a "true" structural model operating in the population from which the sample was drawn *and* assuming that V is uncorrelated with both X_1 and X_2. These assumptions mean, if they are "true," that c_1 and c_2 (in 4.19) can be estimated by slopes.

These two slopes can then be computed as in the experimental case:

$$c_0 = P(Y = 1 \mid X_1 = 0 \text{ and } X_2 = 0) = \overline{Y}_{X_1=0,X_2=0} = .833 \tag{4.20a}$$

$$\begin{aligned} c_1 &= P(Y = 1 \mid X_1 = 1 \text{ and } X_2 = 0) - c_0 \\ &= \overline{Y}_{X_1=1,X_2=0} - c_0 \\ &= .000 \end{aligned} \tag{4.20b}$$

$$\begin{aligned} c_2 &= P(Y = 1 \mid X_1 = 0 \text{ and } X_2 = 1) - c_0 \\ &= .250 - .833 = -.583 \end{aligned} \tag{4.20c}$$

Problem Indicate why no interaction term is needed to fit a linear model to the data in Table 4.4.

The results clearly support the spurious alternative model (diagram B in Figure 4.3) in which there is no direct causal relation between threat and authoritarianism: the association between the two is explained by education's correlation with social threat (people with more education live in less threatening neighborhoods) and education's negative effect on authoritarianism (people with more education also are more likely to develop democratic personalities). Thus, through the technique of computing partial slopes, cross-sectional data have been used to test the relative plausibility of two causal models, and one of the two has been seen to be more plausible. However, there is a catch: interpretation of the slopes as measures of causal relations was *assumed*—not demonstrated. Nothing about the design of the survey or the results provides any evidence that contradicts a reverse causation explanation of the findings (changes in Y produce changes in X_1 or X_2). Nor is there any evidence that the factors represented by V do not produce a spurious association between X_2 and Y—or even *mask* a true causal relation between X_1 and Y. In short, the information provided about causation is indeed limited in this type of study. In fact, about all that can be said is that *if* the slopes are indicative of causal relations, then model A in Figure 4.3 can be rejected in favor of model B. But there are so many plausible alternatives to model B that no "experimentlike" statements about causation are warranted.

Problem Regress Y on X_1, Y on X_2 and X_1 on X_2 (using the data in Table 4.4). Notice that the first slope is equal to the product of the other two. Does this result resonate with your intuitions about the meaning of a spurious association between two variables?

4.6 MODEL FITTING VERSUS ESTIMATION

Thus far in this chapter we have focused upon the construction of linear equation models that reproduce the probability differentials observed in a sample of observations on dichotomous variables. It is important to understand this procedure of model fitting, but researchers are usually interested in a *population* of persons. Thus, we must be concerned with the question of whether or not the sample statistics that result from model fitting are good estimates of *parameters* for population models.

Suppose, for example, that the following causal model is "true" for Americans aged 20 to 70:

$$Y = b_0^* + b_1^* X + T^*$$

where the asterisks indicate population parameters, and X and Y are perfectly measured dichotomous variables. Now, suppose that a random sample of 500 persons is drawn from that population and a *sample estimate* is computed as outlined in this chapter:

$$Y = b_0 + b_1 X + T$$

Two concerns now arise about b_0 and b_1:

Upon repeated replications, would b_1 vary about b_1^*? More concisely, would the expected value of b_1 be equal to b_1^*? If b_1 does have this property, then b_1 is termed an *unbiased* estimator of b_1^*.

Does the method for computing b_1 make maximal use of the information in the sample such that the sample statistic is likely to be as close to the parameter as possible? In other words, it would be desirable if the sampling variance of b_1 about b_1^* were a minimum. It is possible, for instance, to prove mathematically that, under many circumstances, the ordinary least squares (OLS) estimators that compose conventional regression procedures are both unbiased and best (in the sense of minimum sampling variance) among all possible linear estimators. Such estimators are said to be BLUE (best linear unbiased estimate).

It may seem to the reader that only an unbiased estimator could have minimum variance. However, this may not be the case, especially when independent variables are highly correlated (a condition called *multicolinearity*).

Problems with Dichotomous Dependent Variables This discussion of estimation raises a problem with the use of linear equation models where the dependent variable is dichotomous. The main problem is that the criterion of additivity (no interaction) is inappropriate in many circumstances. There are also some more subtle and less important problems with linear equation models for dichotomous variables. Appendix B presents an approach that solves many of these problems: log linear models. It is hoped, therefore, that the reader will treat the preceding examples in the intended spirit: as simple demonstrations of how to apply linear models to systematic observations without the need for tedious calculations.

The reader is urged to review the application of linear models to dichotomous data by attempting to concoct data sets that have certain relationships "built in." Then obtain the regression coefficients using the procedures outlined in equation 4.4 and 4.15. The crucial insight that should result concerns the way in which the regression coefficients reflect the expected change in the dependent variable (Y) that is associated, or perhaps caused by, a change in an independent variable (X). This same logic will be generalized to variables with many levels, rather than just two, in the subsequent chapter, and a sound understanding at this point will greatly facilitate your progress. Can you start with a linear equation involving dichotomous variables?: First, assign values to the regression coefficients and then try to generate a data set that would fit the equation. Can you concoct a set of results analogous to Table 4.4c in which X_1 and X_2 interact such that the effect of social threat on the likelihood of authoritarianism is less for people with more education?

4.7 DIGRESSION: STANDARDIZED MEASURES OF ASSOCIATION

For the reader who enjoys puzzles, the following discussion of standardized measures of association will be both fun and useful in understanding the following chapters.

Suppose that two dichotomous variables (X and Y) are related in some population such that the following contingency table applies:

		Y	
		0	1
X	0	a	b
	1	c	d

where a, b, c, and d are the frequencies of persons falling in each of the four categories defined by the two levels of the two dichotomous variables. If a and d are larger than c and b, we would say that X and Y are positively correlated; if c and b are larger than a and d, we say that X and Y are negatively correlated. This assignment of negative and positive sign may be seen intuitively and through the computation of the regression coefficient resulting from the regression of Y on X. Recall that the slope of Y on X is given by

$$\begin{aligned}\Delta p_{YX} &= P(Y = 1 | X = 1) - P(Y = 1 | X = 0)\\ &= d/(c + d) - b/(a + b)\\ &= (ad - bc)/(a + b)(c + d)\end{aligned}$$

revealing that if the product ad is larger than the product bc, then the slope will be positive, and vice versa.

It may now be of interest to know the relationship between this regression coefficient and the correlation coefficient for X and Y. Recall that the correlation between two dichotomous variables is known as the phi coefficient (after the Greek letter phi that is usually used to denote it). It just so happens that the phi coefficient is related to the regression coefficient in an interesting manner. The correlation between X and Y is the multiplicative mean of the two regression coefficients that result from regression of Y on X and the regression of X on Y. The regression of Y on X was given above, and the regression of X on Y is given by the analogous equation:

$$\begin{aligned}\Delta p_{XY} &= P(X = 1 | Y = 1) - P(X = 1 | Y = 0)\\ &= d/(b + d) - c/(a + c)\\ &= (ad - bc)/(b + d)(a + c)\end{aligned}$$

The correlation coefficient is then defined as the square root of the product of the two regression slopes:

$$\begin{aligned}\text{phi} &= (\Delta p_{YX}\Delta p_{XY})^{1/2}\\ &= (ad - bc)/[(a + b)(c + d)(a + c)(b + d)]^{1/2}\end{aligned}$$

Notice that each of the regression coefficients will equal the correlation only if the product of the row marginals equals the product of the column marginals. This relationship between the slopes and the correlation coefficient will also hold when we move to more continuous variables. The usefulness of the correlation coefficient stems from its independence from differences between the variances of the two correlated variables—since it is this difference that gives rise to differences between the two regression coefficients. In this sense, then, the correlation gives the slope that would occur if the two variables had identical variances, and this is part of what occurs when variables are standardized.

The correlation is only one of many possible measures of association between two dichotomous variables. Two others that deserve note, for distinct reasons, are Yule's Q and the odds ratio. Yule's Q has the same numerator as the phi coefficient, but a different denominator:

$$Q = (ad - bc)/(ad + bc)$$

Q tends to be larger than phi, although both vary between -1 and $+1$. The important difference between Q and phi is their respective criterion for a perfect relationship (yielding a value of $+1$ or -1). Q will be equal to one if X is a necessary condition for Y or if X is a sufficient condition for Y (i.e., if any of the four cells is empty). Phi, however, can equal one only if X is both a necessary *and* sufficient condition for Y (i.e., both a and d, or b and c, must be zero). The choice between Q and phi may therefore stem from philosophical positions concerning the appropriate basis for causal inference. It may also be of interest to note that the original debate between the British statisticians Pearson and Yule over which measure was appropriate had a bit of an ideological twist. Pearson argued that a measure similar to the phi was most appropriate, since it gave results more consistent with his position as a eugenicist (a somewhat progressive political position at the turn of the century). Yule, on the other hand, opposed the eugenics movement and attempted to sway his statistical colleagues toward use of his Q statistic. The actual extent to which this apparently scientific debate was shaped and fueled by ideological differences between the two men remains, of course, difficult to ascertain from the historical record.

The final measure of association between two dichotomous variables that will be mentioned at this point is the *odds ratio:*

$$OR_{YX} = (d/c)/(b/a) = ad/bc$$

where d/c is the odds that Y is one when X is one, and b/a is the odds that Y is one when X is zero. The odds ratio ranges from zero to infinity with a value of one when there is no relationship between X and Y (i.e., when the phi or Q statistics are equal to zero). As shall be demonstrated in Appendix B, the odds ratio—or the log of the odds ratio—is probably the best measure of association for causal analysis of dichotomous or polychotomous variables.

Problem Concoct several two-by-two contingency tables and compute the various measures of association for each. Can you predict from the pattern of frequencies the approximate value of the association measures?

5

The Simple Regression Model

5.1 INTRODUCTION

The previous chapters presented a fairly nonmathematical introduction to how linear equation models can be applied to research problems. In order to develop these topics further and in a more precise manner, it will be necessary to give a more mathematical treatment of linear models and their estimation. It is my hope that the previous chapters have provided a cognitive map that will enhance the reader's understanding of what is to come.

The most crucial concepts for the reader to have gleaned are those involved in the interpretation of linear model coefficients: as associations of change (regression slopes) and causal effects (structural coefficients). Thus, the next section of this chapter summarizes these concepts.

5.2 STRUCTURAL EQUATIONS AND REGRESSION EQUATIONS

As we move toward actually estimating the magnitude of causal effects it becomes important to distinguish between *structural* equations and *regression* equations. Most simply, structural equations describe causal relations and can be estimated with regression equations if

1a. specified causal variables are independent of unspecified causes (as in a true experiment) *or* all important causal variables have been specified (statistically controlled);

1b. changes in the dependent variable are not causing changes in the independent variables;

2. causal variables are measured without error *or* the measurement error can be estimated (as in multiple indicator models—the multitrait multimethod strategy);

3. the measurement error in the causally dependent variables is symmetrically distributed about the true values of the variables (i.e., positive and negative errors tend to cancel one another over measurements of different people);

4. all variables in the model are either dichotomous or *linearly* interrelated after appropriate transformations (Note: contrary to common belief, linear models *are* applicable to investigation of many nonlinear systems, but basic variables must first be transformed to achieve linearity. For example, multiplicative and power functions can be linearized by logarithmic transformations. Some nonlinear relationships, however, are difficult to "linearize.")

All the above conditions are necessary for a causal interpretation of the coefficient b_1 in the following *regression* equation:

$$Y = b_0 + b_1 X + S$$

where b_1 is the amount that Y would change (for each person or other unit of analysis) if X were increased by one unit and *all other factors influencing Y remained unchanged.*

5.3 SOLVING FOR A STRUCTURAL COEFFICIENT OR SLOPE

The goal of this section is to explore the algebraic relations between the variances and covariances among variables and the coefficients of linear models.

Suppose that a particular structural model holds for a population of persons:

$$Y = b_0^* + b_1^* X_1 + T^* \tag{5.1}$$

where the asterisks indicate *population* parameters, and all variables are measured on ratio scales (including dichotomies) without error. To simplify the subsequent algebra, subtract the expectation (mean) of each side of equation 5.1 from each side:

$$Y - E(Y) = b_0^* + b_1^* X_1 + T^* - E(b_0^* + b_1^* X_1 + T^*)$$

which may be rewritten as

$$\begin{aligned} Y - \overline{Y} &= b_1^*[X_1 - E(X_1)] + [T^* - E(T^*)] \\ &= b_1^*(X_1 - \overline{X}_1) + T^* \end{aligned} \tag{5.2}$$

if we arbitrarily scale T^* such that $E(T^*) = 0$.

The trick to solving for b_1^* in terms of the variances and covariances of X_1 and Y in equation 5.2 arises from the definition of covariance given in Chapter 3 (equation 3.6). Both sides of equation 5.2 are multiplied by $(X_1 - \overline{X}_1)$ and the expectation is taken:

$$\begin{aligned} E[(Y - \overline{Y})(X_1 - \overline{X}_1)] &= E\{(X_1 - \overline{X}_1)[b_1^*(X_1 - \overline{X}_1) + T^*]\} \\ &= b_1^* E(X_1 - \overline{X}_1)^2 + E[T^*(X_1 - \overline{X}_1)] \end{aligned} \tag{5.3}$$

From the definitions of variance and covariance, it should be apparent that equation 5.3 can be rewritten as:

$$\sigma_{YX_1} = b_1^* \sigma_{X_1X_1} + \sigma_{X_1T^*} \tag{5.4}$$

where the sigmas are *population* variances or covariances.

We can now attempt to solve equation 5.4 for b_1^* as follows:

$$b_1^* = (\sigma_{YX_1} - \sigma_{X_1T^*})/\sigma_{X_1X_1} \tag{5.5}$$

The structural coefficient has now been expressed in terms of variances and covariances, but one of the covariances cannot be estimated: the term $\sigma_{X_1T^*}$. This covariance presents a problem since T^* is unmeasured (or unobserved).

The experimental solution to this "problem" is random manipulation of X_1 so that the covariance of X_1 and all other variables (except those influenced by variation in X_1) is, on the average, zero. Then equation 5.5 can be reduced to:

$$b_1^* = \sigma_{YX_1}/\sigma_{X_1X_1} \tag{5.6}$$

since it is safe to assume that $\sigma_{X_1T^*} = 0$ when X_1 is randomly manipulated. The population coefficient b_1^* can then be estimated from a sample by b_1 where

$$b_1 = S_{YX_1}/S_{X_1X_1} \tag{5.7}$$

The conditions in which b_1 is a good estimator of b_1^* will be discussed in a later section.

Now, what if the design of a study does not warrant the assumption that X_1 and T^* are uncorrelated? In such instances we can define the population *slope* c_1^* in the following manner:

$$c_1^* = \sigma_{YX_1}/\sigma_{X_1X_1} = b_1^* + \sigma_{X_1T^*}/\sigma_{X_1X_1} \tag{5.8}$$

In other words, if it cannot be assumed that $\sigma_{X_1T^*}$ is zero, then the slope c_1^* is defined as a function of the structural coefficient b_1^* and a bias due to the covariance of X_1 and T^*. Equation 5.8 is easily derived from equation 5.5 by rearranging terms.

The reader may be puzzled, initially, by equations 5.6 and 5.8, since b_1^* seems to be defined differently in the two equations. This is the case because equation 5.6 applies when it can be assumed that X_1 is uncorrelated with all other factors influencing Y, and equation 5.8 applies when this assumption is unwarranted. Notice that equation 5.8 reduces to equation 5.6 when $\sigma_{X_1T^*}$ is zero.

The most obvious formula for estimating c_1^* from a sample is:

$$c_1 = S_{YX_1}/S_{X_1X_1} \tag{5.9}$$

which is identical to equation 5.7. Thus, the distinction between an estimate of a slope and a structural coefficient does not necessarily rely upon computational procedures; the distinction flows from design assumptions.

Problem Use equation 5.9 to solve for the slope of Y regressed on X_1—using the data in Table 4.3. Note that when a variable is dichotomous and scored zero or one,

$$S_{XX} = N_0N_1/(N_0 + N_1)^2$$

where N_0 is the number of people with a score of zero and N_1 is the number of people with a score of one on X. Similarly, when X and Y are both dichotomous and scored zero or one:

$$S_{XY} = (Nd - N_XN_Y)/N^2$$

where d is the number of people scoring one on both X and Y, N_X is the number of people scoring one on X, N_Y is the number of people scoring one on Y and N is the sample size. To check yourself: $S_{XX} = .25$, $S_{XY} = .10$

Residuals and Disturbances The point of this section is to distinguish between structural *disturbances* and regression *residuals*. We will use the label disturbance to refer to the unmeasured factors that cause change in the dependent variable of a structural model (e.g., T^* in equation 5.1). But suppose that nonexperimental data are of interest and a slope is defined using equation 5.8. We can then write the population regression equation as:

$$Y = c_0^* + c_1^* X_1 + U^* \tag{5.10}$$

where the scores U^* are labeled the *residuals* of regressing Y on X_1. If we solve for c_1^* as we solved for b_1^* in equations 5.2 and 5.3:

$$(Y - \bar{Y}) = c_1^*(X_1 - \bar{X}_1) + (U^* - \bar{U}^*)$$

and

$$E(Y - \bar{Y})(X_1 - \bar{X}_1) = c_1^* E(X_1 - \bar{X}_1)^2 + E(X_1 - \bar{X}_1)(U^* - \bar{U}^*)$$
$$\sigma_{YX_1} = c_1^* \sigma_{X_1X_1} + \sigma_{X_1U^*}$$

Thus, it should be clear that U^* is a set of scores (one for each person) that satisfies the condition

$$\begin{aligned}\sigma_{X_1U^*} &= \sigma_{YX_1} - c_1^* \sigma_{X_1X_1} \\ &= \sigma_{YX_1} - (\sigma_{YX_1}/\sigma_{X_1X_1})\sigma_{X_1X_1} \\ &= 0\end{aligned}$$

In order to see just what these residual scores consist of, solve equation 5.10 for U^*

$$\begin{aligned}U^* &= Y - c_0^* - c_1^* X_1 \\ &= Y - (c_0^* + c_1^* X_1)\end{aligned} \tag{5.11}$$

and substitute the right side of equation 5.1 for Y:

$$U^* = (b_0^* - c_0^*) + (b_1^* - c_1^*)X_1 + T^* \tag{5.12}$$

Equation 5.11 defines the residual as the difference between Y and the so-called predicted value of $Y(c_0^* + c_1^* X_1)$. The predicted values of Y are usually labeled Y' and defined by the straight line equation

$$Y' = c_0^* + c_1^* X_1$$

Construing the regression model in this manner gives rise to an alternative way of defining the slope c_1^* and the intercept c_0^*: they define a line such that the mean of U^* is zero and the variance of U^* is a minimum. That is, if the scores X and Y are plotted as points in two-dimensional space, the c_0^* and c_1^* define a line such that the average squared distance of points from the line (measured along a direction parallel with the Y-axis) is the least possible for any straight line. For this reason, slopes defined as in equations 5.8 and 5.9 are described as resulting from the method of ordinary least squares (OLS), since the slopes are obtained that minimize the sum of the squared residuals.

Equation 5.12 makes it clear that when b_1^* equals c_1^*, the residual is equivalent to the disturbance (plus a constant). However, in general, the residuals overstate or understate the disturbances.

5.4 THE CALCULUS OF ORDINARY LEAST SQUARES

The OLS solution for the line of "best fit" in two (or more) dimensional space is elegant in its simplicity and power. It is summarized here for those who would like to know how differential calculus is used in curve fitting applications.

Suppose that X and Y are continuous variables and that N points in two-dimensional space have been defined by pairs of scores (X_i, Y_i). Straight lines in this space are defined by equations of the form

$$Y_i' = d_0 + d_1 X_i \tag{5.13}$$

The question of concern here is whether or not there are unique values of d_0 and d_1, call them c_0 and c_1, such that

$$\sum_{i=1}^{N} (Y_i - Y_i')^2 = \Sigma(Y_i - d_0 - d_1 X_i)^2 \tag{5.14}$$

is a minimum. (There are alternative criteria for a "best fitting" line, but the criterion in equation 5.14 has numerous advantages.)

If the sum in equation 5.14 is defined as $F(d_0, d_1)$, then our goal is to find values of d_0 and d_1 that minimize the function F. The standard approach to such a problem is to conceive of F, d_0, and d_1 as dimensions in three-dimensional space. Then, given a set of scores (X_i, Y_i),

$$F(d_0, d_1) = \sum_{i=1}^{N} (Y_i - d_0 - d_1 X_i)^2 \tag{5.15}$$

defines a surface that may have a minimum point where the partial derivatives of F are equal to zero. First, expand equation 5.15:

$$F(d_0, d_1) = \Sigma(Y^2 + d_0^2 + d_1^2 X^2 - 2d_0 Y - 2d_1 XY + 2d_0 d_1 X) \tag{5.16}$$

The partial derivatives can then be written as:

$$\partial F/\partial d_1 = 2d_1 \Sigma X^2 - 2\Sigma XY + 2d_0 \Sigma X$$

$$\partial F/\partial d_0 = 2d_0 N - 2\Sigma Y + 2d_1 \Sigma X$$

This yields two equations with two unknowns if we set each partial equal to zero [which defines the point $F(c_0, c_1)$ where $F(d_0, d_1)$ is a minimum]:

$$c_1 \Sigma X^2 + c_0 \Sigma X = \Sigma XY \tag{5.17a}$$

$$c_1 \Sigma X + c_0 N = \Sigma Y \tag{5.17b}$$

Equations (5.17) can then be solved for c_0 and c_1 in terms of observed values of X and Y:

$$c_1 = (\Sigma X \Sigma Y - N \Sigma XY)/[(\Sigma X)^2 - N \Sigma X^2] \tag{5.18a}$$

$$c_0 = (\Sigma X \Sigma XY - \Sigma Y \Sigma X^2)/[(\Sigma X)^2 - N \Sigma X^2] \tag{5.18b}$$

Problem Show that equation 5.18a is equal to S_{XY}/S_{XX} and that equation 5.18b is equal to $(\overline{Y} - c_1 \overline{X})$.

Notice that the idea of estimating a population model from a sample model does not enter into the logic of the OLS procedure. Nevertheless, it can be shown that OLS coefficients derived from a random sample are BLUE (best linear unbiased estimates) relative to the coefficients for the population from which the sample was drawn. Thus, the intuitively appealing methods of estimation given in equations 5.7 and 5.9—developed by simply substituting sample variances and covariances for population variances and covariances—turn out to be very useful estimates of population parameters.

There are some further assumptions, however, that are necessary for OLS estimates to be BLUE. A sufficient set of such assumptions is the following: (1) the dependent variable is continuous and measured on an interval scale; (2) the independent variable is also of an interval type and values are *fixed* upon replications; (3) the disturbances or residuals are independent of the fixed values of the independent variable. The effect of violating one or more of these assumptions is best measured by simulation studies, in which data from populations with known characteristics are sampled. Large numbers of samples are generated under conditions of minor and major assumption violation. The reader with an interest in very precise estimation should consult this literature in order to gauge the impact of a suspected violation (e.g., Hanushek and Jackson, 1977). In general, however, OLS estimates from large samples are *robust* with respect to violations of the fixed effects assumption (that different values of independent variables are like fixed levels in an experimental design) and of the requirement of a continuous dependent variable. (*Logit* or *probit* analysis procedures are most appropriate when excessive distortions in OLS estimates are suspected from the use of a dichotomous dependent variable. See Appendix B for a discussion of *log-linear models*.)

5.5 AN ALTERNATIVE APPROACH TO MINIMIZING RESIDUALS

The method of fitting a straight line to a set of points that was developed in the previous section involves partial derivatives and may be less than clear to many readers. Thus, this brief section outlines an alternative way of seeing that defining the slope of Y on X as we have (i.e., $c_1 = S_{YX}/S_{XX}$) generates residuals with the least possible variance.

Suppose that we have a set of N observations on variables X and Y. Now, if Y is expressed as a linear function of X and a residual U, then

$$Y_i = c_0 + c_1 X_i + U_i \qquad (i = 1, 2, \ldots, N) \tag{5.19}$$

Summing both sides of equation 5.19 over all N persons and dividing by N gives

$$\Sigma Y_i/N = \Sigma(c_0 + c_1 X_i + U_i)/N$$

or

$$\overline{Y} = c_0 + c_1\overline{X} + \overline{U} \tag{5.20}$$

Subtracting equation 5.20 from equation 5.19 gives

$$Y_i - \overline{Y} = c_1(X_i - \overline{X}) + (U_i - \overline{U}) \tag{5.21}$$

Now, the OLS procedure gave values of c_0 and c_1 such that $\Sigma(U_i)^2$ is the least possible. We found that this OLS solution was the same as that that follows from assuming that X and U are uncorrelated. That is, we multiply both sides of equation 5.21 by $(X_i - \overline{X})$ and sum over the N persons:

$$\Sigma(Y_i - \overline{Y})(X_i - \overline{X}) = c_1\Sigma(X_i - \overline{X})^2 + \Sigma(X_i - \overline{X})(U_i - \overline{U}) \quad (5.22)$$

Dividing both sides of equation 5.22 by N gives

$$S_{YX} = c_1 S_{XX} + S_{XU} \quad (5.23)$$

Recall now that OLS indicated that c_1 equals S_{YX}/S_{XX} when S_{UU} is minimized. Plugging this value for c_1 into equation 5.23 implies that S_{XU} is zero.

We had come to the conclusion that S_{XU} should be zero when X was an experimental manipulation (and U would be considered a disturbance). But it is not immediately clear that choosing a set of residuals U_i such that $S_{XU} = 0$ would necessarily mean that those residuals had the least possible variance. This can be demonstrated, however, with a bit of algebra.

Our goal is to express the variance of the residuals S_{UU} as a function of S_{XU} and demonstrate that S_{UU} is a minimum when S_{XU} is zero. We begin by squaring equation 5.21, summing over N persons, and dividing by N. This gives

$$S_{YY} = c_1^2 S_{XX} + 2c_1 S_{XU} + S_{UU}$$

Solving for S_{UU} gives

$$S_{UU} = S_{YY} - c_1^2 S_{XX} - 2c_1 S_{XU} \quad (5.24)$$

If equation 5.23 is solved for c_1 and substituted into equation 5.24 we obtain

$$\begin{aligned} S_{UU} &= S_{YY} - S_{XX}\left(\frac{S_{YX} - S_{XU}}{S_{XX}}\right)^2 - 2\left(\frac{S_{YX} - S_{XU}}{S_{XX}}\right)S_{XU} \\ &= S_{YY} - (S_{XY}^2/S_{XX}) + (S_{XU}^2/S_{XX}) \end{aligned} \quad (5.25)$$

Since the third term in equation 5.25 is necessarily positive, it should be clear that S_{UU} is a minimum when S_{XU} is zero. Furthermore, when S_{XU} is zero, equation 5.25 can be rewritten as

$$S_{YY} = (S_{XY}^2/S_{XX}) + S_{UU}$$

which can be rewritten as

$$S_{YY} = c_1^2 S_{XX} + S_{UU} \quad (5.26)$$

Equation 5.26 indicates that the OLS solution for c_1 yields a linear model with residuals such that the variance of Y is *partitioned* into a portion associated with $X(c_1^2 S_{XX})$ and a portion uncorrelated with $X(S_{UU})$. We will return to this useful outcome later in this chapter, but it indicates that minimizing S_{UU} is equivalent to maximizing the portion of S_{YY} associated with X.

Problem Notice that both c_0 and $\overline{U}$ were not taken into account in the demonstration that defining U such that S_{XU} is zero minimizes S_{UU}. According to equation 5.20 we have

$$
\begin{aligned}
c_0 &= \overline{Y} - c_1\overline{X} - \overline{U} \\
&= \overline{Y} - \overline{X}S_{XY}/S_{XX} - \overline{U} \qquad \text{(when } S_{XU} = 0\text{)}
\end{aligned}
$$

Is there a unique solution for c_0? (Hint: $c_0 + \overline{U}$ is uniquely defined and $\overline{U}$ must be chosen such that ΣU_i^2, is a minimum. Can you demonstrate that $\overline{U}$ must be zero?)

5.6 AN EXAMPLE

Before we proceed further to specify the various regularities of OLS linear models it will probably be helpful for most readers to think in terms of a numerical example.

Suppose that the structural model

$$Y = b_0 + b_1X + T \tag{5.27}$$

is known to hold for a population of 16 persons where Y = authoritarianism score (possible range: 0 to 20), X = zero (low threat) or one (normal threat), T = disturbance, $b_0 = 7.5$, and $b_1 = 3.0$. The data for this hypothetical population are given in Table 5.1. Notice that each authoritarianism score can be reproduced by substituting appropriate values for X and T into equation 5.27, with $b_0 = 7.5$ and $b_1 = 3.0$.

The reader should consider the data in Table 5.1 very carefully, noticing that the structural coefficient b_1 is not equal to the difference

$$\overline{Y}_{X=1} - \overline{Y}_{X=0} = 4.0$$

That is, the average difference in authoritarianism between the low and normal threat groups is greater than the causal effect of changing from one threat condition to another. Recall that the structural model reflects *causal change:* b_1 is the change in Y caused by a change in X from zero (low threat) to one (normal threat).

Thus, we must conclude that the difference of four F-scale points between the F-scale means in the two threat conditions ($\overline{Y}$ is 7.5 when $X = 0$ and 11.5 when $X = 1$) must be due to more than the effect of a change in X. Since Y is completely determined by X and T, this means that the unmeasured (and usually unknown—except in concocted examples such as this one) factors T must have been confounded with differences in threat (X). This is to be expected in nonexperiments and can be seen in Table 5.1 by noting that T tends to be larger when X is one than when X is zero. The extent to which X and T are confounded is indexed by their covariance ($S_{XT} = .25$).

We already know that b_1 is 3.0, but we could solve for it using equation 5.5 above and substituting the appropriate values from Table 5.1:

$$
\begin{aligned}
b_1 &= (S_{YX} - S_{XT})/S_{XX} \\
&= (1.0 - .25)/.25 \\
&= 3.0
\end{aligned}
$$

Similarly, b_0 can be obtained with the equation

$$
\begin{aligned}
b_0 &= \overline{Y} - b_1\overline{X} - \overline{T} \\
&= 9.5 - 3.0(.5) - .5 \\
&= 7.5
\end{aligned}
$$

Table 5.1 Hypothetical data for 16 persons. See text for explanation of variables. The disturbance T^* has a nonzero covariance with X and thus a larger variance than the residuals U.

Person	Y	X	T^*	U
P_1	3	0	−4.5	−4.5
P_2	6	0	−1.5	−1.5
P_3	6	0	−1.5	−1.5
P_4	6	0	−1.5	−1.5
P_5	9	0	1.5	1.5
P_6	9	0	1.5	1.5
P_7	9	0	1.5	1.5
P_8	12	0	4.5	4.5
P_9	7	1	−3.5	−4.5
P_{10}	10	1	−.5	−1.5
P_{11}	10	1	−.5	−1.5
P_{12}	10	1	−.5	−1.5
P_{13}	13	1	2.5	1.5
P_{14}	13	1	2.5	1.5
P_{15}	13	1	2.5	1.5
P_{16}	16	1	5.5	4.5
Mean	9.5	.5	.5	0.0
Variance	10.75	.25	7	6.75
$S_{YX} = 1.0$	$S_{XT} = .25$	$S_{XU} = 0$		
$S_{YU} = 6.75$	$S_{YY/X} = 6.75$			

These numbers can perhaps make more sense if T is given a substantive interpretation. Suppose that T is primarily an index of political powerlessness and that people in higher threat conditions ($X = 1$) tend to have less political power ($\overline{T} = 1$) than people in low threat conditions ($X = 0$ and $\overline{T} = 0$). Thus the difference in authoritarianism between the low and normal threat conditions (4 points) represents both the effect of threat (3 points: b_1) *and* the effect of political powerlessness (T) that covaries with threat conditions (1 point).

Regression Up to this point we have dealt with a structural model in which causal relations were known. This is a very unrealistic set of circumstances, especially when we assume we know the effects of unmeasured factors (the disturbance T). The more usual situation with nonexperimental designs is that T is truly unknown and we must settle for computation of the regression slope and resulting residuals. This procedure is equivalent to fitting a line to points defined by the Y and X scores according to the best fit criterion of OLS.

We can write the regression equation as

$$Y = c_0 + c_1 X + U$$

where Y and X are as given in Table 5.1. The solution for c_1 is given by equation 5.8:

$$\begin{aligned} c_1 &= S_{YX}/S_{XX} \\ &= 1.0/.25 \\ &= 4.0 \end{aligned}$$

Similarly, from equation 5.17b we know that

$$
\begin{aligned}
c_0 &= \overline{Y} - c_1\overline{X} \\
&= 9.5 - 4(.5) \\
&= 7.5
\end{aligned}
$$

Thus, the regression of authoritarianism on social threat in Table 5.1 yields the following equation:

$$Y = 7.5 + 4X + U \tag{5.28}$$

And equation 5.28 can be rewritten to give the residuals for each case

$$U = Y - 7.5 - 4X$$

Notice that these residuals are very different from the disturbances T. The disturbances were real factors influencing authoritarianism; the residuals are simply a set of scores that satisfy the requirements of OLS (i.e., using any values for c_0 and c_1 other than those in equation 5.28 will result in residuals with a larger variance).

If we now compare the slope to the structural coefficient, we can see that the slope is biased upwards relative to the structural coefficient. This bias is said to result from *specification error* in the regression equation. That is, an important variable or variables that are both correlated with X and causally related to Y have been left out of the model. Thus, if we specify that S_{XT} is zero, we would be misspecifying the model.

Problem If X in Table 5.1 was a randomly manipulated variable and the cases constituted a *sample,* then how would you interpret the difference between c_1 and b_1? Why might S_{XT} not be zero?

5.7 SOME IDENTITIES FOR LINEAR MODELS

Careful inspection of the numerical example in Table 5.1 reveals some useful relationships between the various variances and covariances among variables in a linear model. Consider those that hold for the OLS regression model.

First, the variance in Y is partitioned by the regression model into two portions

$$
\begin{aligned}
S_{YY} &= c_1^2 S_{XX} + S_{UU} &\quad (5.29a)\\
&= c_1 S_{YX} + S_{UU} &\quad (5.29b)
\end{aligned}
$$

Substituting from Table 5.1 gives

$$
\begin{aligned}
10.75 &= (4)^2(.25) + 6.75 \\
&= 4.0 + 6.75
\end{aligned}
$$

Now, we know that the value 6.75 for S_{UU} is the least possible, and thus the value of 4.0 for $c_1^2 S_{XX}$ is a maximum (since S_{YY} is fixed by the data). The variance in Y has thus been partitioned into the maximum amount attributable to X and the least amount attributable to other factors (that have no covariance with X).

Equations 5.29 can be derived by starting with the standard regression model expressed in mean deviations:

$$(Y - \overline{Y}) = c_1(X - \overline{X}) + U \tag{5.30}$$

Squaring both sides of equation 5.30 and averaging over N persons gives

$$\Sigma(Y - \overline{Y})^2/N = [c_1^2\Sigma(X - \overline{X})^2/N] + (\Sigma U^2/N) + [2c_1\Sigma U(X - \overline{X})/N]$$

which may be rewritten as

$$S_{YY} = c_1^2 S_{XX} + S_{UU}$$

since S_{XU} is zero. (Note that $c_1^2 S_{XX} = c_1 S_{YX}$, since $c_1 = S_{YX}/S_{XX}$). And the term $c_1^2 S_{XX}$ is simply the variance of c_1X.

Second, the variance of the residuals, the covariance of Y and U, and the conditional (or *within group*) variance of Y are all equal. That is

$$S_{UU} = S_{YU} = S_{YY|X}$$

and all are 6.75 in Table 5.1. The definitions of S_{UU} and S_{YU} are straightforward, and the conditional variance of Y is defined as the variance of Y when X is held constant. (The conditional variance can also be construed as variability about the OLS line of best fit.)

Problem Compute the variance of Y when X is zero in Table 5.1. Then compute the variance of Y when X is one. A weighted sum of these two variances defines the conditional variance of Y if Y and X are linearly related and the two variances are approximately equal (homoscedasticity).

These equalities can be derived from equation 5.30 by multiplying both sides of the equation by U and averaging:

$$\Sigma(Y - \overline{Y})U/N = [c_1\Sigma(X - \overline{X})U/N] + \Sigma U^2/N$$

or

$$\begin{aligned} S_{YU} &= S_{XU} + S_{UU} \\ &= S_{UU} \end{aligned}$$

It can be seen that the conditional variance is also equal to S_{UU} by noticing that when X is held constant, only U is varying—thus accounting for all the variance in Y when X is not changing.

Third, we can define the proportion of variance in Y that is explained by X as

$$r_{YX}^2 = c_1^2 S_{XX}/S_{YY}$$

or

$$r_{YX}^2 = (S_{YY} - S_{UU})/S_{YY} = 1 - [S_{UU}/S_{YY}]$$

These two expressions for the squared correlation coefficient suggest two interpretations. The first is that r^2 is proportional to the square of the magnitude of the slope; and the second is that r^2 is inversely proportional to the variance of the residuals.

For the data in Table 5.1, we can compute r_{YX}^2 as

$$\begin{aligned} r_{YX}^2 &= (S_{YY} - S_{UU})/S_{YY} \\ &= (10.75 - 6.75)/10.75 \\ &= .37 \end{aligned}$$

We also could have computed r_{YX} in the manner described in Chapter 3:

$$\begin{aligned} r_{YX} &= S_{YX}/S_XS_Y \\ &= 1.0/(10.75)(.25) \\ &= .61 \end{aligned}$$

Problem Contrast how the structural equation 5.27 partitions the variance in Y with the partitioning found for the regression model in equation 5.29a. (Hint: show that

$$S_{YY} = b_1^2S_{XX} + 2b_1S_{XT} + S_{TT})$$

How does the variance of b_1X contrast with the variance of c_1X for the data in Table 5.1?

5.8 SAMPLING AND STANDARD ERROR

In the previous two sections of this chapter we have dealt with the data from Table 5.1 as if they constituted a population. In most scientific applications, however, we select a *sample* from a population with the hope of generalizing our findings to the population. In other words, a researcher wants some indication of whether or not a replication of his study is likely to give similar results. (In the context of Table 5.1 this means that if those 16 persons compose a sample from a population, then what kind of confidence can we have that including observations on all the people in the population would yield a value of c_1 close to that observed in the sample: $c_1 = 4.0$). The best way to determine whether or not a relationship between variables (represented by coefficients in a linear equation model) is replicable is to replicate the finding with a new sample. However, it would be helpful to have some indication from a first sample as to how close the sample estimates of population coefficients are likely to be.

As the reader might expect, making inferences about what *would* happen *if* a study were replicated is usually a risky endeavor. However, there is a very special circumstance under which we can have considerable confidence that a sample statistic (i.e., the sample mean $\overline{X}$) has a very specific and well understood relationship to its population parameter (i.e., the population mean μ). That special circumstance results from the procedure of *random sampling*—selecting a sample of N persons from a population such that all possible samples of N persons are equally likely to be selected.

Intuitively, the random sample seems consistent with the goal of obtaining a representative sample of a population. But there is another more subtle and literally astonishing consequence of random sampling. This consequence has been expressed formally by one of the most extraordinary theorems in the history of mathematics: the *central limit theorem.* This theorem simply states that the distribution of means of random samples of N persons from a population approach normality as N increases.

In other words, the sampling distribution of $\overline{X}$ is approximately normal *regardless* of the shape of the distribution of X. In equation form,

$$Z = \frac{(\overline{X} - \mu)}{\sigma_X/\sqrt{N}} \tag{5.31}$$

defines a normal variable with zero mean and unit variance; where $\overline{X}$ is a random sample mean, μ is the population mean, σ_{XX} is the population variance, and N is the number of persons in the sample.

Equation (5.31) means that all the regularities associated with the bell-shaped normal curve can be used to make statements about $\overline{X}$. For example, defining the *standard error of the mean as*

$$\sigma_{\overline{X}} = \sigma_X/\sqrt{N}$$

we know that, when N is large (say over 100), $\overline{X}$ will range between $\mu - \sigma_{\overline{X}}$ and $\mu + \sigma_{\overline{X}}$ for 68 percent of random samples with N participants. Furthermore, $\overline{X}$ will range between μ plus and minus two standard errors for 95 percent of samples. These familiar characteristics of a normal variable would be of enormous help to any investigator who wants information concerning the closeness of sample means to population means.

There is a crucial catch, however. There are few research situations in which the population variance is known. Thus, we must substitute a sample estimate for the population standard error:

$$\hat{\sigma}_{\overline{X}\overline{X}} = \hat{\sigma}_{XX}/N = S_{XX}/(N - 1) \tag{5.32}$$

The reader may be tempted to substitute this sample estimate of the standard error of the mean into equation 5.31 to obtain a standard normal variate. Unfortunately, the variate that actually results from such a substitution does not have the nice properties of a normal distribution unless sample size is large. In fact, the resulting ratio

$$t = \frac{\overline{X} - \mu}{S_X/\sqrt{N - 1}} \tag{5.33}$$

has well defined properties *only* if the original distribution of X scores is normal. When such normality holds, t is distributed according to student's famous t-distribution with $(N - 1)$ degrees of freedom.

The requirements of randomness and normality are troublesome. Serious efforts to obtain random samples are usually restricted to opinion polls and variables are frequently non-normal in distribution. But the problem of nonrandomness is usually the most serious, since the expense involved in obtaining a random sample is so often prohibitive *and* since deviations from normality can be compensated by transformations or simply increasing the sample size. The message of simulation studies is that the t-test is quite robust in the face of nonnormality if sample size exceeds 50.

Problem Can you show that the variance of a sampling distribution of means ($\sigma_{\overline{Y}\overline{Y}}$) for samples of size N (drawn randomly from a population with variance σ_{YY}) is equal to σ_{YY}/N?

Solution: Recall that the definition of population variance is as follows

$$\sigma_{YY} = E[(Y - \mu)^2]$$

that is, the expected squared deviation of Y from μ. The variance of sample means is similarly defined

$$\sigma_{\overline{Y}\overline{Y}} = E[(\overline{Y} - \mu)^2] \tag{5.34}$$

Since $\overline{Y}$ is the average of N scores, equation 5.34 can be rewritten as

$$\begin{aligned} \sigma_{\overline{Y}\overline{Y}} &= E[(1/N)(Y_1 + Y_2 + \ldots + Y_N) - \mu]^2 \\ &= (1/N)^2 E(Y_1 + Y_2 + \ldots + Y_N - N\mu)^2 \\ &= (1/N)^2 E[(Y_1 - \mu) + (Y_2 - \mu) + \ldots + (Y_N - \mu)]^2 \\ &= (1/N)^2 E[(Y_1 - \mu)^2 + \ldots + (Y_N - \mu)^2 \\ &\quad + 2(Y_1 - \mu)(Y_2 - \mu) + \ldots] \end{aligned} \tag{5.35}$$

Now, notice that the first N terms in equation 5.35 are each equal to the variance of Y and that the subsequent terms are analogous to covariances between one sampled person and another. It is at this point that the requirement of a random sample becomes crucial: each Y score sampled must be independent of all others so that the covariance terms in equation 5.35 are zero. Then equation 5.35 becomes

$$\begin{aligned} \sigma_{\overline{Y}\overline{Y}} &= (1/N)^2 N\sigma_{YY} \\ &= \sigma_{YY}/N \end{aligned}$$

Then, since the expectation of $NS_{YY}/(N - 1)$ is σ_{YY}, it follows that

$$\begin{aligned} E[S_{YY}/(N - 1)] &= \sigma_{YY}/N \\ &= \sigma_{\overline{Y}\overline{Y}} \end{aligned} \tag{5.36}$$

Therefore, $S_{YY}/(N - 1)$ is an unbiased sample estimate of $\sigma_{\overline{Y}\overline{Y}}$ regardless of whether or not Y is normally distributed.

Problem Take the population of scores (1, 2, 3). The variance of this population is $\sigma_{YY} = 2/3$. Drawing samples of two scores from this population, there are six possible samples if scores are drawn *without* replacement (1,2; 2,1; 1,3; 3,1; 2,3; 3,2) and nine possible samples if scores are drawn *with* replacement (1,1; 1,2; 2,1; 1,3; 3,1; 2,2; 2,3; 3,2; 3,3). Under which of these sampling procedures does equation 5.36 hold? Hints: There are two approaches to this problem. One is to compute the variance of all possible sample means to obtain $\sigma_{\overline{Y}\overline{Y}}$ directly; note that the result should be σ_{YY}/N. The other approach is to compute $\hat{\sigma}_{\overline{Y}\overline{Y}} = NS_{YY}/(N - 1)$ for each possible sample and see which sampling procedure satisfies the assumption that the average value of $\hat{\sigma}_{\overline{Y}\overline{Y}}$ over all samples is $\sigma_{\overline{Y}\overline{Y}}$.

Problem Suppose that a researcher randomly selects adults from a community and then, in order to save expense, selects a second person from the household of each person in the original sample. When analyzing these data, which assumption would be violated when the standard error of a mean is estimated from the total sample?

Standard Error of a Regression Coefficient Our primary concern here is to find a method for estimating standard errors of an OLS slope such that we can assign probability statements to sample estimates of population slopes. (The same arguments will apply to estimates of structural coefficients when they are estimated by OLS.) Suppose, for example, that c_1 is a sample estimate of the population slope c_1^*. Then we would like to come up with a number D such that the interval from $(c_1 - D)$ to $(c_1 + D)$ has a 95 percent probability of containing c_1^*. Such an interval is called a 95 percent confidence interval.

Confidence intervals can be computed in several ways, but we will develop a procedure based upon the standard error of a slope and the assumption that the t-distribution can be used to assign probability statements to the standard error. Keep in mind that the standard error of a slope can be estimated whether the distribution assumptions required for the assignment of probability statements hold or not. But the standard error has no simple probabilistic interpretation unless the normality assumption holds.

We begin with the usual population model

$$Y = c_0^* + c_1^*X + U^* \tag{5.37}$$

where X is uncorrelated with U^*. Selecting a random sample of N persons from this population and applying OLS procedures yields

$$Y = c_0 + c_1X + U$$

where U and X are uncorrelated. Very few assumptions are necessary for the expected value of c_1 to be c_1^* (c_1 unbiased). However, an estimate can be unbiased and yet very poor in the sense that it is unlikely to be very close to the population value for any one sample. So we would like to obtain from a single sample the best information about how close our estimate is to the population value. In order to achieve this goal, however, several very restrictive constraints must be placed upon the distributions of Y, X, and U^*.

A sufficient set of such constraints is the following:

1. the values of X are fixed and measured without error;
2. the variance of U^* at each level of X ($\sigma_{UU/X}$) is the same at each level of X ($\sigma_{UU} = \sigma_{UU/X}$);
3. for a random sample of size N, $U_1^*, U_2^*, \ldots, U_N^*$ are mutually independent and normally distributed with variance σ_{UU} and zero mean.

The general gist of these constraints is that, in the population, the residuals must spread about the conditional mean of Y in a bell-shaped distribution that has the same shape at each level of X.

Notice in Table 5.1 that our concocted example has such characteristics; the residuals have the same variance when X is zero as when X is one, and their mean is zero at each level of X. Furthermore, the residuals have a roughly bell-shaped distribution.

With these assumptions we can proceed to estimate the standard errors of the regression coefficients and assign probability statements to them. We first define the variance of the sample slope c_1 as

$$\sigma_{c_1c_1} = E(c_1 - c_1^*)^2 \tag{5.38}$$

that is, the variability of c_1 about c_1^* that would occur with a large number of replications (and thus a large number of estimates c_1). The simplest way to expand equation 5.38 is to start with the population model (equation 5.37) and apply it to data from a random sample. First, subtract the mean of Y from Y and the mean of X from X to obtain

$$Y_i - \overline{Y} = (c_0^* - \overline{Y} + c_1^*\overline{X}) + c_1^*(X_i - \overline{X}) + U_i^* \tag{5.39}$$

where $i = 1, \ldots, N$ for a random sample of N persons from a population. (In the remainder of this section, for notational ease, U_i^* will be expressed as V_i). Now multiply both sides of (5.39) by $(X_i - \overline{X})$ and sum over the N cases to obtain

$$\begin{aligned} S_{YX} = {} & [\Sigma(c_0^* - \overline{Y} + c_1^*\overline{X})(X_i - \overline{X})/N] \\ & + [c_1^*\Sigma(X_i - \overline{X})^2/N] + \Sigma V_i(X_i - \overline{X})/N \end{aligned}$$

which reduces to

$$S_{YX}/S_{XX} = c_1^* + [\Sigma V_i(X - \overline{X})/NS_{XX}] \tag{5.40}$$

Notice that the left side of equation 5.40 is simply the sample estimate c_1. Thus, we can rewrite equation 5.40 as

$$c_1 - c_1^* = \Sigma V_i(X - \overline{X})/NS_{XX}$$

Squaring and taking the expectation gives

$$\begin{aligned} E(c_1 - c_1^*)^2 &= E[V_1(X_1 - \overline{X}) + \ldots + V_N(X_N - \overline{X})]^2/N^2S_{XX}^2 \\ &= E[V_1^2(X_1 - \overline{X})^2 + \ldots + V_N^2(X_N - \overline{X})^2 \\ &\quad + 2V_1V_2(X_1 - \overline{X})(X_2 - \overline{X}) + \ldots]/N^2S_{XX}^2 \end{aligned} \tag{5.41}$$

We can now capitalize upon the specification that the values of X are fixed across samples and thus can be factored out of the expectations. So equation 5.41 becomes

$$\begin{aligned} \sigma_{c_1c_1} = {} & [(X_1 - \overline{X})^2E(V_1^2) + \ldots + (X_N - \overline{X})^2E(V_N^2) \\ & + 2(X_1 - \overline{X})(X_2 - \overline{X})E(V_1V_2) + \ldots]/N^2S_{XX}^2 \end{aligned} \tag{5.42}$$

Now recall that each V_i has the same variance and they are mutually independent so that they have zero covariance. Thus (5.42) can be written as

$$\begin{aligned} \sigma_{c_1c_1} &= E(V_i^2)\Sigma(X_i - \overline{X})^2/N^2S_{XX}^2 \\ &= \sigma_{VV}/NS_{XX} \end{aligned} \tag{5.43}$$

Equation 5.43 defines the variance of c_1 in terms of the variance of the population residuals, the sample size, and the variance of X. It will turn out that the sample residuals can be used to obtain an unbiased estimate of σ_{VV}, and this estimate can be used to estimate $\sigma_{c_1c_1}$.

We would now like to have a formula similar to equation 5.43 for $E(c_0 - c_0^*)$. In a manner similar to that applied above, it can be shown that

$$\sigma_{coco} = (1 + \overline{X}^2/S_{XX})\sigma_{VV}/N \tag{5.44}$$

Notice that σ_{coco} becomes larger as $\overline{X}$ becomes different from zero; this is the case because fewer data points are located near the *Y*-axis when the mean of *X* is different from zero—making estimation of c_0^* more risky. Also notice that in both equation 5.43 and equation 5.44, the variance of the sample estimate decreases as N increases—which agrees with the intuition that larger random samples should provide estimates closer to population parameters than smaller samples.

In order to obtain estimates of $\sigma_{c_1c_1}$ and σ_{coco} from a sample we need an estimate of σ_{VV}. This can be obtained from the sample residuals U_i as follows:

$$S_{UU} = \Sigma U_i^2/N$$

and it can be shown that for random samples

$$E(S_{UU}) = (N - 2)\sigma_{VV}/N$$

Thus, an unbiased estimate of σ_{VV} is $NS_{UU}/(N - 2)$. This gives sample estimates of $\sigma_{c_1c_1}$ and σ_{coco} by substitution into equations 5.43 and 5.44. Taking the square root in each instance gives the estimated error of c_0 and c_1:

$$\text{est std err of } c_1 = S_{c_1} = \sqrt{S_{UU}/(N - 2)S_{XX}} \tag{5.45}$$

$$\text{est std err of } c_0 = S_{c0} = \sqrt{(1 + \overline{X}^2/S_{XX})S_{UU}/(N - 2)} \tag{5.46}$$

Suppose now that the data in Table 5.1 are from a random sample of 16 persons from a specified population. Recall that the regression line (equation 5.28) fit to those data was

$$Y = 7.5 + 4X + U$$

where $c_0 = 7.5$ and $c_1 = 4$. We can now use equations 5.44 and 5.45 to obtain estimated standard errors of these sample coefficients:

$$S_{c_1} = \sqrt{6.75/(.25)(16 - 2)} = 1.39$$

$$S_{c0} = \sqrt{(1 + .25/.25)6.75/(16 - 2)} = 0.98$$

Furthermore, since we have assumed that U^* is normally distributed,

$$t_1 = (c_0 - c_0^*)/S_{c0}$$

and

$$t_0 = (c_0 - c_0^*)/S_{c0}$$

are distributed as the *t*-distribution with $(N - 2)$ degrees of freedom.

Thus, to compute 95 percent confidence intervals we use a calculator or table to determine the value of t $(df = 14)$ such that

$$\text{Prob}\ (t_1 > t) = .025$$

The value $t = 2.145$ satisfies this probability statement and can be used to compute the confidence intervals as follows:

$$c_1 \pm t_1 S_{c_1} = 4.0 \pm 2.145(1.39) = 4.0 \pm 2.98$$

$$c_0 \pm t_0 S_{c0} = 7.5 \pm 2.145(.98) = 7.5 \pm 2.10$$

The appropriate interpretation of these intervals is that the probability that any interval based on a sample of the same size will contain c_1^* (or c_0^*) is .95.

The traditional hypothesis test is accomplished by testing whether or not the 95 percent confidence interval includes zero. That is, we assume that c_1^* is zero; thus

$$t_{\text{obs}} = c_1 / S_{c_1}$$

is distributed according to the t-distribution if the null hypothesis ($c_1^* = 0$) is true. A significance test for c_0 is accomplished in an analogous manner:

$$t_{\text{obs}} = c_0 / S_{c0}$$

If the observed value of t exceeds the critical value of $t(t_{\text{crit}})$ such that

$$P(t_{\text{obs}} > t_{\text{crit}}) \leq .025 \tag{5.47}$$

then the observed value of c_0 or c_1 is said to be significantly different from zero at the 5 percent level for a two-tailed test and at the .025 level for a one-tailed test. Note that if t_{obs} is negative, then equation 5.47 should be written

$$P(t_{\text{obs}} < t_{\text{crit}}) \leq .025$$

A general rule of thumb for samples with over 60 cases is that a regression coefficient more than twice its standard error is significant at the 5 percent level. Similarly, a 95 percent confidence interval about a slope estimate is defined by a bracket from two standard errors below the slope estimate to two standard errors above the estimate. In most instances, of course, the standard errors should be treated as rough approximations. Approximate standard errors are helpful in assessing the results of a specific study, but in practice *they cannot substitute for independent replication.*

Finally, it should be pointed out that some computer programs for regression give F ratios for regression coefficients rather than t ratios. The former is simply the square of the latter:

$$F(df = 1, N - 2) = t^2(df = N - 2)$$

5.9 PROPERTIES OF ORDINARY LEAST SQUARES ESTIMATES

We have already discussed the method of OLS for estimating population slopes or structural coefficients from sample observations. And given certain assumptions, the OLS estimates have some very desirable properties; i.e., they are the best linear unbiased estimates (BLUE) of population coefficients. The following discussion explains the meaning of these important properties of OLS estimates.

To begin, we assume a population of persons with scores on Y, X, and U^* such that

$$Y_i = c_0^* + c_1^* X_i + U_i^*$$

where the Y_i are observable random variables and the X_i are fixed and known variables. Also, the U_i^* are random variables that are uncorrelated with one another and with the X_i, and they have a mean of zero with variance σ_{UU}. Given this population model, the OLS estimators from a sample

$$c_0 = \overline{Y} - c_1\overline{X}$$

$$c_1 = S_{YX}/S_{XX}$$

are BLUE relative to c_0^* and c_1^*. This result is known as the Gauss-Markoff theorem.

To say that c_1 is an unbiased estimate of c_1^* simply means that as more and more samples are selected from a particular population, the average value of c_1 converges on c_1^*. The specification that c_1 is linear means that c_1 is equal to a linear combination of the Y_i:

$$c_1 = \Sigma a_i Y_i$$

where a_i is some series of constants. With these two restrictions, c_1 is the *best* estimator of c_1^* in the sense that the sampling variance of c_1 about c_1^* is a minimum. That is,

$$E(c_1 - c_1^*)^2 = \sigma_{c_1c_1}$$

is a minimum among all unbiased and linear estimators of c_1^*.

An alternative of OLS to be discussed subsequently, maximum likelihood (ML) estimation, gives estimates with even greater claims. An ML estimator of c_1^* is not only BLUE, but it is the best (minimum variance) of all possible unbiased estimators—linear or nonlinear. The problem with ML, however, is that it applies only to models in which the U^* are normally distributed. Recall that this was a necessary assumption to obtain the probabilistic standard errors in section 5.8. Thus, it is some comfort to know that when the normality assumption holds, the ML estimates of c_0^* and c_1^* are exactly the same as those resulting from OLS.

It should also be mentioned that, under certain circumstances, a biased estimate has smaller squared deviations about the population parameter than does an unbiased estimator. This may sound strange at first, but it should be obvious that the goal of estimation is to be as close as possible to the population value. If an estimate is likely to be closer to the desired parameter, then it may be preferable even though it is actually an estimate of some population value near the desired parameter. This situation arises when we consider multiple regression and the problem of *multicolinearity*.

5.10 A NOTE ON DICHOTOMOUS VARIABLES AND LINEARITY

Many of the examples in the previous sections have involved dichotomous variables. At this point it is necessary to point out the advantages and problems associated with such discrete variables.

Dichotomous Dependent Variables Recall that for OLS estimates of regression coefficients to be BLUE, the residuals must all have the same variance. This means that the conditional variance of Y, $\sigma_{YY/X}$, must be equal for all levels of X. Furthermore, the standard errors are defined in terms of this assumption. Clearly, the variance of a dichotomous variable coded as one or zero has a variance that depends upon the proportion of cases scored as one. Thus, when such a variable is the dependent variable in a linear model this proportion changes from one value of X to the next when Y and X are strongly related. Thus, when Y is dichotomous, the OLS estimates of regression coefficients are not BLUE and the standard errors outlined in Section 5.8 are biased. The extent of these problems is a function of the extent to which the conditional variances of Y are different for different levels of X.

Simulation studies suggest that the distortions in OLS estimators with dichotomous dependent variables are not usually very large. However, there are instances in which such distortions are likely to be considerable, and there are also certainly instances in which the underlying model of differences in probabilities is not appropriate. For these reasons, log linear models are presented in Appendix B. Fortunately, the log linear models are based upon the same logic that underlies our entire discussion of linear models.

Dichotomous Independent Variables The example in Section 5.6 involves the regression of an interval scale variable that ranges from three to 16 onto a dichotomous variable with range (0, 1). Nothing that we have developed concerning the OLS estimators or their probabilistic standard errors is violated by this example. In fact, it is intuitively consistent with the requirement that levels of X are fixed values—not random variables. Furthermore, the use of a dichotomous independent variable has enabled us to avoid discussion of linearity, the next topic in this section. The convenience of dichotomous variables is that they are necessarily interval scales and they necessarily have linear relationships with other variables.

Linearity Suppose that we have a population of persons with scores on Y and *three* levels of X. Let X range over the values (1, 2, 3). The conditional means of Y can be defined as

$$\overline{Y}_{X=1} = M_1;\ \overline{Y}_{X=2} = M_2;\ \overline{Y}_{X=3} = M_3$$

Now, to say that the simple relationship between Y and X is linear is to say that $(M_2 - M_1)$ equals $(M_3 - M_2)$. Or, alternatively, if we plot $\overline{Y}_X$ against X, the three means should define a straight line. Furthermore, the slope and intercept of that line should be the same as the OLS line fit to all pairs of (X, Y) scores.

The hypothesis that two population variables are linearly related can be tested with sample data. However, the details of the procedure depend upon multiple regression and will be described in subsequent chapters.

5.11 STANDARDIZED REGRESSION

When variables are measured without error it can be useful to standardize variables by transforming them into Z scores before computing regression coefficients. The resulting coefficients are then not expressed in the natural units of measurement. For example, c_1 was the increase in Y associated with a one unit increase in X and is referred to as a *metric* coefficient. When regressions are performed with standardized variables, the resulting coefficients are interpreted in terms of standard deviations: how many *standard deviations* Y increases with a one standard deviation increase in X.

To see how the standardized regression model (sometimes called *normalized*) relates to the more usual *un*standardized model, we will begin with the latter and transform it into the former. We start with the usual model

$$Y_i = c_0 + c_1X_i + U_i$$

with the usual assumptions. In order to transform Y_i and X_i into Z scores, we must subtract their respective means from each score and divide by their respective standard deviations. First, then

$$Y_i - \overline{Y} = c_0 - \overline{Y} + c_1\overline{X} + c_1(X_i - \overline{X}) + U_i$$

which can be reduced further by recalling that $c_0 = \overline{Y} - c_1\overline{X}$. This gives

$$Y_i - \overline{Y} = c_1(X_i - \overline{X}) + U_i$$

We now divide both sides of the equation by S_Y, the sample standard deviation of Y, and multiply $(X_i - \overline{X})$ by S_X/S_X to obtain

$$\frac{(Y_i - \overline{Y})}{S_Y} = c_1\left(\frac{S_X}{S_Y}\right)\left(\frac{X_i - \overline{X}}{S_X}\right) + \left(\frac{U_i}{S_Y}\right) \tag{5.48}$$

Now, note that the term on the left side of equation 5.48 is the Z score for Y (Z_Y), while the Z score for X is included in the first term on the right-hand side. Thus, equation 5.48 can be rewritten as

$$Z_{Yi} = (c_1S_X/S_Y)Z_{Xi} + (U_i/S_Y)$$

If we then define β_1 as (c_1S_X/S_Y), we have

$$Z_{Yi} = \beta_1Z_{Xi} + (U_i/S_Y) \tag{5.49}$$

The standardization procedure can even be carried further by multiplying the last term in equation 5.49 by (S_U/S_U):

$$Z_{Yi} = \beta_1Z_{Xi} + (S_UU_i/S_US_Y)$$

Noting that the mean of U_i is zero and defining (S_U/S_Y) as β_U gives

$$Z_{Yi} = \beta_1Z_{Xi} + \beta_UZ_{Ui} \tag{5.50}$$

These standardized coefficients (often referred to as *beta weights*) then have some interesting properties.

The first such property can be deduced from multiplying equation 5.49 by Z_{Xi}, summing over the sample and dividing by N:

$$(1/N)\Sigma Z_{Yi}Z_{Xi} = \beta_1(1/N)\Sigma Z^2_{Xi} + \beta_U(1/N)\Sigma Z_{Xi}Z_{Ui} \tag{5.51}$$

The left-hand term in equation 5.51 is simply the correlation between Y and X; $(1/N)\Sigma Z^2_X$ is the sample variance of Z_X, which is, by definition, one; $(1/N)\Sigma Z_{Xi}Z_{Ui}$ is the correlation between Z_X and Z_U, which by OLS is zero. Thus, equation 5.51 reduces to

$$r_{YX} = \beta_1$$

revealing that, in the simple case of one independent variable, the slope resulting from a regression of variables measured in standard deviations is simply the correlation between the two variables.

In order to elucidate the meaning of β_U we can follow an analogous procedure: multiply equation 5.49 by Z_{U_i}, sum over the N cases, and divide by N to obtain

$$r_{YU} = \beta_1 r_{XU} + \beta_U r_{UU}$$

Now, since r_{XU} is zero by OLS and r_{UU} is one by definition,

$$r_{YU} = \beta_U$$

Thus, β_U is the correlation between Y and the residuals.

Yet another insight into the meaning of β_1 and β_U can be obtained by squaring equation 5.50, summing, and dividing by N. The result is

$$r_{YY} = \beta_1^2 r_{XX} + \beta_U^2 r_{UU} + 2\beta_1\beta_U r_{XU}$$

which reduces to

$$1 = \beta_1^2 + \beta_U^2 \tag{5.52}$$

Thus, the unit variance of Z_Y has been partitioned into a component associated with X (β_1^2) and a residual component (β_U^2). Expression (5.52) also demonstrates that the absolute value of β_1 cannot be greater than one.

The reader should note, however, that the partition of variance in equation 5.52 is for the *sample* and does not provide unbiased estimates of population proportions. To see this, we can write

$$\beta_U^2 = S_{UU}/S_{YY}$$

Now, we know that

$$E(S_{UU}) = (N - 2)(\sigma_{UU}/N)$$
$$E(S_{YY}) = (N - 1)(\sigma_{YY}/N)$$

Thus, it is tempting to write

$$E(\beta_U^2) = (N - 2)\sigma_{UU}/(N - 1)\sigma_{YY} \tag{5.53}$$

and conclude that β_U^2 must be multiplied by $(N - 1)/(N - 2)$ to obtain an unbiased estimate of σ_{UU}/σ_{YY}. However, the equality in equation 5.53 does not necessarily hold (because the expectation of a ratio of random variables is not necessarily equal to the ratio of the expectations). Thus, we can define an *adjusted* R_{YX}^2 as

$$\text{adj } R_{YX}^2 = 1 - (N - 1)\beta_U^2/(N - 2)$$

but it must be kept in mind that the adjusted R^2 is not an unbiased estimator of the population R^2. (Usually denoted as ρ^2, rho squared).

The advantage of the standardized form of a regression model is that the *magnitude* of relationships appears to be reflected in a standardized format. However, there is a problem that arises with the use of standardized regression coefficients. The problem is that they are not invariant with respect to changes in the range of X values that are observed or with respect to changes in the amount of measurement error in Y, *even though* the *structural model* or population slopes underlying the observations remain unchanged.

Imagine, for example, that an increase of one unit along a dimension of social threat produces a two-unit increase in authoritarianism. This structural relationship would be reflected in an unstandardized regression model for observations involving two, three, or more levels of X. However, the standardized coefficient would increase in size as the range of X levels included in the observations increased. This can be seen by expressing β_1 as in equation 5.48:

$$\beta_1 = c_1 S_X / S_Y$$

Clearly, as S_X increases, β_1 increases even though c_1 remains unchanged. (The reader may also wonder if increases in S_Y offset the changes in S_X. This would be the case only if X and Y were perfectly correlated.)

Thus, in most research situations the unstandardized regression model is preferable for between study comparisons. (The invariance of these coefficients in the face of changes in measurement error in Y will be discussed in Chapter 9.)

Problem Plot the following data on a graph. Then compute the metric and standardized regression coefficients for the first four cases. Finally, compute the metric and standardized coefficients for all six cases. Notice how increasing the range of X alters β_1 but not c_1.

Case no.	1	2	3	4	5	6
X	0	0	1	1	2	2
Y	2	4	3	5	4	6

5.12 SUMMARY OF FORMULAS FOR SIMPLE REGRESSION

A. For Regression of Y on X: $Y_i = c_0 + c_1 X_i + U_i$

1. Metric coefficients

$$c_1 = S_{YX}/S_{XX}$$
$$c_0 = \overline{Y} - c_1\overline{X}$$

2. Standard errors of metric coefficients

$$S_{c1} = \sqrt{S_{UU}/(N-2)S_{XX}}$$
$$S_{c0} = \sqrt{(1 + \overline{X}^2/S_{XX})S_{UU}/(N-2)}$$

3. Standardized coefficients: $Z_Y = \beta_1 Z_X + \beta_U Z_U$

$$\beta_1 = c_1 S_X/S_Y = (S_{YX}/S_{XX})(S_X/S_Y) = S_{YX}/S_Y S_X = r_{YX}$$

$$\beta_U = S_U/S_Y = (1 - r^2_{YX})^{1/2}$$

$$\beta_1^2 + \beta_U^2 = 1$$

B. Adjusted *R*-square

$$\text{adj } R^2_{YX} = 1 - [(N-1)\beta_U^2/(N-2)]$$
$$= 1 - [(N-1)(1 - R^2_{YX})/(N-2)]$$

6
Multiple Regression

6.1 INTRODUCTION

The next logical step in the development of linear regression models is generalization of the simple model to one involving two or more independent variables. (In a subsequent chapter we shall generalize even further to a *multivariate regression* model that involves multiple *de*pendent variables as well.) As was demonstrated in Chapter 4, failure to control for other variables in a causal system, or failure to include interactions among independent variables, can leave important relationships undetected or misrepresented. Dichotomous variables were used to elucidate how complex and subtle relationships among variables can be modeled with linear equations. The goal of the current chapter is to develop and demystify the analytic procedures for estimating structural coefficients or slopes from sample data and to develop a concrete understanding of how the coefficients should be interpreted.

The step from simple to multiple regression models is the first step toward a statistical methodology that can be used to analyze complex models involving numerous correlated and interacting variables. Given the appropriate conditions, these models enable a researcher to untangle relationships that have been obscured by the complexities of real human systems. Interestingly, the multivariate technique of multiple regression is a whole that is made up of parts, and the parts are simple (or bivariate) regression coefficients or structural coefficients. In one sense, this whole is greater than the sum of its parts. This is especially true when a large number of variables are involved: it is virtually beyond the integrative capacity of the human mind to "see" in the bivariate regressions among numerous variables what can be "seen" with the aid of multiple regression coefficients. Nevertheless, multiple regression relies completely on the relative magnitudes of simple regression coefficients among the variables involved. And this raises an important issue that must be stressed once again before we proceed with the details of multiple regression methods.

That issue is measurement error. That is, when we study social psychological systems we usually settle for *indicators* of hypothetical constructs that we know are only rough estimates of a person's "true" location on the hypothetical dimension. As will be demonstrated subsequently, *random* measurement error in an independent variable tends to reduce (attenuate) the absolute magnitude of both metric and standardized *bivariate* regression coefficients. Random measurement error in a dependent variable attentuates standardized (but not metric) simple regression coefficients. This means that random measurement error can lead to an understatement of relationships in bivariate regression models—which is problematic—but, with large samples, does not impair judgment concerning whether a relationship is inverse, near zero, or positive. However, when numerous variables are used in a multiple regression and each is measured with a different (and substantial) amount of random error, then the differential attenuations of the bivariate regression coefficients tend to pyramid and give erroneous information about the relative magnitudes of multiple regression coefficients.

A reasonable conclusion for the reader to draw at this point would be that since social scientists almost always measure with considerable error, it is probably not very fruitful to apply multivariate statistical methods to social psychological problems. Fortunately, however, methods have been developed for taking account of measurement error and adjusting coefficients accordingly. These procedures require multiple and method divergent indicators of hypothetical dimensions and will be discussed in Chapter 9. In the meantime, we will develop the method of multiple regression, assuming that independent variables are measured without error.

6.2 AN EXAMPLE WITH NO INTERACTION

It is often difficult to imagine just what is meant by a multiple regression model, so we will begin with a concrete example and then proceed with mathematical derivation of estimation formulas.

Imagine an *additive* causal system in which the effects of social threat and years of education on authoritarianism can be added together for individuals. That is

$$Y = b_0^* + b_1^* X_1 + b_2^* X_2 + T^* \tag{6.1}$$

where Y is an F scale score, X_1 is an errorless measure of social threat, and X_2 is an errorless measure of educational attainment (in years). Furthermore, T^* represents all other factors that determine Y (perhaps including random measurement error in the F scale), and the coefficients with asterisks are structural parameters defining the causal relationship between Y and the independent variables (X_1 and X_2). If X_1 is increased by one unit and all other relevant variables are held constant, then Y will increase by b_1^* for any person in the specified population. Similarly, an isolated one-unit increase in X_2 (i.e., one more year of education) will produce a b_2^* increase in authoritarianism.

If we add the specification that the disturbances T^* are uncorrelated with social threat and education in the population, then the structural coefficients in equation 6.1 can also be construed as average or expected effects of changes in X_1 and X_2. This means that the expected value of Y, given particular values of X_1 and X_2, is

$$\begin{aligned} Y_{X_1X_2} &= E(Y|X_1 \text{ and } X_2) \\ &= b_0^* + b_1^* X_1 + b_2^* X_2 \end{aligned} \tag{6.2}$$

Thus, we could also interpret b_1^* as the increase in the expected or mean value of Y that is caused by a one-unit increase in X_1. This latter interpretation is more in line with the logic of causal inference that characterizes science, than was the previous interpretation concerning individuals: we seldom make causal statements about specific instances; rather, causality in the instance is inferred from an aggregate of observations.

Now, suppose that observations of Y, X_1, and X_2 for a sample of 18 persons gives rise to the data in Table 6.1. Is it now possible to use OLS regression methods to obtain reasonable estimates of the known structural parameters? It is possible if one of the following two conditions holds:

The independent variables X_1 and X_2 are randomly manipulated (as in a true experiment) such that they correlate with the disturbance T^* only by change or;

The independent variables included in the model theoretically exhaust the systematic determinants of Y, leaving only random (and thus uncorrelated) fluctuations in Y to be defined as the disturbance T^* (as in the econometric tradition), or other systematic determinants are uncorrelated with X_1 and X_2.

Table 6.1 Hypothetical data for an additive causal model where Y, X_1, and X_2 are the observed measures of authoritarianism, social threat, and years of education, respectively.

Case	X_1	X_2	Y	= b_0^*	$+ b_1^*X_1$	$+ b_2^*X_2$	$+ T^*$	T	$\hat{Y}$
1	1	12	7	27	2	−18	−4	−3.6	10.6
2	1	12	15	27	2	−18	+4	4.4	10.6
3	2	12	8	27	4	−18	−5	−4.7	12.7
4	2	12	16	27	4	−18	+3	3.3	12.7
5	3	12	10	27	6	−18	−5	−4.9	14.9
6	3	12	12	27	6	−18	−3	−2.9	14.9
7	3	12	15	27	6	−18	0	0.1	14.9
8	3	12	18	27	6	−18	+3	3.1	14.9
9	3	12	20	27	6	−18	+5	5.1	14.9
10	1	16	0	27	2	−24	−5	−5.2	5.2
11	1	16	2	27	2	−24	−3	−3.2	5.2
12	1	16	5	27	2	−24	0	−0.2	5.2
13	1	16	8	27	2	−24	+3	2.8	5.2
14	1	16	10	27	2	−24	+5	4.8	5.2
15	2	16	4	27	4	−24	−3	−3.4	7.4
16	2	16	13	27	4	−24	+6	5.6	7.4
17	3	16	5	27	6	−24	−4	−4.5	9.5
18	3	16	13	27	6	−24	+4	3.5	9.5
Mean	2	14	10.06	27	4	−21	.06	.00	10.06
Variance	.78	4	29.9	0	3.11	9	15.5	15.3	14.6
S.D.	.88	2	5.47	0	1.76	3	3.94	3.91	3.82

A. True model: $Y = b_0^* + b_1^*X_1 + b_2^*X_2 + T^* = 27 + 2X_1 - 1.5X_2 + T^*$

B. OLS estimate: $Y = b_0 + b_1X_1 + b_2X_2 + T = 24.51 + 2.14X_1 - 1.34X_2 + T$

C. OLS predicted scores: $\hat{Y} = b_0 + b_1X_1 + b_2X_2$

Note: The coefficients b_j^* are structural parameters (that are usually not known in research situations, but are given in this hypothetical example), whereas the coefficients b_j are OLS estimates of those parameters based upon this sample of 18 cases. Similarly, T^* is the true disturbance for the population model and T is the OLS residual that serves as an estimate of the disturbance.

Given that one of these conditions holds for the model in Table 6.1, we can assume that the population covariances $\sigma_{X_1T^*}$ and $\sigma_{X_2T^*}$ are zero and proceed to apply OLS regression methods to the estimation of the structural parameters. But before we do this, let us look more closely at the causal model assumed to underlie the interrelations of these data.

To begin with, note in Table 6.1 that b_1^* is +2 and b_2^* is −1.5. This means that increasing social threat by one unit *increases* the expected F-scale score by two points, whereas a one-year increase in education causes a 1.5 point *decrease* in the expected F-scale. Furthermore, these two influences are additive (no interaction) in that the effect on Y of change in one independent variable is simply added to the effect of change in the other independent variable. And it is crucial to keep in mind that b_1^* (structural) describes the extent to which changes in X_1 influence Y if *all other* influences on Y are held constant, while $c_{YX_1}^*$ (slope) reflects the observed change in the *average* value of Y *if* X_2 *is held constant*. An analogous interpretation of b_2^* applies to the effect of X_2 on Y. (Note that the statements concerning "unit" changes are assumed to apply throughout the full ranges of all the variables. This means that unit changes must have the same meaning at any interval in the range. In other words, interval scale measures are assumed.)

The parameter b_0^* is the population intercept: the expected value of Y when X_1 and X_2 are zero. Since Y and X_1 are not ratio scales (the zero point is arbitrary and, in this case, undefined) b_0^* has no special significance in the model.

In practice, of course, the structural parameters are not known (nor are the true disturbances T^*). But it is instructive in this example to see just what kind of causal process is discoverable with OLS regression. Note that once the structural model has been specified, a "simulation" can be produced by simply concocting values of X_1, X_2, and T^* (or Y) and substituting into the equation:

$$Y = 27 + 2X_1 - 1.5X_2 + T^* \tag{6.3}$$

However, if the structural parameters are to be estimable by OLS, then we must place some restrictions on the values of X_1, X_2, and T^* that we select. The most important of these restrictions is that T^* be uncorrelated with X_1 and X_2 *in the population of scores*. Also, the expected value of T^* at each combination of levels of X_1 and X_2 must be zero if a good estimate of b_0^* is desired. But the reader should keep in mind that *samples* from a population will evidence correlations between the disturbance and the independent variables merely by chance. Such is the case in Table 6.1.

Note in Table 6.1 that each F-scale score Y is completely determined by equation 6.3 and that each component of the score ($b_1^*X_1$, $b_2^*X_2$, and T^*) is assumed to be a (sometimes unobservable) "true" determinant of the F-scale score. We can also then say that variability in Y (across persons) can be attributed to variability in these causal components.

Table 6.1 gives both the true structural model and the OLS estimate based on the sample of 18 cases. The OLS procedures give regression coefficients b_0, b_1, and b_2 that are used as estimates of b_0^*, b_1^*, and b_2^*, and these in turn give residuals T that serve as estimates of T^*. The question of concern here is thus whether or not reasonable estimates are obtained and how they are obtained. Before seeing how the regression coefficients are computed, contrast the population and estimated models:

$$Y = 27 + 2X_1 - 1.5X_2 + T^* \text{ (population)} \quad (6.4a)$$

$$Y = 24.5 + 2.14X_1 - 1.34X_2 + T \text{ (estimate)} \quad (6.4b)$$

$$\text{SE}(b_0) = 8.96$$

$$\text{SE}(b_1) = 1.24$$

$$\text{SE}(b_2) = 0.55$$

$$R\text{-square} = .485$$

Clearly, the estimates in equation 6.4b are not misleading concerning the population parameters in equation 6.4a—the signs of the coefficients are all appropriate and the magnitudes of the estimates differ no more than 10 or 11 percent from the parameters. Furthermore, inspection of T in Table 6.1 reveals that it is very close to T^* in every instance.

The intercorrelations of the various variables in Table 6.1 and the simple regression coefficients among Y, X_1, and X_2 are presented in Table 6.2. These statistics are the building blocks of the multiple regression statistics. It is especially important to note that the statistics in Table 6.2 are analogous to results obtained from a *single* sample selected from a population in which equation 6.4a holds. Thus, we would expect the sample statistics to differ from the population values because of sampling error. This effect of sampling, however, should be distinguished from *biasing* effects of OLS estimation of regression coefficients due to a population correlation between the disturbances (T^*) and one or both of the independent variables.

Several meaningful characteristics of the data and model described in these two tables should be noted:

1. The sample correlation between X_2 and T^* ($r_{X_2T^*}$) is not zero (it is .071). With a sample size of only 18, this correlation could easily be much larger, since the zero correlation between these variables applies to a large population. The deviation of this

Table 6.2 Correlation matrix and bivariate regression coefficients for the hypothetical data presented in Table 6.1. See that table for a description of the variables.

	Y	X_1	X_2	T^*	T
Y	1.0				
X_1	.530	1.0			
X_2	−.619	−.378	1.0		
T^*	.681	.00	.071	1.0	
T	.715	.00	.00	.997	1.0
$\hat{Y}$	.699	.761	.888	−.049	.00

A. $Y = c_1 + c_{YX1}X_1 + u_1$; $c_{YX1} = 3.29$

B. $Y = c_2 + c_{YX2}X_2 + u_2$; $c_{YX2} = -1.69$

C. $X_1 = c_3 + c_{X1X2}X_2 + u_3$; $c_{X1X2} = -.167$

D. $X_2 = c_4 + c_{X2X1}X_1 + u_4$; $c_{X2X1} = -.857$

correlation from zero means that the OLS estimates cannot be equal to the population coefficients. (The zero correlation between X_1 and T^* is rather unlikely and occurs in this example only because the data are contrived.)

2. However, T is *un*correlated with both X_1 and X_2. This turns out to be the result of an OLS approach to parameter estimation: values of b_0, b_1, and b_2 are chosen so that T has this property. Since T^* was *almost* uncorrelated with X_1 and X_2, the OLS procedure produces estimates of b_1^*, b_2^*, and T^* that are quite close to population values. To the extent that T^* *is* correlated in the population with the independent variables, the OLS estimates would be more biased as estimates of the true *structural* parameters.

3. Even though Y is, in truth (a constructed truth albeit), determined by $b_1^*X_1$, $b_2^*X_2$, and T^*, the sample variance in Y ($S_{YY} = 29.9$) is *not* equal to the sum of the variances of these three components. In fact, the sum of these three variances is only 27.6 ($3.11 + 9 + 15.5$), or 92 percent of the total variance of Y. The main reason for this is the positive correlation between social threat and years of education. That is, when two variables are positively correlated, the variance of their sum is greater than the sum of the two variances. Thus, the variance of $(b_1^*X_1 + b_2^*X_2)$ is greater than the sum of the variances of $b_1^*X_1$ and $b_2^*X_2$, if X_1 and X_2 are positively correlated. (The direction of this inequality between the variance of a sum and the sum of variances is reversed for negatively correlated variables.)

4. The variance of Y *is* equal to the sum of the variance of T and the variance of $\hat{Y}$ $(b_0 + b_1X_1 + b_2X_2)$: $29.9 = 15.3 + 14.6$ (see Table 6.1). This is true because the OLS procedure produces values of T that are uncorrelated with both X_1 and X_2. OLS thus partitions the variability in Y into that portion associated with X_1 *and* X_2 and that portion uncorrelated with the independent variables. Note, however, that the variance of Y is still not equal to the sum of the variances of b_1X_1, b_2X_2, and T if X_1 and X_2 are correlated.

5. Notice that the sample variance of T is not equal to the sample variance of T^*. This will usually be the case and serves as a reminder of the distinction between the two variables.

6. The sample mean of T^* is not zero. This distorts the estimate of b_0^*, the intercept. Such distortion is likely to occur, even though the population mean of T^* is zero. Since the intercept is often of little importance, zero means for the disturbances are often not crucial.

7. The simple regression of Y on X_1 gives an inflated impression of the causal influence of changes in social threat on authoritarianism. The value of c_{YX_1} is 3.29, whereas b_1 is only 2.14 (and $b_1^* = 2$). The reason for this is that completing more years of education decreases authoritarianism *and* is associated with lower threat conditions. Thus, the tendency for people in high threat conditions to be high in authoritarianism (as measured by c_{YX_1}) results not only because of the direct causal link between the two variables, but also because people in high threat conditions tend to have less education, and having less education increases authoritarianism. Another way of saying this is that the regression of Y on X_1 (c_{YX_1}) results from both a direct causal effect of X_1 on Y (b_1^*) and because of effects of X_2 on Y that are correlated with X_1 ($c_{YX_1}^* - b_1^*$).

The untangling of these two sources of covariation between Y and X_1 is the goal of our estimation efforts. ($c^*_{YX_2}$ is also inflated relative to b_2 for conversely analogous reasons: $c_{YX_2} = -1.69$, whereas $b_2 = -1.34$.)

8. Finally, we should note that the model says nothing about the causal relations between social threat and education. The model makes no predictions concerning the magnitude of the correlation between X_1 and X_2 and, more importantly, does not specify any causal explanation of the observed correlation between these variables. These two variables are thus said to be *exogenous* in the model.

Multiple Correlation

There are two ways in which to approach the idea of *multiple* correlation. Suppose, for example, that we have observed one left-hand variable (Y) and two right-hand variables (X_1, X_2). If we use OLS regression procedures to minimize the variance of the residuals (e) in the linear equation

$$Y = c_0 + c_1X_1 + c_2X_2 + e$$

then the multiple correlation between Y and the pair of X variables is given by the relation

$$R^2_{Y \cdot X_1X_2} = 1 - (S_{ee}/S_{YY})$$

where S_{ee} is the variance of the OLS residuals and S_{YY} is the variance of the observed variable Y. In other words, the multiple correlation of Y with X_1 and X_2 is the proportion of variance in Y that is *not* attributed to the residuals. This definition can also be stated in positive terms: the multiple correlation of Y with X_1 and X_2 is the proportion of variance in Y that is attributed to the linear combination ($c_0 + c_1X_1 + c_2X_2$). In fact the R-square in this instance is the square of the simple correlation between Y and this linear combination of X_1 and X_2.

It is important to keep in mind that R-square is a descriptive characteristic of relationships *in a sample.* In the case of small samples, this means that the observed R-square may be used in the hope of making some statements about a randomly sampled population. In other words, we should be concerned in small sample cases about the relationship between the sample R-square and the R-square in the population from which the sample was drawn:

$$\text{population R-square} = 1 - (\sigma_{e^*e^*}/\sigma_{YY})$$

where e^* is the residual obtained from the population equation

$$Y = c^*_0 + c^*_1X_1 + c^*_2X_2 + e^*$$

where e^* is uncorrelated with X_1 and X_2 and has a mean of zero. It turns out that the sample R-square is biased upward relative to the population R-square. Thus, a correction procedure, resulting in the so-called *adjusted* R-square, may be applied to the sample R-square in an effort to compensate for this bias. The rationale for the adjusted R-square and procedures for computation are given in the final section of this chapter.

When X_1 and X_2 are uncorrelated with one another, then the multiple correlation of Y with X_1 and X_2 is simply the sum of the squares of the simple correlations between Y and X_1 and between Y and X_2. However, as the correlation between X_1 and X_2 increases, this equality tends not to hold.

Problem Use the sample variances of T and Y for the data in Table 6.1 to compute the multiple correlation of Y with X_1 and X_2. Compare this R-square with the sum of $r^2_{YX_1}$ and $r^2_{YX_2}$ (see Table 6.2) and explain the difference. (Hint: the extent to which X_1 and X_2 are correlated is the extent to which R^2 is less than the sum of the simple correlations.)

6.3 CAUSAL DIAGRAMMING

Now that some familiarity with the model presented in Tables 6.1 and 6.2 has been achieved, we can proceed to develop a diagrammatic device that will accomplish two goals: (1) provide visually meaningful representation of the linear structural equation model and (2) provide a simple algorithm for generating equations that allow estimation of the structural coefficients b_1^* and b_2^* from the sample regression coefficients c_{YX_1}, c_{YX_2}, $c_{X_2X_1}$, $c_{X_1X_2}$.

Figure 6.1 is a causal diagram of the structural model developed in Section 6.2. The conventions that have been observed in constructing the diagram are simple:

1. Straight arrows are drawn from causal to affected variables.
2. Curved and bi-directional arrows are drawn connecting variables that are correlated, but not causally related in a specified manner.
3. Pairs of variables that are *not* correlated cannot be linked by following any series of directed arrows (e.g., there is no route from T^* to X_1, without tracing backwards on one of the arrows).

Given the causal diagram in Figure 6.1, we are in a position to develop some rules concerning the interrelations of the variables with the special goal of untangling the causal effects from the simple regression coefficients. The rule is that the coefficient

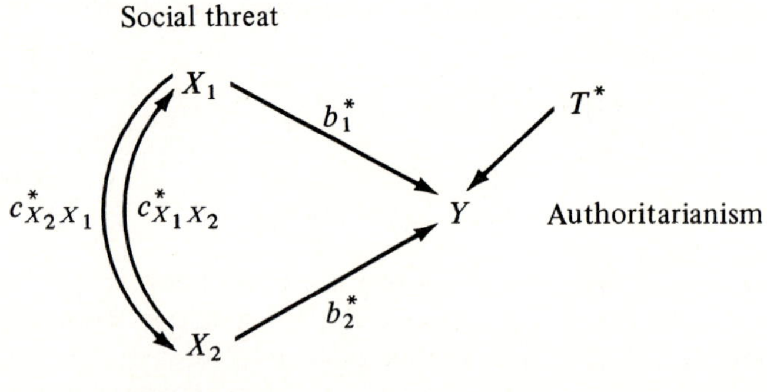

Figure 6.1 A path diagram of the structural equation model given in equation 6.3.

resulting from the regression of any variable in the causal diagram on a variable previous in the causal chain is equal to the sum of the various paths from the previous variable to the subsequent variable; furthermore, the numerical value of each path is equal to a product of the coefficient associated with each arrow along the path. Thus, for the causal diagram in Figure 6.1, the regression of Y on X_1 gives $c^*_{YX_1}$. By the rules of path diagrams, we can express this coefficient as equal to the sum of two paths: one directly from X_1 to Y (b^*_1) and one path involving two arrows (X_1 to X_2 and X_2 to Y) and thus a product of two coefficients ($c^*_{X_2X_1}$ and b^*_2). Finally, then, we can write:

$$c^*_{YX_1} = b^*_1 + c^*_{X_2X_1}b^*_2 \tag{6.5}$$

Equation 6.5 essentially partitions the association between Y and X_1 into a direct causal component (b^*_1 or how a one-unit change in X_1 affects Y if other variables are held constant) and a correlational component ($c^*_{X_2X_1}b^*_2$). It may be the case that changes in X_1 actually cause changes in X_2 so that some or all of this correlational component is an *indirect* causal link between X_1 and Y but this is not specified in the model defined by the equation in Table 6.1.

Analogously, the regression of authoritarianism on education can be partitioned as follows:

$$c^*_{YX_2} = b^*_2 + c^*_{X_1X_2}b^*_1 \tag{6.6}$$

If we now substitute sample values in equations 6.5 and 6.6 we have

$$c_{YX_1} = b_1 + c_{X_2X_1}b_2 \tag{6.7a}$$

$$c_{YX_2} = c_{X_1X_2}b_1 + b_2 \tag{6.7b}$$

which is a system of two linear equations with b_1 and b_2 unknown (the simple regression coefficients are known since they can easily be computed from the sample data). Solving equations 6.7 for b_1 and b_2 gives

$$b_1 = (c_{YX_1} - c_{X_2X_1}c_{YX_2})/(1 - c_{X_1X_2}c_{X_2X_1}) \tag{6.8a}$$

$$b_2 = (c_{YX_2} - c_{X_1X_2}c_{YX_1})/(1 - c_{X_1X_2}c_{X_2X_1}) \tag{6.8b}$$

and these formulas can be further simplified by noting that $c_{X_1X_2}c_{X_2X_1}$ is equal to $r^2_{X_1X_2}$. It should be clear that equations 6.8 allow computation of b_1 and b_2 from the simple regression coefficients among the three variables.

If we now return to the numerical example in Tables 6.1 and 6.2 we can write equations 6.7 as

$$3.29 = b_1 + b_2(-.857)$$

$$-1.69 = b_1(-.167) + b_2$$

Solving these equations for b_1 and b_2 gives

$$b_1 = 2.14\ (b^*_1 = 2)$$

$$b_2 = -1.34\ (b^*_2 = -1.5)$$

which are unbiased OLS estimates of b_1^* and b_2^* if T^* is uncorrelated with X_1 and X_2 in the population from which the sample was drawn. Equations 6.7 can now be written as

$$\begin{aligned} c_{YX_1} &= 3.29 = 2.14 + (-1.34)(-.857) \\ &= 2.14 + 1.15 \end{aligned} \tag{6.9a}$$

$$\begin{aligned} c_{YX_2} &= -1.69 = 2.14(-.167) + (-1.34) \\ &= -.35 - 1.34 \end{aligned} \tag{6.9b}$$

for the sample data in Table 6.1. Note that equation 6.9a partitions the association between authoritarianism and social threat (c_{YX_1}) of 3.29 into a component of direct causation equal to 2.14 (or 65 percent of c_{YX_1}) and a correlated cause component of 1.15 (the remaining 35 percent of c_{YX_1}). These numbers mean that from the sample data we would estimate that an increase of one unit of social threat would, on the average, produce a 2.14 unit increase in F-scale scores if years of education were held constant (not allowed to vary). This estimate is quite close to the structural parameter (b_1^*) of 2.0 that was used to generate these sample data and, it is hoped, will serve as a rather intuitive demonstration of how the causal diagram can be used to disentangle the structural coefficients (estimated by b_1 and b_2) from the observed bivariate regression coefficients (c_{YX_1}, c_{YX_2}, $c_{X_1X_2}$, $c_{X_2X_1}$).

To further clarify the meaning of b_1 (and b_2), recall that coefficients in linear regression models with multiple independent variables are *partial* regression coefficients. And partial coefficients are equivalent to *conditional* regression coefficients. Thus, if we regress the Y scores on the X_1 scores for the first nine cases in Table 6.1, where X_2 is always equal to 12, we obtain the conditional regression of Y on X_1 *given that* $X_2 = 12$:

$$(Y|X_2 = 12) = 8.52 + 2.11X_1 + U_5 \tag{6.10}$$

Similarly, if we regress Y on X_1 for the last nine cases, where X_2 is constant, we have:

$$(Y|X_2 = 16) = 3.06 + 2.17X_1 + U_6 \tag{6.11}$$

These two conditional slopes obtained in equations 6.10 and 6.11 can be labeled $c_{YX_1|X_2=12}$ and $c_{YX_1|X_2=16}$, respectively. Note that the two conditional slopes are approximately equal and are both very close to the partial slope (b_1) of 2.14 obtained from the multiple regression that was developed from our heuristic analysis of the causal diagram in Figure 6.1. In fact, in this instance, the partial slope b_1 is simply the average of the two conditional slopes. So it should now be clear that describing b_1 as the direct effect of X_1 on Y *(controlling for X_2)* is equivalent to saying that b_1 is the average regression coefficient that would occur when data were included only from persons with identical scores on other variables that both affect Y and are correlated with X_1. For example, the partial slope b_1 reflects the effect of increased social threat on authoritarianism with respect to social threat for people with 12 years of education and change in authoritarianism with respect to social threat for people with 16 years of education.

Finally, we can depict the results of the estimation procedure in a causal diagram as in Figure 6.2.

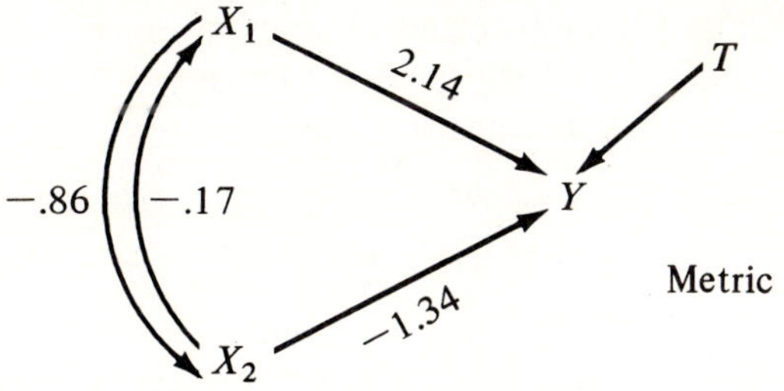

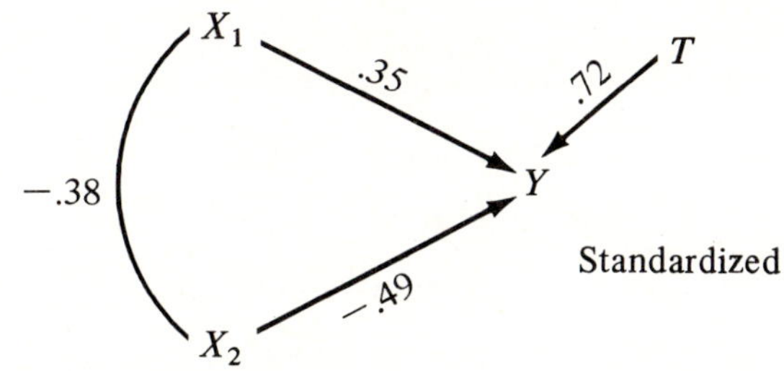

Figure 6.2 Path diagram of estimated equation model in metric and standardized form.

We are now ready to proceed to a more formal derivation of multiple regression coefficients and their properties. But, before doing so, I should repeat a warning concerning the interpretation of slopes as structural coefficients. In the above example we were given a causal system and our task was to untangle the causal relations from the bivariate regression coefficients. The reader must be careful not to construe this exercise as indicating that causal relations (structural coefficients) can be untangled from any set of bivariate regressions. On the contrary, attribution of causal characteristics to partial regression coefficients must stem from design assumptions, not from the application of any statistical procedure. In the experimental tradition, the crucial design assumption is that randomization and manipulation produce uncorrelated disturbances. In the econometric tradition, complete model specification or observation over time is assumed to produce uncorrelated disturbances or such correlations are explicitly modeled. Neither of these assumptions is made more or less true by any mathematical procedures applied to the data.

6.4 ALGEBRAIC DERIVATION OF REGRESSION COEFFICIENTS

Algebraic derivation of partial regression coefficients is essentially analogous to the procedures described for the simple regression model in Chapter 5. We begin with a model involving two independent variables:

$$Y = b_0^* + b_1^* X_1 + b_2^* X_2 + T^* \tag{6.12}$$

where equation 6.12 holds in some specified population. Further manipulation is simplified if the following algebraic rearrangements are used to transform the variables into deviations from their respective means (expected values):

$$Y - \overline{Y} = (b_0^* - \overline{Y} + b_1^*\overline{X}_1 + b_2^*\overline{X}_2) + b_1^*(X_1 - \overline{X}_1) + b_2^*(X_2 - \overline{X}_2) + T^* \tag{6.13}$$

where a bar over a variable indicates its population mean.

[A subtle point of clarification is in order here. Note that the population means of the observed variables (which are the expectations of the variables) are indicated by a horizontal line over a variable. $\overline{X}_1$ is thus a parameter that does not vary, and its expectation is itself.

In other contexts, the notation $\overline{X}_1$ is used to denote a *sample* mean. In such an instance, $\overline{X}_1$ is a statistic, and the expected value of $\overline{X}_1$ *across samples* is the population mean of X_1. A more cumbersome notation provides greater clarity:

$$\begin{aligned} X_1 &= \text{a random variable defined for population Q} \\ m(X_1) &= \text{a sample mean of } X_1 \\ m^*(X_1) &= E(X_1) = \text{population mean of } X_1 \end{aligned}$$

It can then be shown that the long run average of sample means is the population mean. That is

$$E[m(X_1)] = m^*(X_1)]$$

Equation 6.13 can be simplified considerably by noticing that the first term is zero if $E(T^*)$ is zero. (To see this, take the expectations of both sides of equation 6.12.) Note that, in general, adding constants to variables changes the intercept, but leaves the slopes unchanged. We may then write equation 6.13 as

$$Y - \overline{Y} = b_1^*(X_1 - \overline{X}_1) + b_2^*(X_2 - \overline{X}_2) + T^* \tag{6.14}$$

In order that we may solve for b_1^* and b_2^* in terms of simple indices of bivariate relationships, the usual procedure of multiplying by independent variables is followed. First, equation 6.14 is multiplied by $(X_1 - \overline{X}_1)$, and then the expectation over the population is taken:

$$\begin{aligned} E[(Y - \overline{Y})(X_1 - \overline{X}_1)] &\\ = b_1^*E(X_1 - \overline{X}_1)^2 &+ b_2^*E(X_1 - \overline{X}_1) + b_2^*[(X_1 - \overline{X}_1)(X_2 - \overline{X}_2^*)] \\ &+ E[T^*(X_1 - \overline{X}_1)] \end{aligned}$$

which may be rewritten as

$$\sigma_{YX_1} = b_1^*\sigma_{X_1X_1} + b_2^*\sigma_{X_1X_2} + \sigma_{X_1T^*} \tag{6.15}$$

if $E(T^*) = 0$. If equation 6.14 is similarly multiplied by $(X_2 - \overline{X}_2)$ and expectations taken, we have

$$\sigma_{YX_2} = b_1^*\sigma_{X_1X_2} + b_2^*\sigma_{X_2X_2} + \sigma_{X_2T^*} \tag{6.16}$$

if $E(T^*) = 0$.

Now, since Y, X_1, and X_2, are observed variables, we could compute the variances and covariances involving these variables and substitute them into equations 6.15 and 6.16. However, the last covariance in each of these equations involves the unobservable T^*. Thus, the covariances $\sigma_{X_1T^*}$ and $\sigma_{X_2T^*}$ must be assigned some values to allow solution for b_1^* and b_2^*; the values assigned in OLS are zero. The reader should keep in mind that if equation 6.12 is a structural model, it may or may not be appropriate to specify that T^* is uncorrelated with X_1 and X_2. However, if b_1^* and b_2^* are OLS slopes, then these slopes are actually *defined* by the values that result from the zero correlation specification.

Incorporating these OLS assumptions into equations 6.15 and 6.16 gives what are referred to as the *normal equations:*

$$\sigma_{YX_1} = b_1^*\sigma_{X_1X_1} + b_2^*\sigma_{X_1X_2} \tag{6.17a}$$

$$\sigma_{YX_2} = b_1^*\sigma_{X_1X_2} + b_2^*\sigma_{X_2X_2} \tag{6.17b}$$

And the normal equations can be solved for the coefficients:

$$b_1^* = (\sigma_{YX_1}\sigma_{X_2X_2} - \sigma_{YX_2}\sigma_{X_1X_2})/(\sigma_{X_1X_1}\sigma_{X_2X_2} - \sigma^2_{X_1X_2}) \tag{6.18a}$$

$$b_2^* = (\sigma_{YX_2}\sigma_{X_1X_1} - \sigma_{YX_1}\sigma_{X_1X_2})/(\sigma_{X_1X_1}\sigma_{X_2X_2} - \sigma^2_{X_1X_2}) \tag{6.18b}$$

The coefficients b_1^* and b_2^* can now be computed in terms of the variances of, and covariances among, the observable variables (although population parameters are seldom available).

Equations 6.18 are not readily interpretable, and their form does not appear to be the same as the equations derived from the path diagram in Section 6.3. This problem can be remedied by a simple transformation of the normal equations (equations 6.17): divide the first equation by $\sigma_{X_1X_1}$ and the second by $\sigma_{X_2X_2}$ to obtain

$$c^*_{YX_1} = b_1^* + b_2^*c^*_{X_1X_2} \tag{6.19a}$$

$$c^*_{YX_2} = b_1^*c^*_{X_2X_1} + b_2^* \tag{6.19b}$$

where $c^*_{YX_1}$, for example, is the population slope of Y regressed on X_1. Equations 6.19 can then be solved to give

$$b_1^* = (c^*_{YX_1} - c^*_{X_1X_2}c^*_{YX_2})/(1 - r^{*2}_{X_1X_2}) \tag{6.20a}$$

$$b_2^* = (c^*_{YX_2} - c^*_{X_2X_1}c^*_{YX_1})/(1 - r^{*2}_{X_1X_2}) \tag{6.20b}$$

Equations 6.19 and 6.20 are more readily interpreted: b_1^* is the direct effect of change in X_1 on change in Y, whereas $c^*_{YX_1}$ includes both that direct effect and a component due to the correlation between X_1 and another factor influencing Y (i.e., X_2).

All that now remains to be done is to find the OLS value of b_0^*. This is accomplished by taking the expectation of both sides of equation 6.12:

$$E(Y) = \overline{Y} = b_0^* + b_1^*\overline{X}_1 + b_2^*\overline{X}_2$$

if $E(T^*) = 0$. Solving for b_0^*, we have

$$b_0^* = \overline{Y} - b_1^*\overline{X}_1 - b_2^*\overline{X}_2 \tag{6.20c}$$

Notice that equation 6.20c indicates that b_0^* will be zero if the mean of each variable is zero.

Matrix Form of Derivation When more than two independent variables are involved in a regression equation it is convenient to portray the algebra of derivation in terms of matrices. Suppose that we define $\underline{Y}$, $\underline{X}_1$, and $\underline{T}^*$ as vectors of N scores (i.e., columns of numbers), for some finite population of persons. Then the model (equation 6.12) could be portrayed as

$$\underline{Y} = (\underline{1}, \underline{X}_1, \underline{X}_2) \begin{bmatrix} b_0^* \\ b_1^* \\ b_2^* \end{bmatrix} + \underline{T}^*$$

where $\underline{1}$ is a vector of N ones. With the following definitions:

$$\underline{X} = (\underline{1}, \underline{X}_1, \underline{X}_2)$$

$$\underline{b}^* = \begin{bmatrix} b_0^* \\ b_1^* \\ b_2^* \end{bmatrix}$$

we can rewrite the model in its most compact form:

$$\underset{(N \times 1)}{\underline{Y}} = \underset{(N \times 3)}{\underline{X}} \underset{(3 \times 1)}{\underline{b}^*} + \underset{(N \times 1)}{\underline{T}^*} \tag{6.21}$$

where the numbers in parentheses indicate, respectively, the number of rows and columns in each matrix.

One advantage of the matrix form is its generality for regression models involving any number of independent (and, in subsequent chapters, dependent) variables. For example, if a model with p independent variables is specified, then letting $q = p + 1$ allows such a model to be written as

$$\underset{(N \times 1)}{\underline{Y}} = \underset{(N \times q)}{\underline{X}} \underset{(q \times 1)}{\underline{b}^*} + \underset{(N \times 1)}{\underline{T}^*} \tag{6.22}$$

If the reader is familiar with the rules of matrix algebra (see Appendix A), then solution of equation 6.22 for the vector of coefficients $\underline{b}^*$ is quite straightforward. For convenience and without a loss of generality, suppose that each variable has a mean of zero so that b_0^* is zero. Then, multiplying both sides of equation 6.22 by the transpose of $\underline{X}$ gives

$$\underset{(q \times N)}{\underline{X}'} \underset{(N \times 1)}{\underline{Y}} = \underset{(q \times N)}{\underline{X}'} \underset{(N \times q)}{\underline{X}} \underset{(q \times 1)}{\underline{b}^*} + \underset{(q \times N)}{\underline{X}'} \underset{(N \times 1)}{\underline{T}^*} \tag{6.23}$$

This matrix equation can be better understood if written with expanded matrices as follows

$$N \underset{(q \times 1)}{\begin{bmatrix} Y \\ \sigma_{YX_1} \\ \sigma_{YX_2} \\ \cdot \\ \cdot \\ \cdot \\ \sigma_{YX_p} \end{bmatrix}} = N \underset{(q \times q)}{\begin{bmatrix} 1 & X_1 & X_2 & & X_p \\ X_1 & \sigma_{X_1X_1} & \sigma_{X_1X_2} & \cdots & \sigma_{X_1X_p} \\ X_2 & \sigma_{X_2X_1} & \sigma_{X_2X_2} & \cdots & \sigma_{X_2X_p} \\ \cdot & & & & \\ \cdot & & & & \\ \cdot & \cdots & & & \cdots \\ X_p & \sigma_{X_pX_1} & \sigma_{X_pX_2} & \cdots & \sigma_{X_pX_p} \end{bmatrix}}$$

$$\times \begin{bmatrix} b_0^* \\ b_1^* \\ b_2^* \\ \cdot \\ \cdot \\ \cdot \\ b_p^* \end{bmatrix} + N \begin{bmatrix} E(T^*) \\ \sigma_{X_1T^*} \\ \sigma_{X_2T^*} \\ \cdot \\ \cdot \\ \cdot \\ \sigma_{X_pT^*} \end{bmatrix} \quad (6.24)$$

$$(q \times 1) \qquad\qquad (q \times 1)$$

If we can now specify that T^* is uncorrelated with each independent variable and has a mean of zero, then the last vector in equation 6.24 is simply a vector of zeros and can be ignored. Once this vector is deleted, the remaining matrices define the *normal equations* that may be solved for the regression coefficients in terms of variances and covariances among the observable variables:

$$\begin{aligned} \sigma_{YX_1} &= b_1^*\sigma_{X_1X_1} + b_2^*\sigma_{X_1X_2} + \ldots + b_p^*\sigma_{X_1X_p} \\ \sigma_{YX_2} &= b_1^*\sigma_{X_2X_1} + b_2^*\sigma_{X_2X_2} + \ldots + b_p^*\sigma_{X_2X_p} \\ &\cdot \\ &\cdot \\ &\cdot \\ \sigma_{YX_p} &= b_1^*\sigma_{X_pX_1} + b_2^*\sigma_{X_pX_2} + \ldots + b_p^*\sigma_{X_pX_p} \end{aligned} \quad (6.25)$$

The actual solution of these normal equations is tedious without the aid of a computer when $p > 2$. But with a computer capable of inverting large matrices, the solution is achieved with standard programming techniques. (Even handheld calculators can solve sets of simultaneous equations, such as 6.25, when $p = 6$ or more!) The matrix solution for the OLS coefficients follows from multiplying both sides of equation 6.23 by the inverse of $(\underline{X}'\underline{X})$—i.e., the inverse of N times the covariance matrix of the independent variables:

$$(\underline{X}'\underline{X})^{-1}(\underline{X}'\underline{Y}) = (\underline{X}'\underline{X})^{-1}(\underline{X}'\underline{X})\underline{b}^* = \underline{b}^* \quad (6.26)$$

since the product of a matrix and its inverse is the identity matrix.

6.5 ESTIMATION OF REGRESSION COEFFICIENTS FROM SAMPLE DATA

Most applications of multiple regression do not involve population data. Rather, a researcher has usually (at best) selected a random sample from some population and on the basis of those data, seeks to make generalizations about the population. In other words, she uses the sample data to estimate population regression coefficients.

As in the case of the simplest regression model discussed in Chapter 5, sample estimates of multiple regression coefficients are BLUE (best linear unbiased estimate) if the OLS method of estimation is applied and certain assumptions hold. This means that we can simply substitute sample variances and covariances into the normal equations 6.17 or 6.25 and obtain BLUE estimates of $\underline{b}^*$. For example,

$$b_1 = (S_{YX_1}S_{X_2X_2} - S_{YX_2}S_{X_1X_2})/(S_{X_1X_1}S_{X_2X_2} - S^2_{X_1X_2})$$
$$= (c_{YX_1} - c_{X_1X_2}c_{YX_2})/(1 - r^2_{X_1X_2})$$

The assumptions that are sufficient for b_1 and b_2 to be BLUE estimates of b_1^* and b_2^* are the following:

1. For every combination of levels of X_1 and X_2, T^* must be distributed with zero mean and the same variance.
2. T^* is an independent random variable for each observation. That is, disturbances or residuals are uncorrelated with one another and uncorrelated with X_1 and X_2.

The central idea here is that each observation of a Y score is an independent replication of a random selection process. Thus, scores obtained from husbands and wives, for example, could not be included as distinct observations in a regression estimate since factors influencing Y, but unrelated to X_1 and X_2, would be more similar between mates than between different families.

It is important at this point to raise two issues: one, that violations of the above assumptions often introduce inconsequential bias or inflation of sampling variance; two, that the most serious estimation problems of a practical sort are unrelated to these standard assumptions. In a sense, the entire issue of whether or not an estimate is BLUE fades in importance as the following real questions are addressed by researchers:

1. What is the functional relationship between the variables of interest? Is it linear? Does the form of the relationship change as levels of the variables change?
2. How much measurement error is included in the observed variables? Are these errors correlated across variables?
3. How representative is my sample?
4. Have important variables been left out of the model so that observed relationships are distorted or misleading?

It is my opinion that these sources of bias are of such import to behavioral scientists that concern about violation of purely statistical assumptions is often misplaced. Of course, researchers should avoid such violations when feasible.

Estimating Standard Errors

The standard errors of estimates of multiple regression coefficients are obtained in a manner analogous to that outlined in Chapter 5 in the bivariate case (see the end of this chapter for formulas). However, there is an added factor to be considered in the case of multiple independent variables. This additional consideration is the intercorrelation between one independent variable and other independent variables. It turns out that the standard error of a partial regression coefficient tends to increase as the correlation between it and other independent variables increases. At a very intuitive level, the reason for this increase is that it is very difficult to partial out the variance uniquely associated with each of two tightly correlated variables. Common sense would suggest that more observations would be needed to distinguish the independent effects of two highly correlated influences on a dependent variable, and common sense is congruent with theoretical developments in mathematical statistics.

When independent variables in a set are related to one another in such a way that one, or more, of the variables is essentially a linear combination of the remaining variables, then we have the problem of *multicollinearity*. In the case of only two variables, multicollinearity occurs when the two variables are very strongly correlated. With complete multicollinearity, estimates of model parameters cannot be computed. However, as we approach multicollinearity, the standard error of OLS estimates becomes quite large—leaving the estimates as almost useless.

Interactive Models

A detailed discussion of interactive models is presented in Chapter 7. However, a few general comments are appropriate here. In general, interactions between independent variables are incorporated into linear models via the formation of new variables, which are in some way a *product* of two or more independent variables (or transformations of those variables). These "product" variables produce a model in which the effect of one independent variable depends upon the value of one or more other independent variables. However, one must be very careful in the construction of these product variables to avoid very strong correlations between the products and the variables multiplied together. These strong correlations can be eliminated through the use of alternative schemes for coding dummy variables and through the subtraction of means from variables.

An important point to keep in mind when forming interaction variables is that linear equations with product variables are capable of modeling only a certain class of interactive patterns (unless dozens of dummy variables are used). When two variables are multiplied together to form an interactive term, the equation is then capable of modeling an interaction in which the effect of X_1 on Y increases *linearly* as X_2 increases, for example. If the effect of X_1 on Y increases as X_2 increases up to a point and then levels off, the linear model will not fit this process without further adjustments. Therefore, the fact that inclusion of an interactive term in a linear equation model fails to produce a significant coefficient does not necessarily mean that there is no interaction of any type.

A Numerical Example

Now suppose that a survey is conducted in which three new variables are measured: annual income (Y), years of education completed (X_1), and years of work experience (X_2). Our primary concern here is with how much the income of a person seems to be boosted with the addition of one year of formal education. The results of the survey are as follows:

Respondent:	1	2	3	4	5	6	7	8	9	10	11	12	13	14	15	16	17	18
Educ X_1:	12	14	14	16	16	16	12	12	14	14	16	16	12	12	12	14	14	16
Exper X_2:	2	2	2	2	2	2	6	6	6	6	6	6	10	10	10	10	10	10
Income Y:	8	6	12	7	10	13	9	15	10	16	17	11	13	16	19	14	20	18

Use the REGRESSION subprogram of SPSS to analyze these data. Fit them to the following models:

$$Y = c_0 + c_1X_1 + e \tag{Eq1}$$

$$Y = d_0 + d_1X_2 + e \tag{Eq2}$$

$$Y = f_0 + f_1X_1 + f_2X_2 + e \tag{Eq3}$$

Contrast the estimate of c_1 with f_1 and d_1 with f_2. Why are there these differences? Plot these data in a way that shows why the partial slopes are different from the simple slopes.

An SPSS control file that will perform these analyses is as follows:

```
RUN NAME          REGRESS INCOME ON EDUCATION AND WORK EXPER
VARIABLE LIST     RESP,EDUC,EXPER,INCOME
INPUT MEDIUM      CARD
N OF CASES        18
INPUT FORMAT      FREEFIELD
TASK NAME         REGRESS INCOME ON YRS OF EDUC AND YRS OF WRK EXPER
REGRESSION        VARIABLES=INCOME,EDUC,EXPER/
                  REGRESSION=INCOME WITH EDUC(2)RESID=0/
                  REGRESSION=INCOME WITH EXPER(2)RESID=0/
                  REGRESSION=INCOME WITH EDUC,EXPER(2)RESID=0/
STATISTICS        1,2,6
READ INPUT DATA
1 12 2 8   2 14 2 6   3 14 2 12   4 16 2 7   5 16 2 10
6 16 2 13   7 12 6 9   8 12 6 15   9 14 6 10   10 14 6 16
11 16 6 17   12 16 6 11   13 12 10 13   14 12 10 16   15 12 10 19
16 14 10 14   17 14 10 20   18 16 10 18
BREAKDOWN          TABLES=INCOME BY EDUC BY EXPER/
                    INCOME BY EXPER BY EDUC/
STATISTICS         ALL
FINISH
```

Can you now concoct a new set of values for the variable Y so that education (X_1) and work experience (X_2) interact? They are already correlated, but interaction is distinct from correlation. This topic will be addressed in great detail in Chapter 7.

Formulas Associated with Multiple Regression

Suppose that the linear model

$$Y = b_0^* + b_1^*X_1 + b_2^*X_2 + u^* \tag{6.27}$$

holds in some population. The coefficients b_1^* and b_2^* may be structural coefficients or simply slopes. The asterisks indicate population parameters or, in the case of u^*, population values. Now, OLS estimates of population parameters based upon observation of a sample are given by

$$b_1 = (c_{YX_1} - c_{X_1X_2}c_{YX_2})/(1 - r^2_{X_1X_2}) \tag{6.28a}$$

$$b_2 = (c_{YX_2} - c_{X_2X_1}c_{YX_1})/(1 - r^2_{X_1X_2}) \tag{6.28b}$$

$$b_0 = \overline{Y} - b_1\overline{X}_1 - b_2\overline{X}_2 \tag{6.28c}$$

where r indicates a sample correlation and c indicates a simple slope (e..g, $c_{YX_1} = S_{YX_1}/S_{X_1X_1}$, where S_{YX_1} is the sample covariance of Y and X_1, and $S_{X_1X_1}$ is the sample variance of X_1). The OLS estimate of equation 6.27 is then

$$Y = b_0 + b_1X_1 + b_2X_2 + u \tag{6.29}$$

The sample multiple correlation between Y and the right-hand variables is then given by

$$R^2_{Y \cdot X_1X_2} = (b_1S_{YX_1} + b_2S_{YX_2})/S_{YY}$$

for the case of two right-hand variables. For the more general case

$$R^2 = 1 - (S_{uu}/S_{YY}) \tag{6.30}$$

where S_{uu} is the variance of the residuals resulting from the regression of Y on any number of right-hand variables and R^2 is the square of the multiple correlation coefficient between Y and these right-hand variables. The multiple correlation coefficient is simply the correlation between Y and the linear combination of the right-hand variables defined by the OLS coefficients. In the case of two right-hand variables, the multiple correlation R is the correlation between Y and $(b_0 + b_1X_1 + b_2X_2)$. It should also be noted that the OLS estimates of b_1^* and b_2^* maximize this correlation relative to any other estimates (and minimize S_{uu}).

The sample multiple correlation (or its square) is biased upwards in small samples. This bias is analogous to the bias in the sample variance relative to the population variance. The proportion of variance in Y that is explained by the right-hand variables in the population will tend to be less than that reflected in R-squares from samples from that population. In fact, as the number of right-hand variables in a multiple regression equation (p) approaches the sample size (N), R-square approaches one—even if the right-hand variables are completely random. Thus, it may be useful in some circumstances to estimate the population R-square rather than simply compute the sample R-square. One approach to such estimation is to replace the sample variances in equation 6.30 with unbiased estimates of corresponding population variances:

$$\text{est } \sigma_{u^*u^*} = NS_{uu}/(N - p - 1)$$

and

$$\text{est } \sigma_{YY} = NS_{YY}/(N - 1)$$

Thus, we can estimate the population R-square with the so-called *adjusted* R-square:

$$\begin{aligned} \text{adj } R^2 &= 1 - \frac{\text{est } \sigma_{u^*u^*}}{\text{est } \sigma_{YY}} \\ &= 1 - \frac{NS_{uu}/(N - p - 1)}{NS_{YY}/(N - 1)} \\ &= R^2 + (1 - R^2)\{1 - [(N - 1)/(N - p - 1)]\} \\ &= 1 - [(N - 1)(1 - R^2)/(N - p - 1)] \end{aligned} \tag{6.31}$$

where R^2 is as defined in equation 6.30. It should be emphasized, however, that the adjusted R-square in equation 6.31 does not have the ideal properties we associate with the estimate of the population variances.

After obtaining estimates of partial slopes from a sample, it is of interest to estimate the standard errors of those estimates. In the case of only two right-hand variables (as in equation 6.29), the best estimates of the standard errors of b_1 and b_2 are

$$SE(b_1) = (S_u/S_{X_1})w \tag{6.32a}$$

and

$$SE(b_2) = (S_u/S_{X_2})w \tag{6.32b}$$

where

$$w = [1/(N - 3)(1 - r^2_{X_1X_2})]^{1/2}$$

(Note that the standard error becomes smaller as the standard deviation of X_1 becomes larger and as the sample size increases. Conversely, the standard error increases as the correlation between X_1 and X_2 increases. The lessons to be learned from these observations are the following: more precise estimates are derived from designs that increase the range of an independent variable, designs that minimize correlations among independent variables, and designs that employ larger samples.) If u^* is distributed normally, or if sample size is greater than 40, then an approximate 95 percent confidence interval about the estimate b_1 is formed by the interval

$$b_1 - 2\ \mathrm{SE}(b_1) \quad \text{to} \quad b_1 + 2\ \mathrm{SE}(b_1)$$

A t-test for the significance of a regression coefficient against the null hypothesis that the population coefficient is zero is obtained by forming the ratio

$$t = b_1/\mathrm{SE}(b_1) \qquad df = N - p - 1$$

More generally, if Y is regressed on p right-hand variables, then the null hypothesis that k regression coefficients are zero in the population is tested by the F statistic

$$F = \frac{(R_t - R_r)/k}{(1 - R_t)/(N - p - 1)} \qquad df = k,\ (N - p - 1) \tag{6.33}$$

where

R_t = total R-square resulting from the regression of Y on all p right-hand variables

R_r = restricted R-square resulting from the regression of Y on $p - k$ right-hand variables

In the case where only one right-hand variable is constrained to zero in the null hypothesis ($k = 1$), then equation 6.33 gives the square of the t ratio. Since the t ratio is the ratio of the coefficient estimate to its standard error, equation 6.33 allows computation of estimated standard errors for regression slopes from any multiple regression involving p right-hand variables.

Equation 6.33 can also be used to test the null hypothesis that the population R-square is zero. The trick here is to note that this null hypothesis is equivalent to one in which all coefficients of right-hand variables are presumed zero in the population model. In the case of equation 6.29, for example, the null hypothesis that

$$b_1^* = b_2^* = 0$$

can be tested with equation 6.33 as follows:

$$F = \frac{R^2_{Y \cdot X_1 X_2}/2}{(1 - R^2_{Y \cdot X_1 X_2})/(N - 3)} \qquad df = 2, (N - 3)$$

This is also a test of the null hypothesis that the population R-square is zero.

Although most scientific work should focus on metric regression coefficients, it is often instructive to standardize variables and compute standardized regression coefficients (or "beta weights"). If we denote the standardization of a variable by the notation

$$Z_Y = (Y - \overline{Y})/S_Y$$

then the standardized form of equation 6.27 becomes

$$Z_Y = \beta_1^* Z_{X_1} + \beta_2^* Z_{X_2} + \beta_u^* Z_u^*$$

Sample estimates of the beta weights are then easily obtained from estimates of the metric coefficients or from the correlations among the variables:

$$\beta_1 = b_1 S_{X_1}/S_Y = (r_{YX_1} - r_{X_1X_2} r_{YX_2})/(1 - r^2_{X_1X_2})$$
$$\beta_2 = b_2 S_{X_2}/S_Y = (r_{YX_2} - r_{X_1X_2} r_{YX_1})/(1 - r^2_{X_1X_2})$$
$$\beta_u = r_{Yu} = (1 - \beta_1 r_{YX_1} - \beta_2 r_{YX_2})^{1/2}$$

For the data in Table 6.1, $\beta_1 = .35$, $\beta_2 = -.49$, and $\beta_u = .72$. The multiple correlation between Y and X_1, X_2 is unchanged by standardization of variables and is given by

$$R^2{}_{Y \cdot X_1 X_2} = \beta_1 r_{YX_1} + \beta_2 r_{YX_2}$$
$$= (r^2_{YX_1} + r^2_{YX_2} + 2 r_{YX_1} r_{YX_2} r_{X_1X_2})/(1 - r^2_{X_1X_2})$$

And this latter result suggests an approach to computing the proportion of variance in Y that is *uniquely* explained by X_1 (or by X_2):

$$\text{proportion of variance in } Y \text{ uniquely explained by } X_1 = R^2_{Y \cdot X_1 X_2} - r^2_{YX_2}$$

The proportion of variance in Y uniquely explained by X_2 is similarly computed.

Finally, there are two types of special correlations that are conventionally defined: partial and semi-partial. If

$$Y_1 = a_0 + a_1 X_1 + a_2 X_2 + u_1$$

and

$$Y_2 = c_0 + c_1X_1 + c_2X_2 + u_2$$

where a_1, a_2, c_1, and c_2 are OLS estimates, then

$$r_{Y_2Y_1 \cdot X_1X_2} = \text{partial correlation between } Y_1 \text{ and } Y_2 \text{ controlling for (partialling) } X_1 \text{ and } X_2$$
$$= r_{u_1u_2}$$

Thus, the partial correlation is the correlation between the residuals of Y_1 regressed on X_1 and X_2 and the residuals of Y_2 regressed on these same variables. If all relationships are linear, then this partial correlation is an average of conditional correlations. Similarly,

$$r_{Y_1(Y_2 \cdot X_1X_2)} = \text{semi-partial correlation between } Y_1 \text{ and } Y_2 \text{ partialling } X_1 \text{ and } X_2 \text{ from } Y_2 \text{ only}$$
$$= r_{Y_1u_2}$$

and

$$r_{Y_2(Y_1 \cdot X_1X_2)} = r_{Y_2u_1}$$

In the simplest case of partialling only one X variable, these equations reduce to

$$r_{Y_1Y_2 \cdot X} = (r_{Y_1Y_2} - r_{Y_1X}r_{Y_2X})/[(1 - r^2_{Y_1X})(1 - r^2_{Y_2X})]^{1/2}$$
$$r_{Y_1(Y_2 \cdot X)} = (r_{Y_1Y_2} - r_{Y_1X}r_{Y_2X})/(1 - r^2_{Y_2X})^{1/2}$$
$$r_{Y_2(Y_1 \cdot X)} = (r_{Y_1Y_2} - r_{Y_1X}r_{Y_2X})/(1 - r^2_{Y_1X})^{1/2}$$

A close look at these last two equations reveals that beta weights are similar to semi-partial correlations.

The similarity and the difference between a semi-partial correlation and a standardized partial regression coefficient (beta weight) can be seen quite clearly by considering the following regression equations:

$$Z_Y = \beta_1Z_X + \beta_2Z_W + e_1 \tag{6.34}$$

and

$$Z_X = \beta_3Z_W + e_2 \tag{6.35}$$

where Z indicates a standardized variable (mean of zero and variance of one), and e_1 and e_2 are OLS residuals. β_3 is simply the correlation between X and W, while β_1 and β_2 are partial beta weights.

Now, the semi-partial correlation between Y and X partialling W from X is the correlation between Y and e_2:

$$r_{Ye_2} = S_{Ye_2}/S_YS_{e_2} = S_{Z_Ye_2}/S_{e_2}$$

If we multiply equation 6.35 by Z_Y and average over cases, then we see that

$$S_{Z_Ye_2} = r_{YX} - r_{XW}r_{YW}$$

And if we square both sides of equation 6.35 and average over cases, we find that

$$S_{e_2e_2} = 1 - r^2_{XW}$$

Thus, we have

$$r_{Y(X \cdot W)} = (r_{YX} - r_{XW}r_{YW})/(1 - r^2_{XW})^{1/2}$$

for the semi-partial correlation between Y and X with W partialled from X.

The partial beta weight β_1 is now defined in a slightly different manner. Instead of the correlation between Y and e_2, β_1 is the slope of Z_Y regressed on e_2. Thus,

$$\begin{aligned}\beta_1 &= S_{Z_Ye_2}/S_{e_2e_2} \\ &= (r_{YX} - r_{XW}r_{YW})/(1 - r^2_{XW})\end{aligned}$$

Note that this equation differs from that given above for the semi-partial correlation only in that the denominator is a variance rather than a standard deviation. This means that the semi-partial correlation and its corresponding beta weight will have the same sign, but may differ in magnitude. Note also that although the semi-partial correlation must, like all correlations, range between -1 and $+1$, the partial beta weight has no such bounds. The partial beta weight is the slope of a standardized variable regressed on a residual that has a variance less than or equal to one, and this means that the beta weight may take on most any value if X and W are correlated.

REFERENCES AND SUGGESTED READING

Bock, R. Darrel. *Multivariate statistical methods in behavioral research.* New York: McGraw-Hill, 1975.

Hanushek, E. A., and Jackson, J. E. *Statistical methods for social for social scientists.* New York: Academic Press, 1977.

Kleinbaum, D. G., and Kupper, L. L. *Applied regression analysis and other multivariate methods.* North Scituate, Mass.: Duxbury Press, 1978.

Mood, A. M., and Graybill, F. A. *Introduction to the theory of statistics* (2nd ed.). New York: McGraw-Hill, 1963.

7

Univariate Analysis of Variance

7.1 Introduction

Now that we have developed and used a linear model for the analysis of relationships between a dependent and several *in*dependent variables, we may progress to the application of such a model to what has been known as *analysis of variance.* Analysis of variance is often abbreviated as Anova (or ANOVA) and has been associated with the analysis of experiments that involve nominal scale independent variables and continuous (or almost continuous) dependent variables. The very name of this procedure—analysis of *variance*—may be confusing to the reader, since it appeared that the procedures described as multiple regression (in the previous chapter) were based upon an analysis of variance and covariance. In fact, this confusion indicates a prevalent misconception about the relationship between ANOVA and multiple regression.

The essence of that misconception is the belief that ANOVA supplies information about causation, whereas regression techniques only enable an investigator to make predictions or assess correlation. As has been stressed throughout this book, however, the truth of the matter is that causal attribution stems from design considerations rather than statistical procedures, and the linear regression model is the basis for virtually all the statistical methods used by social and behavioral scientists—including ANOVA. As we will also see, the analysis of *co*variance (ANCOVA) is a special application of linear regression to the analysis of experiments in which both manipulated and non-manipulated independent variables are of interest.

Two approaches to the application of regression to analysis of experiments will be demonstrated. The first approach is very much contiguous with the way in which dummy variables were used in Chapter 4. The only important difference between the procedures described there and the ones presented here concerns the definition of the interaction between independent variables. The second method is presented for those readers who use ANOVA on a fairly regular basis to analyze complex experiments. This second approach is based upon the linear regression model known as *regression about the origin.* This procedure involves modification of the linear model that we have worked with up to this point and will be of primary interest to experimenters who employ "factorial designs" involving unusual patterns of *a priori* (i.e., planned) comparisons.

The reader should be aware that the procedures described in this chapter are not applicable to studies involving repeated observations of the same person (or other unit of analysis). This is part of what is meant by the adjective *univariate* in the title of this chapter: the models described are useful when only one dependent variable is observed. When more than one dependent variable is observed in an experiment, we describe the data as *multivariate* and use multivariate regression for analysis. Some confusion arises at this point, however, since traditional presentations of ANOVA have presented so-called *repeated measures* ANOVA procedures as special cases of univariate ANOVA. This unfortunate categorization has obscured the multivariate nature of repeated measures designs. As it turns out, there are some instances in which repeated measures designs are best analyzed by a univariate model, but they are not the rule, and rather troublesome assumptions must be made for the univariate analysis to provide accurate significance tests. In most instances, the more appropriate approach to the analysis of repeated measures designs is the multivariate regression model presented in Chapter 8.

7.2 ONE-WAY ANALYSIS OF VARIANCE

The one-way analysis of variance applies to research situations in which the effect of one independent variable on one dependent variable is of interest. What distinguishes a situation suited to ANOVA from one suited to the type of regression model described in Chapter 6 is that the independent variable is not assumed to be an interval scale variable. For example, if we could manipulate level of social threat such that ten groups of people could be subjected to conditions of threat that corresponded to ten equally separated points along a social threat dimension, then we could regress each person's level of authoritarianism on the almost continuous independent variable of social threat. The resulting structural coefficient would reflect the extent to which we would expect a person to increase in authoritarianism with a one-unit increase in social threat (this specification includes the assumption that the structural relationship between these variables is linear).

Unfortunately, however, social and behavioral scientists have been remiss in the development of such interval scales as the hypothetical one described above. Thus, the more usual situation is one in which a researcher is confident that two experimental conditions involve more social threat than some third condition, but no specific information about the relative magnitude of social threat in the three conditions is available. For example, imagine an experiment in which the independent variable A has three levels corresponding to three "conditions." The three conditions are labeled as follows:

A_1: Control condition
A_2: First type of increased threat condition
A_3: Second type of increased threat condition

These three conditions are termed *levels* of the independent variable A. If participants in the study are randomly assigned to these three conditions, then we have a true randomized experiment. And our primary concern in such a study is with the extent, if any, of the effects of these experimental manipulations on a dependent variable such as authoritarianism.

The data from such an experiment would consist of an F-scale score for each person and information on which experimental condition was applied to each person in the study. The secret to the application of multiple regression to the analysis of these data is in the way that the information about experimental condition is coded with *dummy variables*. These variables are called "dummies" because they are constructed to define group membership rather than position on a continuum. (There are numerous alternative codings that can be used to introduce group membership information into a regression equation. The dummy variable scheme is used here for purposes of contiguity with earlier discussions of dichotomous variables and subsequent discussions of regressions about the origin.) One scheme for defining group membership with dummy variables is as follows:

$$X_1(\mathrm{i}) = \begin{array}{l} 0 \text{ if person } i \text{ is in condition } A_1 \text{ or } A_3 \\ 1 \text{ if person } i \text{ is in condition } A_2 \end{array}$$

$$X_2(\mathrm{i}) = \begin{array}{l} 0 \text{ if person } i \text{ is in condition } A_1 \text{ or } A_2 \\ 1 \text{ if person } i \text{ is in condition } A_3 \end{array}$$

Now that these two dummy variables have been defined, we are in a position to perform an analysis of variance. Notice that two dummy variables were necessary to define the three categories of participants; in general, the number of dummy variables required to define an independent variable is one less than the number of levels in the independent variable. The ANOVA for such an experiment would derive from estimation of the following multiple regression equation:

$$Y(i) = b_0 + X_1(i) + b_2X_2(i) + e(i) \tag{7.1}$$

where X_1 and X_2 are defined for each person i as described above, and Y is an observed dependent variable.

The usefulness of equation 7.1 for the analysis of an experiment is more apparent if we evaluate the expected value of both sides of the equation under different values of the dummy variables (which correspond to different experimental conditions):

$$E[Y(i)\,|\,X_1 = X_2 = 0] = b_0 \tag{7.2a}$$

$$E[Y(i)\,|\,X_1 = 1, X_2 = 0] = b_0 + b_1 \tag{7.2b}$$

$$E[Y(i)\,|\,X_1 = 0, X_2 = 1] = b_0 + b_2 \tag{7.2c}$$

From these three equations it should be apparent that the mean of Y for the first level of A (A_1) is equal to the intercept b_0. Furthermore, the regression coefficient b_1 is equal to the difference between the mean of Y in condition A_2 and condition A_1, and the coefficient b_2 is the difference between the mean of Y for A_3 and A_1. If the condition A_1 is a "control" condition and A_2 and A_3 are experimental conditions, then the usual standard errors of the regression coefficients b_1 and b_2 can be used to test the significance of the effects of conditions A_2 and A_3, respectively.

7.3 HYPOTHESIS TESTING IN ANOVA

In Chapter 6 we focused upon the use of approximate standard errors to obtain confidence intervals for regression coefficients. For ANOVA, the regression coefficients are equal to differences between means for different levels of one (for a one-way ANOVA) or more independent variables. This means that the standard errors obtained in a regression analysis can serve as standard errors of differences between means. For example, if we consider the experiment described above (where the independent variable A had three levels), then the standard errors of b_1 and b_2 provide estimates of the standard errors of the differences between the mean for the control group and the mean of Y for the two-threat conditions:

$$S(b_1) = \text{std error } (Y_2 - Y_1)$$

$$S(b_2) = \text{std error } (Y_3 - Y_1)$$

where Y_1, Y_2, and Y_3 are the means of Y in conditions A_1, A_2, and A_3, respectively.

Now, what is the standard error of a difference between two means? The idea is quite simple and is analogous to the meaning of the standard error of a regression coefficient or any other *statistic* that is derived from observation of a sample from some population. To clarify the meaning of a standard error in the context of our current example, suppose that the hypothetical experiment involving a control (A_1) and two experimental conditions (A_2 and A_3) is replicated many times. That is, suppose that 30 persons are randomly selected from a defined population. A *randomization* procedure is then applied to assign persons to experimental conditions, and observations of authoritarianism are made. All of this yields three means, one for each condition. We may now compute the difference between the mean for condition A_2 and condition A_1. This difference between means would be equal to the regression coefficient b_1 *for this sample.*

Now suppose that we draw another sample of 30 persons from the same population and follow exactly the same procedure. We now have a second sample value for the difference between the mean for persons in condition A_2 and condition A_1 (we also have two values for any other difference between groups that we may care to assess). Repeating this procedure indefinitely, we could generate a long list of differences between group means. That list of differences would constitute a *sampling distribution* of differences between means and that sampling distribution would have a mean and standard deviation. The mean of that distribution would be the *population* difference between people in condition A_1 and A_2, and the standard deviation of that distribution would be the standard error of the difference between means (or, equivalently, the standard error of the regression coefficient b_1). The approximate standard error of b_1 that we compute with a multiple regression program is then an estimate of the standard deviation of the sampling distribution of differences between means.

The usefulness of the standard error, of course, derives from its connection with probabilistic statements about the relationship between an observed sample value of a statistic and the population parameter which it estimates. Those probabilistic statements have been described as *confidence intervals* and should be treated as approximate statements that do not substitute for independent replication in most instances.

The computation of standard errors in ANOVA follows the same logic as was developed in Chapter 6. The standard error of b_1 can then be used to compute a t-statistic by forming the ratio

$$t = b_1/S(b_1) \qquad df = 1, N - p - 1$$

where N is the number of persons in the entire experiment and p is the number of right-hand variables in the multiple regression equation (see equation 7.1 above). When there are three levels of an independent variable, we must define two dummy variables—thus p would be equal to two in this instance. This t-statistic can then be used in a classic Fisherian test of a null hypothesis. The null hypothesis in this instance would be that there is no *average* difference in F-scale scores between persons subjected to conditions A_1 and A_2—in the population. In other words, the observed difference between the means for these two groups is due to chance variation. The probability level for the t-statistic is then the probability that we would obtain a mean difference as large (or larger) than the one obtained if the null hypothesis were true for the population.

Planned and *Post Hoc* Comparisons The above discussion of a test for a null hypothesis raises an issue that is central to the ANOVA procedure. Notice that we focused on testing a null hypothesis concerning the mean F-scale difference between group A_2 and group A_1. We could also consider differences between A_1 and A_3, or between A_2 and A_3. The crucial problem raised by the mutiplicity of possible comparisons of means is that the probability level associated with the significance test of each difference between means applies only to that significance test. Thus, when many means are involved (as would be the case if our independent variable had, for example, six levels), the significance level of each test is deceptive. The deception can be seen quite clearly by an analogy. Suppose that an experiment involved an independent variable with 20 experimental conditions and one control condition. If each of the 20 experimental means were compared with the control mean, and a 5 percent significance level were adopted *for each comparison,* then we would clearly expect that one or more of these comparisons would be significant even if the null hypotheses were exactly true. But, suppose that our theory had predicted that one particular experimental condition would produce a significant difference from the control condition. Such an instance is a *planned comparison,* and under this circumstance the logic of the significance test is appropriate. In such an instance, then, we could use the multiple regression procedure outlined above and the t-statistic for the appropriate regression coefficient to test our null hypothesis.

But now suppose that we have no *a priori* basis for planning specific comparisons between means. Can we test more general null hypotheses concerning a group of means? For example, can we test the null hypothesis that the population F-scale means are identical for our three experimental conditions? The answer is yes and the procedure for such hypothesis testing is quite straightforward once we learn a procedure for testing null hypotheses involving more than one regression coefficient.

F-tests for Groups of Regression Coefficients The problem at hand is how to test the null hypothesis that several regression coefficients have population values of zero *simultaneously*. In other words, can we obtain a rough probability of the likelihood of obtaining several sample slopes or structural coefficients if the population values of those slopes or structural coefficients are all zero? (In an ANOVA situation such a null hypothesis would be equivalent to the assumption that several population means were equal.) The answer, of course, is yes, and the procedure is quite straightforward.

Suppose that we have a regression model with p right-hand variables, and our null hypothesis is that k ($k < p + 1$) of the p regression coefficients are zero in the population. An F-test for this null hypothesis is obtained by regressing Y on the p right-hand variables to obtain the total R-squared, and then deleting the k variables hypothesized to have zero slopes and obtaining the regression of Y on the remaining ($p - k$) right-hand variables. The latter regression yields a *restricted* (or *constrained*) R-squared (restricted under the null hypothesis). The difference between the total and restricted R-squares then reflects the proportion of dependent variable variance that is uniquely associated with the deleted variables. For large samples, this difference is distributed as a chi-square with k degrees of freedom and can be used to compute an F ratio as follows:

$$F = \frac{[R(t) - R(r)]/k}{[1 - R(t)]/(N - p - 1)} \qquad df = k, N - p - 1 \tag{7.3}$$

where

$R(t)$ = total R-square for the regression of Y on p right-hand variables

$R(r)$ = restricted R-square for the regression of Y on ($p - k$) right-hand variables

N = sample size

Equation 7.3 is a very useful tool for developing significance tests in the analysis of experiments since, as we shall see, it allows the testing of very general null hypotheses. It should be noted that the F-statistic is a ratio of chi-square statistics and, therefore, is distributed as the ideal F-distribution only if certain criteria are met. For example, if the residuals for the full model (including p right-hand variables) are independent of one another (i.e., each person in the sample is independently sampled from the population) and normally distributed with zero mean and equivalent variances, then F in equation 7.3 is distributed as the F distribution with k and ($N - p - 1$) degrees of freedom when the null hypothesis is true. For sample sizes greater than 50 or 60, the F-statistic is quite robust under violation of the normality and equal variance assumptions, but not so robust under violation of the independence assumption. The assumption of independence is particularly violated in models involving repeated measurements on the same individual or other unit of analysis as we shall see in Chapter 11, which means that methods other than OLS regression may be necessary where assessment of change is part of a research design.

An Example In order to demonstrate the use of equation 7.3 in the analysis of an experiment, suppose that the experiment described above (involving a control condition and two experimental manipulations of social threat) is conducted and the following data result:

$$Y = \text{F-scale scores observed after treatment}$$

$$X_1(i) = \begin{array}{l} 1 \text{ if person } i \text{ is in condition } A_2 \\ 0 \text{ otherwise} \end{array}$$

$$X_2(i) = \begin{array}{l} 1 \text{ if person } i \text{ is in condition } A_3 \\ 0 \text{ otherwise} \end{array}$$

i	Y	X_1	X_2	
1	7	0	0	
2	9	0	0	
3	10	0	0	Group A_1
4	11	0	0	
5	13	0	0	
6	10	1	0	
7	12	1	0	
8	13	1	0	Group A_2
9	14	1	0	
10	16	1	0	
11	9	0	1	
12	11	0	1	
13	12	0	1	Group A_3
14	13	0	1	
15	15	0	1	

It is clear from inspection of these data that the F-scale scores tend to be larger in the threat conditions (A_2 and A_3) than in the control condition (A_1). The next question that presents itself is the probability that such results would occur if the experimental manipulations would actually have no effect in the population from which the sample was drawn. If this probability is sufficiently low, then we would have greater confidence that this result is unlikely to be a "fluke" that could not be replicated.

There are two approaches to hypothesis testing that may be applied to this situation. The first involves *planned (or a priori) comparisons* and the second involves *a posteriori* or *exploratory* comparisons.

(1) *Planned comparisons.* Suppose that we made two predictions about the outcome of this experiment. In other words, this study was actually composed of two experiments: the first experiment stimulated the prediction that Y_2 (mean F-scale for condition A_2) would be greater than Y_1, and the second experiment involved the prediction that Y_3 would be greater than Y_1. Each of these predictions can be used to generate a null hypothesis:

$$\text{first } H_0\text{:} \quad Y_2^* - Y_1^* = 0$$

$$\text{second } H_0\text{:} \quad Y_3^* - Y_1^* = 0$$

where the asterisk indicates a population mean and H_0 stands for *null hypothesis*. If we fit the above data to the multiple regression equation 7.1, then these two null hypotheses about differences between means can also be stated in terms of the population values of the regression coefficients:

$$\text{first } H_0\text{:} \quad b_1^* = 0 \tag{7.4a}$$

$$\text{second } H_0\text{:} \quad b_2^* = 0 \tag{7.4b}$$

If we consider these two null hypotheses as bases for tests of planned comparisons, then we are considering each as a separate probabilistic event, and the *joint* probability of *both* regression coefficients being different from zero is not a concern. Rather, we are treating the comparison of A_2 with A_1 as a separate event from the comparison of A_3 with A_1, and our concern is with the probability of each difference between means given each null hypothesis.

In order to test the two null hypotheses stated in equations 7.4, we begin by fitting the data above to a regression model like that in equation 7.1. The regression of Y on X_1 and X_2 yields the follwoing results:

$$\begin{aligned} Y &= b_0 + b_1X_1 + b_2X_2 + e \\ &= 10 + 3X_1 + 2X_2 + e \end{aligned} \tag{7.5}$$

$$[\text{R-square} = .28,\ r(Y, X_1) = .4,\ r(Y, X_2) = .1]$$

Notice that the intercept (b_0) is equal to the mean for condition A_1 ($Y_1 = 10$), b_1 is equal to the mean difference between conditions A_2 and A_1 ($Y_2 - Y_1 = 3$), and b_2 is equal to the mean difference between conditions A_3 and A_1 ($Y_3 - Y_1 = 2$). These coefficients have this relationship to the group means on the F-scale because of the way in which the dummy variables were coded; as we shall see subsequently, there are many different schemes for coding these dummy variables, and each scheme produces a different relationship between the regression coefficients and the group means.

We may now proceed to use the results of this multiple regression to perform hypothesis tests. In order to test the first null hypothesis in equation 7.4, we constrain b_1 to be zero by regressing Y on X_2 alone:

$$Y = 11.5 + .5X_2 + e \tag{7.6}$$

$$(\text{R-square} = .01)$$

and this result may now be used in equation 7.3 to obtain

$$\begin{aligned} F &= \frac{(.28 - .01)/1}{(1 - .28)/(15 - 2 - 1)} \qquad df = 1, 15 - 2 - 1 \\ &= 4.5 \end{aligned}$$

These results indicate that 28 percent of the sample variance in Y is explained by mean differences between the three conditions of the study. The probability of obtaining the observed value of b_1 ($Y_2 - Y_1 = 3$) if b_1^* is zero (the null hypothesis) is estimated by the probability that the F ratio would be greater than the observed 4.5. This probability is obtained from a calculator or table for 1 and 13 degrees of freedom to be .055

(5.5 percent). Whether or not this probability is sufficiently small to warrant rejection of the null hypothesis would depend upon the significance level adopted by the investigator.

We may now proceed to test the second of our null hypotheses: that condition A_3 has no population effect relative to the control condition on the average F-scale score (i.e., $b_2^* = 0$). In order to obtain the F ratio for this null hypothesis we must constrain b_2 to be zero by regressing Y on X_1 alone:

$$Y = 11 + 2X_1 + e$$
$$(\text{R-square} = .16)$$

which may be used with equation 7.3 to obtain

$$F = \frac{(.28 - .16)/1}{(1 - .28)/(15 - 2 - 1)} \qquad df = 1,\ 15 - 2 - 1$$
$$= 2.0 \qquad (\text{Prob} = .183)$$

Thus if we had adopted a significance level of .10, we would be motivated to reject the first null hypothesis and accept the second.

(2) *Unplanned comparisons.* Suppose now that we had no specific hypotheses about how the means of our three conditions would differ. (This may sound odd to the reader: What research situation would involve such a lack of specific predictions? I sympathize with such reactions, but there are many researchers who prefer an exploratory strategy.) One approach to such a situation is to test the null hypothesis that the population means are equal for all three of the conditions in our hypothetical study of social threat and authoritarianism:

$$H_0\colon Y_1^* = Y_2^* = Y_3^*$$

If we subtract Y_1^* from each of these three means we obtain the logically equivalent null hypothesis that

$$H_0\colon Y_2^* - Y_1^* = Y_3^* - Y_1^* = 0$$

Recalling the definitions of the regression coefficients in equation 7.1, it should now be clear that this more general null hypothesis about the equivalence of means is also equivalent to

$$H_0\colon b_1^* = b_2^* = 0$$

It should now be apparent that this more general null hypothesis (associated with the ANOVA tradition) is simply a combination of the two null hypotheses generated by our previous planned comparisons. Our goal is thus to determine the likelihood of *both* the observed value of b_1 and the observed value of b_2 if b_1^* and b_2^* are both zero.

Once again we rely upon equation 7.3 to obtain an F ratio to test this new null hypothesis:

$$F = \frac{(.28 - .00)/2}{(1 - .28)/(15 - 2 - 1)} \qquad df = 2,\ 15 - 1 - 1$$
$$= 2.33 \qquad (\text{Prob} = .139)$$

Thus, if we had not made a specific prediction about mean differences, we would be unable to reject the null hypothesis that there are no population differences among the three F-scale means. Note that the zero subtracted from the total R-square is the restricted R-square that results from the regression of Y on only a constant term (b_0); in other words, zero is the R-square that results from the deletion of X_1 *and* X_2 from equation 7.1.

This approach raises a bit of a problem if no planned comparisons were specified. The problem becomes apparent if we consider the ANOVA null hypothesis that b_1^* and b_2^* are zero. If this null hypothesis *is* rejected, what alternative hypothesis is suggested? The logical alternative to the null hypothesis is that b_1^* *and* b_2^* are not *both* zero. But this alternative does not specify the way in which these two parameters are "not zero." Perhaps only one of the two parameters is nonzero, or perhaps the observed data are sufficient to reject both null hypotheses. The procedures developed to answer these questions about specific (i.e., pairwise) comparisons of means are known as methods for unplanned (or *post hoc,* or *a posteriori*) comparisons.

Before discussing the procedures for unplanned comparisons in more detail, it is important to clarify the probabilistic distinction between the planned and unplanned comparisons. The probabilities associated with the planned comparisons of Y_1 with Y_2 and Y_1 with Y_3 are as follows: suppose that b_1 is a variable that is equal to the difference between Y_2 and Y_1 ($Y_2 - Y_1$) for different random samples from a given population while b_1' is a value of b_1 for a particular sample. Then

$$\text{Prob}(b_1 > b_1' \mid b_1^* = 0) = a_1$$

and similarly,

$$\text{Prob}(b_2 > b_2' \mid b_2^* = 0) = a_2$$

where the researcher's hypothesis is directional. These two probabilities, a_1 and a_2, are defined separately: they are linked only in the sense that they are estimated from the same regression equation (and it should be noted that this strategy increases the degrees of freedom, and therefore the statistical power, of the significance tests). a_1 is the probability that a sample of the given size would yield a difference between Y_2 and Y_1 that is as large as (or larger than) the observed difference (b_1) if the population difference (b_1^*) is zero. The probability a_2 is similarly defined. In contrast, the probability associated with the null hypothesis that all three population means are equal is

$$\text{Prob}[(b_1, b_2) > f(b_1', b_2') \mid b_1^* = b_2^* = 0] = a_3$$

where $f(b_1, b_2)$ = some distance function that defines a distance of the point (b_1, b_2) from the origin (b_1^*, b_2^*)—or (0, 0)—in a two-dimensional space. Notice that this probability a_3 is defined in terms of a two-dimensional region that is beyond a particular distance from the origin of that region. The point (b_1^*, b_2^*) or (0, 0) defines the origin (center point) of a plane, whereas b_1 and b_2 are sides of a right triangle. The probability associated with a significance test of the null hypothesis that b_1' and b_2' have occurred by chance from a population in which $b_1^* = b_2^* = 0$ is then the probability of observing two regression coefficients with at least as great a distance from the origin as b_1' and b_2'.

[The distinction between a_1 or a_2 and a_3 is similar to that involving individual tosses of a coin and a series of tosses. Whereas the probability of tossing a head with a fair coin is 0.5 on *each* toss, the probability of obtaining two heads on two specified tosses is 0.25. Furthermore, the probability of obtaining one head on either of two tosses is 0.75. This latter situation is analogous to the one involving unplanned pairwise comparisons.

Problem Suppose that an experiment involves the tossing of a sample of coins from three populations (A_1, A_2, and A_3). Two coins are sampled from each population, and each coin is tossed once. If the outcome of a toss is a head, then that coin receives a score of 1; if the outcome of a toss is a tail, then the sampled coin receives a score of zero. The means for the three samples are Y_1, Y_2, and Y_3, respectively. The corresponding population means are Y_1^*, Y_2^*, and Y_3^*. There are four possible outcomes for each sample (HH, HT, TH, and TT) and three possible means (1.0, 0.5, and 0.0). Considering all three samples jointly, there are 64 possible outcomes. Figure out the following probabilities:

$$\text{Prob}[(Y_2 - Y_1) > 0 \,|\, \text{coins in } A_1 \text{ and } A_2 \text{ fair}] = a_1$$
$$\text{Prob}[(Y_3 - Y_1) > 0.5 \,|\, \text{coins in } A_1 \text{ and } A_3 \text{ fair}] = a_2$$
$$\text{Prob}[(Y_2 - Y_1) > 0 \text{ and } (Y_3 - Y_1) > 0.5 \,|\, \text{all coins fair}] = a_3$$
$$\text{Prob}[([Y_2 - Y_1] > 0 \text{ and } [Y_3 - Y_1] > 0.5) \text{ or } ([Y_2 - Y_1] > 0.5 \text{ and } [Y_3 - Y_1] > 0) \,|\, \text{all coins fair}] = a_4$$

The value a_2 is the probability of two heads on the sample from A_3 and two tails on the sample from A_1—given that all the coins in the two populations are fair. This is the probability of incorrectly rejecting the null hypothesis (all coins in A_1 and A_3 fair) when the null hypothesis is true *and* results in the opposite direction (more tails in sample from A_3) are not considered a basis for rejection (directional hypothesis testing). However, the value a_4 is the probability that *either* Y_2 or Y_3 would be this deviant from Y_1—given the null hypothesis of fair coins. Finally, two times a_4 would be the probability of obtaining differences as large as those giving rise to a_1 and a_2 if deviations in either direction were considered a basis for rejecting the null hypothesis. Notice that a_4, and especially $2 \times a_4$, is larger than a_2. (Answers: $a_1 = 20/64$; $a_2 = 4/64$; $a_3 = 3/64$; $a_4 = 5/64$.)]

This brings us to the problem of unplanned pairwise comparisons of means. Here the question concerns the probability that b_1 would be observed given that b_1^* is zero *and* given that we are also going to test a null hypothesis for b_2. The intuitive point here is that the probability of b_1 being as different from zero as it is, if the null hypothesis is true, must be greater than a_1 if b_1 is only one of several coefficients that are being considered (just as the probability of a head on an unspecified toss in a series must be greater than the probability of a head on a specified toss).

Numerous schemes have been developed to deal with the problem of adjusting probabilities like a_1 and a_2 when unplanned pairwise comparisons are made among several means. Such methods as the Scheffe test, the Newman-Keuls test, and Duncan's multiple range test are discussed in most statistics texts and are available in packaged computer programs. SPSS, for example, includes in its routine entitled ONEWAY no less than seven options for *a posteriori* contrasts. The lack of consensus on approaches to unplanned comparisons reflects, in my opinion, the vagaries of unplanned scientific investigation and the penchant of methodologists to proliferate schemes for splitting hairs. My own advice for researchers who find themselves in a situation calling for unplanned comparisons is to use the most conservative method—the Scheffe test. Since significance tests tend to overstate the replicability of findings in the social and behavioral sciences in any case, the use of a conservative test and a reduction of attention to the finer points of proliferating procedures seems the best tack.

In the spirit of the above discussion, then, we will not delve into the complexities of unplanned comparisons.

The Advantage of More Observations Before we leave the one-way ANOVA, it is important to point out the advantage of combining all the groups in a one-way design to perform a *priori* significance tests. The advantage over pairwise t-tests is greater statistical *power*. The term power is used in statistics to refer to the likelihood that a null hypothesis will be rejected when it is false.

$$(1 - \text{beta}) = \text{Prob}(H_0 \text{ rejected} \mid H_1 \text{ is true})$$

where (1 − beta) is the power of a significance test and H_1 is an alternative to the null hypothesis. The problem with computation of this probability is that some alternative hypothesis must be stated explicitly before beta can be estimated. In the context of the linear models that have been developed in this book, specification of an H_1 is equivalent to specification of the magnitude of a structural coefficient (or slope) or an R-square. For example, we could estimate the probability of rejecting the null hypothesis that three group means are equal if the "true" R-square for the population is .09. (The details for such computations are given in J. Cohen, 1969.) From a table in Cohen (1969) we would find that for a significance level of .05 the value of (1 − beta) increases with sample size. For a sample of 15 (1 − beta) is only .19. This means that if a sample of 15 persons is drawn from three populations with mean differences such that R-square is .09, then the probability of obtaining a sample R-square that is significant at the .05 level is .19. However, if our sample size increases to 50, then (1 − beta) increases to .57; and an N of 100 yields a power of .87. Clearly, then, the likelihood of statistical significance increases considerably with sample size.

The power of a significance test may be contrasted with the significance level of such a test (alpha):

$$\text{alpha} = \text{Prob}(\text{reject } H_0 \mid H_0 \text{ true})$$

Thus, alpha is the probability of making an incorrect inference from observations, whereas (1 − beta) was the probability of making a correct inference. The probability beta is more analogous to alpha in that it is also the likelihood of an incorrect inference:

$$\text{beta} = \text{Prob}(\text{accept } H_0 \mid H_1 \text{ is true})$$

The probabilities alpha and beta are then referred to as the likelihoods of *type one* and *type two* errors in inference, respectively. Alpha is the probability that H_0 will be rejected when it is true, whereas beta is the probability that H_0 will be accepted when some alternative H_1 is true (and H_0 is false).

The importance of each of these types of error varies with research setting. For example, an epidemiologist searching for environmental carcinogens would be less concerned with type one errors than with type two errors. In other words, the error of claiming that a chemical is carcinogenic when it is not is preferable to the error of incorrectly claiming that a chemical is safe. Even sophisticated researchers often fall into the logical trap of assuming that minimization of type one error is always paramount. In fact, however, the relative importance of these two types of error is closely linked with the values of the investigator. If one is more concerned with the profits that may be lost from the banning of a product than with the incidence of cancer, then one may demand "solid proof" that the product causes cancer (i.e., require a very small alpha level). However, if one is more concerned with human welfare than profits, then type one errors are less worrisome than type two errors—meaning that higher alpha levels should be tolerated.

We are now prepared to return to the issue that motivated this discussion of power and types of errors. That issue was the increased power of planned comparisons using equations 7.1 and 7.3 relative to pairwise t-tests. If we return to our numerical example involving three conditions of social threat, the advantage of the multiple regression approach is easy to see. Recall that the mean difference between the first two groups was reflected in the regression coefficient b_1 and that the null hypothesis $b_1^* = 0$ yielded an F ratio of 4.5 with $df = 1{,}12$. The probability of this F ratio is .055 if H_0 is true and the assumptions about the distributions of the residuals hold. Suppose now that instead of testing this null hypothesis in this way we excluded group three from our analysis and regressed Y only on X_1:

$$\begin{aligned} Y &= c_0 + c_1X_1 + e \\ &= 10 + 3X_1 + e \\ &\quad (\text{R-square} = .36) \end{aligned}$$

Notice that c_1 has the same value as b_1; both are equal to the difference between the means of groups A_2 and A_1. And if we compute the F ratio under the null hypothesis that $c_1^* = 0$, we find that

$$\begin{aligned} F &= \frac{(.36 - 0.0)/1}{(1 - .36)/(10 - 1 - 1)} \qquad df = 1, 8 \\ &= 4.5 \quad (\text{Prob} = .067) \end{aligned}$$

The F ratio turns out to be the same, but the probability of the observed F is greater (.067 versus .055; see equation 7.6). The reason for this is the larger number of observations when all three groups are considered together. The logic of this difference is consistent with your intuition in the sense that more observations give us greater confidence in our estimates of population parameters. Thus, when more groups of people are incorporated into an analysis, the estimate of the residual variance is more accurate and produces more powerful significance tests than when pairs of groups are compared without the information from the other groups.

A Disclaimer Now that we have gone to some length to discuss the disadvantages of pairwise t-tests in the analysis of one-way experiments, I will close this section with the disclaimer that pairwise t-tests are probably the most convenient and most easily interpreted approach to such analyses in many situations. The increased power of the multiple regression method is not great in most instances, and simple t-tests may be computed directly from the means and standard deviations of each group in the various levels of the independent variable. If some *a priori* t-test is "almost significant" according to some inflexible alpha level, then the multiple regression equation may be applied to "pool" error terms. But for most purposes, and especially if a researcher does not adhere to the fetish of the .05 level, the pairwise t-tests supply all the inferential information needed; and the correlation between the dummy variable defining the two groups and the dependent variable supplies the descriptive information concerning the magnitude of the relationship.

The real usefulness of the multiple regression approach to the analysis of experiments becomes apparent when we apply it to *factorial* designs. In this instance, the exploration of interactions and increases in power due to reduced residual variance can be of considerable importance.

7.4 ANALYSIS OF FACTORIAL EXPERIMENTS

Factorial designs are used in experiments involving more than one independent variable. It is important at this point to distinguish between an independent variable and the dummy variables that define that independent variable. For example, the one-way design described in the previous section included only one independent variable—even though two dummy variables were used to define it (the dummy variables are then referred to as "right-hand" variables to distinguish them from independent variables). Thus, a factorial design is not one that requires more than one dummy variable to analyze; rather, it is one in which an independent variable with two or more levels is *crossed* with one or more additional independent variables to ascertain the interacting effects of all these variables on one dependent variable.

An Example Suppose that the experiment in the previous section was expanded so that the effects of both social threat (A) *and* knowledge (B) on authoritarianism (Y) can be assessed. The independent variable called "knowledge," and labeled "B," is added to the experiment because of our suspicion (following Solomon Asch) that knowledge of social laws enables people to transcend those laws. Thus, the independent variable B has two levels:

B_1: persons are not informed of the effect of social threat on authoritarianism

B_2: participants *are* informed of the effect of social threat on authoritarianism

We may now construct a crossed design with six "cells":

	A_1	A_2	A_3	
B_1	Y_{11}	Y_{21}	Y_{31}	$k = 1$
B_2	Y_{12}	Y_{22}	Y_{32}	$k = 2$
	$j = 1$	$j = 2$	$j = 3$	

where Y_{jk} are the cell means for level j of independent variable A and level k of independent variable B. Different people are randomly placed in each of the six possible combinations of levels of A and B. It should be pointed out that there are certain advantages of having an equal number of people in each cell (this will make all our dummy variables uncorrelated), but the method of analysis described in the following is applicable to *unbalanced* designs (unequal cell sizes) as well as balanced designs (with equal cell sizes).

Two approaches to analysis of the results of such an experiment will be presented. The first approach is referred to as the planned comparison approach, the second approach the classic ANOVA approach. As we shall see, the only difference between the two approaches is the type of null hypothesis that may be *easily* tested. The practical difference between the two approaches is the way in which the dummy variables are coded. The planned comparison approach employs the 0/1 coding scheme we have used thus far. The classic ANOVA approach requires a slight alteration in the way that the dummy variables are coded.

Planned Comparison Approach The planned comparison approach to analysis of the above experiment would follow from the hypothesis that increased social threat will have a greater effect on F-scale scores if people are not informed of the relationship between these two variables. In terms of the cell means from the factorial design, this means that we would expect the difference $(Y_{21} - Y_{11})$ to be greater than the difference $(Y_{22} - Y_{12})$, and we would expect $(Y_{31} - Y_{11})$ to be greater than $(Y_{32} - Y_{12})$. In order to explore these hypotheses, we need to construct a linear model with coefficients that correspond to these comparisons. The population parameters associated with these coefficients can then be assumed zero under a null hypothesis, and the null hypothesis can be subjected to the usual significance test.

Our hypotheses can be explored by specifying the following linear model

$$Y(i) = b_0^* + b_1^* X_1 + b_2^* X_2 + b_3^* X_3 + b_4^*(X_1 X_3) + b_5^*(X_2 X_3) + e^*(i) \quad (7.7)$$

where we define the right-hand dummy variables in the usual way

$$Y(i) = \text{F-scale score for person } i \text{ (dependent variable)}$$

$$X_1(i) = \begin{cases} 1 \text{ if person } i \text{ is in } A_2 \\ 0 \text{ otherwise} \end{cases}$$

$$X_2(i) = \begin{cases} 1 \text{ if person } i \text{ is in } A_3 \\ 0 \text{ otherwise} \end{cases}$$

$$X_3(i) = \begin{cases} 1 \text{ if person } i \text{ is in } B_2 \\ 0 \text{ otherwise} \end{cases}$$

Note that the interactions in equation 7.7 have been formed by multiplying the dummy variables for A by the dummy variable defining B. Furthermore, the levels of A and B have been defined in a rather straightforward manner that is consistent with our previous practice.

Our next task is to see just how the regression coefficients in equation 7.7 relate to our hypotheses about the effects of social threat and knowledge on authoritarianism. One way to demonstrate the relationship between the coefficients and our hypotheses is to rewrite equation 7.7 in the following (sample) form:

$$Y = (b_0 + b_3X_3) + (b_1 + b_4X_3)X_1 + (b_2 + b_5X_3)X_2 + e \tag{7.8}$$

In this form, it should be clear that we have essentially two regression equations in one. When $X_3 = 0$ (no knowledge condition), equation 7.8 reduces to

$$Y = b_0 + b_1X_1 + b_2X_2 + e \tag{7.9}$$

which is analogous to equation 7.1 used to analyze a one-way ANOVA. In fact, the coefficients in equation 7.9 are equivalent to what would follow from a one-way ANOVA of those persons in level B_1 of this experiment. The coefficient b_1 reflects the mean difference between condition A_2B_1 and condition A_1B_1, and the coefficient b_2 is equal to the difference between the mean in condition A_3B_1 and the control condition A_1B_1. Now, when $X_3 = 1$, a number must be added to each coefficient to make the equation consistent with the mean differences between analogous groups in level B_2. Therefore, the coefficients b_4 and b_5 are the differences between the effects of social threat in the no-knowledge condition (B_1) and the knowledge condition (B_2).

These relationships among the cell means and the coefficients can be seen more clearly if we take the expectation of both sides of equation 7.8 under all the possible combinations of values of the dummy variables (which, in turn, define all of the cells of the factorial design):

Cell	X_1	X_2	X_3	$E(Y\|X_1, X_2, X_3)$	$E(b_0 + b_1X_1 \ldots + e)$
A_1B_1	0	0	0	Y^*_{11}	b^*_0
A_2B_1	1	0	0	Y^*_{21}	$b^*_0 + b^*_1$
A_3B_1	0	1	0	Y^*_{31}	$b^*_0 + b^*_2$
A_1B_2	0	0	1	Y^*_{12}	$b^*_0 + b^*_3$
A_2B_2	1	0	1	Y^*_{22}	$b^*_0 + b^*_1 + b^*_3 + b^*_4$
A_3B_2	0	1	1	Y^*_{32}	$b^*_0 + b^*_2 + b^*_3 + b^*_5$

We are now in a position to state our hypotheses more explicitly and see how they correspond to the regression coefficients.

The hypothesis that the effect of threat condition A_2 is greater without knowledge than with knowledge is equivalent to stating that $(Y^*_{21} - Y^*_{11})$ is greater than $(Y^*_{22} - Y^*_{12})$. Translating these mean differences into functions of the coefficients gives

$$Y^*_{21} - Y^*_{11} = (b^*_0 + b^*_1) - b^*_0$$

$$Y^*_{22} - Y^*_{12} = (b^*_0 + b^*_1 + b^*_3 + b^*_4) - (b^*_0 + b^*_3)$$

where, as usual, the asterisk indicates a population value. Now, if we specify the null hypothesis that these two mean differences are equal, then subtracting the first equation from the second will give us an expression that should be zero under the null hypothesis that $(Y^*_{21} - Y^*_{11})$ is equal to $(Y^*_{22} - Y^*_{12})$. And the result of such subtraction is

$$b^*_4 = 0$$

Thus, we can test one of our hypotheses by constraining b_4 to be zero [by deleting the X_1X_3 term from the regression equation (equation 7.8) and using the relationship (equation 7.3) to obtain an F ratio].

The null hypothesis that threat condition A_3 has the same effect in level B_1 as in level B_2 is similarly tested by the constraint

$$b^*_5 = 0$$

This latter hypothesis is tested by excluding the X_2X_3 term from equation 7.7 to obtain an F ratio. We could also test the two of these null hypotheses jointly by deleting both the interaction terms from the model.

Classic ANOVA Model The classic ANOVA approach focuses on the testing of *main effects* along with interactions. Main effects are defined as means of levels of one independent variable *across* levels of other independent variables. For example, in the above factorial design, the main effect of A is defined as the differences among the following three marginal means:

$$(Y^*_{11} + Y^*_{12})$$
$$(Y^*_{21} + Y^*_{22})$$
$$(Y^*_{31} + Y^*_{32})$$

Now, if the reader uses the expectations of the coefficients in equation 7.7 to evaluate these three sums of means, it will be apparent that the null hypothesis that they are all equal does not generate a testable set of restrictions on the coefficients (at least not testable with the usual regression programs such as SPSS). That is, there is no set of coefficients that, when set equal to zero, correspond to the main effects null hypothesis.

This situation is remedied by an alternative scheme for coding dummy variables. The scheme is identical to that used for the planned comparisons approach except that the *first* level of each independent variable is coded as -1 instead of zero. Thus, for the factorial design above we specify

$$Y = c^*_0 + c^*_1Z_1 + c^*_2Z_2 + c^*_3Z_3 + c^*_4(Z_1Z_3) + c^*_5(Z_2Z_3) + e(i) \quad (7.10)$$

where

$$Y(i) = \text{F-scale score for person } i$$

$$Z_1(i) = \begin{cases} -1 \text{ if person } i \text{ in } A_1 \\ 1 \text{ if person } i \text{ in } A_2 \\ 0 \text{ otherwise} \end{cases}$$

$$Z_2(i) = \begin{cases} -1 & \text{if person } i \text{ in } A_1 \\ 1 & \text{if person } i \text{ in } A_3 \\ 0 & \text{otherwise} \end{cases}$$

$$Z_3(i) = \begin{cases} -1 & \text{if person } i \text{ in } B_1 \\ 1 & \text{if person } i \text{ in } B_2 \\ 0 & \text{otherwise} \end{cases}$$

Using equation 7.10, the null hypothesis for the main effect of A becomes

$$c_1^* = 0$$

$$c_2^* = 0$$

These two restrictions may then be used with equation 7.3 to obtain F ratios for the main effect of A. The main effect of B may similarly be tested by the constraint

$$c_3^* = 0$$

and the interaction between A and B is tested by the constraints

$$c_4^* = 0$$

$$c_5^* = 0$$

The problem with the use of this approach to ANOVA is that it is rather cumbersome to demonstrate why the coding scheme allows the testing of main effects. And this means that most users will not have a good understanding of how the procedure actually works—a situation that maximizes the likelihood of incorrect application. For this reason an alternative method for the computation of ANOVA statistics is presented in Section 7.5 of this chapter. The alternative computation method is much easier to understand, but it cannot be performed with the computer programs that are restricted to *ordinary* least squares (OLS) regression (e.g., the SPSS subprogram REGRESSION, version 8). The alternative computation method requires a least squares program that allows user specified constraints on the values of regression coefficients (including the intercept). Although such programs are readily available (e.g., the BMD and TSP packages include such programs), there may be some readers who wish to perform an ANOVA using a "tried and true" OLS program. Thus, the following discussion outlines the reasons why equation 7.10 enables one to test for main effects of the independent variable A.

The goal now is to see how the classic ANOVA coding scheme enables one to specify a main effects null hypothesis. As noted above, the main effects null hypothesis for the independent variable A is that

$$(Y_{11}^* + Y_{12}^*) = (Y_{21}^* + Y_{22}^*) = (Y_{31}^* + Y_{32}^*) \tag{7.11}$$

In words, this null hypothesis states that for each level of the independent variable A (i.e., A_1, A_2, and A_3) the average (across levels of B) population mean of Y is the same. When the numbers of participants in all cells of the experiment are equal, this null hypothesis is equivalent to the statement that the population means of Y are equal under the three conditions of the A variable. However, when cell sizes are not equal, it is important to specify that the null hypothesis refers to an average (or sum) of population means.

In order to see how the constraint that c_1^* and c_2^* are zero is equivalent to the null hypothesis of no main effect for A, we simply apply the familiar procedure of finding the expectation of Y under different fixed values of the dummy variables. Applying this procedure to equation 7.10 gives the following results:

Cell	Z_1	Z_2	Z_3	Cell mean $E(Y\|Z_1, Z_2, Z_3)$	$E(c_0 + c_1Z_1 \ldots + e)$
A_1B_1	-1	-1	-1	Y_{11}^*	$c_0^* - c_1^* - c_2^* - c_3^* + c_4^* + c_5^*$
A_2B_1	1	0	-1	Y_{21}^*	$c_0^* + c_1^* \qquad - c_3^* - c_4^*$
A_3B_1	0	1	-1	Y_{31}^*	$c_0^* \qquad + c_2^* - c_3^* \qquad - c_5^*$
A_1B_2	-1	-1	1	Y_{12}^*	$c_0^* - c_1^* - c_2^* + c_3^* - c_4^* - c_5^*$
A_2B_2	1	0	1	Y_{22}^*	$c_0^* + c_1^* \qquad + c_3^* + c_4^*$
A_3B_2	0	1	1	Y_{32}^*	$c_0^* \qquad + c_2^* + c_3^* \qquad + c_5^*$

where the expected values of equation 7.10 are obtained by substituting the values of the dummy variables for each cell into the equation itself.

Now that the relationship between the cell means and the regression coefficients has been clarified, we can proceed to express the null hypothesis for the main effect of A in terms of those coefficients. This is done by replacing the means in equation 7.11 with equivalent combinations of regression coefficients. For example, the mean Y_{11} is replaced with the expression $(c_0 - c_1 - c_2 - c_3 + c_4 + c_5)$ and so on. Finally, then, equation 7.11 may be rewritten as

$$(c_0 - c_1 - c_2 - c_3 + c_4 + c_5) + (c_0 - c_1 - c_2 + c_3 - c_4 - c_5)$$
$$= (c_0 + c_1 - c_3 - c_4) + (c_0 + c_1 + c_3 + c_4)$$
$$= (c_0 + c_2 - c_3 - c_5) + (c_0 + c_2 + c_3 + c_5)$$

And after applying some basic rules of algebra we have

$$2(-c_1 - c_2) = 2(c_1) = 2(c_2)$$

which implies that c_1 and c_2 are both zero. [If this conclusion is not obvious, note that

$$2(-c_1 - c_2) = 2(c_1)$$

implies that

$$2(c_1) + c_2 = 0$$

Putting this result together with the constraint

$$2(c_1) = 2(c_2)$$

gives

$$3(c_2) = 0 \text{ and/or } 3(c_1) = 0]$$

Although the persistent reader can surely follow the above algebra, it is difficult to achieve an enduring intuitive sense of how the cell means relate to the regression coefficients. The reader who is concerned about this difficulty can take heart, however, since the method of computation presented in Section 7.5 involves a much less complex cognitive map. Nevertheless, the above algebra demonstrates that the main effects null hypothesis of classic ANOVA is equivalent to the null hypothesis that

$$c_1^* = c_2^* = 0$$

And this null hypothesis can be tested with an OLS regression program using equation 7.3.

We are now ready to see how the dummy variable coding scheme used for equation 7.10 also enables us to test the classic ANOVA null hypothesis for the interaction between the factors A and B. Once again, we begin by defining the interaction null hypothesis in terms of population means and then proceed to find the regression coefficients that correspond to the mean differences that define the interaction.

An interaction between two independent variables is evidenced when the effect of one variable is dependent upon the level of the other variable that is involved. [The term *specification* is used by some authors instead of interaction. The reason for this nomenclature is that the effect of one independent variable upon the dependent variable can be stated only if the level of one or more other variables is specified. For example, it might be found that statements about the effect of social threat on authoritarianism are meaningful only if we can specify whether or not the participants have knowledge of the effect of social threat (without knowledge).] In the context of our example, an interaction between the independent variable A and the independent variable B would mean that the effect of social threat (A) on authoritarianism (Y) would be different for people with (B_2) and without (B_1) knowledge of the relationship between social threat and authoritarianism. Alternatively, the interaction between A and B could be stated as the finding that the effect of knowledge of the relationship between social threat and authoritarianism on F-scale scores would be different for different levels of social threat. Thus, if a researcher found that participants with knowledge of the relationship showed less of an increase in authoritarianism when subjected to social threat, then this would be described as an interaction between social threat and social knowledge as these two variables affect authoritarianism. (It is important to distinguish this notion of interaction from the concept of *reciprocal causation.* This latter concept will be discussed in Chapter 11 on longitudinal designs; it means that changes in one variable produce changes in a second variable, and that change in the second variable produces change in the first variable. The label "interaction" is easily confused with this concept of reciprocal causation.)

For our 3×2 ANOVA design, we can define the effects of social threat on authoritarianism for each level of B. The null hypothesis of *no interaction* in the population can then be stated by equating the effects of A under the two levels of B. The effects of the conditions A_1, A_2, and A_3 on Y *in level one of* B are reflected in the mean differences between persons in the three cells A_1B_1, A_2B_1, and A_3B_1. These are the participants that were not informed of the relationship between social threat and authoritarianism prior to the study. The *effects* of A in this level of B are therefore

$$Y_{21} - Y_{11} = \text{effect 1}$$
$$Y_{31} - Y_{11} = \text{effect 2}$$
$$Y_{31} - Y_{21} = \text{effect 3}$$

and similarly for level B_2

$$Y_{22} - Y_{12} = \text{effect 4}$$
$$Y_{32} - Y_{12} = \text{effect 5}$$
$$Y_{32} - Y_{22} = \text{effect 6}$$

Note that these six effects exhaust the set of possible mean differences *within* levels of B. The null hypothesis of no interaction, then, is that in the population

$$\text{effect 1} = \text{effect 4}$$
$$\text{effect 2} = \text{effect 5}$$
$$\text{effect 3} = \text{effect 6}$$

or

$$Y^*_{21} - Y^*_{11} = Y^*_{22} - Y^*_{12}$$
$$Y^*_{31} - Y^*_{11} = Y^*_{32} - Y^*_{12}$$
$$Y^*_{31} - Y^*_{21} = Y^*_{32} - Y^*_{22}$$

If the means in these last three equations are now replaced with the regression coefficients from equation 7.10 that are equal to each mean, then we can determine which constraints on the coefficients are equivalent to the null hypothesis of no interaction. Performing that algebra gives the following results

$$-2(c_4) - c_5 = 2(c_4) + c_5$$
$$-c_4 - 2(c_5) = c_4 + 2(c_5)$$
$$c_4 - c_5 = -c_4 + c_5$$

And these three equations hold if and only if $c_4 = c_5 = 0$. Thus, the null hypothesis of no interaction between the independent variables A and B is equivalent to the statement that

$$c^*_4 = c^*_5 = 0$$

And this null hypothesis may be tested in a straightforward manner using equation 7.3 and an OLS regression program.

Example Computations In order to clarify the computational procedures that are necessary for analysis of a factorial experiment, suppose that the following data resulted from a study designed to assess the effects of social threat and knowledge on authoritarianism. [The data for level B_1 are the same as in the hypothetical example of Section 7.3.]

B_1 No knowledge	B_2 Knowledge	
7	7	
9	8	
10 (10)	(9.2) 9	A_1 Control
11	10	
13	12	
10	6	
12	8	
13 (13)	(9) 8	A_2 First threat manipulation
14	9	
16	10	
	10	
	12	
9	5	
11	7	
12 (12)	(8) 7	A_3 Second threat manipulation
13	7	
15	8	
	10	
	12	

(The mean for each cell is given in parentheses.)

From inspection of the six cell means it is apparent that, on the average, participants in the "no knowledge" conditions (B_1) have higher authoritarianism scores than those participants in the "knowledge" condition (B_2). Furthermore, it appears that the threat manipulations (A_2 and A_3) are associated with increased authoritarianism scores (relative to the control group) only for the "no knowledge" condition. The goal of the ANOVA is to gain some indication of how replicable these patterns of mean differences are. As has been pointed out previously, statistical estimates of replicability are important aids, but they should not be treated with the same respect as actual replication by independent experimenters. The clues concerning replicability are couched in terms of probabilities that observed differences (such as the ones between the means in our hypothetical experiment) would have occurred if some null hypothesis were true in the population from which the observed sample was drawn.

Two approaches to the analysis of these data will be computed: the *planned comparison* approach and the *classic* ANOVA approach.

Planned Comparison Computations The planned comparison approach to the analysis of designed experiments involves the use of a linear model with dummy variables coded as zero or one. This approach is most appropriate when one level (or condition) of each independent variable is a control condition. In our hypothetical experiment level A_1 and level B_1 are control conditions relative to the "treated" or "manipulated" conditions A_2, A_3, and B_2, respectively. The primary concern of such an experiment would be the pattern of differences between the treated groups and the control groups on the dependent variable (F-scale scores).

The linear model specified in equation 7.7 (or equation 7.8) is appropriate for this planned comparison analysis. The data needed for input into an OLS regression program would be as follows:

i	$Y(i)$	X_1	X_2	X_3	$X_1^*X_3$	$X_2^*X_3$	
1	7	0	0	0	0	0	
2	9	0	0	0	0	0	
3	10	0	0	0	0	0	Cell A_1B_1
4	11	0	0	0	0	0	
5	13	0	0	0	0	0	
6	10	1	0	0	0	0	
7	12	1	0	0	0	0	
8	13	1	0	0	0	0	Cell A_2B_1
9	14	1	0	0	0	0	
10	16	1	0	0	0	0	
11	9	0	1	0	0	0	
12	11	0	1	0	0	0	
13	12	0	1	0	0	0	Cell A_3B_1
14	13	0	1	0	0	0	
15	15	0	1	0	0	0	
16	7	0	0	1	0	0	
17	8	0	0	1	0	0	
18	9	0	0	1	0	0	Cell A_1B_2
19	10	0	0	1	0	0	
20	12	0	0	1	0	0	
21	6	1	0	1	1	0	
22	8	1	0	1	1	0	
23	8	1	0	1	1	0	
24	9	1	0	1	1	0	Cell A_2B_2
25	10	1	0	1	1	0	
26	10	1	0	1	1	0	
27	12	1	0	1	1	0	
28	5	0	1	1	0	1	
29	7	0	1	1	0	1	
30	7	0	1	1	0	1	
31	7	0	1	1	0	1	Cell A_3B_2
32	8	0	1	1	0	1	
33	10	0	1	1	0	1	
34	12	0	1	1	0	1	

[The participant numbers (i) would not, of course, be *necessary* input for most regression programs, but they are usually included for identification purposes.]

We may commence our analysis of these data by regressing the dependent variable (Y) on all the right-hand dummy variables. This gives

$$\begin{aligned} Y = {} & \underset{(0.96)}{10.0} + \underset{(1.36)}{3.0X_1} + \underset{(1.36)}{2.0X_2} - \underset{(1.36)}{0.8X_3} \\ & - \underset{(1.85)}{3.2(X_1X_3)} - \underset{(1.85)}{3.2(X_2X_3)} + e \end{aligned} \tag{7.12}$$

$$\text{R-square} = .445 \qquad S(ee) = 128.8/34 = 3.79$$
$$S(YY) = 232/34 = 6.82$$

where $S(ee)$ is the variance of the sample residuals, $S(YY)$ is the sample variance of the dependent variable, and the numbers below the coefficients are estimates of standard errors. [The reader should verify that the coefficients in equation 7.12 reproduce the observed cell means. For example, the mean of Y for cell A_3B_2 was 8. If the appropriate values for the dummy variables are substituted into equation 7.12 we have

$$\begin{aligned} Y_{32} &= 10 + 3(0) + 2(1) - .8(1) - 3.2(0) - 3.2(1) \\ &= 10 + 2 - .8 - 3.2 \\ &= 8 \end{aligned}$$

This procedure may be applied for each of the cells of the experiment.]

Equation 7.12 enables us to draw several conclusions about the results of the hypothetical experiment. First, we note that the dummy variables representing the experimental manipulations account for 44.5 percent of the total variance in the dependent variable. [The reader should keep in mind the fact that the R-square tends to be inflated relative to the proportion of variance in the dependent variable that would be explained by the independent variables *in the population* from which the observed sample was drawn. The inflation is especially great when the sample size is small. In the case of equation 7.12 a (better) estimate of the population R-square is obtained by estimating population variances of Y and e:

$$\text{Est of population variance of } Y = 232/33 = 7.03$$
$$\text{Est of population variance of } e = 128.8/28 = 4.6$$

where the denominator values of 33 and 28 are the degrees of freedom for the sample sum of squared deviations. These population estimates can then be used to estimate the population R-square:

$$\begin{aligned} \text{Est of population R-square} &= (7.03 - 4.60)/7.03 \\ &= .346 \end{aligned}$$

Thus, the sample R-square is inflated by a considerable extent (.445/.346 = 1.29) in this example.] The immediate question raised by this finding is the probability that this R-square would have been observed in a sample of 34 persons from a population in which the R-square is zero. A rough estimate of this probability can be obtained through the application of equation 7.3:

$$\begin{aligned} F &= \frac{(.445 - 0)/5}{(1 - .445)/(34 - 5 - 1)} \qquad df = 5, 28 \\ &= 4.49 \end{aligned}$$

$$\text{Prob}(F > 4.49 \,|\, H_0 \text{ true}) = .004$$

Thus, if the appropriate assumptions hold, regression coefficients large enough to explain at least 44.5 percent of the variance in Y would occur in equation 7.12 only four times out of 1000 if the null hypothesis were true. The null hypothesis in this instance may be stated in terms of the population R-square (that it is zero) or in terms of the population regression coefficients (that they are all zero). Thus, the results of the experiment would be quite unlikely if the null hypothesis were true, so we are motivated to reject the null hypothesis and entertain an alternative hypothesis in which the experimental manipulations have nonzero structural parameters associated with them (i.e., nonzero effects on the dependent variable).

The above result would be most comforting to an experimenter who was hoping to find some effects of her manipulations, but the null hypothesis was rather nonspecific. Furthermore, some of the planned comparisons could be significant even if the total R-square was not. We can now proceed to the more specific planned comparisons to see if any of them are large enough, and significant enough, to use for theoretical purposes.

We begin by looking at each of the coefficients in equation 7.12 to identify the null hypothesis associated with each. Recalling the expected values of each coefficient, we have

$$b_1 = Y_{21} - Y_{11} \tag{7.13a}$$

$$b_2 = Y_{31} - Y_{11} \tag{7.13b}$$

$$b_3 = Y_{12} - Y_{11} \tag{7.13c}$$

$$b_4 = (Y_{22} - Y_{12}) - (Y_{21} - Y_{11}) \tag{7.13d}$$

$$b_5 = (Y_{32} - Y_{12}) - (Y_{31} - Y_{11}) \tag{7.13e}$$

From these relations we see from the null hypothesis that $b_1^* = 0$ means that threat manipulation A_2 has no effect on authoritarianism under condition B_1 (no knowledge). Similarly,

$b_2^* = 0$ implies that threat condition A_3 has no effect under condition B_1

$b_3^* = 0$ implies that knowledge has no effect under condition A_1

$b_4^* = 0$ implies that the effect of threat A_2 is the same under condition B_1 as under condition B_2

$b_5^* = 0$ implies that the effect of threat A_3 is the same under condition B_1 as under condition B_2

An experimenter may have specified the alternatives to any or all of these null hypotheses as *a priori* hypotheses or *planned comparisons*. Significance tests for the planned comparisons can be computed with equation 7.3. Each significance test is obtained by comparing the variance in Y that is explained by a regression model that excludes those coefficients that are constrained to zero under the null hypothesis. The resulting *restricted* R-square is then compared to the total R-square given in equation 7.12.

As an example, suppose that the experimenter's prediction was that (1) there would be no difference between the two groups that did not receive the threat manipulation (cells A_1B_1 and A_1B_2), and (2) the mean F-scale scores in both threat conditions would increase more, relative to the control condition, when there was no knowledge than when there was knowledge. The first of these hypotheses would be supported by acceptance of the null hypothesis $b_3^* = 0$, and the second would be supported by the rejection of the null hypothesis $b_4^* = b_5^* = 0$.

Assuming that $b_3^* = 0$, we proceed to regress Y on all dummy variables *except* X_3. This gives:

$$Y = 9.6 + 3.4X_1 + 2.4X_2 - 4X_1X_3 - 4X_2X_3 + e \tag{7.14}$$

$$\text{R-square} = .438 \qquad S(ee) = 130.4/34 = 3.84$$

Note that the deletion of the X_3 dummy variable had almost no effect on the R-square; the R-square of .445 was reduced by only a small increment to .438. The probability of this observation if $b_3^* = 0$ is estimated with equation 7.3:

$$F = \frac{(.445 - .438)/1}{(1 - .445)/(34 - 5 - 1)} \qquad df = 1, 28$$
$$= .348$$
$$\text{Prob}(F > .348) = .56$$

Thus, the probability of the observed difference between the two control groups (A_1B_1 and A_1B_2) is quite reasonable under the null hypothesis, and there is no reason to reject it. The second planned comparison is dealt with in an analogous manner. We begin by estimating a regression equation involving all dummy variables except those whose corresponding population parameters are constrained to zero under the null hypothesis that $b_4^* = b_5^* = 0$:

$$Y = 11.12 + 1.32X_1 + .32X_2 - 3.04X_3 + e$$

$$\text{R-square} = .368 \qquad S(ee) = 146.7/34 = 4.32$$

In this instance the restricted R-square has been reduced by a more substantial amount. The probability of this reduction under the null hypothesis is again estimated using equation 7.3:

$$F = \frac{(.445 - .368)/2}{(1 - .445)/(34 - 5 - 1)} \qquad df = 2, 28$$
$$= 1.94$$
$$\text{Prob}(F > 1.94) = .16$$

In view of the small sample size, the finding that the observations would have occurred less than two times out of ten (if the null hypothesis were true) might be sufficient to motivate further experimentation. However, by the conventions of the 5 and 10 percent levels of significance, the data are insufficient to reject the null hypothesis. Any success or failure in an attempt to reject a null hypothesis should be considered in an appropriate context, and the most important aspects of that context are sample size and measurement error. When sample size is small or when the dependent variable is measured with considerable error, even substantial theoretical relationships can turn up as not significant. Fortunately, methods for the estimation of measurement error have been developed (and will be presented in Chapter 9), and the effect of sample size on standard errors—and, therefore, significance tests—is compensated by estimates of the relative magnitudes of experimental effects. In other words, although a larger sample for an experiment virtually assures statistical significance of some miniscule effects, the increased sample size will not tend to increase the *magnitude* of the observed relationship. And, conversely, large relationships found with small samples will evidence large effect magnitudes even though the appropriate F ratio is not very large. Procedures for the estimation of effect magnitudes will be presented in Section 7.7, but a hint of how such relative magnitudes are estimated can be gleaned from noting the relationship between the total R-square and the restricted R-square. For intuitive purposes, the difference between these two R-squares can be considered the relative magnitude of the corresponding independent variable's effect on the dependent variable. As will be pointed out in Section 7.7, there are some problems with the use of this statistic as an estimate of an effect magnitude in a population, but it is nevertheless a useful and straightforward description of sample results.

By applying analogous procedures to any other subset of the regression coefficients in equation 7.12 various other planned comparisons or *a priori* hypotheses may be subjected to significance testing. It should also be reiterated that the standard errors of the coefficients could be used to compute t ratios or confidence intervals for any *single* coefficient in equation 7.12. This means that computation of a restricted regression model is not necessary (for the computation of F ratios) when a null hypothesis involves only one regression coefficient. Furthermore, when the assumptions necessary for the probabilistic interpretation of the F ratio are severely violated, an investigator may prefer to report only the regression coefficients and their approximate standard errors.

Now that the procedures for the computation of F-ratios for planned comparisons using an OLS regression program have been presented, we can proceed to a presentation of the computations needed for a classic analysis of variance.

Classic ANOVA Computations At the broadest conceptual level, the crucial difference between the planned comparison and the classic ANOVA approaches to the analysis of experiments is that the classic approach utilizes a standardized set of null hypotheses and a standardized model for the testing of those null hypotheses. This standardization overlaps with the suitability of the classic ANOVA for situations in which no *a priori* hypotheses have been formulated by the experimenter. From a computational perspective, the difference between the planned comparison approach and classic ANOVA is simply the coding of the dummy variables; comparing equation 7.7 (planned comparison) and equation 7.10 (classic ANOVA) reveals the difference in coding. Applying the classic ANOVA coding scheme to the current numerical example yields the following data input for a regression program:

i	$Y(i)$	Z_1	Z_2	Z_3	Z_1Z_3	Z_2Z_3	
1	7	−1	−1	−1	1	1	
2	9	−1	−1	−1	1	1	
3	10	−1	−1	−1	1	1	Cell A_1B_1
4	11	−1	−1	−1	1	1	
5	13	−1	−1	−1	1	1	
6	10	1	0	−1	−1	0	
7	12	1	0	−1	−1	0	
8	13	1	0	−1	−1	0	Cell A_2B_1
9	14	1	0	−1	−1	0	
10	16	1	0	−1	−1	0	
11	9	0	1	−1	0	−1	
12	11	0	1	−1	0	−1	
13	12	0	1	−1	0	−1	Cell A_3B_1
14	13	0	1	−1	0	−1	
15	15	0	1	−1	0	−1	
16	7	−1	0	1	−1	0	
17	8	−1	0	1	−1	0	
18	9	−1	0	1	−1	0	Cell A_1B_2
19	10	−1	0	1	−1	0	
20	12	−1	0	1	−1	0	
21	6	1	0	1	1	0	
22	8	1	0	1	1	0	
23	8	1	0	1	1	0	
24	9	1	0	1	1	0	Cell A_2B_2
25	10	1	0	1	1	0	
26	10	1	0	1	1	0	
27	12	1	0	1	1	0	
28	5	0	1	1	0	1	
29	7	0	1	1	0	1	
30	7	0	1	1	0	1	
31	7	0	1	1	0	1	Cell A_3B_2
32	8	0	1	1	0	1	
33	10	0	1	1	0	1	
34	12	0	1	1	0	1	

The analysis begins with estimation of the full model specified in equation 7.10:

$$\begin{aligned} Y = \underset{(.37)}{10.2} + \underset{(.52)}{.80}Z_1 - \underset{(.52)}{.20}Z_2 - \underset{(.37)}{1.47}Z_3 \\ - \underset{(.52)}{.53}(Z_1Z_3) - \underset{(.52)}{.53}(Z_2Z_3) + e \end{aligned} \tag{7.15}$$

$$\text{R-square} = .445 \qquad S(ee) = 3.79$$

The meaning of the regression coefficients becomes clear if the cell means are inspected. The intercept term (c_0 = 10.2) is the *unweighted* mean of the six cell means. The effect of B (c_3 = −1.47) is the difference between the unweighted mean of the three cell means under condition B_2 and the intercept $[(9.2 + 9 + 8)/3 - 10.2 = -1.47]$. And finally, the effects of the independent variable A ($c_1 = .8$ and $c_2 = -.2$) are the differences between unweighted means of cells under condition A_2 and the intercept and between condition A_3 and the intercept $[(9 + 13)/2 - 10.2 = .8; (12 + 8)/2 - 10.2 = -.2]$. Note that the total R-square is the same as in the planned comparison model (equation 7.12). The real difference between the two approaches is in how they partition the dependent variable variance into components associated with each dummy variable. This difference becomes apparent when we test the null hypotheses of the classic ANOVA approach.

The classic ANOVA null hypotheses are of two types—main effects and interactions. The main effects in our hypothetical experiment are the two independent variables (A and B), and the interaction is the interaction between these two independent variables. The null hypotheses corresponding to each main effect and interaction are as follows:

Source of variation	Null hypothesis
Main effects	
A (social threat)	$c_1^* = c_2^* = 0$
B (knowledge)	$c_3^* = 0$
Interaction	
$A \times B$	$c_4^* = c_5^* = 0$

The F ratios and dependent variable variance associated with each of the main effects and interactions can now be obtained through successive application of an OLS regression program to regression of Y on subsets of the right-hand variables in equation 7.10. The results are as follows:

Null hypothesis for main effect of A

$$Y = 10.2 - 1.44Z_3 - .44(Z_1Z_3) - .60(Z_2Z_3) + e$$

$$\text{R-square} = .395 \qquad S(ee) = 140.3/34 = 4.13$$

$$F = [(.445 - .395)/2]/[(1-.445)/(34 - 5 - 1)] = 1.26 \qquad df = 2, 28$$

$$\text{Prob}(F > 1.26) = .29$$

Null hypothesis for main effect of B

$$Y = 10.1 + .72Z_1 - .28Z_2 - .61(Z_1Z_3) - .61(Z_2Z_3) + e$$

$$\text{R-square} = .137 \qquad S(ee) = 200.1/34 = 5.89$$

$$F = [(.445 - .137)/1]/[(1 - .445)/28] = 15.54 \qquad df = 1, 28$$

$$\text{Prob}(F > 15.54) < .001$$

Null hypothesis for interaction between A and B

$$Y = 10.2 + .77Z_1 - .23Z_2 - 1.52Z_3 + e$$

$$\text{R-square} = .368 \qquad S(ee) = 146.7/34 = 4.31$$

$$F = [(.445 - .368)/2]/[(1 - .445)/28] = 1.94 \qquad df = 2, 28$$

$$\text{Prob}(F > 1.94) = .16$$

The results of such an analysis are usually presented in an ANOVA summary table that includes the source of variation, the degrees of freedom of each source, the so-called sums of squares, and the F ratios. All these numbers are available from the above computations. The degrees of freedom for each source correspond to the number of dummy variables required to specify the independent variable (number of levels minus one), the F ratios are obtained as above, and the *sums of squared deviations* or *sums of squares* are simple functions of the squared residuals. (The term "sums of squares" is often used to refer to "squares about the origin" rather than squares of deviations about the mean. The sum of squares in that case is simply the sum of the squares of each score. The reader should note that the SPSS subprogram ANOVA labels the sum of squared deviations about the mean as the sum of squares.)

The total sum of squares is the sum of squared residuals when only the intercept has been included in the model. Thus,

$$\text{Total sum of squares} = SS(t) = N[S(YY)] = 232$$

The remaining sums of squares are obtained from the sum of squared residuals that result from the four regression equations estimated above.

1. Error sum of squares $= SS(e)$ is the sum of squared residuals when Y is regressed on *all* the dummy variables in the model. Thus

$$SS(e) = 128.8$$

2. The *explained* sum of squares is then the difference between the total sum of squares and the error sum of squares:

$$\begin{aligned} \text{Explained sum of squares} &= SS(t) - SS(e) \\ &= 232 - 128.8 = 103.2 \end{aligned}$$

   This particular sum of squares is not always reported in the ANOVA summary table.
3. The sum of squares associated with each main effect and interaction is then the difference between the sum of squared residuals for the full model and the sum of squared residuals for the restricted model used to test that main effect or interaction:

$$\begin{aligned} \text{Sum of squares for } A = SS(A) &= 140.3 - 128.8 = 11.5 \\ \text{Sum of squares for } B = SS(B) &= 200.1 - 128.8 = 71.3 \\ \text{Sum of squares for } A \times B = SS(A \times B) &= 146.7 - 128.8 = 17.9 \end{aligned}$$

These sums of squares can also be derived from the R-squares obtained from each of the regression equations:

$$\begin{aligned} SS(A) &= SS(t)[R(t) - R(c_1 = c_2 = 0)] \\ &= 232 \times (.445 - .395) = 11.5 \\ SS(B) &= SS(t)[R(t) - R(c_3 = 0)] \\ &= 232 \times (.445 - .137) = 71.3 \\ SS(A \times B) &= SS(t)[R(t) - R(c_4 = c_5 = 0)] \\ &= 232 \times (.445 - .368) = 17.9 \end{aligned}$$

$$SS(e) = SS(t)[1 - R(t)]$$
$$= 232 \times (1 - .445) = 128.8$$

where $R(t)$ is the R-square for the full model (equation 7.12), $R(c_1 = c_2 = 0)$ is the R-square when c_1 and c_2 are constrained to zero by deletion of Z_1 and Z_2 from the regression, and so on. Each of these sums of squares is a *partition* of the total sum of squares; and if each sum of squares is divided by the number of cases ($N = 34$), each is transformed into a component of *sample* variance in Y. Finally, the ratio of each variance component—or sum of squares component—to the total variance of Y is the proportion of sample variance associated with a particular main effect or interaction. This ratio is referred to as *eta squared* in most of the literature on ANOVA. The reader should keep in mind, however, the fact that eta squared is a biased estimate of the proportion of variance explained by an independent variable in the population from which the sample was drawn. A procedure for reducing that bias will be outlined in Section 7.7 of this chapter.

The results of all these computations are usually presented in a summary table of the following form:

ANOVA Summary Table

Source	SS	df	F	P	eta squared
intercept		1	750.2	<.001	
A	11.5	2	1.26	.29	.05[= $R(t) - R(c_1 = c_2 = 0)$]
B	71.3	1	15.54	<.001	.31[= $R(t) - R(c_3 = 0)$]
$A \times B$	17.9	2	1.94	.16	.08[= $R(t) - R(c_4 = c_5 = 0)$]
Error	128.8	28			.56[= $1 - R(t)$]
Total	232.0	34			.99

The ANOVA Summary Table makes it clear that the primary source of *systematic* variance in authoritarianism was the independent variable B (no knowledge/knowledge). This manipulation accounted for 31 percent of the sample variance in Y, while A and the interaction between A and B accounted for 5 and 8 percent, respectively. Furthermore, the F ratio for factor B is an extremely unlikely event if there is no effect of B in the population from which this sample was drawn. The intercept is of little interest in this ANOVA computation, since the zero point of the measurement scale is arbitrary or irrelevant. ANOVA procedures are often applied in situations in which differences between groups are the focus of concern rather than actual magnitudes of measurement scales. The intercept is included in the ANOVA Summary Table, however, to indicate the total partition of the degrees of freedom in a data set generated from 34 observations.

If the reader has not performed the calculations presented above on a calculator or computer, it may not be apparent that there is an unsuspected inconsistency in the sum of squares column of the summary table. The sum

$$SS(A) + SS(B) + SS(A \times B) + SS(e) = 229.5$$

is not equal to the $SS(t)$ of 232.0 listed in the table. The reason for this discrepancy is not rounding error (although the reader will find that many of the arithmetic statements made in the last few paragraphs are the result of more precise arithmetic than that given—i.e., calculations were made with more decimal places than are presented). The reason for the discrepancy is the fact that the number of participants was not uniform for each cell of the experiment. This creates a correlation among the dummy variables used to code the independent variables in the experiment. And when right-hand variables in a regression equation are correlated, the sum of the components of variance attributable to each right-hand variable and the residual will be different from the total variance of the dependent variable.

Digression

When the number of persons in each cell is quite uneven, it is possible for the difference between the total sum of squares and the sum of component sums of squares to be substantial. For the simple case of two right-hand variables, it is straightforward to demonstrate the extent of this difference. If we have the model

$$Y = b_0 + b_1X_1 + b_2X_2 + e$$

where the residual e is uncorrelated with X_1 and X_2, then the proportion of sample variance in Y associated *uniquely* with X_1, X_2, and e is

$$\begin{aligned} \text{variance explained by } X_1 &= R(Y \cdot X_1, X_2) - R(Y \cdot X_2) \\ \text{variance explained by } X_2 &= R(Y \cdot X_1, X_2) - R(Y \cdot X_1) \\ \text{variance explained by } e &= 1 - R(Y \cdot X_1, X_2) \end{aligned}$$

Now, when the correlation between X_1 and X_2 is zero, the sum of these three proportions is one. However, when X_1 and X_2 are correlated, the sum of the three proportions of explained variance is

$$1 - r[b_1(rb_1 + b_2) + b_2(b_1 + rb_2)]$$

where r is the correlation between X_1 and X_2 and b_1 and b_2 are standardized partial slopes. It should be clear that this expression is one when r is zero. Similarly, when b_1 and b_2 are equal to 0.5 and r is 0.2, the sum of the proportions of variance attributed to X_1, X_2, and e would be 0.88—indicating that 12 percent of the association between Y and the two independent variables could not be assigned *uniquely* to either of the two independent variables. Thus, the discrepancy between the sum of the variance components and the total variance is a function of the difference between the total R-square and the sum of the proportions of dependent variable variance uniquely attributed to each subset of right-hand variables that are used to define main effects and interactions; and this difference is due to the fact that the total R-square is based upon variance components that are *covariance* terms as well as terms that are unique to a particular independent variable. This can be seen for the two variable standardized model by squaring both sides of the equation and summing over all cases in the sample:

$$S(YY) = 1 = b_1^2 + b_2^2 + 2(b_1b_2r) + S(ee)$$

where R-square is the variance of $(b_1X_1 + b_2X_2)$ or

$$\text{R-square} = b_1^2 + b_2^2 + 2(b_1b_2r)$$

The last term in this expression for R-square makes it clear that the covariance of X_1 and X_2 (which is a correlation in this standardized case) contributes negatively or positively to the variance in Y explained by the linear combination of the two independent variables. On the other hand, the proportion of variance in Y that is uniquely associated with X_1 is given by

$$R(t) - R(b_1 = 0) = (b_1^2)(1 - r^2)$$

and the proportion of variance explained by X_2 is given by

$$R(t) - R(b_2 = 0) = (b_2^2)(1 - r^2)$$

where r is the correlation between X_1 and X_2. These last two expressions for the proportion of sample variance explained by each independent variable reveal that the difference between the total R-square and the sum of unique components is

$$\begin{aligned} R(t) &- [R(t) - R(b_1 = 0) + R(t) - R(b_2 = 0)] \\ &= R(b_1 = 0) + R(b_2 = 0) - R(t) \\ &= r[b_1(rb_1 + b_2) + b_2(b_1 + rb_2)] \end{aligned}$$

And this difference may be positive or negative (e.g., if $b_1 = b_2 = .5$ and $r = -.2$, then the sum of proportions of variance attributed to X_1, X_2, and e is 1.16). In general, if the regression coefficients in a linear model are positive, then positive correlations between the right-hand variables reduce the sum of components relative to the total R-square, whereas negative correlations between right-hand variables increase the sum of unique components to a level that exceeds the R-square. Fortunately, however, the sign of the correlation between the right-hand variables does not change the F ratios or standard errors for sample regression coefficients. Note that this *does not* mean that the magnitude of the correlation between independent variables has no effect on the magnitude of F ratios. On the contrary, a balanced design is always more powerful than an unbalanced design. This means that if two experiments are identical in every respect, except that one has equal cell sizes and the other does not, then the experiment with equal cell sizes (balanced design) will, on the average, obtain larger F ratios. Thus, although experimenters should strive to obtain balanced data, they should also not exclude cases that produce unbalanced data.

All this attention to proportions of variance explained and to F ratios should not obscure the fact that the structural coefficients are the main focus of any experimental investigation. And the expectation of the sample estimates of the population structural coefficients is unaffected by correlation among the causal variables. Only the standard errors of those coefficients are changed by an unequal distribution of participants. Thus, if we deemphasize the importance of significance tests relative to independent replication, then the disadvantage of unbalancing is less important. Nevertheless, a smaller standard error of any sample estimate is a desirable outcome, since greater confidence in the accuracy of the estimate ensues.

To anticipate the discussion of Section 7.6, the previous paragraphs point to an important lesson concerning measures of the magnitude of a relationship. Any attempt to standardize such measures creates problems when different experiments are compared. A measure such as the proportion of variance explained by an experimental manipulation is affected by a number of theoretically irrelevant factors—including the extent to which the design is balanced. For example, the proportion of sample variance explained by an experimental manipulation could shift from 20 percent to 40 percent by simply reversing a pattern of cell inequalities—even though the magnitude of the structural coefficient did not change. Clearly, then, statistics such as eta squared are useful for inter-experimental comparison only when designs are approximately balanced. The estimates of the structural coefficients are, in most cases, the best basis for comparisons between different experiments.

The reader should have a good grasp at this point of how the R-squares and sums of squared residuals can be used to perform a classic ANOVA on experimental data. However, only a few readers are likely to be comfortable with their understanding of how the dummy variable coding scheme allows tests of main effect and interaction null hypotheses. The following section is included to remedy this situation and to assure that the reader can apply these regression techniques to the analysis of any crossed design—whether balanced or unbalanced, complete or incomplete (a crossed design in which some combinations of crossed levels are not observed).

Use of SPSS Subprogram ANOVA Although the above method for computing an analysis of variance has the virtue of clarity that derives from the use of an OLS program for multiple regression, it is nevertheless more convenient in some instances to use a computer program that generates the appropriate dummy variables automatically. Provided that "option 9" is utilized (and that the version of SPSS available to you is a recent one—version 8 or newer), then the ANOVA subprogram of the Statistical Package for the Social Sciences (Nie et al, 1975) is such a program. The following statements constitute the appropriate SPSS instructions to perform an "unweighted means" analysis of variance of the data used in the example above:

```
RUN NAME          SPSS ANOVA ANALYSIS OF 34 CASE EXAMPLE
VARIABLE LIST     Y,A,B
INPUT MEDIUM      CARD
N OF CASES        34
INPUT FORMAT      FREEFIELD
ANOVA             Y BY A(1,3),B(1,2)/
OPTIONS           9
STATISTICS        ALL
READ INPUT DATA
7 1 1    9 1 1     10 1 1    11 1 1    13 1 1
10 2 1    12 2 1    13 2 1    14 2 1    16 2 1
9 3 1    11 3 1    12 3 1    13 3 1    15 3 1
7 1 2    8 1 2    9 1 2    10 1 2    12 1 2
6 2 2    8 2 2    8 2 2    9 2 2    10 2 2    10 2 2    12 2 2
5 3 2    7 3 2    7 3 2    7 3 2    8 3 2    10 3 2    12 3 2
FINISH
```

The reader should be warned that failure to use "option 9" (specified by the statement "OPTIONS 9" in the SPSS control cards) will lead to erroneous results if cell sizes are substantially unequal. What the SPSS manual refers to as the "classic experimental design approach" (Nie et al., 1975, p. 404) is, in fact, not what Sir Ronald had in mind. If an approach to ANOVA can be identified as classic, it is the one outlined above—the unweighted means approach. Only what the SPSS manual refers to as the "regression approach" (option 9) provides this analysis when cell sizes are unbalanced. Furthermore, versions of SPSS prior to version 8 incorporated programming errors that provided erroneous results even when option 9 was specified.

The moral of this story is that reliance upon a "tried and true" linear regression program for most all analyses is clearly a wise course for most researchers. Any more specialized program, such as the SPSS subprogram ANOVA, should be checked with a reliable regression program to determine whether or not the more specialized program does what its authors claim.

The difference between the default option and option 9 in the SPSS subprogram ANOVA is as follows: in the default option, assessment of main effects is done without controlling for interactions; whereas option 9 partials all interaction effects from other interaction and main effects. A concrete understanding of the difference between these two procedures, and the preferability of the regression approach, can be seen by running the following SPSS control cards:

```
RUN NAME          COMPARISON OF OPTIONS 0 AND 9 ON ANOVA IN SPSS
VARIABLE LIST     Y,A,B
INPUT MEDIUM      CARD
N OF CASES        30
INPUT FORMAT      FREEFIELD
ANOVA             Y BY A(1,2),B(1,2)/
OPTIONS           9
READ INPUT DATA
10 1 1    11 1 1    12 1 1    13 1 1    14 1 1
10 1 1    11 1 1    12 1 1    13 1 1    14 1 1
13 1 2    14 1 2    15 1 2    16 1 2    17 1 2
13 2 1    14 2 1    15 2 1    16 2 1    17 2 1
13 2 1    14 2 1    15 2 1    16 2 1    17 2 1
10 2 2    11 2 2    12 2 2    13 2 2    14 2 2
ANOVA             Y BY A(1,2),B(1,2)/
STATISTICS        ALL
*COMPUTE          X1=2*A-3
*COMPUTE          X2=2*B-3
*COMPUTE          X1X2=X1*X2
REGRESSION        VARIABLES=Y,X1,X2,X1X2/
                  REGRESSION=Y WITH X1,X2(2)/
                  REGRESSION=Y WITH X1,X2,X1X2(2)/
FINISH
```

By inspection of the cell means from this 2×2 factorial design, the reader will see that the appropriate conclusion should be that there are no main effects of A or B—there is only an interaction. However, this result is only obtained with option 9.

Problem Use the output from the two regressions specified above to reconstruct the analysis of variance summary tables given by the subprogram ANOVA. [Help: the F ratio for the main effect of A with the default option uses the first regression to estimate the effect of A, but it uses the second regression to estimate the variance of the residuals.]

7.5 LEAST SQUARES REGRESSION WITH CONSTRAINTS

Thus far in this book there has been an almost exclusive use of the ordinary least squares (OLS) regression model. The OLS model is an intuitively appealing approach to analysis of linear (and "well behaved" non-linear) relationships involving interval scale variables. In the behavioral sciences, however, there is a strong commitment to procedures of experimental design that have their roots in biology—especially the comparison of different methods of treating agricultural crops. These procedures usually involve nominal or ordinal scaled independent variables, and analytic attention is focused upon the presence or absence of a manipulation rather than upon different magnitudes of an independent variable. This special focus on nominal scaled independent variables is also manifest in greater attention to differences between group means at every level of a variable rather than to average differences across many levels. This focus on group means in classic ANOVA is best accommodated by a least squares model that is slightly different from the OLS model. Too much importance should not be attached to the difference in the form of the model, however. In fact, the results of the analysis will be identical to those obtained with OLS, and the intermediate computational steps are equivalent. What is different about the more general least squares model is simply the mechanics of utilitization—the more general model is much more simply applied to the analysis of experiments than is the OLS procedure described in Section 7.4.

A Least Squares Model for ANOVA Recalling the example experiment for this chapter, three levels of the independent variable A were crossed with two levels of B. The results of the experiment are scores for 34 participants grouped into six "cells." Each cell of the experiment is thus associated with a mean on the dependent variable. The ANOVA least squares model for analysis of these data is specified as follows

$$Y(i) = d_1 W_1 + d_2 W_2 + d_3 W_3 + d_4 W_4 + d_5 W_5 + d_6 W_6 + e(i) \quad (7.16)$$

where

$$\begin{aligned}
Y(i) &= \text{F-scale score for person } i \\
W_1(i) &= 1 \text{ if person } i \text{ is in group } A_1B_1 \\
& \ 0 \text{ otherwise} \\
W_2(i) &= 1 \text{ if person } i \text{ is in group } A_2B_1 \\
& \ 0 \text{ otherwise} \\
W_3(i) &= 1 \text{ if person } i \text{ is in group } A_3B_1 \\
& \ 0 \text{ otherwise} \\
W_4(i) &= 1 \text{ if person } i \text{ is in group } A_1B_2 \\
& \ 0 \text{ otherwise} \\
W_5(i) &= 1 \text{ if person } i \text{ is in group } A_2B_2 \\
& \ 0 \text{ otherwise} \\
W_6(i) &= 1 \text{ if person } i \text{ is in group } A_3B_2 \\
& \ 0 \text{ otherwise}
\end{aligned}$$

There are two important differences between the model specified in equation 7.16 and an OLS model for computation of a classic ANOVA (e.g., equation 7.10). The first difference is that equation 7.16 does not include an intercept term. The second, and related, difference is that the dummy variables have been defined in such a way that each dummy variable identifies a cell of the experiment rather than an independent variable or interaction. These differences will mean that each regression coefficient corresponds to a cell mean, and this correspondence will greatly simplify the testing of varied null hypotheses.

Equation 7.16 specifies a *least squares* regression model if the coefficients $d(j)$ are selected such that the sum of squared residuals is a minimum. The residual for the ith case may be written as

$$\begin{aligned} e(i) &= \{Y(i) - [d_1 W_1(i) + \ldots + d_6 W_6(i)]\} \\ &= [Y(i) - d(j)] \end{aligned} \tag{7.17}$$

where the ith case is in the jth cell and $d(j)$ is the jth regression coefficient (for the jth cell of the experiment). Note that equation 7.17 simplifies to the expression $[Y(i) - d(j)]$ because of the way in which the dummy variables are defined. For any given cell of the experiment, only one of the variables $W(j)$ will be non-zero—meaning that all right-hand terms in equation 7.16 will vanish for a given case except for the $d(j)$ associated with the cell containing the ith case. The task of finding the coefficients $d(j)$ that minimize the sum of squared residuals is thus equivalent to finding a number for each cell of the experiment that minimizes the within cell squared deviations:

$$\begin{aligned} \text{sum of } e(i)^2 = {} & \text{sum of } [Y(i) - d_1]^2 \quad (i = 1, \ldots, 5) \\ & + \text{sum of } [Y(i) - d_2]^2 \quad (i = 6, \ldots, 10) \\ & + \ldots \\ & \ldots \\ & + \text{sum of } [Y(i) - d_6]^2 \quad (i = 28, \ldots, 34) \end{aligned}$$

It should be clear that the least squares solution to this equation will assign the six cell means to the six regression coefficients. This is the case since the mean of any set of scores minimizes squared deviations. More formally, the six partial derivatives of this sum of squared residuals (one for each regression coefficient) would be zero where the sum was a minimum. It can be shown that these partial differential equations are zero when

$$\begin{aligned} d_1 &= Y_{11} \quad d_2 = Y_{21} \quad d_3 = Y_{31} \\ d_4 &= Y_{12} \quad d_5 = Y_{22} \quad d_6 = Y_{32} \end{aligned}$$

where Y_{pq} is the mean of Y for the participants in the pth level of factor A and the qth level of factor B.

Now that we have seen how equation 7.16 is defined, we are in a position to apply it to the analysis of experiments. As was pointed out previously, the results of this analysis will not be any different from the results obtained with the OLS procedures in Section 7.4. The advantage of this new computational method, the ANOVA least squares method, will be the simplicity of program preparation and null hypothesis specification.

Computations for ANOVA Least Squares with Constraints In order to demonstrate the use of a least squares program (with constraints other than deletion of variables) for the analysis of experiments, we return to the data given for the 3 × 2 design in Section 7.4. The computations presented in this section could be performed with any packaged computer program for regression or least squares estimation that allows linear constraints on coefficients. [BMD11V in the UCLA BMD series, LSQ in the University of Wisconsin TSP package, and Finn's (1972) MULTIVARIANCE routine are examples of such programs.]

We begin by noting the ease with which null hypotheses about population cell means can be stated in terms of the coefficients in equation 7.16.

Main Effect of A

$$H_0\text{: } Y^*_{11} + Y^*_{12} = Y^*_{21} + Y^*_{22} = Y^*_{31} + Y^*_{32}$$
$$\text{implies } d^*_1 + d^*_4 = d^*_2 + d^*_5 = d^*_3 + d^*_6$$

Main Effect of B

$$H_0\text{: } Y^*_{11} + Y^*_{21} + Y^*_{31} = Y^*_{12} + Y^*_{22} + Y^*_{32}$$
$$\text{implies } d^*_1 + d^*_2 + d^*_3 = d^*_4 + d^*_5 + d^*_6$$

Interaction of A and B

$$H_0\text{: } Y^*_{31} - Y^*_{11} = Y^*_{32} - Y^*_{12}$$
$$Y^*_{21} - Y^*_{11} = Y^*_{22} - Y^*_{12}$$
$$(\text{or } Y^*_{31} - Y^*_{21} = Y^*_{32} - Y^*_{22}, \text{ redundant})$$
$$\text{implies } d^*_3 - d^*_1 = d^*_6 - d^*_4 \text{ and } d^*_2 - d^*_1 = d^*_5 - d^*_4$$

The null hypotheses about population means are immediately transformed into *constraints* involving coefficients, since each coefficient is equal to a cell mean. Thus, the analysis may proceed as outlined in Section 7.4 using equation 7.16 and the constraints outlined above.

Traditional approaches to ANOVA have utilized the concept of a *contrast* among cell means. A contrast is a linear combination of a set of means where the sum of the coefficients defining the linear combination is zero. As an example, we could take the six means resulting from our hypothetical experiment and define a linear contrast L as

$$L = f_1 Y_{11} + f_2 Y_{21} + f_3 Y_{31} + f_4 Y_{12} + f_5 Y_{22} + f_6 Y_{32}$$

where

$$f_1 + \ldots + f_6 = 0$$

The reader should now notice that each of the *constraints* outlined above (one or more for each main effect and interaction) could be stated in terms of a *contrast*. For example, the constraint that factor B has no effect on Y is

$$Y^*_{11} + Y^*_{21} + Y^*_{31} = Y^*_{12} + Y^*_{22} + Y^*_{32}$$

or

$$d^*_1 + d^*_2 + d^*_3 = d^*_4 + d^*_5 + d^*_6$$

This constraint may be rewritten in terms of a linear contrast by simple algebraic rearrangement:

$$L^* = 1(d^*_1) + 1(d^*_2) + 1(d^*_3) + (-1)(d^*_4) + (-1)(d^*_5) + (-1)(d^*_6)$$

where

$$L^* = 0 \text{ (under the null hypothesis of no } B \text{ effect)}$$

(Note that the coefficients of the contrast meet the requirement of summing to zero.) And it turns out that all the constraints defining main effects and interactions in classic ANOVA are linear contrasts as well. Furthermore, the contrast, or contrast*s,* defining any main effect or interaction are *orthogonal* to those contrasts defining other main effects or interactions. The definition of orthogonality is that two contrasts, say L_1 and L_2, are orthogonal if the vector product of the coefficients defining the two contrasts is zero. This means that if the coefficients $f_1, f_2, \ldots, f_6$ define the contrast L_1, and the coefficients $g_1, \ldots, g_6$ define L_2, then L_1 and L_2 are orthogonal if

$$f_1g_1 + f_2g_2 + \ldots + f_6g_6 = 0$$

The intuitive meaning of contrast orthogonality is that the two contrasts do not overlap—each relies upon information that is "at right angles" with the other. In a balanced experimental design (all cell sizes equal), this orthogonality means that the sample variance of the dependent variable can be partitioned into components associated with each subset of orthogonal contrasts (one subset for each main effect and interaction). If these contrasts were not orthogonal with one another, then some of the dependent variable variance *explained* by one contrast would overlap with the variance explained by another contrast. (The computer program BMD11V accepts input in the form of contrasts for computation of classic ANOVA statistics.)

In order to compute the statistics associated with a classic ANOVA, we begin by estimating equation 7.16 without constraints. The result is

$$\begin{array}{l} Y = \underset{(.96)}{10W_1} + \underset{(.96)}{13W_2} + \underset{(.96)}{12W_3} + \underset{(.96)}{9.2W_4} + \underset{(.81)}{9W_5} + \underset{(.81)}{8W_6} + e \end{array} \qquad (7.18)$$

$$\begin{aligned} \text{R-square} &= [S(YY) - S(ee)]/S(YY) \qquad & S(ee) &= 128.8/34 \\ &= [6.82 - 3.79]/6.82 & &= 3.79 \\ &= .445 \end{aligned}$$

Note that the R-square and the sum of squared residuals are the same for equation 7.18 as they were with the OLS regression procedures in Section 7.4. By now specifying the appropriate constraints on the coefficients we can also obtain the other R-squares and sums of residuals necessary to complete a classic ANOVA on the data from our hypothetical experiment.

CONSTRAINT: $d_1 + d_4 = d_2 + d_5 = d_3 + d_6$

yields

$$Y = 10.63W_1 + 12.10W_2 + 12.27W_3 + 9.83W_4 + 8.36W_5 + 8.19W_6 + e_1$$

R-square = .395 $\quad S(e_1,e_1) = 140.3/34 = 4.13$

CONSTRAINT: $d_1 + d_2 + d_3 = d_4 + d_5 + d_6$

yields

$$Y = 8.38W_1 + 11.38W_2 + 10.38W_3 + 10.82W_4 + 10.16W_5 + 9.16W_6 + e_2$$

R-square = .137 $\quad S(e_2,e_2) = 200.1/34 = 5.89$

CONSTRAINT: $d_3 - d_1 = d_6 - d_4$ and $d_2 - d_1 = d_5 - d_4$

gives

$$Y = 11.12W_1 + 12.44W_2 + 11.44W_3 + 8.08W_4 + 9.40W_5 + 8.4W_6 + e_3$$

R-square = .368 $\quad S(e_3,e_3) = 146.7/34 = 4.32$

Note that the R-square and the sum of squared residuals are the same as in the OLS analyses. The reason for this is that the coefficients constrained to be zero in the OLS models placed the same constraints on the OLS model that the above constraints place on the ANOVA least squares model. The results of these computations could now be used to construct an ANOVA summary table—as outlined in Section 7.4 of this chapter. [The reader should verify that the coefficients in the above constrained equations fulfill the constraints. Note that the coefficients obtained under the constraints are close to the cell means when the constraint corresponds to a null hypothesis that is not strongly contradicted by the data (e.g., the null hypothesis for the interaction), but the constrained model reproduces the cell means very poorly when the constraint corresponds to a rejected null hypothesis (e.g., the null hypothesis for the main effect of *B*). The "fit" of each constrained model to the observed data could be assessed by the accuracy with which each constrained model reproduces the observed means. This line of thinking will be developed more systematically when maximum likelihood methods for model fitting are introduced in Chapter 9.]

It should be immediately apparent that any linear (or nonlinear) combination of means could be specified as a constraint and tested as an *a priori* hypothesis. The analog to equation 7.3 is

$$F = \frac{[R(t) - R(r)]/k}{[1 - R(t)]/(N - j)} \qquad df = k, N - j \tag{7.19}$$

where

$R(t)$ = R-square for the unconstrained regression of Y on j dummy variables defining j groups

$R(r)$ = R-square for least squares solution with k constraints (or contrasts)
N = total sample size

In this least squares form, the model is applicable to unbalanced designs—even to designs with incomplete cross classifications. The concept of main effect may or may not apply in incomplete designs, but *a priori* hypotheses can certainly be tested through the specification of constraints. The flexibility of this approach should not, however, encourage researchers to explore their data for apparent patterns of differences and then use the obtained F ratio to estimate the likelihood of a type one error. *A posteriori* hypothesis testing capitalizes on chance factors in a manner that must be counteracted in some systematic way.

A Posteriori Constraints *A posteriori* constraints will yield F ratios that understate the likelihood of the observation if the null hypothesis is true. The advantage of the classic ANOVA approach to *a posteriori* hypothesis testing is that the main effects and interactions usually subsume the more specific *a posteriori* tests that may be of interest. A significant F ratio for the main effect of a factor with four levels, for example, means that the results are sufficient to reject the null hypothesis that the central tendency of the dependent variable is the same under all four conditions. This does not mean, however, that the F ratios obtained for specific comparisons among these four conditions give accurate estimates of type one error probabilities. In the case of a four level factor with population means Y_1^*, Y_2^*, Y_3^*, and Y_4^*, there are many "states of reality" that could lead to a rejection of the main effects null hypothesis. For example, any one of the following could produce a significant F ratio for the main effect:

$$Y_1^* \neq Y_2^*$$
$$Y_1^* \neq Y_3^*$$
$$Y_1^* \neq Y_4^*$$
$$Y_2^* \neq Y_3^*$$
$$Y_2^* \neq Y_4^*$$
$$Y_3^* \neq Y_4^*$$
$$Y_1^* \neq (Y_2^* + Y_3^* + Y_4^*)/3$$
$$Y_2^* \neq (Y_1^* + Y_3^* + Y_4^*)/3$$
$$Y_3^* \neq (Y_1^* + Y_2^* + Y_4^*)/3$$
$$Y_4^* \neq (Y_1^* + Y_2^* + Y_3^*)/3$$
$$(Y_1^* + Y_2^*) \neq (Y_3^* + Y_4^*)$$
$$(Y_1^* + Y_3^*) \neq (Y_2^* + Y_4^*)$$
$$(Y_1^* + Y_4^*) \neq (Y_2^* + Y_3^*)$$

and so on.

For the sake of illustration, suppose that the true state of nature is that $Y_2^* = Y_3^* = Y_4^*$, and Y_1^* is different from the other three, and that this state of nature led to the rejection of the main effects null hypothesis. There are at least three null hypotheses that could be tested with constraints after this significant finding that *should* be accepted:

$$Y_2^* = Y_3^*,\ Y_2^* = Y_4^*, \text{ and } Y_3^* = Y_4^*$$

And yet, if each is tested separately, the probability that one of the three F ratios will be significant (at the 5 percent level, say) is greater than 5 percent. In fact, the probability that at least one of the three F ratios will be significant is approximated by the assymmetric binomial distribution:

$$\begin{aligned}
\text{Prob}(0 \text{ significant } F\text{'s}) &= .8574 \\
\text{Prob}(1 \text{ significant } F) &= .1354 \\
\text{Prob}(2 \text{ significant } F\text{'s}) &= .0071 \\
\text{Prob}(3 \text{ significant } F\text{'s}) &= .0001 \\
\text{Prob}(1 \text{ or more}) &= 1 - .8574 = .1426
\end{aligned}$$

And if the number of levels is expanded beyond three, the probability of at least one significant F for a pairwise comparison of means also expands:

Probability that one or more pairwise comparisons will be significant at the 5 percent level when all population means are equal

$$\begin{aligned}
\text{with 3 means} &= .1426 \\
\text{with 4 means} &= .2649 \\
\text{with 5 means} &= .4012 \\
\text{with 6 means} &= .5367
\end{aligned}$$

Thus, when six or more means are involved, the probability of finding a significant difference between at least two of the means is greater than the probability of *not* finding any significant differences.

Several approaches to the estimation of type one errors for *a posteriori* contrasts have been developed. The SPSS package provides no less than seven alternative approaches to the problem. Unfortunately, none of them have been generalized to the multivariate case (where more than one dependent variable is measured in one experiment; see Chapter 8). For this reason, I am inclined to follow Bock (1975) in his recommendation of the method of "protected" hypothesis testing. As developed by Fisher (1951), this method of testing *a posteriori* constraints involves the use of the usual *a priori* F ratio with the proviso that the classical ANOVA constraint(s) (main effect or interaction effect), which subsumes the *a posteriori* constraint, must be significant at the appropriate level. For example, the *a posteriori* hypothesis that $(Y_{21}^* + Y_{31}^*)/2$ is greater than Y_{11}^* (that the threat manipulations A_2 and A_3 increase authoritatianism in the "no knowledge" condition) is tested via the null hypothesis that

$$Y_{11}^* = (Y_{21}^* + Y_{31}^*)/2$$

or the constraint that

$$d_1^* = (d_2^* + d_3^*)/2$$

or the contrast

$$0 = (2)(d_1^*) + (-1)(d_2^*) + (-1)(d_3^*) + (0)(d_4^*) + (0)(d_5^*) + (0)(d_6^*)$$

If this were an *a priori* hypothesis situation, we would estimate equation 7.16 with the above constraint to obtain

$$Y = 11.67Z_1 + 12.17X_2 + 11.17X_3 + 9.2X_4 + 9X_5 + 8X_6 + e$$

$$\text{R-square} = .355 \qquad S(ee) = 149.6/34 = 4.40$$

Recalling that the R-square for the model with no constraints was .445, we can obtain the F ratio for this null hypothesis using equation 7.19:

$$F = \frac{(.445 - .355)/1}{(1 - .445)/(34 - 6)} \qquad df = 1, 28$$

$$= 4.54 \quad \text{Prob}(F > 4.54 \mid H_0) = .042$$

Thus, if this hypothesis were *a priori,* and a significance level of 5 percent had been chosen, then the null hypothesis would be rejected. However, by the rules of protected significance testing, this null hypothesis would be tested using equation 7.19 only if the main effect of A or the interaction between A and B was significant.

It should be noted that the protected procedure does not correct the probability obtained from the regression equations in any way. There are various procedures for doing so, but, as pointed out previously, none are generalizable to the multivariate case.

Standard Errors for Linear Contrasts Many researchers have found the intricacies of ANOVA and null hypothesis testing to be more trouble than help in their attempts to discover important relationships. A large sample size can render the most miniscule of effects as statistically significant, and practical experience has convinced many researchers that refined procedures for the estimation of type one error probabilities represent a misplaced emphasis. For this reason, researchers may prefer to report the results of experiments in terms of differences between means (linear contrasts) and the standard errors of those differences.

The standard error of an *a priori* contrast among means is easily obtained from the computation of the R-square for a model with no constraints $[R(t)]$ and a model that incorporates a constraint $[R(r)]$ corresponding to the *a priori* contrast:

Standard error of contrast $L = S(L)$

$$S(L) = |L|/[(R(t) - R(r))(N - J)/(1 - R(t))]^{.5} \tag{7.20}$$

where

$$|L| = \text{absolute value of } (f_1 d_1 + \ldots + f_J d_J)$$

$$J = \text{the number of cells in the design}$$

$$N = \text{total sample size}$$

As an example, consider the contrast associated with the main effect of B in our hypothetical experiment. The main effect of the knowledge manipulation in that experiment would be reflected in the difference between the means in the no knowledge condition (d_1, d_2, and d_3—see equation 7.18) and the means in the knowledge condition (d_4, d_5, and d_6):

$$\begin{aligned}&\text{effect of factor } B\\ &= (d_1 + d_2 + d_3)/3 - (d_4 + d_5 + d_6)/3\\ &= (10 + 13 + 12)/3 - (9.2 + 9 + 8)/3\\ &= 2.93\end{aligned}$$

Equation 7.20 can now be used to compute the standard error of this contrast:

$$\begin{aligned}&\text{standard error } [(d_1 + d_2 + d_3)/3 - (d_4 + d_5 + d_6)/3]\\ &= 2.93/[(.445 - .127)28/(1 - .445)]^{.5}\\ &= 2.93/[15.5]^{.5}\\ &= 2.93/3.94\\ &= 0.74\end{aligned}$$

And this standard error can be given a probabilistic interpretation if certain assumptions hold. If the residual is normally distributed, and it has the same variance within each cell of the design (homoscedasiticity), then the interval

$$\begin{aligned}&L - 2S(L) \text{ to } L + 2S(L)\\ &= 2.93 - 2(.74) \text{ to } 2.93 + 2(.74)\\ &= 1.45 \text{ to } 4.38\end{aligned}$$

will contain the value L^* in approximately 95 percent of experiments that randomly sample participants from the population in which L^* is the effect of manipulation B on authoritarianism.

Use of Constrained Least Squares Programs The use of two routines for least squares estimation with constraints will be demonstrated here and in Chapter 8. The first routine is BMD11V from the UCLA Biomedical series of programs (Dixon, 1968). It is an extremely flexible program that applies to multivariate problems (more than one dependent variable) as well as univariate problems. Its use is demonstrated in Chapter 8. The second is the LSQ command from the University of Wisconsin TSP package. These two routines are quite different, and they represent two alternative ways in which such programs may be organized. The reader should be able to understand the procedures for most any program for least squares estimation if he masters these two widely used routines.

The estimates obtained in equation 7.18 can be computed using the following TSP statements:

```
SAMPLE 1 34 $
LOAD Y1 $
7 9 10 11 13    10 12 13 14 16    9 11 12 13 15
7 8 9 10 12   6 8 8 9 10 10 12    5 7 7 7 8 10 12 $
LOAD W1 $
 1 1 1 1 1 0 0 0 0 0 0 0 0 0 0 0
 0 0 0 0 0   0 0 0 0 0 0 0   0 0 0 0 0 0 0 $
```

```
LOAD W2 $
 0 0 0 0 0 1 1 1 1 1 0 0 0 0 0
 0 0 0 0 0   0 0 0 0 0 0 0   0 0 0 0 0 0 0 $
LOAD W3 $
 0 0 0 0 0 0 0 0 0 0 1 1 1 1 1
 0 0 0 0 0   0 0 0 0 0 0 0   0 0 0 0 0 0 0 $
LOAD W4 $
 0 0 0 0 0  0 0 0 0 0  0 0 0 0 0
 1 1 1 1 1  0 0 0 0 0 0 0  0 0 0 0 0 0 0 $
LOAD W5 $
 0 0 0 0 0  0 0 0 0 0  0 0 0 0 0
 0 0 0 0 0  1 1 1 1 1 1 1  0 0 0 0 0 0 0 $
LOAD W6 $
 0 0 0 0 0  0 0 0 0 0  0 0 0 0 0
 0 0 0 0 0  0 0 0 0 0 0 0  1 1 1 1 1 1 1 $
PARAMETERS D1 9 D2 13 D3 13 $
PARAMETER D4 10 D5 10 D6 10 $
FORMULA FA1 Y1 = D1*W1+D2*W2+D3*W3+D4*W4+D5*W5+D6*W6 $
FORMULA FA2 Y1 = (D2-D4+D5)*W1+D2*W2+D3*W3+D4*W4
                            +D5*W5+(D2-D3+D5)*W6$
FORMULA FA3 Y1 = (-D2-D3+D4+D5+D6)*W1+D2*W2+D3*W3+D4*W4
                        +D5*W5+D6*W6$
FORMULA FA4 Y1 = (D2+D4-D5)*W1+D2*W2+D3*W3+D4*W4
                        +D5*W5+(-D2+D3+D5)*W6$
LSQ FA1 $
LSQ FA2 $
LSQ FA3 $
LSQ FA4 $
END $
```

Note that the statement "FORMULA FA1" defines the unconstrained model, whereas "FORMULA FA2" specifies the constraints constituting the null hypothesis of no main effects for the threat condition A. Similarly, FORMULA FA3 and FORMULA FA4 define the constraints for the main effect of knowledge (B) and the interaction between A and B, respectively. The statement "LSQ FA1" instructs TSP to minimize the squared residuals in order to estimate the unknown linear coefficients in the model. The series of LSQ statements provide the necessary information for computation of the ANOVA.

7.6 RANDOM EFFECTS MODELS

Up to this point we have assumed that the independent variables in an experiment are comprised of *fixed* levels. This sort of assumption is appropriate when an independent variable, such as social threat, is either present or not present (increased by manipulation or not manipulated). However, there are other situations in which levels of independent variables are sampled from some population of possible levels. The inferential goal is somewhat different in this latter instance, since the concern of the experimenter is not so much with specific levels of the independent variable as with variation *across levels* of the independent variable. An example of an experimental situation that involves random effects is the assessment of experimenter effects.

Psychologists in many areas became acutely aware of the social dynamics of experimentation with humans during the 1960's. Numerous investigators found that characteristics of experimenters appeared to influence the outcome of the experiment. Thus, in elaborate experimental designs, some researchers would repeat a particular experiment k times utilizing k different experimenters. The k experimenters would then correspond to k levels of an independent variable in the experimental design. The application of ANOVA to data from such experiments requires some modifications of the methods that were developed in Sections 7.4 and 7.5 of this chapter.

As an example, suppose that a simple experiment was designed to demonstrate the effect of a social threat manipulation on authoritarianism. The control condition is labeled A_1 and the manipulation condition is labeled A_2. From a population of possible experimenters, ten persons are selected to serve as experimenters. Each experimenter runs five participants in each of the two conditions. Thus, we have a 2 × 10 factorial design with independent observations in each of 20 cells. Utilizing an ANOVA least squares approach to computations, we would specify the linear model

$$Y(i) = d_1W_1 + d_2W_2 + \ldots + d_{20}W_{20} + e \tag{7.21}$$

where

$$Y(i) = \text{F-scale score for } i\text{th participant}$$

$$Wj(i) = \begin{cases} 1 \text{ if participant } i \text{ is in cell } j \\ 0 \text{ otherwise} \end{cases}$$

Applying the least squares criterion to the residuals would result in a set of coefficients d_j that would be equal to the corresponding cell means.

Now, if we proceeded as in Section 7.4, we would explore the main effect of the different experimenters by inspecting the mean of the dependent variable (across levels of A) within the different levels of B. However, in this instance, the levels of B correspond to randomly (or representatively) selected experimenters, and thus no one level has any particular significance. Rather, the variability between all the experimenters is the focus of concern. In line with this different focus, we use a different index of the main effect of B. Rather than using the differences between means *per se,* a variance function of those differences is used: the variance of the means at different levels of B. If experimenter number one conducts the participants in cells d_1 and d_2, experimenter number two corresponds to cells d_3 and d_4, and so on, then

$$\begin{aligned} h_1 &= (d_1 + d_2)/2 \\ h_2 &= (d_3 + d_4)/2 \\ &\vdots \\ h_j & \\ &\vdots \\ h_{10} &= (d_{19} + d_{20})/2 \end{aligned}$$

The h_j are the marginal means that have been sampled. The population from which these ten means have been sampled has variance Var(h). And this population parameter can be estimated from the experimental results:

$$E[NS(hh)/(N - 1)] = \text{Var}(h)$$

where E signifies the expectation of the expression in the brackets over an infinite number of replications of the experiment and N is the total number of persons in the sample. The null hypothesis for the main effect of B may now be stated as

$$\text{Var}(h) = 0$$

Now, the question that obviously presents itself is how this slightly different construction of the factor B affects the computation of type one error probabilities. The key to answering this question stems from a closer look at the F-ratio that has been used to test null hypotheses with fixed effects.

Equations 7.3 and 7.19 are expressed in terms of R-squares for constrained and unconstrained least squares models. For the ANOVA least squares model, the F ratio associated with k constraints was given in equation 7.19:

$$F = \frac{[R(t) - R(r)]/k}{[1 - R(t)]/(N - J)} \qquad df = k, N - J$$

Now, if the residuals from the unconstrained model are e and the residuals from the constrained model are e_1, then equation 7.19 may be rewritten as

$$F = \frac{[S(e_1,e_1) - S(e,e)]/k}{S(e,e)/(N - J)} \qquad df = k, N - J \tag{7.22}$$

where S connotes sample variance, k is the number of constraints, N is the total sample size, and J is the number of cells in the experimental design (J is also the number of coefficients in the ANOVA least squares model). Now, an F ratio is a ratio of two independent chi-square statistics (each divided by its appropriate degrees of freedom), and a chi-square statistic with m degrees of freedom is the sum of the squares of m scores selected independently from a normal distribution of scores. The degrees of freedom for the F ratio are then the degrees of freedom for the chi-square in the numerator of the F ratio and the degrees of freedom for the chi-square in the denominator of the F ratio. The expression in equation 7.22, thus, is distributed as the theoretical F distribution with k and $(N - J)$ degrees of freedom if $[S(e_1,e_1) - S(e,e)]$ is the sum of the squares of k normally distributed scores and $S(e,e)$ is the sum of the squares of $(N - J)$ normally distributed scores. The denominator and numerator of equation 7.22 can be conceptualized as averages of, respectively, k and $N - J$ squared scores selected from two normal distributions of scores. Finally, if the two normal distributions from which the squared scores are sampled—one normal distribution for the numerator scores and one for the denominator scores—have equal variances, then the resulting F ratio is distributed as the *central* F distribution (as opposed to the *non*central F distribution) that is tabulated in most statistics texts.

We are now in a position to look at equation 7.22 and determine when the numerator and denominator may be treated as sums of squared scores from normal distributions with equal variances. The conclusion of this foray is going to be that when all independent variables are comprised of fixed effects, then any F ratio of the form in equation 7.22 will be distributed as the central F distribution when the null hypothesis being tested is true in the population from which the sample was drawn (i.e., the constraint(s) placed upon the model hold in the population). However, where some independent variables are fixed and others are random, this is not necessarily the case. In fact, in a two-factor design (say $A \times B$), when A is fixed and B is random, the constraints for the main effect of A produce an F ratio that is distributed as a noncentral F distribution when the null hypothesis is true. This means that the difference between the variances of the residuals $[S(e_1,e_1) - S(e,e)]$ has a different expected value (over replications) than the unconstrained residuals $[S(e,e)]$. For this reason, then, it is necessary to use an equation other than that in 7.22 when random effects are included in an experimental design.

An intuitive grasp of why equation 7.22 yields values distributed as the central F distribution when only fixed effects are involved can be gleaned from the observation that $S(e_1,e_1)$ will be larger than $S(e,e)$ only because of random factors when the constraint holds in the population. That is, the only random component in the population model is the disturbance (which is estimated by the sample residual). Thus, the only reason that $S(e_1,e_1)$ is not equal to $S(e,e)$ when the constraint holds in the population is because relaxing the constraint allows sample values to adjust to random fluctuations in means in which population expectations are identical. In this sense, then, the extent to which the residuals increase in size when the model is constrained will be equal to the extent of randomness in the true disturbance. This can be stated more formally in terms of the expectations of the numerator and denominator of equation 7.22 when the constraint holds in the population:

$$E[S(e_1,e_1) - S(e,e)] = k \operatorname{Var}(e^*)/N \qquad (7.23a)$$

and

$$E[S(e,e)] = (N - J) \operatorname{Var}(e^*)/N \qquad (7.23b)$$

where $\operatorname{Var}(e^*)$ is the population variance of the "true" disturbances—the only source of randomness in a fixed effects model. [In the traditional jargon of ANOVA, the expression 7.23a reflects "between group" differences, whereas expression 7.23b reflects "within group" differences. The extent to which the ratio of these expressions (divided by k and $(N - J)$, respectively) is distributed as the theoretical density is a function of the truth of the null hypothesis *and* the independence of the between group and within group variation. Thus, one way to violate the assumptions of the usual F-test is to have data in which the variability of scores within cells is related to the magnitude of cell means. For example, if the within cell variance increases as the cell mean increases, then equation 7.23a will not be independent of equation 7.23b. Some transformation of the data should be applied in such instances to eliminate this dependence.] If the constraint does not hold in the population, then a term must be added to the right side of equation 7.23a; this additional term reflects the extent to which the constraint *does not* hold in the population and can be responsible for F ratios that are quite unlikely under the assumption that the null hypothesis holds.

Now, the line of reasoning applied above must be modified when we consider an experimental design involving a fixed effects independent variable (A) crossed with a random effects independent variable (B). In this case, we might expect some *sample results* to deviate from true population constraints *not only* because of random disturbances, *but also because of random sampling of levels* of the random effects independent variable. This turns out to be the case with the constraints used to specify the main effects of the fixed effects independent variable. From equation 7.21 the main effects null hypothesis for factor A could be stated in terms of the following constraint:

$$d_1 + d_3 + \ldots + d_{19} = d_2 + d_4 + \ldots + d_{20} \tag{7.24}$$

This constraint states that the average cell mean in the control condition (odd-numbered cells) is equal to the average cell mean in the manipulated condition (even-numbered cells). Now, if this constraint holds in the population, then sample results will deviate from equation 7.24 because of random disturbances *and* because there may be an interaction between A and B. In other words, differences between control and treatment groups may be a function of which experimenter is involved, and, therefore, the observed deviation from equation 7.24 would depend upon the variability among experimenters even though the main effect of A across all experimenters in the population was zero (as stated in the constraint 7.24). (If there were no interaction between A and B, then these considerations would be irrelevant. But in most instances where random effects are involved, this lack of interaction is implausible.) More formally, this difference between the sources of variability in an experimental design involving only fixed effects and a design involving both fixed and random effects (called a "mixed model") can be stated in terms of the expectation of the numerator of equation 7.22. When constraint 7.24 holds for equation 7.21 in the population of participants and experimenters, then we have

$$E[S(e_1,e_1) - S(e,e)] = k \operatorname{Var}(e^*)/N + kN^2 \operatorname{Var}(d^*) \tag{7.25}$$

where

$$\operatorname{Var}(d^*) = E\{[(d_2^* - d_1^*) - (d_4^* - d_3^*)]^2\}$$

Thus, when a fixed effects independent variable is crossed with a random effects independent variable, the constraint(s) corresponding to the fixed effects produces an increase in the variance of the residuals that is a function of both $\operatorname{Var}(e^*)$ and $\operatorname{Var}(d^*)$—when the null hypothesis of no main effect for the fixed effects factor holds. [Note that $\operatorname{Var}(d^*)$ is the variance of the contrast that defines the interaction between A and B. The null hypothesis of no interaction between A and B would be stated in terms of the constraint that $(d_2 - d_1) = (d_4 - d_3) = (d_6 - d_5) = \ldots = (d_{20} - d_{19})$. This constraint could be stated in terms of the nine contrasts

$$(-1)d_1 + (1)d_2 + (1)d_3 + (-1)d_4 = 0$$
$$(-1)d_3 + (1)d_4 + (1)d_5 + (-1)d_6 = 0$$
$$\vdots$$
$$(-1)d_{17} + (1)d_{18} + (1)d_{19} + (-1)d_{20} = 0$$

The variance $\operatorname{Var}(d^*)$ is then the variance of these contrasts in the population.]

The task that remains is to find a denominator that has the same expectation as in equation 7.25, since the expectation of $S(e,e)$ is the same in the mixed model as in the fixed model (see equation 7.23b). If some function of the observations does have this same expectation under the null hypothesis, then a ratio could be formed that has the characteristics of the central F-distribution. It can be shown that the appropriate expectation stems from the interaction constraint. If e_2 is the residual under the interaction constraint, then when the null hypothesis of no main effect of A is true,

$$E[S(e_2,e_2) - S(e,e)]/k_2 = E[S(e_1,e_1) - S(e,e)]/k$$

where k_2 is the number of constraints necessary to specify the interaction null hypothesis (nine in the case of a 2×10 factorial design). This means that the null hypothesis of no main effects for the fixed effect independent variable A can be tested with the F ratio

$$F = \frac{[S(e_1,e_1) - S(e,e)]/k}{[S(e_2,e_2) - S(e,e)]/k_2} \qquad df = k,\ k_2 \tag{7.26}$$

or

$$F = \frac{[R(t) - R(r_1)]/k}{[R(t) - R(r_2)]/k_2} \qquad df = k,\ k_2 \tag{7.27}$$

where $R(r_1)$ is the R-square under the constraint of no main effect of A, and $R(r_2)$ is the R-square under the constraint of no interaction between A and B.

An alternative conceptualization of the distinction between random and fixed effects is suggested by the degrees of freedom for the main effect of the fixed effects independent variable. Note in equations 7.26 and 7.27 that the degrees of freedom for the F ratio are k and k_2—the number of constraints necessary to specify the main effects of A and the interaction null hypotheses, respectively. Unlike the fixed effects model, the degrees of freedom for the F ratio appear to be independent of the number of participants within each cell of the experiment. Instead, the degrees of freedom for the denominator are a function of the number of levels of the random factor (and the number of levels in the fixed effects factor), since more levels require more constraints. The insight to be gained from this observation is that the inference to be made about the main effect of A (when B is random) is made concerning the population of levels of B. In other words, a rejection of the null hypothesis for the main effect of A supports the alternative hypothesis that an effect of A holds over the population of experimenters from which the actual experimenters were drawn. Clearly, this is a broader conclusion than that that stems from the fixed effects model—i.e., that the significant effect of A holds for those particular (fixed) levels of B that have been included in the experiment. Thus, it is intuitively appealing that the larger numbers of participants *within* each cell would not affect inferences concerning the effect of A over the population of levels of the independent variable B. Only inclusion of more levels of B would increase assurance that any observed effects of A would hold for the entire population of levels of B.

Some General Principles We now have two equations for the computation of F ratios in the analysis of experiments with qualitative independent variables. The first, equation 7.22, uses the variance of the residuals as the denominator; the second, equation 7.26, uses the increase in the residual variance under the interaction constraint(s) as the denominator. In general, equation 7.22 should be used to test all main effects and interactions when all independent variables in an experimental design are fixed. In a mixed model (A fixed, B random), equation 7.22 is also appropriate for all F ratios *except* the one for the main effect of the fixed factor; in this instance, equation 7.26 should be used to obtain an F ratio. In a *random model* (A random, B random), equation 7.22 is appropriate for testing the interaction and equation 7.26 is used for both main effects.

When more than two independent variables are included in an experiment, determination of the appropriate procedure for computing F ratios is more complex and can present problems. For example, a three-factor experiment with two random factors does not contain adequate information for an exact test of the null hypothesis for the main effect of the fixed effects main effect. Similarly, a design involving three random factors cannot be used to obtain exact tests of any of the main effects. In my experience, however, this is not a major block to the use of random effects models in the social and behavioral sciences, since most designs that include random effects include only one such factor. One of the most frequent uses of random factors is the analysis of program effectiveness *across* institutions. An investigator may, for instance, be concerned with the activity patterns of persons in group therapy. If several different types of groups were formed within a number of treatment centers, and the treatment centers were randomly selected from some population of treatment centers, then the different treatment centers could be analyzed as levels of a random factor. Similarly, studies of educational processes may assign students within schools to various treatment conditions. If the schools studied were randomly (or representatively) selected from a larger population of schools, then each school could comprise a level of a random factor. In such instances, it is rare for these random factors to be crossed with another random factor. (It is not unusual, however, for random factors to be "nested" within another random factor (rather than crossed). For example, classrooms within schools may be selected for study on a random basis. The "classroom variable" would then be nested within the "school variable" rather than crossed with it. The problems with obtaining exact F ratios do not arise in this case.)

A generalization of the random effects model that may be of some utility to researchers is the *finite random effects* model. Up to this point we have assumed that the levels of a random factor are sampled from a large population. There may be cases, however, in which, for example, five schools are selected from a population of 15 schools in a district. This situation gives rise to a slightly different statistical model, since the assumption of sampling *with replacement* is inappropriate. The notion of finite random effects also gives rise to a parsimonious conceptualization of fixed, finite random, and random effects. From this perspective, fixed effects are finite random effects in which the entire population of levels has been included in the experiment. Finite random effects result from sampling without replacement from small populations of levels. And random effects arise from the sampling of levels from large or infinite populations. A computer program that makes good use of this parsimony is BMD08V (Dixon, 1968). It is one of the most versatile programs for Anova that has been developed. The user is able to specify a wide range of crossed and nested designs with fixed, finite random and random factors. The limitations of the program, however, are often prohibitive; it requires equal cell sizes and applies only to the univariate case. Nevertheless, when a complex experimental design is used and the limitations are not problematic, BMD08V greatly simplifies the rather complex calculations and choices of error terms.

7.7 ESTIMATING THE MAGNITUDE OF EXPERIMENTAL RELATIONSHIPS

The realization that even miniscule experimental effects can be statistically significant with sufficient sample sizes has encouraged many investigators to be more concerned with measures of effect magnitude than significance levels. The structural equation approach adopted in this book is consonant with this concern, since the major focus of this approach is the estimation of structural (causal) parameters and their standard errors. As has already been stated in Section 7.4, however, the *relative* magnitude of experimental relationships is also of interest.

There are essentially two ways in which the relative magnitude of an independent variable's influence on a dependent variable can be summarized. The first is to standardize the structural coefficients to obtain *beta weights* (see Chapter 5). Beta weights reflect the relative magnitude of a causal relationship in the sense that all causal changes are expressed in terms of standard deviations of the independent and dependent variables. This means that the influence of different independent variables on the dependent variable can be compared without distortion due to differing units of measurement. Also, since the absolute value of standardized coefficients should range from zero to one, the magnitude of the coefficient communicates in percentage terms: a standardized coefficient of 0.3 means that an increase of one standard deviation in the independent variable produces a change in the dependent variable equal to 30 percent of a standard deviation of the dependent variable. And this characteristic of the standardized coefficients is closely linked with the inverse relationship between the absolute value of these coefficients and the extent to which variability has been attributed to the residual in the model. As the absolute value of the standardized coefficients increases, the standard deviation of the residual decreases.

Before discussing the second approach to the estimation of *relative* effect magnitudes (variance components), it is important to point to the shortcomings of standardized coefficients as communicators of the relative *importance* of a causal relationship. The first drawback of standardized coefficients is that they are a function of the range of the independent variable that is included in a study. Suppose, for example, that a study of health determinants found that an expenditure of one dollar per person on health education reduced subsequent health expenditures by six dollars per year, and that an expenditure of two dollars reduced subsequent health costs by 12 dollars per year. This structural relationship would clearly be of importance, since the savings are so much larger than the costs. However, if medical costs were regressed on the health education expenditure independent variable (and other control variables as well), the magnitude of the standardized coefficient would probably be miniscule, since the differences of 6 and 12 dollars would be a very small proportion of the variability (between persons or families) in health care expenditures. If a much larger range of health education expenditures was included in the study (say, from 1 to 80 dollars per person), and the structural relationship remained linear, then health education expenditures would explain a large percentage of the individual differences in health care expenditure. This example underscores an important characteristic of social and behavioral science that must be considered when the relative magnitude of experimental effects is of concern: investigators are usually restricted by practicalities or moral considerations to weak manipulations that represent only a small portion of the potential range of the independent variable. This tends to mean, then, that theoretically and practically important experimental manipulations in the social and behavioral sciences may be associated with small standardized coefficients in multiple regression equations.

The other drawback of standardized coefficients is that they are attenuated by measurement error in the dependent variable (they are also attenuated by measurement error in the independent variable, but the ANOVA model assumes that there is such error only in the dependent variable). Again, this drawback is particularly relevant in the social and behavioral sciences, since measures of dependent variables are usually rather rough. The unstandardized coefficients have the virtue of being unaffected by either random measurement error in the dependent variable or restriction of range in the independent variable (although estimation of the structural coefficients *is* affected by these factors, since each increases the standard error of estimation). Thus, a greater focus on the unstandardized coefficients is clearly in order in experimental research in the social and behavioral sciences. (It should be pointed out that the measurement models presented in Chapter 9 provide a solution to the distortion of standardized coefficients by measurement error in the dependent variable. However, these models cannot ameliorate the problem created by restricted range in the independent variable.)

Variance Components The second procedure for estimation of effect magnitudes is the estimation of variance components in ANOVA models. In the case of balanced designs, there is a close link between the variance components of fixed effects models and the standardized coefficients of the linear model used to obtain the F ratios for the ANOVA. Although general procedures for the estimation of these variance components have been developed for fixed, finite random, and random models with crossed and nested factors (see Dwyer, 1974), only the simplest case will be presented here.

Suppose that an experimental design involves two independent variables (A and B) and one dependent variable. The variables A and B each have only two fixed levels, and the four cells of the experiment each contain n observations (i.e., $N = 4n$). The results of this experiment could then be analyzed with an ANOVA model of the following form:

$$Y(i) = b_0 + b_1X_1 + b_2X_2 + b_3(X_1X_2) + e(i) \tag{7.28}$$

where

$$\begin{aligned} Y(i) &= \text{score of the } i\text{th participant on the dependent variable} \\ X_1(i) &= -1 \text{ if } i\text{th participant in } A_1 \\ &\quad\;\; 1 \text{ if } i\text{th participant in } A_2 \\ X_2(i) &= -1 \text{ if } i\text{th participant in } B_1 \\ &\quad\;\; 1 \text{ if } i\text{th participant in } B_2 \end{aligned}$$

The coefficients in equation 7.28 would be sample estimates of the population parameters in equation 7.29:

$$Y = b_0^* + b_1^* X_1 + b_2^* X_2 + b_3^*(X_1 X_2) + e^* \tag{7.29}$$

If the model 7.29 holds for the population from which the sample of participants was drawn, and e^* is uncorrelated with X_1, X_2, and $(X_1 X_2)$, then

$$\begin{aligned} \text{Var}(Y) &= E[Y - b_0^*]^2 \\ &= E[b_1^* X_1 + b_2^* X_2 + b_3^*(X_1 \times X_2) + e^*]^2 \\ &= b_1^{*2}\,\text{Var}(X_1) + b_2^{*2}\,\text{Var}(X_2) + b_3^{*2}\,\text{Var}(X_1 \times X_2) + \text{Var}(e^*) \\ &= b_1^{*2} + b_2^{*2} + b_3^{*2} + \text{Var}(e^*) \end{aligned} \tag{7.30}$$

[Note that $\text{Var}(Y)$ simplifies to equation 7.30 because the variances of X_1, X_2, and $(X_1 \times X_2)$ are all one, and the covariances among these dummy variables are zero—due to the classic ANOVA coding scheme and equal cell sizes.] Thus, according to the population model, the variance of the dependent variable is partitioned into *variance components* associated with A(b_1^* squared), B (b_2^* squared), and $A \times B$ interaction (b_3^* squared), and the disturbance [$\text{Var}(e^*)$]. These components sum to equal the total variance of Y in the population and may be interpreted as the relative contribution of each source of variation to differences in the dependent variable.

The task at hand is to develop unbiased estimates of the population parameters in equation 7.30 from the sample estimates in equation 7.28. In the case of fixed effects and homoscedasticity, this task is quite straightforward. Since the dummy variables are uncorrelated with one another, it can be shown that

$$E[b_1^2] = b_1^{*2} + \text{Var}(e^*)/N \tag{7.31a}$$

$$E[b_2^2] = b_2^{*2} + \text{Var}(e^*)/N \tag{7.31b}$$

$$E[b_3^2] = b_3^{*2} + \text{Var}(e^*)/N \tag{7.31c}$$

$$E[S(e,e)] = (N - J)\,\text{Var}(e^*)/N \tag{7.31d}$$

where N is the size of the total sample and J is the number of cells in the experimental design ($J = 4$ in the case of a 2×2 factorial design). Note that equations 7.31 make intuitive sense in that the term $\text{Var}(e^*)/N$ would be the variance of the mean of samples of N cases of the disturbance (e^*)—i.e., the variance of e^* divided by N is the variance of the sampling distribution of the mean of e^* for samples of size N. Thus, equation 7.31a, for example, stems from the fact that the variance of b_1, the sample estimate of b_1^*, is

$$\begin{aligned} \text{Var}(b_1) &= E[(b_1 - b_1^*)^2] \\ &= E[b_1^2 + b_1^{*2} - 2(b_1)(b_1^*)] \\ &= E[b_1^2] - b_1^{*2} \end{aligned}$$

which may be rearranged as

$$E[b_1^2] = b_1^{*2} + \text{Var}(b_1)$$
$$= b_1^{*2} + \text{Var}(e^*)/N$$

since the dummy variables in a balanced design are uncorrelated (see Chapter 6).

Equations 7.31 can now be solved for the population variance components in terms of the expected values of the sample statistics. We begin by solving equation 7.31d for $\text{Var}(e^*)$ and then using the result to solve the first three equations (7.31a to 7.31c) for the squares of b_1^*, b_2^*, and b_3^*:

$$\text{estimate of } (b_1^{*2}) = b_1^2 - NS(e,e)/(N - J) \quad (7.32a)$$

$$\text{estimate of } (b_2^{*2}) = b_2^2 - NS(e,e)/(N - J) \quad (7.32b)$$

$$\text{estimate of } (b_3^{*2}) = b_3^2 - NS(e,e)/(N - J) \quad (7.32c)$$

$$\text{estimate of Var}(e^*) = NS(e,e)/(N - J) \quad (7.32d)$$

These four equations can then be used to compute unbiased estimates of the population variance components associated with each main effect and interaction (and error) term in the model 7.29. For slightly unbalanced designs, these equations would provide adequate, although biased, estimates. However, as the correlation between the dummy variables becomes nontrivial, equations 7.32 will not provide useful information.

At this point, it is tempting to sum the variance components [which equals $NS(Y,Y)/(N - 1)$] and divide the sum into each component to obtain the *proportion* of variance in Y that is explained by each component of the model. The result of yielding to this temptation is an estimate of a population parameter that is analogous to the sample statistic *eta squared* that was described in Section 7.4 of this chapter. Thus, the ratio of the variance components is preferable to eta squared in the same way that the adjusted R-square is preferable to the sample R-square (see Chapter 6). [The reader should recall that the adjusted R-square differs from the usual R-square only in that estimates of population parameters are used:

$$\text{R-square} = 1 - [S(ee)/S(YY)]$$

whereas

$$\text{adjusted R-square} = 1 - [NS(ee)/N - p - 1)]/[NS(YY)/(N - 1)]$$

The estimates of the variance components refer to population parameters, whereas eta squared refers to characteristics of the sample. However, as with the adjusted R-square, the ratios of the variance component estimates are not necessarily unbiased, and there are no straightforward procedures for estimation of the standard errors of those ratios. Thus, the estimation of variance components is recommended in reports of experiments where the magnitude of the estimates of the structural coefficients is insufficient to communicate the "importance" of the relationship; the reporting of variance component proportions is recommended as an attempt to summarize the relative magnitudes of the various effects—but the latter should be interpreted with caution, since all the drawbacks associated with standardized regression coefficients are also applicable to these proportions. [Hays (1970) refers to these ratios of variance components as *omega squared*. A general method for the computation of variance components for different experimental designs is presented in Dwyer (1974). Also, the ANOVA computer program BMD08V provides estimates of variance components as part of standard output.]

The reader should be warned that estimates of population variance components, unlike sample values of eta squared, may be negative. This arises when a main effect or interaction is smaller than would be expected when the null hypothesis for that main effect or interaction is true (i.e., the F ratio is less than one). This event will occur roughly one-half the time when the null hypothesis is true. Thus, negative values for these estimates are no problem, and they should be reported as such.

7.8 ANALYSIS OF COVARIANCE (ANCOVA)

In the case of true experiments in which variability in the independent variables is randomized (i.e., determined by some random process), the goal of *analysis of covariance* is reduction of the amount of "noise" in the observation of the dependent variable. This is accomplished through the inclusion of *covariates* in the linear model that is presumed to hold in the population and inclusion of observed covariates in the multiple regressions used to construct F-tests and/or standard errors for main effects and interactions. A covariate is an observed variable that has not been randomized; typical covariates are age, sex, years of education—i.e., variables that tend to be related to most individual differences, but that cannot be manipulated or randomly assigned. If the covariate(s) is sufficiently correlated with the dependent variable *after controlling for the randomized independent variables,* then the inclusion of the covariate in the model increases the total R-square. (The notion of net relationships after controlling for third variables should be one that is easily construed by the reader of this book. In the traditional jargon of ANOVA, this notion is referred to as a *within group* relationship. That is, for a covariate to increase the power of ANOVA F-tests, the covariate must have an adequate correlation with the dependent variable within the cells of the experiment.) This increase in the R-square is indicative of a reduction in the variance of the residuals, and this reduction is indicative of a general reduction in the amount of "noise" or unexplained variance in the dependent variable. And since the standard error of the estimates of the structural coefficients *decrease* as R-square increases, the inclusion of covariates can markedly increase confidence in parameter estimates.

The reader may recall from Chapter 6 that the inclusion of additional variables in a linear model involving non-randomized variables can have several different results. For example, a relationship that apparently arises from a direct causal link can be "explained away" by inclusion of the appropriate "third variable." In true experiments, however, the randomization of independent variables assures that any covariate will be correlated with the independent variables only by chance; and the probability of such correlations being substantial becomes quite small as sample size increases above 40 or 50.

Fortunately, the regression approach to ANOVA allows a straightforward generalization from ANOVA to ANCOVA. Traditional presentations of ANOVA (in terms of mean squares and marginal means) were favored because they enabled the user to perform calculations on a mechanical calculator. Since the widespread use of electronic computers has rendered this concern obsolete, however, there is no need to perpetuate the belief that ANCOVA is a rather mysterious topic. The regression approach to the analysis of experimental data reveals ANCOVA to be little more than the inclusion of additional right-hand variables in a multiple regression program. This ease of mechanical setup should not lull the reader into complacency concerning the knotty conceptual problems that arise in research designs other than randomized experiments; subsequent discussions of non-equivalent control group designs will focus upon these problems with interpretation of covariance analysis.

A final word of caution should be digested before proceeding to the details of ANCOVA computations. If a covariate is poorly measured, or only slightly correlated with the dependent variable within experimental cells for some other reason, then the inclusion of the covariate can reduce the power of F-tests through losses in degrees of freedom. This loss of power will not be substantial, however, unless the number of poorly correlated covariates reduces the residual degrees of freedom from above 30 to below 30.

Computational Example In order to demonstrate how easily an ANOVA model is generalized to an ANCOVA model, we will add a covariate to the hypothetical study concerning the effects of social threat and scientific knowledge on authoritarianism. The goal here will be to find a variable that has some observable value for each participant in the experiment and that should be correlated with the F-scale scores within cells of the experiment. (We would not expect large correlations *between* the covariate and mean differences between cells because those differences are presumably due to the experimental manipulations that were *randomized*.) A likely candidate for such a covariate would be years of education, since the inverse correlation between years of education and the F-scale is well established (see Adorno et al., 1950). Suppose now that all the 34 participants in the experiment have reported their educational background and that a new right-hand variable (X_7) is defined

$$X_7(i) = \text{years of education completed by the } i\text{th participant in the experiment}$$

Note that there is no statistical distinction between X_7 and any of the dummy variables used to define the independent variables in the experiment. The difference between X_7 and the dummy variables stems from design considerations: the dummy variables represent *randomized* variables that can be correlated with variables other than those they influence causally only by chance; in contrast, the covariate X_7 is an observed variable that may be correlated with any other variable for a host of reasons. Thus, we proceed by simply adding X_7 to the ANOVA regression model that is being used to analyze the experimental results. If we were using the classic ANOVA coding scheme $(-1,0,1)$ and an OLS program, then the ANOVA model in equation 7.10 would become the following ANCOVA model:

$$\begin{aligned} Y(i) = c_0 + c_1Z_1 + c_2Z_2 + c_3Z_3 + c_4(Z_1 \times Z_2) \\ + c_5(Z_2 \times Z_3) + c_6X_7 + e(i) \end{aligned} \tag{7.33}$$

where Z_1, Z_2, and Z_3 are defined as in equation 7.10, and X_7 is the education variable defined above. It should be noted that this ANCOVA regression model is identical to the ANOVA model in equation 7.10 except for the inclusion of education (X_7) as an additional right-hand variable in the linear equation.

Now, suppose that the education scores for the 34 participants in the experiment lined up with the F-scale scores as follows:

	B_1 No knowledge			B_2 Knowledge		
i	Y	X_7	i	Y	X_7	
1	7	12	16	7	12	
2	9	16	17	8	16	
3	10	12	18	9	12	A_1 Control
4	11	8	19	10	8	
5	13	12	20	12	12	
6	10	12	21	6	12	
7	12	16	22	8	16	
8	13	12	23	8	12	
9	14	8	24	9	12	A_2 First threat manipulation
10	16	12	25	10	12	
			26	10	8	
			27	12	12	
11	9	12	28	5	12	
12	11	16	29	7	16	
13	12	12	30	7	12	
14	13	8	31	7	12	A_3 Second threat manipulation
15	15	12	32	8	12	
			33	10	8	
			34	12	12	

We are now in a position to perform F-tests for the main effects and interaction null hypotheses. The procedure followed is the same as that outlined in Section 7.4 except that the covariate X_7 is included in each regression.

The calculations begin with the estimation of the population parameters for the ANCOVA model using the regression equation 7.33. The result is

$$\begin{aligned} Y = {} & \underset{(1.79)}{13.3} + \underset{(.50)}{.77Z_1} - \underset{(.50)}{.13Z_2} - \underset{(.36)}{1.5Z_3} \\ & - \underset{(.50)}{.50(Z_1 \times Z_3)} - \underset{(.50)}{.60(Z_2 \times Z_3)} - \underset{(.14)}{.25X_7} + e \end{aligned} \tag{7.34}$$

$$\text{R-square} = .502 \qquad S(ee) = 115.6/34 = 3.40$$

This result should be compared with equation 7.15 (the regression of Y on the same independent variables, but without the covariate), which is repeated here

$$\begin{aligned} Y = {} & \underset{(.37)}{10.2} + \underset{(.52)}{.80Z_1} - \underset{(.52)}{.20Z_2} - \underset{(.37)}{1.47Z_3} \\ & - \underset{(.52)}{.53(Z_1Z_3)} - \underset{(.52)}{.53(Z_2Z_3)} + e \end{aligned}$$

$$\text{R-square} = .445 \qquad S(ee) = 3.79$$

A comparison of equations 7.34 (ANCOVA) and 7.15 (ANOVA) raises several issues:

(1). The residual for the ANCOVA model has a smaller variance than the ANOVA model (3.40 versus 3.79). This occurs because the covariate explains some of the variance in Y that was unexplained by the experimental manipulations. But the addition of any new variable to a regression model will decrease the variance of the residual; the important question is whether or not the decrease is sufficient to increase the power of the F-tests. This will be determined when we compute the F ratios for the main effects and interactions using the ANCOVA model.

(2). The R-square is larger for the ANCOVA model than for the ANOVA model (.502 versus .445). This result is simply the converse of the reduction in the variance of the residual.

(3). The regression coefficients are roughly the same in both models. This occurs because the fabricated data were constructed such that the covariate is correlated with the dependent variable only within the cells of the experiment. In other words, the mean of the covariate within each cell is the same (12 years of education), which is what would occur *on the average* in randomized experiments. Since the covariate has an almost zero correlation with the dummy variables used to define the main effects and interactions, the inclusion of the covariate has little influence on the magnitude of the regression coefficients.

(4). The standard errors of all the coefficients (except the intercept) are all slightly smaller in the ANCOVA model than in the ANOVA model. This finding means that the inclusion of the covariate has produced a slight increase in the estimated precision of the sample estimates. The increased precision has occurred because the ANCOVA model has successfully reduced the proportion of uncertainty in the data (although only slightly). This reduction in the magnitude of standard error means that we can expect the ANCOVA F ratios for main effects and interactions to have smaller probabilities than the F-ratios obtained from the ANOVA.

We can now proceed to test the null hypotheses associated with the main effects and interaction by the now familiar application of equation 7.3:

Main effect of A; H_0: $c_1^* = c_2^* = 0$

$$Y = 13.36 - 1.50Z_3 - .41(Z_1 \times Z_3) - .66(Z_2 \times Z_3) - .26X_7 + e_1$$

$$\text{R-square} = .454 \qquad S(ee) = 3.72$$

$$\begin{aligned} F &= \{[R(t) - R(c_1 = c_2 = 0)]/2\}/\{[1 - R(t)]/(N - 7)\} \\ &= [(.502 - .454)/2]/[(1 - .502)/27] \\ &= 1.29 \qquad df = 2, 27 \end{aligned}$$

$$\text{Prob}(F > 1.29) = .29$$

Main effect of B; H_0: $c_3^* = 0$

$$Y = 12.73 + .69Z_1 - .22Z_2 - .58(Z_1 \times Z_3) - .67(Z_2 \times Z_3) - .22X_7 + e_2$$

$$\text{R-square} = .181 \qquad S(ee) = 5.59$$

$$\begin{aligned} F &= [(.502 - .181)/1]/[(1 - .502)/27] \\ &= 17.40 \qquad df = 1, 27 \end{aligned}$$

$$\text{Prob}(F > 17.40) < .001$$

Interaction of A and B; H_0: $c^*_4 = c^*_5 = 0$

$$Y = 13.07 + 75Z_1 - .17Z_2 - 1.55Z_3 - .24X_7 + e_3$$

$$\text{R-square} = .419 \qquad S(ee) = 3.96$$

$$F = [(.502 - .419)/2] \;/\; [(1 - .502)/27]$$
$$= 2.23 \qquad df = 2, 27$$
$$\text{Prob}(F > 2.23) = .13$$

And the following table summarizes the results of the ANCOVA and ANOVA analyses:

ANCOVA/ANOVA Comparison

Source	*SS*	*df*	*F*	Prob	eta squared
Intercept		1	55.04(750.2)	<.001(<.001)	
A	11.0(11.5)	2	1.29(1.26)	.29 (.29)	.05(.05)
B	74.4(71.3)	1	17.39(15.54)	<.001(<.001)	.32(.31)
A×*B*	19.1(17.9)	2	2.23(1.94)	.13 (.16)	.08(.08)
Covariate	13.2	1(0)	3.29	.08	.06
Error	115.6(128.8)	27(28)			.50(.56)
Sum	233.3(229.5)	34			1.01(.99)
Total	232.0(232.0)	34			

Note. The numbers in parentheses result from the ANOVA model and are presented for comparison with the ANCOVA results.

The immediate conclusion to be drawn from the comparison of the ANCOVA and ANOVA models is that the inclusion of the covariate in the linear model increased the magnitudes of the F-ratios slightly, but did not affect the portion of sample variance in the dependent variable that was associated with main effects and interactions. [This increase in the magnitudes of the F ratios is counteracted, to some degree, by the deletion of one degree of freedom from the error term—a degree of freedom given to the covariate. The impact of this loss is almost equal to the gain due to the reduction in the variance of the residual: Note that the probability of the observations under the interaction null hypothesis is only slightly less (.13 versus .16) under the ANCOVA model than under the ANOVA model.] If the covariate had been more strongly correlated with the dependent variable *within cells,* then the ANCOVA F ratios would have been considerably larger than the ANOVA F ratios. However, if, by chance, the covariate was correlated with the independent variables, then the main effects and interaction F ratios could be considerably smaller for the ANCOVA than the ANOVA.

The analysis of covariance can also be computed using the ANOVA least squares model outlined in Section 7.5. For example, the ANCOVA model for the above data could be analyzed with a regression model of the following form:

$$Y = d_1W_1 + \ldots + d_6W_6 + d_7X_7 + e \tag{7.35}$$

where the W_j ($j = 1, \ldots, 6$) are defined as in equation 7.16. The various null hypotheses are then tested through the use of constraints in a least squares estimation procedure. The form of equation 7.35 suggests an alternative conceptualization of ANCOVA. Recall that the ANOVA least squares method uses a 0/1 coding scheme for dummy variables such that only one term in the model is nonzero for any particular participant. With the addition of the covariate X_7 in the model, the model becomes a group of six regression equations where the coefficients $d_1, \ldots, d_6$ are the intercepts and d_7 is the slope of Y regressed on X_7. There is a regression equation for each cell of the experiment, and the constraints (or null hypotheses) apply to the intercepts of the different regression equations.

This protrayal of the ANCOVA model as a series of regressions raises several important points concerning ANCOVA. First, note that the slope d_7 is the same for every cell of the experiment. This specification constitutes an assumption that the population slopes are equal in the populations defined by the different cells. And this assumption is essentially that the effects of the independent variables do not *interact* with the covariate (e.g., the effect of increased social threat on authoritarianism does not depend upon the level of education of the participant). If this assumption is inaccurate, then the results of the ANCOVA can be misleading without the inclusion of terms for interactions between the independent variables and the covariate. In the case of equation 7.35, the hypothesis that the slope of Y on X_7 is different for any particular cell of the experiment could be explored by adding a multiplicative term to the model. For example, suppose that the relationship between Y and X_7 is greater in the knowledge condition (B_2) than in the control condition (B_1). Recalling that d_1, d_2, and d_3 are associated with B_1, whereas d_4, d_5, and d_6 are associated with B_2, we have

$$Y = d_1W_1 + \ldots + d_6W_6 + d_7X_7 + d_8(W_1 + W_2 + W_3)X_7 + e \quad (7.36)$$

which may be rewritten as

$$Y = d_1W_1 + \ldots + d_6W_6 + X_7[(d_7 + d_8(W_1 + W_2 + W_3)] + e$$

In this form, then, the slope of X_7 is d_7 when $(W_1 + W_2 + W_3)$ equals zero (condition B_2) and is $(d_7 + d_8)$ when $(W_1 + W_2 + W_3)$ is equal to one (B_1). This specification allows the slope of Y regressed on X_7 to be different for the two levels of B. Whether or not the difference between the two slopes is statistically reliable can be assessed by the standard error of the coefficient d_8.

Equation 7.36 raises a bit of a problem, however, since the within cell regression lines are no longer assumed to be parallel. The problem is that the intercepts are now arbitrarily changed by the choice of the zero point of X_7. My recommendation in such instances, therefore, is to transform covariates to mean deviations before including them in an ANCOVA. Subtracting the mean from covariates before analysis also maintains the interpretation of $d_1, \ldots, d_6$ in equation 7.35 as cell means—except that they are now *adjusted* cell means. They are adjusted so that the between cell differences in the dependent variable are uncorrelated with between cell differences in the covariate. This adjustment should be small in true experiments, since between cell differences in the dependent variable should be related to the covariate only by chance.

It is also possible to allow interactions between independent variables and covariates when using OLS regression programs and the classic ANOVA coding scheme (see equation 7.10).

7.9 CONCLUDING REMARKS

The goal of this chapter has been to present computational procedures for the analysis of variance and covariance as special applications of linear regression models. This approach to ANOVA and ANCOVA liberates researchers from the task of learning the algebra of ANOVA as an almost distinct area of statistics. Once linear regression models are understood, ANOVA is simply the application of regression to the case of an almost continuous dependent variable regressed on dummy variables that code levels of ordinal or nominal scale independent variables and their interactions. And the analysis of covariance is accomplished by simply adding other independent variables to the dummy variables in an ANOVA model. There is no computational distinction between independent variables and covariates; rather, the distinction is usually one that stems from design considerations. For example, randomized independent variables in an experiment are usually distinguished from passively observed covariates; but this distinction stems from the distinction between randomization and passive observation and is relevant to substantive interpretation rather than procedures of statistical analysis.

Another advantage of the linear model approach to ANOVA computations is that the assumptions sufficient for good estimation of effects and useful estimates of confidence intervals are precisely those that apply to regression models generally. In short (see Chapters 5 and 6 for complete discussions of this topic), the estimates of effects are unbiased if the residuals have zero mean in each cell of the design, and the confidence intervals (and significance tests) are accurate if the residuals are normally distributed with equal variances within each cell (homoscedasticity) or sample sizes are large. Furthermore, the estimates of standard errors derived from populations that violate normality and/or homoscedasticity assumptions are quite robust.

REFERENCES AND SUGGESTED READING

Adorno, T. W., Frenkel-Brunswick, E., Levinson, D. J., and Sanford, R. N. *The authoritarian personality*. New York: Harper, 1950.

Bock, R. D. *Multivariate statistical methods in behavioral research*. New York: McGraw-Hill, 1975.

Cohen, J. *Statistical power analysis for the behavorial sciences*. New York: Academic Press, 1969.

Cohen, J., and Cohen, P. *Applied multiple regression/correlation analysis for the behavioral sciences*. Hillsdale, N.J.: Lawrence Erlbaum, 1975.

Dixon, W. J. (ed.). *BMD biomedical computer programs*. Los Angeles: UCLA, 1968.

Dwyer, J. H. Analysis of variance and the magnitude of effects: A general approach. *Psychological Bulletin*, 1974, *81*, 731–737.

Finn, J. D. *MULTIVARIANCE: Univariance and multivariance analysis of variance, covariance and regression*. Chicago: National Education Resources, 1972.

Fisher, R. A. *Design of experiments* (8th ed.). New York: Hafner, 1966.

Hays, W. L. *Statistics for the social sciences*. New York: Holt, Rinehart and Winston, 1973.

Kirk, R. E. *Experimental design: Procedures for the behavioral sciences*. Belmont, Calif.: Brooks/Cole, 1968.

Kleinbaum, D. G., and Kupper, L. L. *Applied regression analysis and other multivariable methods*. North Scituate, Mass.: Duxbury Press, 1978.

Lindquist, E. F. *Design and analysis of experiments in psychology and education.* Boston: Houghton Mifflin, 1953.

Nie, N. H., Hull, C. H., Jenkins, J. G., Steinbrenner, K., and Bent, D. H. *Statistical package for the social sciences (SPSS)* (2nd ed.). New York: McGraw-Hill, 1975.

Overall, J., and Klett, C. J. *Applied multivariate analysis.* San Francisco: McGraw-Hill, 1972.

Winer, B. J. *Statistical principles in experimental design* (2nd ed.). New York: McGraw-Hill, 1971.

8

Multivariate Regression and Multivariate ANOVA

8.1 INTRODUCTION

The reader of the first seven chapters of this book has been exposed to correlational and structural uses of linear models involving one left-hand (dependent) variable. The versatility of the linear model and least squares estimation procedures should be apparent. In fact, however, many—if not most—research problems in the social and behavioral sciences could only be approached rather crudely if the cafeteria of available quantitative methods were restricted to models with one dependent variable that was influenced by one or more independent variables. The intuitively more realistic models would involve a blurring of the distinction between independent and dependent variables, when change in X may cause a change in Y over one time period, but a change in Y produces a change in X over some subsequent time period. Also, realistic models of human systems are likely to include large numbers of reciprocally causal variables.

The presentation of a quantitative methodology adequate to the task of testing such realistic models is the goal of this book. A general multivariate regression model is our next step toward that plateau in the evolution of quantitative methods that was discussed in the preface. Multivariate regression models allow multiple dependent variables to be included in a procedure of hypothesis testing. What will need to be added to this regression model is the capacity to incorporate models of measurement error and reciprocal causation into the general model. These additions will require a slightly different approach to model fitting (maximum likelihood rather than least squares), but least squares procedures will be adequate for the multivariate regression model presented in this chapter.

There are two basic types of research situations that can make use of a multivariate regression model. The first is the experiment with multiple dependent variables. This application of the model is a straightforward generalization of the methods developed in Chapter 7. The second area of application is to correlational problems involving two sets of distinct variables. For example, if a number of measures of personality dimensions have been administered to persons in different occupations, a multivariate regression model could be used to explore the relationships between these two variable domains (personality dimensions versus occupations).

Advantages of Multivariate Models The advantages of the multivariate model over successive application of the now familiar univariate model can be important, but in many situations the multivariate analysis provides little additional information relative to a series of univariate analyses. There are two distinct advantages of the multivariate approach: (1) As the number of dependent variables in an experiment (or left-hand variables in a correlational study) increases, the likelihood that larger (but random) relationships will be observed also increases. The multivariate model enables a researcher to estimate the overall likelihood of a type one error in such situations. (2) Important relationships between left- and right-hand variables may involve linear combinations of the left-hand variables (as well as of the right-hand variables). The multivariate model draws attention to such relationships through the covariance of residuals.

The logic concerning the increased likelihood of a type one error with multiple dependent variables is analogous to that applied to multiple comparisons in a one-way analysis of variance. In the one-way ANOVA with numerous groups, the likelihood that at least one pair of groups will differ is considerably greater than the probability that a particular pair of groups will differ—when the null hypothesis of no population difference is true. Similarly, the probability that two groups will differ on any one of a battery of dimensions is greater than the probability that they will differ on a particular measure if the null hypothesis of no population differences on any of the measures is true. The multivariate model aids in this respect in that it enables the researcher to test null hypotheses concerning *all measures* jointly. It also allows the estimation of confidence intervals (or multidimensional "confidence regions") for more than one dimension, once again, *jointly*.

The capacity of the multivariate model to detect relationships between right-hand variables and linear combinations of the left-hand variables is, in my experience, a less useful advantage of the multivariate model over the univariate model. Nevertheless there may be research situations in which this capacity is of importance. One can imagine, for example, that the difference between two personality dimensions is more highly correlated with occupational membership than is either dimension individually. If this were the case, then the multivariate model would draw attention to this fact by producing a much larger F-ratio for a multivariate null hypothesis than for the two corresponding univariate null hypotheses. The prospect of discovering such relationships is tantalizing; however, in my opinion, the likelihood of such discoveries is not great with most measurement instruments available in the social and behavioral sciences.

We are now prepared to specify a multivariate regression model and explore its characteristics. The discussion will begin with a description of the general model as a generalization of the univariate model. The application of the general model to multivariate ANOVA (MANOVA) will then be illustrated.

8.2 A GENERAL MULTIVARIATE REGRESSION MODEL

Harking back to Chapter 6, the univariate linear model was specified by an equation of the form

$$Y = b_0^* + b_1^* X_1 + b_2^* X_2 + \ldots + e^* \tag{8.1}$$

where X_1, X_2, . . . are interval scale variables with fixed values, and the e^* are independent of one another with zero mean and variance Var(e^*) at each combination of levels of the right-hand variables. If the model 8.1 holds in some population, then the application of least squares procedures to data from random samples will yield estimates of b_0^*, b_1^*, and b_2^* that are BLUE (*b*est, *l*inear *u*nbiased *e*stimates). Furthermore, if the sample size is large or e^* is distributed normally, then probabilistically interpretable estimates of standard errors can also be computed.

In order to specify the multivariate generalization of this linear model, it is convenient to change some of the notational conventions. The content of the multivariate regression model will be a series of regression equations of the form in equation 8.1, where each equation has a *different* left-hand variable, but the *same* right-hand variables. This series of equations is best written in a form that is congenial with matrix algebra. Suppose, for example, that a sample of N persons is drawn from a population of mental hospital patients. Each person is then administered a battery of q psychological tests. Each of the q tests is labeled as a variable Y_k where k ranges from 1 to q. If we then use the letter i as the participant index, which ranges from 1 to N, then the score of the ith participant on the kth psychological test would be $Y_k(i)$. If the entire set of scores for the N participants on the q tests were arranged in a rectangle, in which rows corresponded to different participants and columns corresponded to different psychological tests, then we would have a rectangular matrix of scores with N rows and q columns. This matrix is denoted as Y and may be depicted as follows:

$$\underline{Y} = \begin{bmatrix} Y_1(1) & Y_2(1) & \ldots & Y_k(1) & \ldots & Y_q(1) \\ Y_1(2) & Y_2(2) & \ldots & Y_k(2) & \ldots & Y_q(2) \\ \vdots & & & & & \\ Y_1(i) & Y_2(i) & \ldots & Y_k(i) & \ldots & Y_q(i) \\ \vdots & & & & & \\ Y_1(N) & Y_2(\mathrm{N}) & \ldots & Y_k(N) & \ldots & Y_q(N) \end{bmatrix} \tag{8.2}$$

Now suppose that p background variables (e.g., education, occupation, family data) are also measured for each of the N persons in this sample of mental hospital patients. If the index j ranges from 1 to p, then the score of the ith participant on the jth background variable would be denoted $X_j(i)$. And the N by p matrix of $N \times p$ scores on these background variables could be organized as follows:

$$\underline{X} = \begin{bmatrix} X_1(1) & X_2(1) & \ldots & X_j(1) & \ldots & X_p(1) \\ X_1(2) & X_2(2) & \ldots & X_j(2) & \ldots & X_p(2) \\ \vdots & & & & & \\ X_1(i) & X_2(i) & \ldots & X_j(i) & \ldots & X_p(i) \\ \vdots & & & & & \\ X_1(N) & X_2(N) & \ldots & X_j(N) & \ldots & X_p(N) \end{bmatrix} \tag{8.3}$$

where the score in the ith row and the jth column is the score of the ith participant on the jth background variable. It should also be noted that when an intercept term is to be included in the linear model fit to these data, it is necessary for the variable X_1 to have the value one for all participants. This means that if six background variables were actually included in the model, then p, the number of X variables, would be equal to seven, since the dummy variables X_1 would be added to the matrix. When the multivariate model is used to perform an analysis of variance, and the ANOVA least squares model is used, then the dummy variable equal to one for all cases should not be included.

The task before us now is specification of a linear model that relates the scores in Y to the scores in X. From a knowledge of univariate linear models, the logical way to proceed would be to regress each of the Y variables on the group of X variables—since the design of the data collection suggests that the psychological variables may be functions of the background variables. Thus, we might specify the following population models:

$$\begin{aligned} Y_1(i) = {} & b^*_{11}X_1(i) + b^*_{21}X_2(i) + \ldots + b^*_{j1}X_j(i) \\ & + \ldots + b^*_{p1}X_p(i) + e^*_1(i) \end{aligned} \quad (8.4a)$$

$$\begin{aligned} Y_2(i) = {} & b^*_{12}X_1(i) + b^*_{22}X_2(i) + \ldots + b^*_{j2}X_j(i) \\ & + \ldots + b^*_{p2}X_p(i) + e^*_2(i) \end{aligned} \quad (8.4b)$$

$$\vdots$$

$$\begin{aligned} Y_k(i) = {} & b^*_{1k}X_1(i) + b^*_{2k}X_2(i) + \ldots + b^*_{jk}X_j(i) \\ & + \ldots + b^*_{pk}X_p(i) + e^*_k(i) \end{aligned} \quad (8.4c)$$

$$\vdots$$

$$\begin{aligned} Y_q(i) = {} & b^*_{1q}X_1(i) + b^*_{2q}X_2(i) + \ldots + b^*_{jq}X_j(i) \\ & + \ldots + b^*_{pq}X_p(i) + e^*_q(i) \end{aligned} \quad (8.4d)$$

where the kth equation model depicts the kth psychological measure as a linear function of the p background variables and a disturbance term. Construction of a mental map of this model should be enhanced through its graphic depiction in Figure 8.1 as a path diagram. [Recall that $X_1(i)$ is equal to one for all participants; this means that the coefficients $b^*_{11}, b^*_{12}, \ldots, b^*_{1k}, \ldots, b^*_{1q}$ are the intercept terms for the q equations. These coefficients do not enter into the path diagram.]

Equations 8.4 may be written much more compactly in matrix form:

$$\underline{Y} = \underline{X}\,\underline{B}^* + \underline{E}^* \quad (8.5)$$

where $\underline{B}^*$ is the $(p \times q)$ matrix of population coefficients

$$\underline{B}^* = \begin{bmatrix} b^*_{11} & b^*_{12} & \ldots & b^*_{1k} & \ldots & b^*_{1q} \\ b^*_{21} & b^*_{22} & \ldots & b^*_{2k} & \ldots & b^*_{2q} \\ \vdots & & & & & \\ b^*_{j1} & b^*_{j2} & \ldots & b^*_{jk} & \ldots & b^*_{jq} \\ \vdots & & & & & \\ b^*_{p1} & b^*_{p2} & \ldots & b^*_{pk} & \ldots & b^*_{pq} \end{bmatrix} \quad (8.6)$$

and $\underline{E}^*$ is the $(N \times q)$ matrix of population disturbances

$$\underline{E}^* = \begin{bmatrix} e_1^*(1) & e_2^*(1) & \dots & e_k^*(1) & \dots & e_q^*(1) \\ e_1^*(2) & e_2^*(2) & \dots & e_k^*(2) & \dots & e_q^*(2) \\ \vdots & & & & & \\ e_1^*(i) & e_2^*(i) & \dots & e_k^*(i) & \dots & e_q^*(i) \\ \vdots & & & & & \\ e_1^*(N) & e_2^*(N) & \dots & e_k^*(N) & \dots & e_q^*(N) \end{bmatrix} \tag{8.7}$$

Now, if the model 8.5 holds in a population, and if each column of $\underline{E}^*$ has the usual properties of independence, zero mean, and constant variance within each combination of levels of the X variables, then the matrix $\underline{B}^*$ may be estimated with least squares regression techniques—and those estimates (the matrix $\underline{B}$) will be BLUE. Furthermore, if the sample size is large, or if the disturbances are jointly normal in distribution, then probabilistically interpretable confidence intervals (and corresponding F ratios for null hypotheses) for elements of $\underline{B}$ (the least squares estimate of $\underline{B}^*$) may be calculated from the sample data.

What distinguishes this multivariate model from a series of univariate models is not the values of the estimates ($\underline{B}$) of the model coefficients ($\underline{B}^*$). The estimates obtained from a multivariate regression will be identical to those that would be obtained from regressing each Y variable on the group of X variables—the usual univariate multiple regression. Rather, what the multivariate model achieves is the capacity to calculate confidence intervals, or test null hypotheses, involving coefficients from all the univariate equations. For example, it would be possible to test the null hypothesis that all the slopes in $\underline{B}^*$ are zero; that is, that all of the observed coefficients relating psychological measures to background characteristics are the result of random fluctuations.

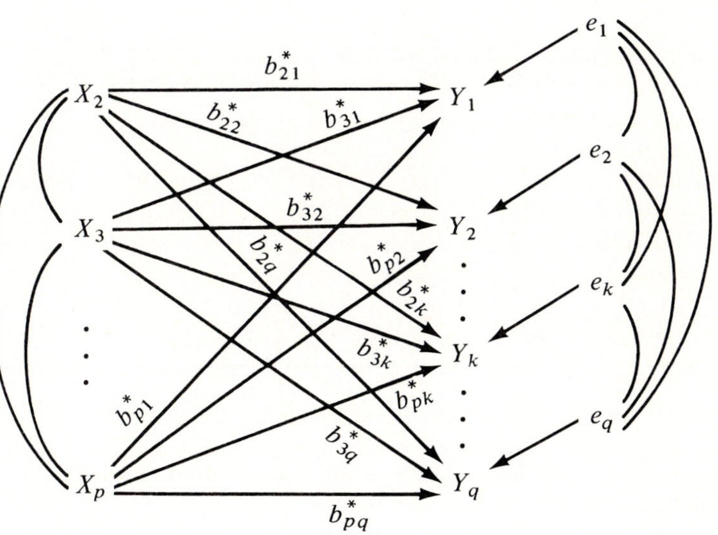

Figure 8.1 Path diagram of a multivariate regression model (see equations 8.4).

When the number of Y variables is large, this could be a useful bit of information to have in hand before proceeding to interpret individual coefficients that would be significant if they were the only ones measured or if they had been singled out in an *a priori* hypothesis.

A Less General Example The above equations are written in a very general form that is rather confusing at first encounter. It should be helpful to reduce this problem to a more manageable size. Suppose, for instance, that our design is reduced to the case of two psychological tests and two background variables:

$Y_1(i)$ = score for the ith person on a scale measuring Ego Strength

$Y_2(i)$ = score for the ith person on a scale measuring Alienation

$X_1(i)$ = 1 (this is a dummy variable for the intercept)

$X_2(i)$ = number of years of education completed by the ith person

$X_3(i)$ = occupational prestige score of father of ith person

Now suppose that the following model holds for the population of persons sampled

$$Y_1(i) = b^*_{11} + b^*_{21}X_2(i) + b^*_{31}X_3(i) + e^*_1(i) \qquad (8.8a)$$

$$Y_2(i) = b^*_{12} + b^*_{22}X_2(i) + b^*_{32}X_3(i) + e^*_2(i) \qquad (8.8b)$$

where the variable X_1 has been deleted since it is equal to one. The coefficients in equation 8.8a could be estimated by the usual procedures of OLS regression from a random sample. The coefficients in equation 8.8b could be similarly estimated.

The model (8.8) may be rewritten as

$$\underline{Y} = \underline{X}\,\underline{B}^* + \underline{E}^* \qquad (8.9)$$

or

$$\begin{bmatrix} Y_1(1) & Y_2(2) \\ Y_1(2) & Y_2(2) \\ \vdots & \\ Y_1(i) & Y_2(i) \\ \vdots & \\ Y_1(N) & Y_2(N) \end{bmatrix} = \begin{bmatrix} 1 & X_2(1) & X_3(1) \\ 1 & X_2(2) & x_3(2) \\ & \vdots & \\ 1 & X_2(i) & X_3(i) \\ & \vdots & \\ 1 & X_2(N) & X_3(N) \end{bmatrix} \begin{bmatrix} b^*_{11} & b^*_{12} \\ b^*_{21} & b^*_{22} \\ b^*_{31} & b^*_{32} \end{bmatrix} + \begin{bmatrix} e^*_1(1) & e^*_2(1) \\ e^*_1(2) & e^*_2(2) \\ \vdots & \\ e^*_1(i) & e^*_2(i) \\ \vdots & \\ e^*_1(N) & e^*_2(N) \end{bmatrix}$$

where $\underline{Y}$ and $\underline{E}^*$ are $(N \times 2)$ rectangular matrices, $\underline{X}$ is an $(N \times 3)$ rectangular matrix, and $\underline{B}^*$ is a (3×2) rectangular matrix (i.e., $p = 3$ and $q = 2$). In matrix form, the least squares estimates of $\underline{B}^*$ are obtained from the sample regression

$$\underline{Y} = \underline{X}\,\underline{B} + \underline{E} \qquad (8.10)$$

where $\underline{B}$ and $\underline{E}$ are sample estimates of the corresponding matrices in the population model. The least squares solution for $\underline{B}$ is

$$\underline{B} = (\underline{X}'\underline{X})^{-1}(\underline{X}'\underline{Y}) \qquad (8.11)$$

where the resulting residuals [$e_1(i)$ and $e_2(i)$] are uncorrelated with the X variables, but may be correlated with one another [i.e., $S(e_1,e_2)$, the covariance of e_1 and e_2, may be non-zero]. Equation 8.11 is simply a shorthand way of writing the *normal equations* that were developed in Chapter 6. The first expression on the right side of equation 8.11 is analogous to the inverse of the covariance matrix of the X variables, and the second expression in parentheses is analogous to the YX covariance matrix. The difference between this equation and the *univariate* normal equations is that the matrix $(\underline{X}'\underline{Y})$ contains covariances (or cross products) between *several* Y variables and the X variable. Thus, equation 8.11 is essentially a cluster of univariate normal equations. (As specified above, these matrices are cross product matrices that become covariance matrices when variables are transformed to deviations about their respective means.)

Now suppose that data on a sample had been gathered and that the following values of $\underline{B}$ were obtained:

$$Y_1 = 1 + 2X_2 + .5X_3 + e_1$$

$$Y_2 = 59.9 - .8X_2 - 1X_3 + e_2$$

or

$$\underline{B} = \begin{bmatrix} 1.0 & 59.9 \\ 2.0 & -.8 \\ .5 & -1.0 \end{bmatrix}$$

Now, if we wanted to obtain an F ratio for the null hypothesis that $b^*_{21} = b^*_{31} = 0$, we could do so using the methods developed in Chapter 6. However, suppose that before proceeding with any specific *a posterori* hypothesis testing, we would like to reject the null hypothesis that

$$b^*_{21} = b^*_{31} = b^*_{22} = b^*_{32} = 0$$

In other words, before investigating specific coefficients, an investigator may want to know the probability of observing these sample values *as a group* if all the corresponding population parameters are zero. In a manner somewhat analogous to the procedures used for testing constraints in univariate models, this multivariate null hypothesis can be tested by placing the constraint

$$b_{21} = b_{31} = b_{22} = b_{32} = 0$$

on the regression model and noting the deterioration in fit (as the reduction in the R-square for constrained models was the focus in the univariate case). The deterioration of fit is measured by a function of the residuals, but the function is a bit more complex than in the univariate case (its computation is best left to electronic devices).

This discussion of hypothesis testing brings us to the next topic of interest: the estimate of type one error probabilities in the multivariate case.

8.3 MULTIVARIATE HYPOTHESIS TESTING

The real advantage of the multivariate regression model over the univariate model is the potential for hypothesis tests involving several dependent variables. The procedure for such hypothesis testing is to formulate a null hypothesis the rejection of which supports the research hypothesis. This null hypothesis is then translated into one or more constraints that can be placed on the model. If those constraints result in a deterioration of the "fit" of the model that is greater than would be expected by chance (one out of 20 times, say), then the null hypothesis may be rejected in favor of an alternative model.

The reader will recall that the measure of how well a univariate model fits the data was the R-square, which was an inverse function of the variance of the residual. In the multivariate case, it might seem reasonable to use the *sum* of the several R-squares as the criterion of the overall fit of the model. Then the constraints could be placed on the model and a new sum of the R-squares computed. In fact, this line of reasoning is quite close to the more exact methods that have been developed by mathematical statisticians (especially Wilks, 1932). And those more exact methods reduce to a function of the R-squares for each equation when the residuals are uncorrelated between equations. But such a procedure would not be appropriate in general because it does not take account of the information that is usually of particular interest in research designs involving multiple dependent variables—that is, the correlation among the dependent variables and the correlation among the residuals. Thus, the index for comparing the fit of an unconstrained model with a constrained model turns out to be a function of the covariance matrix of the residuals. This matrix will be labeled $\underline{S}(EE)$ and may be depicted as follows:

$$\underline{S}(EE) = (\underline{E}'\underline{E})/N \tag{8.12}$$

or

$$\underline{S}(EE) = \begin{bmatrix} S(e_1e_1) & S(e_1e_2) & \dots & S(e_1e_k) & \dots & S(e_1e_q) \\ S(e_2e_1) & S(e_2e_2) & \dots & S(e_2e_k) & \dots & S(e_2e_q) \\ \vdots & & & & & \\ S(e_ke_1) & S(e_ke_2) & \dots & S(e_ke_k) & \dots & S(e_ke_q) \\ \vdots & & & & & \\ S(e_qe_1) & S(e_qe_2) & \dots & S(e_qe_k) & \dots & S(e_qe_q) \end{bmatrix}$$

where

$$\begin{aligned} S(e_1e_2) &= ([e_1(1) \times e_2(1)] + [e_1(2) \times e_2(2)] \\ &\quad + \dots + [e_1(N) \times e_2(N)]/N \\ &= \text{the sample covariance of } e_1 \text{ and } e_2 \end{aligned}$$

and so on. Note that the diagonal elements of this matrix are covariances of residuals from the same equation; the diagonals are thus variances of residuals from the respective equations and are closely linked to the R-square for each of the q equations in the multivariate model. In fact, if all the dependent variables are standardized (zero mean, unit variance), then the kth diagonal element of $\underline{S}(EE)$ is equal to (1 − R-square) for the kth equation in the model (equation 8.4).

It may help the reader to conceptualize the origins of $\underline{S}(EE)$ if it is recalled that there is a residual score generated for each case in a multiple regression analysis.

Estimation of each of the equations in model 8.4, then, would generate N sample residuals. The covariances and variances of these residuals across the N cases would constitute the elements of $\underline{S}(EE)$.

The way in which $\underline{S}(EE)$ is transformed into an index of the fit of a model is through the computation of the *determinant* of the matrix. The determinant of a 2 × 2 matrix is defined as follows:

$$\text{if } \underline{H} = \begin{bmatrix} a & b \\ c & d \end{bmatrix}$$

then the determinant of $\underline{H}$ is given by $\det(\underline{H}) = ad - bc$. For a 3 × 3 matrix

$$\text{if } \underline{H} = \begin{bmatrix} a & b & c \\ d & e & f \\ g & h & i \end{bmatrix}$$

then

$$\det(\underline{H}) = a(ei - fh) - b(di - fg) + c(dh - eg)$$

Thus, when a multivariate model includes two Y variables ($q = 2$) the index of fit of the model is

$$\det[\underline{S}(EE)] = S(e_1e_1) \times S(e_2e_2) - S(e_1e_2)^2 \tag{8.13}$$

The meaning of equation 8.13 is clarified to some extent if the case of standardized dependent variables is considered. When Y_1 and Y_2 have unit variance and zero mean, then equation 8.13 may be written as

$$\det[\underline{S}(EE)] = (1 - R_1^2) \times (1 - R_2^2) - S(e_1e_2)^2 \tag{8.14}$$

where

R_1^2 = R-square for the first Y variable in the model

R_2^2 = R-square for the second Y variable in the model

In this form, it should be clear that the determinant of the residual covariance matrix becomes smaller as the R-squares for the individual equations become larger. In the extreme case, the determinant is zero when the two R-squares are equal to one. Thus, as the R-squares for the individual equations in 8.4 increase, the value of the determinant will decrease. But the value of the determinant is also a function of the covariance(s) of the residuals. As equation 8.14 makes clear, the determinant also decreases in magnitude as the magnitude of the covariance increases. Identification of these two sources of variation in the multivariate index of model "fit" suggests the following: the multivariate model treats unexplained (or "unfitted") variance in the dependent variables as that which is not accounted for by the independent variables *or the intercorrelation of the dependent variables.* In other words, the multivariate model provides the opportunity to reduce the amount of variability attributed to "noise" (and thus the standard errors of estimation) when the various dependent variables are correlated with one another. The logic behind these procedures is somewhat analogous to that behind the analysis of covariance, in which non-causal variables were added to a model to reduce the variance of the residual (the unexplained variation of the dependent variable).

Comparing the Fit of Two Models

Now that an index of the fit of a multivariate linear model to sample data has been described, we are prepared to outline procedures for comparing two models. The purpose of this comparison will be to test null hypotheses. One of the two models being compared will be *un*constrained, while the other will have constraints corresponding to the appropriate null hypothesis. If the fit of the constrained model is much worse than the unconstrained model, the constraints can then be rejected as plausible hypotheses—and the unconstrained model survives the test. (It should be pointed out that the model treated as unconstrained in one comparison may actually be constrained relative to a less constrained model. Thus, the labels constrained and unconstrained are relative in meaning.)

As an introduction to the testing of constraints in multivariate linear models, consider the case in which the null hypothesis is the most complete. That is, the case of the null hypothesis that all coefficients in the model (except the intercepts) are zero:

$$H_0: \underline{B}^* = \begin{bmatrix} b^*_{11} & b^*_{12} & \cdots & b^*_{1q} \\ 0 & 0 & \cdots & 0 \\ 0 & 0 & \cdots & 0 \\ \vdots & & & \\ 0 & 0 & \cdots & 0 \end{bmatrix}$$

which is equivalent to the null hypothesis that all the R-squares for the q equations in 8.4 are zero (jointly). In a manner analogous to the procedures outlined in Chapter 7, we proceed by estimating a multivariate linear model that incorporates the constraints specified in the null hypothesis. That is, we constrain all the coefficients in $\underline{B}$ to be zero, and compute the residuals for each equation in the model. In this case, in which only the intercept terms remain, the residuals are simply equal to the original dependent variables transformed to mean deviation scores. Thus, the covariance matrix of the residuals for the constrained model will simply be the covariance matrix of the dependent variables.

The next step in the procedure is to calculate the determinant of the covariance matrix for the residuals of the *unconstrained* model and the determinant of the covariance matrix of the residuals for the *constrained* model so that the fit of the two models to the sample data may be compared. This comparison is accomplished by forming a ratio of these two determinants. The ratio is then referred to as Wilk's lambda or the multivariate *U-statistic*. In the case of the null hypothesis that all population coefficients are zero, this ratio would be

$$U = \det[\underline{S}(EE)]/\det[\underline{S}(YY)]$$

where $\underline{S}(YY)$ is the sample covariance matrix of the dependent variables in the model 8.4. For the simplest case of two dependent variables, this ratio may be rewritten as

$$U = [S(e_1e_1)S(e_2e_2) - S(e_1e_2)^2]/[S(Y_1Y_1)S(Y_2Y_2) - S(Y_1Y_2)^2] \quad (8.15)$$

Note that U in equation 8.15 would be one if the sample residuals were equal to the mean deviations of the Y variables (sample coefficients all equal to zero), and U would be zero if the variances of the two residuals were both zero (indicating R-squares of one for both equations). Furthermore, U would decrease in size if the covariance of the residuals were larger than the covariance of Y_1 and Y_2 [i.e., $S(e_1e_2)$ greater in absolute magnitude than $S(Y_1Y_2)$]. Thus, as the fit of the constrained model deteriorated relative to the unconstrained model, the value of the U-statistic would approach zero. Conversely, when the two models fit the data equally well, the U-statistic would be closer to one.

In order to describe the U-statistic more generally, let $\underline{B}_1$ be the sample regression coefficients for model 8.4, and let $\underline{B}_2$ be the regression coefficients for model 8.4 with constraints. Similarly, let $\underline{E}_1$ be the matrix of residuals for the unconstrained model, while $\underline{E}_2$ is the matrix of residuals for the constrained model. Thus,

$$\underline{Y} = \underline{X}\underline{B}_1 + \underline{E}_1 \tag{8.16a}$$

and

$$\underline{Y} = \underline{X}\underline{B}_2 + \underline{E}_2 \tag{8.16b}$$

where equation 8.16a is the unconstrained model and equation 8.16b is the constrained model. We can now define Wilk's lambda (or U-statistic) as

$$U = \det[\underline{S}(E_1E_1)]/\det[\underline{S}(E_2E_2)] \tag{8.17}$$

where $\underline{S}(E_1E_1)$ and $\underline{S}(E_2E_2)$ are the covariance matrices for the residuals of the unconstrained and the constrained models, respectively. Furthermore, if the constraints placed upon equation 8.16a to obtain 8.16b meet an *estimability criterion* (which is usually met with logically sensible constraints), then U has a distribution over random samples that may be used to estimate the probability of the sample observations under the assumption that the null hypothesis (represented by the constrained model) is true in the population. The function describing the distribution of U has three degree of freedom parameters:

s = the number of Y variables included in the null hypothesis (ranges from 1 to q)

r = the number of constrained parameters (ranges from 1 to $p \times q$)

$n = N - p$

Thus, the null hypothesis that all the coefficients in $\underline{B}^*$ are zero (except the intercepts) would be tested with a U-statistic with q, $[q(p - 1)]$, and $(N - p)$ degrees of freedom. If the probability that the observed value of U (or one smaller) would occur when the null hypothesis were true was small enough, then the constrained model would be rejected. The alternative hypothesis, that some of the observed coefficients are estimates of nonzero population coefficients, would then be supported. If a less global null hypothesis was of interest (i.e., one involving fewer than q dependent variables and/or fewer than $p \times q$ constraints), then a U-statistic with r, s, and n degrees of freedom would be computed.

An Approximation of the U-statistic Although equation 8.17 is elegantly simple, the actual distribution of the U-statistic is difficult to present in a table. For this reason,

Rao (1959) developed a procedure whereby the observed U-statistic can be transformed into an F ratio—and the probability of this F ratio (or one greater) can be used as an approximation of the probability of the observed U-statistic (or one smaller) if the null hypothesis is true.

The approximate F-statistic is computed from the observed U-statistic (with s, r, and n degrees of freedom) as follows:

$$\text{approx } F = [(1 - u)/u][h/rs] \qquad df = rs, h \tag{8.18}$$

where

$$\begin{aligned} u &= U^{1/t} \\ t &= [(r^2s^2 - 4)/(r^2 + s^2 - 5)]^{.5} && \text{if defined} \\ &= 1 && \text{if } r^2 + s^2 = 5 \\ h &= [N - p - (s - r + 1)/2]t - (rs/2) + 1 \end{aligned}$$

(It should be noted that this approximation is exact if r or s is 1 or 2.) Fortunately, this approximation is computed for the user of most multivariate regression programs (e.g., BMD11V and MULTIVARIANCE).

Depicting Constraints in Matrix Format For ease of presentation in the rest of this chapter and for the use of multivariate regression programs, it is helpful to express the constraints stemming from the null hypothesis in matrix format. We begin with the matrix of population coefficients $\underline{B}^*$ in model 8.4. For the unconstrained model, it is assumed that none of the elements of $\underline{B}$ (the sample estimate of $\underline{B}^*$) are constrained. This means that no statements about specific values for elements of $\underline{B}^*$ have been specified. For a constrained model, however, a null hypothesis specifies that certain elements of $\underline{B}^*$ have certain values (such as zero) or that certain linear combinations (i.e., *contrasts*) of elements of $\underline{B}^*$ have certain values. This null hypothesis can then be translated into constraints that are placed on the sample estimates of $\underline{B}^*$. The sample estimates under the constraints are the matrix $\underline{B}_1$.

In Chapter 7, the constraints to be placed on regression coefficients were stated in terms of equalities between a coefficient and a constant (such as zero) or contrasts among two or more coefficients. For continuity with developments in later chapters, a model parameter (i.e., element of $\underline{B}^*$) that is fixed to some specific value under the null hypothesis is referred to as a *fixed parameter;* parameters that are constrained by virtue of a linear contrast are referred to as *constrained parameters.* Thus, the null hypothesis that a particular element of $\underline{B}^*$ is zero would make that element a fixed parameter of the constrained model. Similarly, the null hypothesis that two elements of $\underline{B}^*$ were equal would make them constrained (or contrasted) parameters. Both these types of constraints can be summarized efficiently through the specification of three additional matrices—$\underline{A}$, $\underline{C}$, and $\underline{D}$ where

$$\underline{D} = \underline{A}\,\underline{B}^*\underline{C} \tag{8.19}$$

expresses the null hypothesis and

$$\underline{D} = \underline{A}\,\underline{B}_1\underline{C} \tag{8.20}$$

expresses the constraints placed upon the least squares estimates of the null hypothesis model (i.e., $\underline{B}^*$ with fixed and/or contrasted elements).

The matrix $\underline{A}$ defines which X variables are to be involved in the constraints—or, more precisely, $\underline{A}$ specifies the particular rows of $\underline{B}^*$ that are to have fixed and/or constrained elements. The matrix $\underline{A}$, thus, has r rows and p columns; when the multivariate regression model is applied to the analysis of variance, using the ANOVA least squares approach, the number of rows in $\underline{A}$ indicates the number of linear contrasts to be tested; but for OLS type regression models, the number r is usually equal to the number of X variables *in each equation* of model 8.4 that have fixed coefficients under the null hypothesis.

The matrix $\underline{C}$ in equations 8.19 and 8.20 specifies which of the equations in the model 8.4 will be involved in the null hypothesis. For example, a multivariate regression model with five dependent variables may be the basis for a null hypothesis involving only three of the five linear equations in the model. The number of *columns* in the matrix $\underline{C}$ (i.e., s) is the number of the linear equations in equations 8.4 that are included in the null hypothesis.

Finally, the matrix $\underline{D}$ defines the numerical values that are assigned to fixed and constrained elements of $\underline{B}^*$. In most instances, $\underline{D}$ is a matrix of zeros—but not necessarily. The matrix $\underline{D}$ will have r rows and s columns, indicating that the total number of constraints is rs. In other words, each element of $\underline{D}$ is the left-hand side of an equation that expresses a constraint or a part of a null hypothesis. The reader who is not accustomed to thinking in terms of matrices may find this specification of the constraints a bit perplexing. However, with the following examples and a bit of practice, I have found that most researchers can master the use of these matrices.

To demonstrate the use of matrices to specify constraints in a more concrete way, recall the model presented in equations 8.8. That model included two personality measures as dependent variables and two background variables as independent variables (and a dummy variable for the intercept). For the sake of generality, suppose that the number of dependent measures was increased to three and the number of background variables was increased to three—meaning that four X variables will be included in the model, since the dummy variable X_1 is needed to estimate an intercept. The population model could then be written as

$$Y_1 = b^*_{11} + b^*_{21}X_2 + b^*_{31}X_3 + b^*_{41}X_4 + e^*_1 \tag{8.21a}$$

$$Y_2 = b^*_{12} + b^*_{22}X_2 + b^*_{32}X_3 + b^*_{42}X_4 + e^*_2 \tag{8.21b}$$

$$Y_3 = b^*_{13} + b^*_{23}X_2 + b^*_{33}X_3 + b^*_{43}X_4 + e^*_3 \tag{8.21c}$$

or more compactly,

$$\underset{(N \times 3)}{\underline{Y}} = \underset{(N \times 4)}{\underline{X}}\underset{(4 \times 3)}{\underline{B}^*} + \underset{(N \times 3)}{\underline{E}^*} \tag{8.22}$$

where

$$\underline{B}^* = \begin{bmatrix} b^*_{11} & b^*_{12} & b^*_{13} \\ b^*_{21} & b^*_{22} & b^*_{23} \\ b^*_{31} & b^*_{32} & b^*_{33} \\ b^*_{41} & b^*_{42} & b^*_{43} \end{bmatrix} \tag{8.23}$$

Notice that the columns of equation 8.23 correspond to the coefficients of the separate linear equations in equations 8.21. Similarly, the rows of equation 8.23 correspond to the coefficients for a particular X variable. Column 1 of equation 8.23 includes the coefficients for the regression of the first personality measure on the three background measures and row 3 the three coefficients relating the second background measure to each of the three personality measures.

Now, the task at hand is to see how various null hypotheses concerning $\underline{B}^*$ in equation 8.23 can be expressed with the matrices $\underline{A}$, $\underline{C}$, and $\underline{D}$. Consider the following three examples:

Null Hypothesis Number 1 All the coefficients in $\underline{B}^*$ are zero except for the intercepts. In matrix form

$$H_0\colon \underline{B}^* = \begin{bmatrix} b^*_{11} & b^*_{12} & b^*_{13} \\ 0 & 0 & 0 \\ 0 & 0 & 0 \\ 0 & 0 & 0 \end{bmatrix}$$

This first null hypothesis specifies that nine of the 12 elements in $\underline{B}^*$ are zero—only the intercepts are free parameters. The null hypothesis could also be stated in the form

$$H_0\colon b^*_{21} = 0,\ b^*_{22} = 0, \ldots,\ b^*_{43} = 0$$

Finally, this null hypothesis could be stated in the form

$$\underline{D} = \underline{A}\,\underline{B}^*\underline{C}$$

or

$$\begin{bmatrix} 0 & 0 & 0 \\ 0 & 0 & 0 \\ 0 & 0 & 0 \end{bmatrix} = \begin{bmatrix} 0 & 1 & 0 & 0 \\ 0 & 0 & 1 & 0 \\ 0 & 0 & 0 & 1 \end{bmatrix} \begin{bmatrix} b^*_{11} & b^*_{12} & b^*_{13} \\ b^*_{21} & b^*_{22} & b^*_{23} \\ b^*_{31} & b^*_{32} & b^*_{33} \\ b^*_{41} & b^*_{42} & b^*_{43} \end{bmatrix} \begin{bmatrix} 1 & 0 & 0 \\ 0 & 1 & 0 \\ 0 & 0 & 1 \end{bmatrix}$$

If the reader performs the above matrix multiplication, it will be clear that this last equation expresses the null hypothesis. The "trick" in this procedure, of course, is to figure out which values should be placed in the three *hypothesis matrices* ($\underline{A}$, $\underline{C}$, and $\underline{D}$). Some more examples should assist the reader in ferreting out the cognitive principles involved.

Null Hypothesis Number 2 The three coefficients relating X_2 (the first background measure) to the three personality measures are zero. That is

$$H_0\colon \underline{B}^* = \begin{bmatrix} b^*_{11} & b^*_{12} & b^*_{13} \\ 0 & 0 & 0 \\ b^*_{31} & b^*_{32} & b^*_{33} \\ b^*_{41} & b^*_{42} & b^*_{43} \end{bmatrix}$$

or

$$b^*_{21} = 0,\ b^*_{22} = 0,\ \text{and}\ b^*_{23} = 0$$

And this null hypothesis may be restated in terms of the hypothesis matrices as follows

$$\underline{D} = \underline{A}\,\underline{B}^*\underline{C}$$

or

$$[0 \quad 0 \quad 0] = [0 \quad 1 \quad 0 \quad 0] \begin{bmatrix} b^*_{11} & b^*_{12} & b^*_{13} \\ b^*_{21} & b^*_{22} & b^*_{23} \\ b^*_{31} & b^*_{32} & b^*_{33} \\ b^*_{41} & b^*_{42} & b^*_{43} \end{bmatrix} \begin{bmatrix} 1 & 0 & 0 \\ 0 & 1 & 0 \\ 0 & 0 & 1 \end{bmatrix}$$

Null Hypothesis Number 3 $b^*_{32} = b^*_{42}$. Although this particular constraint would probably not make sense in the current example, there are instances when a null hypothesis involves the equality of two coefficients (especially when the ANOVA least squares approach to ANOVA is used—this procedure will be demonstrated in Section 8.5 of this chapter). This null hypothesis may be stated as

$$H_0\colon \underline{B}^* = \begin{bmatrix} b^*_{11} & b^*_{12} & b^*_{13} \\ b^*_{21} & b^*_{22} & b^*_{23} \\ b^*_{31} & b^*_{32} & b^*_{33} \\ b^*_{41} & \underline{b}^*_{32} & b^*_{43} \end{bmatrix}$$

Note that the equating of b^*_{32} and b^*_{42} has been expressed by using the same label (b^*_{32}) for both coefficients. In terms of the hypothesis matrices, this null hypothesis is stated as

$$\underline{D} = \underline{A}\,\underline{B}^*\underline{C}$$

or

$$[0] = [0 \quad 0 \quad 1 \quad -1]\underline{B}^* \begin{bmatrix} 0 \\ 1 \\ 0 \end{bmatrix}$$

Note here that the elements of $\underline{A}$ are used to subtract one element of $\underline{B}^*$ from another, $\underline{C}$ is used to select that difference from the second equation in the model, and $\underline{D}$ is used to set that difference to zero (as specified by the null hypothesis that $b^*_{32} = b^*_{42}$ or $b^*_{32} - b^*_{42} = 0$).

The General Form of Wilk's Lambda (U-statistic) Now that the use of hypothesis matrices has been outlined, the procedure for comparing the fit of a constrained and unconstrained multivariate regression model may be stated in quite general terms. If the model

$$\underline{Y} = \underline{X}\,\underline{B}^* + \underline{E}^*$$

holds for some population (and the disturbances satisfy the usual assumptions of independence, zero mean, and homoscedasticity within equations), *and* the null hypothesis

$$\underline{D} = \underline{A}\,\underline{B}^*\underline{C}$$

also holds, then

$$U = \det[\underline{C}'\underline{S}(EE)\underline{C}]/\det[\underline{C}'\underline{S}(E_1E_1)\underline{C}] \qquad (8.24)$$

is distributed as Wilk's lambda with s, r, and n degrees of freedom where

$\underline{S}(EE)$ = the covariance matrix for the sample residuals using $\underline{B}$ to estimate $\underline{B}^*$

$\underline{S}(E_1E_1)$ = the covariance matrix for sample residuals using $\underline{B}_1$ (the constrained least squares coefficients) to estimate $\underline{B}^*$

r = the number of rows in the hypothesis matrix $\underline{A}$

s = the number of columns in the hypothesis matrix $\underline{C}$

$n = N - p$ (N = total sample size; p = number of X variables in the model—including intercept)

$\underline{C}'$ = the transpose of $\underline{C}$ (i.e., rows and columns are interchanged)

This U-statistic then ranges from zero to one. A value close to one indicates that the constrained model (null hypothesis) fits the sample data about as well as the unconstrained model. A value of U that is closer to zero indicates that the fit of the constrained model is poor relative to the unconstrained model. If the constraints hold in the population (the null hypothesis) and the sample size is large, or the population residuals/disturbances are multivariate normal in distribution, then the probability of observing a U-statistic as small as (or smaller than) the observed one is approximated by Rao's approximate F ratio (equation 8.18).

The conclusion to all this would be a decision concerning the constrained and unconstrained $\underline{B}_1$ and $\underline{B}$, respectively. If U is sufficiently small, then the likelihood of the actual observations is so small under the null hypothesis that the constraints corresponding to that null hypothesis must be rejected as plausible alternatives to the observed values of the coefficients (or their observed contrasts).

8.4 A NONEXPERIMENTAL EXAMPLE OF MULTIVARIATE REGRESSION

The first three sections of this chapter have presented the multivariate regression model in abstract terms. Thus, it is important to reiterate that, from the estimation point of view, the multivariate model has added little to the (univariate) multiple regression model. The multivariate model with q dependent variables yields the same sample estimates of population coefficients as a series of q multiple regressions—one for each dependent variable regressed on the same set of p independent variables. Furthermore, the standard errors of individual regression coefficients in the multivariate model are the same as those that would be obtained from the univariate multiple regression including the coefficient. [This fact can be verified by using equation 8.24 to test the null hypothesis that one coefficient in $\underline{B}^*$ is zero. Rao's approximate F ratio (equation 8.18) is exact in this instance, and the resulting F ratio will be equal to the square of the tested coefficient divided by its standard error. This standard error will be the same as would have been obtained from the application of procedures outlined in Chapter 6 or Chapter 7.] The only real advantages of the multivariate model stem from the opportunity to test null hypotheses involving coefficients from different linear equations.

One context in which the capabilities of the multivariate model are of interest is the nonexperimental study of numerous variables. When many multiple regression analyses have been extracted from a data set, an investigator begins to wonder about the extent to which he is capitalizing on chance variation. The multivariate model provides rough estimates of the likelihood that a series of linear regression results would have occurred if all (or some subset) of the corresponding population parameters were zero. The following empirical example has been contrived to demonstrate such an application of the multivariate regression model. Since it involves only 20 cases and four observed variables, the computations are easier to follow and may be checked with a standard program in less than an hour.

Intergenerational Mobility

One of the most studied topics in American social science is the relationship between status measures for parents and status measures for their children. Researchers have hoped to shed light on the question of how much a person's social origin is associated with adult social position and on the question of how that association comes about. A typical approach in this tradition is to sample a population of men and obtain information concerning the educational and financial position of both the respondent and the respondent's father. Suppose that the data in Table 8.1 resulted from such a random sample of 20 men.

Table 8.1 Hypothetical scores for 20 respondents on two dependent (Y_1 and Y_2) and two independent variables (X_1 and X_2).

Respondent	X_1	X_2	X_3	Y_1	Y_2	Y_2'
1	1.000	4.000	3.200	9.000	14.36	14.36
2	1.000	4.000	3.200	5.000	8.36	8.36
3	1.000	5.000	4.800	12.00	15.84	9.84
4	1.000	5.000	4.800	8.000	9.84	15.84
5	1.000	6.000	4.000	11.00	15.40	15.40
6	1.000	6.000	4.000	7.000	9.40	9.40
7	1.000	7.000	5.600	14.00	16.88	10.88
8	1.000	7.000	5.600	9.000	10.88	16.88
9	1.000	8.000	3.600	11.00	15.48	15.48
10	1.000	8.000	3.600	7.000	9.48	9.48
11	1.000	9.000	6.400	15.00	17.92	11.92
12	1.000	9.000	6.400	11.00	11.92	17.92
13	1.000	10.00	4.400	13.00	16.52	16.52
14	1.000	10.00	4.400	9.000	10.52	10.52
15	1.000	11.00	6.000	15.00	18.00	12.00
16	1.000	11.00	6.000	11.00	12.00	18.00
17	1.000	12.00	5.200	14.00	17.56	17.56
18	1.000	12.00	5.200	10.00	11.56	11.56
19	1.000	13.00	6.800	17.00	19.04	13.04
20	1.000	13.00	6.800	13.00	13.04	19.04

Note. Y_1 = respondent's education, Y_2 = respondent's income, X_1 = dummy variable, X_2 = father's education, X_3 = father's income

The data for the 20 respondents consist of scores on four variables:

Y_1 = the respondent's education in years

Y_2 = the respondent's annual income in \$thousands

X_1 = 1 (dummy variable for intercept)

X_2 = father's education in years completed

X_3 = father's annual income when respondent was 15 in \$thousands

The problem of measurement error in such instances is almost overwhelming (models developed in Chapter 10 can be used to attack these problems), but we shall proceed as if the measures are valid.

The univariate regression approach to such a problem would be to regress Y_1 on X_1 and X_2 and regress Y_2 on X_1 and X_2. An investigator may interpret the resulting coefficients as estimates of structural coefficients in a causal model. However, the more likely interpretation would be in terms of slopes.

If we estimate a multivariate linear model with these data, the resulting estimates of model coefficients will be identical to those resulting from two multiple regression analyses. We begin by specifying the multivariate regression model for the population:

$$Y_1 = b^*_{11}X_1 + b^*_{21}X_2 + b^*_{31}X_3 + e^*_1 \tag{8.25a}$$

$$Y_2 = b^*_{12}X_1 + b^*_{22}X_2 + b^*_{32}X_3 + e^*_2 \tag{8.25b}$$

where equations 8.25 hold for every person in the population and (e^*_1, e^*_2) are well behaved. Note that this model does not address the relationship between the respondent's education (Y_1) and the respondent's income (Y_2). Only the coefficients relating the respondent's characteristics to the father's characteristics are addressed by this model. The least squares estimate of 8.25 from the data in Table 8.1 is:

$$\begin{aligned} Y_1 = \underset{(2.24)}{1.67} + \underset{(.24)}{.28}X_2 + \underset{(.59)}{1.40}X_3 + e_1 \end{aligned} \tag{8.26a}$$

$$\text{R-square} = .54$$

$$\begin{aligned} Y_2 = \underset{(3.25)}{8.00} + \underset{(.34)}{.20}X_2 + \underset{(.86)}{.80}X_3 + e_2 \end{aligned} \tag{8.26b}$$

$$\text{R-square} = .17$$

In matrix format, equations 8.26 may be written as

$$[Y_1 \quad Y_2] = [X_1 \quad X_2 \quad X_3] \begin{bmatrix} 1.67 & 8.00 \\ .28 & .20 \\ 1.40 & .80 \end{bmatrix} + [e_1 \quad e_2] \tag{8.27}$$

or

$$\underline{Y} = \underline{X}\,\underline{B} + \underline{E}$$

Now we notice in equations 8.26 that only one of the four slope estimates is larger than twice its standard error. This fact may motivate an investigator to wonder how likely the occurrence of all four of these sample coefficients would be if the corresponding population coefficients were all zero. An estimate of this probability can be obtained from the fitting of a constrained model (where the four slopes are constrained to zero) to the sample data and comparing the fit of the constrained model to that of the unconstrained model in equation 8.27.

We begin by calculating the covariance matrix for the residuals of the unconstrained model:

$$\underline{S}(EE) = \begin{bmatrix} S(e_1e_1) & S(e_1e_2) \\ S(e_2e_1) & S(e_2e_2) \end{bmatrix} = \begin{bmatrix} 4.27 & 6.15 \\ 6.15 & 9.00 \end{bmatrix} \tag{8.28}$$

$S(e_1e_1)$ = the sample variance of the residuals resulting from the regression of Y_1 on X_1, X_2, and X_3 (equation 8.26a)

$S(e_2e_2)$ = the sample variance of the residuals resulting from the regression of Y_2 on X_1, X_2, and X_3 (equation 8.26b)

$S(e_1e_2)$ = $S(e_2e_1)$ = the *covariance* of the sample residuals from the two regressions

Now we proceed to estimate $\underline{B}^*$ with the constraints implied by the null hypothesis and then compare the covariance matrix of the residuals with that obtained in equation 8.28.

The null hypothesis is that

$$b^*_{21} = b^*_{22} = b^*_{31} = b^*_{32} = 0$$

or

$$\underline{B}^* = \begin{bmatrix} b^*_{11} & b^*_{12} \\ b^*_{21} & b^*_{22} \\ b^*_{31} & b^*_{32} \end{bmatrix} = \begin{bmatrix} b^*_{11} & b^*_{12} \\ 0 & 0 \\ 0 & 0 \end{bmatrix} \tag{8.29}$$

The null hypothesis stated in equation 8.29 could also be stated in terms of hypothesis matrices as follows:

$$H_0\colon \underline{D} = \underline{A}\,\underline{B}^*\underline{C}$$

$$H_0\colon \begin{bmatrix} 0 & 0 \\ 0 & 0 \end{bmatrix} = \begin{bmatrix} 0 & 1 & 0 \\ 0 & 0 & 1 \end{bmatrix} \begin{bmatrix} b_{11} & b^*_{12} \\ b^*_{21} & b^*_{22} \\ b^*_{31} & b^*_{32} \end{bmatrix} \begin{bmatrix} 1 & 0 \\ 0 & 1 \end{bmatrix} \tag{8.30}$$

(Notice that the matrix $\underline{A}$ is used to select the second and third rows of $\underline{B}^*$, while $\underline{C}$ is used to select both columns of $\underline{B}^*$.)

Now that the null hypothesis has been stated, the constraints that define the constrained model can be easily defined. The constrained model may be written as

$$\underline{Y} = \underline{X}\,\underline{B}_1 + \underline{E}_1$$

or

$$[Y_1 \quad Y_2] = [X_1 \quad X_2 \quad X_3] \begin{bmatrix} b_{11} & b_{12} \\ b_{21} & b_{22} \\ b_{31} & b_{32} \end{bmatrix} + [e_3 \quad e_4] \tag{8.31}$$

where

$$\underline{B}_1 = \begin{bmatrix} b_{11} & b_{12} \\ b_{21} & b_{22} \\ b_{31} & b_{32} \end{bmatrix} = \begin{bmatrix} b_{11} & b_{12} \\ 0 & 0 \\ 0 & 0 \end{bmatrix}$$

or

$$\begin{bmatrix} 0 & 0 \\ 0 & 0 \end{bmatrix} = \begin{bmatrix} 0 & 1 & 0 \\ 0 & 0 & 1 \end{bmatrix} \begin{bmatrix} b_{11} & b_{12} \\ b_{21} & b_{22} \\ b_{31} & b_{32} \end{bmatrix} \begin{bmatrix} 1 & 0 \\ 0 & 1 \end{bmatrix}$$

The estimation procedure resulting in equation 8.31 is quite straightforward in this instance, since only the intercept terms are nonzero in the constrained model. Fitting equation 8.31 to the data in Table 8.1 with least squares methods gives

$$\underline{B}_1 = \begin{bmatrix} 11.05 & 13.70 \\ 0.00 & 0.00 \\ 0.00 & 0.00 \end{bmatrix} \tag{8.32}$$

where the two intercept terms (the first row of equation 8.32) are the respective means of Y_1 (respondent's education) and Y_2 (respondent's annual income).

Now that the constrained coefficients $\underline{B}_1$ (equation 8.32) have been obtained, the residuals for the constrained model—and their covariances—can be computed. The result is

$$\underline{S}(E_1E_1) = \begin{bmatrix} S(e_3e_3) & S(e_3e_4) \\ S(e_4e_3) & S(e_4e_4) \end{bmatrix} = \begin{bmatrix} 9.25 & 9.21 \\ 9.21 & 10.89 \end{bmatrix} \tag{8.33}$$

where the variances and covariances of the residuals are actually equal to the variances and covariance of the dependent variable (since all slopes are constrained to zero).

We are now in a position to compute a U-statistic from $\underline{S}(EE)$ and $\underline{S}(E_1E_1)$, using equation 8.24:

$$U = \det[\underline{C}'\underline{S}(EE)\underline{C}]/\det[\underline{C}'\underline{S}(E_1E_1)\underline{C}]$$

Since for the particular null hypothesis at hand, $\underline{C}$ is the identity matrix, it is unnecessary to actually perform the pre- and post-multiplication of the residual covariance matrices by $\underline{C}'$ and $\underline{C}$ (multiplication of a matrix by an identity matrix does not change the original matrix). For the reader who has difficulty thinking in terms of matrix algebra, the pre- and post-multiplication procedure simply accomplishes the task of selecting the rows and columns of the residual covariance matrices that correspond to dependent variables included in the null hypothesis. Our current null hypothesis involves all the dependent variables in the model (both Y_1 and Y_2), so all the rows and columns in the residual covariance matrices are included in the computation of the U-statistic. Thus, for the null hypothesis in equation 8.29

$$U = \det \begin{bmatrix} 4.27 & 6.15 \\ 6.15 & 9.00 \end{bmatrix} \Big/ \det \begin{bmatrix} 9.25 & 9.21 \\ 9.21 & 10.89 \end{bmatrix} \tag{8.34}$$

$$= .589/15.818 = .037 \qquad df = 2, 2, 17$$

where the degrees of freedom (s, r, n) are 2,2,17 because two of the dependent variables in the model (s) are included in the null hypothesis, two of the independent variables in the model (r) are included in the null hypothesis, and the model includes three (p) X variables and 20 cases ($N = 20$; $n = N - p$).

Before proceeding to transform this U-statistic to an F-ratio, some aspects of equation 8.34 should be noted. First, the variance of the residuals increases with the constraints for both Y_1 and Y_2, but the increase is substantial in the case of Y_1 and small in the case of Y_2 (an increase from 4.27 to 9.25 compared with an increase from 9.00 to 10.89). This observation is consistent with the R-squares that were obtained in equations 8.26. Second, the covariance of the residuals increases under the constraints (from 6.15 to 9.21), but the correlation between the residuals actually decreases

$$\text{Corr}(e_1,e_2) = 6.15/(2.07 \times 3.00) = .99$$

$$\text{Corr}(e_3,e_4) = 9.21/(3.04 \times 3.30) = .92$$

This means that "within level" correlation of the residuals is greater for the unconstrained than the constrained model, which means that the unconstrained model involves less noise than the constrained model in terms of the relationships between the independent variables and the dependent variables *and* in terms of the correlation among the residuals. We now proceed to computation of Rao's F-approximation to see how likely these reductions would be if the null hypothesis were true.

Rao's F-approximation is obtained by substituting the observed value of U and the degrees of freedom for U into equation 8.18:

$$F = 33.45 \qquad df = 4,\ 32$$

$$\text{Prob}(F > 33.45) < 10^{-10}$$

Thus, the probability of the observations in Table 8-1 being obtained from a random sample of a population in which the null hypothesis (equation 8.30) holds is extremely small. This result could hardly be anticipated from inspection of the individual regression coefficients and their standard errors. Recall that only one of the four slopes was greater than twice its standard error—and yet the observed value of all four slopes is an extremely unlikely event under the null hypothesis. This means that if we considered the four sample slopes to be a point in four-dimensional space, then the 95 percent confidence region about this point would clearly exclude the point (0,0,0,0). [Methods for estimating such multidimensional confidence intervals are given in Bock (1975).] The reason for this extreme discrepancy between the individual and the joint confidence regions is the (unrealistically) high correlation between the residuals in the unconstrained model. Thus, a researcher who had obtained data, such as those in Table 8.1, and who was hesitant to interpret the one significant slope out of the four computed, would be assured by the multivariate analysis that the one significant slope was part of a nonzero matrix of slopes.

An important point should be made at this juncture. It is clearly possible, in fact likely, that highly correlated residuals could lead to a rejection of the multivariate null hypothesis when none of the individual coefficients in the model were significant (none were larger than twice their standard error). This situation arises when some *linear combination* of the dependent variables would have a significant R-square with the X variables in the model—even though none of the R-squares for the separate dependent variables would be significant. For this reason, one recent book on multivariate statistics (Harris, 1975) has urged the use of an alternative to Wilk's lambda. The alternative test statistic is called the *greatest characteristic root* (gcr) and is related to Wilk's lambda:

$$U = [1/(1 + g_1][1/(1 + g_2)] \dots [1/(1 + g_s]$$

where $g_1, g_2, \dots, g_s$ are the s roots of the equation

$$\det\{\underline{C}'\underline{S}(EE)\underline{C} - g\underline{C}'[\underline{S}(E_1E_1) - \underline{S}(EE)]\underline{C}\} = 0$$

The gcr is then the largest of the s roots of this equation [max(g)]. Computation of the gcr is rather complex, but once it is obtained, tables are available in Harris (1975) for critical values of {max(g)/[1 + max(g)]} at the .01 and .05 levels.

Bock (1975) has provided tables for a *Generalized F-statistic.* The generalized F-statistic gives results equivalent to those of the gcr test, but it has the virtue of reducing to the usual univariate F ratio when the number of dependent variables involved in the null hypothesis is one. Bock (1975, Section 3.4.11) also provides a brief review of studies concerning the relative power of the U-statistic (which he calls the *likelihood-ratio criterion*) and the gcr test. He concludes that the one method is more powerful in some circumstances, whereas the other prevails in other circumstances. The choice not to belabor these alternatives in the current context stems from a preference for continuity of presentation and a desire not to proliferate complexities where it is unlikely that any practical benefits will ensue. The careful use of the U-statistic is thus recommended in multivariate regression analyses when *a posteriori* inspection of numerous coefficients raises the problem of capitalization upon chance variation. The finding that a set of coefficients is jointly distinct from a multi-dimensional zero point provides some protection for conclusions drawn from the separate standard error of a particular coefficient. [The case in which the multivariate null hypothesis is rejected, but none of the individual coefficients exceed twice their standard error, would lead to the conclusion that the independent variables are related to some linear combination of the dependent variables to a greater extent than any individual dependent variable. The instances in which this information is useful are not great in number—in my experience. On the other hand, the case in which some of the coefficients are significant, but the multivariate null hypothesis cannot be rejected, is more straightforward. This outcome communicates to the investigator that the joint results of all the regressions are not unlikely if the population slopes (or structural coefficients) are all zero. In this instance, the investigator would be uneasy about the interpretation of individual coefficients that appear to deviate from zero according to *a priori* hypothesis testing procedures—unless the hypothesis was truly *a priori*.]

Testing a Less General Null Hypothesis

The test of the null hypothesis in equation 8.30 led to the conclusion that the observed coefficients were an extremely unlikely event if

$$\underline{B}^* = \begin{bmatrix} b^*_{11} & b^*_{12} \\ 0 & 0 \\ 0 & 0 \end{bmatrix}$$

But from inspection of the individual coefficients and their standard errors (equations 8.26), it would appear that father's income (X_2) is primarily responsible for the relationship between the Y variables and the X variables. Thus, it may be of interest to test the multivariate null hypothesis that the population coefficients for father's education are zero:

$$H_0: b^*_{21} = b^*_{22} = 0$$

or

$$\underline{D} = \underline{A}\,\underline{B}^*\underline{C}$$

where

$$[0 \quad 0] = [0 \quad 1 \quad 0]\underline{B}^* \begin{bmatrix} 1 & 0 \\ 0 & 1 \end{bmatrix}$$

Since the covariance matrix for the residuals of the unconstrained model were developed for the previous null hypothesis, we can proceed directly to estimation of the constrained model for this new null hypothesis concerning the slopes of Y_1 and Y_2 on X_1.

The constrained model is obtained by essentially deleting X_1 from the sample model and regressing the two dependent variables on X_2 alone:

$$\underline{Y} = \underline{X}\,\underline{B}_1 + \underline{E}_1$$

where

$$\underline{B}_1 = \begin{bmatrix} b_{11} & b_{12} \\ 0 & 0 \\ b_{31} & b_{32} \end{bmatrix}$$

The result is

$$Y_1 = 1.69 + 0.00X_2 + 1.87X_3 + e_5 \quad (\text{R-square} = .50)$$
$$Y_2 = 8.02 + 0.00X_2 + 1.14X_3 + e_6 \quad (\text{R-square} = .16)$$

or

$$\underline{B}_1 = \begin{bmatrix} 1.69 & 8.02 \\ 0.00 & 0.00 \\ 1.87 & 1.14 \end{bmatrix}$$

Comparing these results to equations 8.26 reveals that the R-squares declined only slightly under the constraints of this second null hypothesis. However, recall that multivariate hypothesis testing relies upon the covariance of the residuals as well.

The next task is to compute the U-statistic for the null hypothesis. This computation requires the covariance matrix of the residuals with and without the constraints placed upon the sample estimates of the population model:

$$\underline{S}(E_1E_1) = \begin{bmatrix} S(e_5e_5) & S(e_5e_6) \\ S(e_6e_5) & S(e_6e_6) \end{bmatrix} = \begin{bmatrix} 4.63 & 6.40 \\ 6.40 & 9.18 \end{bmatrix}$$

If we now apply equation 8.24 [using $\underline{S}(EE)$ obtained in equation 8.28 above] we find that

$$U = \det \begin{bmatrix} 4.27 & 6.15 \\ 6.15 & 9.00 \end{bmatrix} \Big/ \det \begin{bmatrix} 4.63 & 6.40 \\ 6.40 & 9.18 \end{bmatrix} \tag{8.35}$$
$$= .589/1.453 = .406 \qquad df = 2,\ 1,\ 17$$

If this value of U is now used in equation 8.18 to obtain Rao's F-approximation, we have

$$F = 11.72 \qquad df = 2,\ 16$$
$$\text{Prob}(F > 11.72 \mid H_0) = .00073$$

Once again, the multivariate null hypothesis has been rejected soundly, even though neither of the coefficients constrained to zero was significant separately. This discrepancy occurs because the confidence region about (b_{21},b_{22}) is much more restricted than the separate confidence intervals about b_{21} and b_{22}. This leaves the investigator with the somewhat ambiguous conclusion that b^*_{21} and b^*_{22} are not *both* zero, but the 95 percent confidence interval for *each* coefficient includes zero. Clearly, a replication of the study is needed in such a situation before any confidence could be placed in the conclusion that a father's education had a significant net relationship with a son's education or income.

The Case of Uncorrelated Residuals

It is instructive to expand this nonexperimental example by changing one of the Y variables slightly. The new Y variables will produce the same sample coefficients as the data in Table 8.1, but the residuals for the two linear equations will be correlated only slightly. The effect of this change on the multivariate hypothesis tests will be dramatic.

We begin by concocting new values for the variable Y_2—the respondent's annual income. These new values for respondent's income are labeled Y_2' in Table 8.1. These fictitious values were constructed so that $\text{Cov}(e_1 e_2')$ will be close to zero for the unconstrained model.

We may now proceed to estimate $\underline{B}$ in equation 8.24 with the new data. The result is the same as before (8.26a,b)

$$\underset{}{Y_1} = \underset{(2.24)}{1.68} + \underset{(.23)}{.28X_2} + \underset{(.59)}{1.40X_3} + e_1$$

$$\text{R-square} = .54$$

$$Y_2' = \underset{(3.25)}{8.00} + \underset{(.34)}{.20X_2} + \underset{(.86)}{.80X_3} + e_2'$$

$$\text{R-square} = .17$$

But the covariance matrix for the residuals is quite different when Y_2' is used in place of Y_2:

$$\underline{S}(EE) = \begin{bmatrix} S(e_1e_1)S(e_1e_2') \\ S(e_2'e_1)S(e_2'e_2') \end{bmatrix} = \begin{bmatrix} 4.27 & -.15 \\ -.15 & 9.00 \end{bmatrix} \tag{8.36}$$

If the matrix in equation 8.36 is compared with that in equation 8.28, it is clear that the variances of the residuals have not changed (found in the diagonals of 8.28 and 8.36); but the covariance between the residuals has dropped from 6.15 for e_1 and e_2 (Y_1 and Y_2) to -0.15 for e_1 and e_2' (Y_1 and Y_2'). The obvious effect of this reduction in the covariance of the residuals will be an increase in the magnitide of the determinant of $\underline{S}(EE)$—and this may mean an increase in the U-statistic when multivariate null hypotheses are tested.

We begin the hypothesis testing procedure for these new data by assuming the model in equation 8.25 as our alternative hypothesis. The sample estimate of 8.25 is 8.26 (the same as with Y_1 and Y_2), but the covariance matrix of the residuals is given in 8.36 for Y_1 and Y_2'. The first multivariate null hypothesis tested with these data is the same as that in equation 8.30—that the four population slopes in the model are zero:

$$\underline{B}^* = \begin{bmatrix} b_{11} & b_{12} \\ 0 & 0 \\ 0 & 0 \end{bmatrix}$$

To test this null hypothesis, we estimate $\underline{B}^*$ by minimizing the squared residuals with constraints. That is, we find the matrix $\underline{B}_1$ such that

$$\underline{Y} = \underline{X}\,\underline{B}_1 + \underline{E}_1$$

where the diagonal elements of $S(E_1E_1)$ are minimums and where

$$\begin{bmatrix} 0 & 0 \\ 0 & 0 \end{bmatrix} = \begin{bmatrix} 0 & 1 & 0 \\ 0 & 0 & 1 \end{bmatrix} B_1 \begin{bmatrix} 1 & 0 \\ 0 & 1 \end{bmatrix}$$

(i.e., $b_{21} = b_{22} = b_{31} = b_{32} = 0$).

The result of this constrained estimation is

$$Y_1 = 11.05 + 0.00X_1 + 0.00X_2 + e_3$$

$$Y_2' = 13.70 + 0.00X_1 + 0.00X_3 + e_4'$$

which gives

$$\underline{S}(E_1E_1) = \begin{bmatrix} S(e_3e_3) & S(e_3e_4') \\ S(e_4'e_3) & S(e_4'e_4') \end{bmatrix} = \begin{bmatrix} 9.25 & 2.91 \\ 2.91 & 10.89 \end{bmatrix} \tag{8.37}$$

And this result (equation 8.37) should be compared with the result in equation 8.33. Note that the covariance of Y_1 and Y_2 was 9.21, whereas the covariance of Y_1 and Y_2' is only 2.91. We are now in a position to calculate the U-statistic, using equation 8.24:

$$U = \det \begin{bmatrix} 4.27 & -.15 \\ -.15 & 9.00 \end{bmatrix} \Big/ \det \begin{bmatrix} 9.25 & 2.91 \\ 2.91 & 10.89 \end{bmatrix}$$

$$= 38.39/92.19 = .416 \qquad df = 2, 2, 17$$

This value of .416 for the U-statistic contrasts with the value of .037 obtained in equation 8.34—indicating that the alternative and null hypotheses are much less divergent relative to the sample data when the residuals for the unconstrained model are uncorrelated. Transforming this new value of U with Rao's approximation (equation 8.18) gives

$$F = 4.40 \qquad df = 4, 32$$
$$\text{Prob}(F > 4.40 \,|\, H_0) = .006$$

which is considerably more modest than the F ratio of 33.45 that was obtained when the residuals were so highly correlated. Nevertheless, even with uncorrelated residuals, the multivariate null hypothesis of zero population slopes is rejected, thus clearing the way for interpretation of individually significant coefficients (b_{31} in this case).

8.5 MULTIVARIATE REGRESSION AND THE ANALYSIS OF EXPERIMENTS

The multivariate regression model is actually of more use in the analysis of experiments than in quasi-experiments. This tends to be the case for two reasons: (1) the model includes a strict distinction between independent and dependent variables that is appropriate in most experiments and *in*appropriate in many quasi-experiments; (2) the model includes no provisions for dealing with measurement error in the independent variables. The problem of measurement error is more crucial in quasi-experiments because estimates of structural coefficients are often dependent upon procedures of statistical control, and statistical control is an erratic procedure when independent variables are strongly correlated and measured with error. Most experiments, on the other hand, involve independent variables that are almost orthogonal (uncorrelated) and that are often presumed to be measured without random error. (There are, of course, many experiments in which this is not the case. The methods outlined in Chapters 9 and 10 may be adapted to such experiments—especially those experiments involving presumptions concerning indirectly observable mediators of manipulated effects.) The reader should recall that random measurement error in the *de*pendent variable of a linear model (or, at least, measurement error that is uncorrelated with the independent variables) does not alter the expected value of sample estimates of metric (i.e., unstandardized) structural coefficients—it only affects the standard error of those estimates. This means that random measurement error in the dependent variable(s) of an experiment can be counteracted by increases in sample size. In most quasi-experiments, however, the lack of distinction between independent and dependent variables, together with substantial correlations among variables, necessitates the inclusion of a measurement model if interpretable estimates of structural parameters are to be obtained. Thus, the following experimental examples are provided to demonstrate the straightforward generalization of ANOVA to Multivariate ANOVA (or MANOVA).

Traditional discussions of ANOVA have included a distinction between experiments with multiple dependent variables and *repeated measures* experimental designs. The first of these two types of design was to be analyzed with a multivariate regression model like the one outlined in Section 8.2, whereas the second category of designs was, with some trepidation, to be analyzed with special models specifically developed for repeated measures designs. The approach taken in this chapter, however, is more parsimonious; the multivariate regression model is applied to both repeated measures designs *and* designs with multiple dependent variables. The reasons for this choice include more than Occam's Razor, however. The special procedures developed for repeated measures designs involve an assumption that is often dubious—that persons do not interact with treatments. The multivariate model, as applied to these designs, does not require this assumption to yield interpretable F-tests of null hypotheses. Furthermore, in many instances, the Multivariate F-tests are more powerful than the univariate repeated measures F-tests.

But, alas, some repeated measures designs are not amenable to a multivariate regression analysis. These are designs in which the number of persons included in the experiment is too small relative to the number of levels of the independent variables. The basis for multivariate null hypothesis testing is the U-statistic (see Section 8.3), and U has three degree of freedom parameters (s, r, and n). The last of these three parameters is equal to the number of cases or participants in the sample (N) minus the number of X variables in the model (p). Thus, as p becomes large relative to N, the number of coefficients that can be included in a null hypothesis becomes very limited or is zero. In such instances, specialized repeated measures analyses would be needed (see Kirk, 1968).

Before launching into the more detailed examples, it may be useful to delineate the distinction between (1) *multiple dependent variable* and (2) *repeated measures* experimental designs. The first of these two categories is distinguished—in traditional ANOVA jargon—by the fact that participants in the experiment are a factor of the experiment that is "nested" within all the combinations of levels of the independent variable(s). Thus the mean of each cell of the experiment is based upon a different set of persons, and the multiple dependent variables do not constitute levels of one or more independent variables. The second category of experimental designs, repeated measures, is distinguished by the "crossing" of participants or cases with one or more independent variables. In other words, persons are not nested within all levels of the independent variables, and one group of persons is the source of observations for more than one cell of the experiment.

As an example, suppose that a study of the effects of threat and social science knowledge on authoritarianism involved three levels of the social threat variable (e.g., control, medium, and high threat) and three levels of the knowledge variable (e.g., control, medium, and high levels of knowledge). If this experiment were "completely crossed," then all levels of A (threat) would be crossed with all levels of B (knowledge), and observations for all nine such cells would be gathered. (If some of the possible nine cells were not included in the experiment, then the design would be described as "incomplete.") If nine different samples of persons were observed in the nine different cells of the experiment, then the cases (subjects) factor would be *nested* within both A and B. Finally, if more than one measure of authoritarianism were included in the experiment as dependent variables, then the design would be a multiple dependent variable design with participants nested within all levels of the independent variables.

But now suppose that the experimental design was altered in the following manner. Three samples of persons (S_1, S_2, and S_3) are drawn from some population, and each of the three groups is assigned to only one of the levels of B. Now, rather than divide each of these groups into three groups, one for each level of A, each group is subjected to all three levels of A. That is, the authoritarianism of S_1, S_2, and S_3 is assessed in a control condition, under medium threat, and under high threat. Thus, repeated measures of each group have constituted the variation in the levels of one of the independent variables (A). We would then say that participants are nested within B and crossed with A—a "split-plot" design, or a two-way factorial design with repeated measures on only one of the two factors. Finally, if more than one dependent variable was observed in the experiment, then the design would be referred to as a repeated measures design with multiple dependent variables.

Fortunately, the multivariate regression model presented in Section 8.2 is applicable to all these designs, and there is no longer any reason to delineate each and every possible combination of crossed and nested designs (see, for example, Kirk, 1968). In this model, both repeated measures and multiple dependent variables are treated as different Y variables in the model 8.4, while the independent variables are defined by (dummy) X variables—as outlined in Chapter 7. Thus, although the actual calculations require an electronic computer, the use of a multivariate linear model for the analysis of experiments eliminates the necessity to master the jargon and nearly endless computational formulas of ANOVA and MANOVA. Instead, an understanding of the basic procedures for testing hypotheses with a regression program is the conceptual building block for the analysis of most experimental results.

The following examples address the case of multiple dependent variables first and the case of repeated measures designs second.

An Experiment with Multiple Dependent Variables

The numerical example in this section will be an expansion of the example used in Chapter 7. Recall that the example in Chapter 7 involved two *in*dependent variables—social threat (A) and social science knowledge (B)—and only one *de*pendent variable—the F-scale as a measure of authoritarianism. The independent variable A had three levels (control, medium threat, and high threat—A_1, A_2, and A_3), and the independent variable B had two levels (control and informed concerning the relationship between social threat and authoritarianism—B_1 and B_2). If a different random sample of participants is manipulated and observed in each of the six possible combinations of treatment conditions (A_1B_1, A_2B_1, A_3B_1, A_1B_2, A_2B_2, A_3B_2), then this design is a two-way design with persons nested within both of the completely crossed independent variables.

The goal of such an experiment would be the testing of certain hypotheses about the effect of increased social threat on authoritarianism and how that effect may interact with knowledge of previous studies concerning the relationship. Now, suppose that the same experiment was conducted, but that a second dependent variable was included: racial prejudice (or ethnocentrism). For theoretical reasons, we might expect social threat to increase both authoritarianism and racial prejudice. Furthermore, it is known that measures of prejudice are correlated with measures of authoritarianism, and this covariation may increase the inferential power of multivariate hypothesis tests.

The main reason for doing a multivariate analysis of data from an experiment such as this one is to estimate the probability of *joint* type one errors. In other words, when several dependent variables are obtained from one set of participants, the probability that any one of those dependent variables will evidence a sizable chance relationship with an experimental manipulation is greater than the probability that a particular dependent variable would do so. The MANOVA thus adds to the univariate analysis of *each* dependent variable a multivariate hypothesis test; for example, the univariate F ratio for the main effect of A may be computed for each dependent variable, whereas a multivariate F ratio may be obtained for the null hypothesis that there is no population main effect of A for *any* of the dependent variables.

The data from this hypothetical experiment are presented in Table 8.2. Note that the scores for Y_1 (authoritarianism) are the same as in Chapter 7; the only change in this experiment is that a second dependent variable, Y_2 (prejudice), has been observed. The prejudice scores show roughly the same pattern of group differences as the authoritarianism scores, except that the interaction between A and B is more apparent with Y_2 than with Y_1 (i.e., the difference between the effect of increased threat under condition B_1 and condition B_2 is more apparent with the prejudice scores than with the authoritarianism scores).

Before we outline the procedures for the MANOVA computations, consider the results of the analysis as presented in Table 8.3. The table includes both the results of the two univariate ANOVA (one for Y_1 and one for Y_2) and the MANOVA (for Y_1 and Y_2 jointly). The univariate results for Y_1 are the same as those obtained in Chapter 7. The univariate results for Y_2 are comparable to those for Y_1, except that the interaction between A and B is significant at the .05 level for Y_2 ($F = 3.99$; $df = 2{,}28$; $\text{Prob}(F > 3.99 \mid H_0) = .030$) and not for Y_1 ($F = 1.95$; df $= 2{,}28$; $\text{Prob}(F > 1.95 \mid H_0) = .163$). This raises the question: How likely is it that the interaction would be significant for *either* Y_1 or Y_2 if *both* interactions are zero in the population from which the sample was drawn? This question is addressed by the multivariate F ratio, which is given below the univariate F ratios (for each main effect and interaction). The F ratio of 2.20 ($df = 4{,}54$) gives a probability of 0.081. This means that the two observed interactions between A and B (one with Y_1 and one with Y_2), or ones greater, would occur in about 8 percent of experiments even if the population interactions were zero. In other words, if an investigator had adopted a 5 percent alpha level in a traditional Fisherian scenario of hypothesis testing, then the null hypothesis of no interaction (for both dependent variables) could not be rejected. And the univariate analyses would not even be considered—even though one of them gave significant results for the interaction. From a more practical point of view, the multivariate result communicates to the investigator a context within which the univariate results can be interpreted—and that context is the likelihood that *overall* results (i.e., across multiple dependent variables) are due to chance variation.

The reader should also note the relationship between the univariate F ratios and the multivariate F ratios for the main effects of A and B. The "multivariate context" is

Table 8.2 Scores on two dependent variables for a factorial experimental design with two independent variables (A and B).

Case	Dummy	Y_1	Y_2	Condition
1	100000	7.0	14.0	A_1 B_1
2	100000	9.0	22.0	
3	100000	10.0	20.0	
4	100000	11.0	18.0	
5	100000	13.0	26.0	
6	010000	10.0	20.0	A_2 B_1
7	010000	12.0	28.0	
8	010000	13.0	26.0	
9	010000	14.0	24.0	
10	010000	16.0	32.0	
11	001000	9.0	18.0	A_3 B_1
12	001000	11.0	26.0	
13	001000	12.0	24.0	
14	001000	13.0	22.0	
15	001000	15.0	30.0	
16	000100	7.0	14.0	A_1 B_2
17	000100	8.0	20.0	
18	000100	9.0	18.0	
19	000100	10.0	16.0	
20	000100	12.0	24.0	
21	000010	6.0	12.0	A_2 B_2
22	000010	8.0	20.0	
23	000010	8.0	20.0	
24	000010	9.0	18.0	
25	000010	10.0	16.0	
26	000010	10.0	16.0	
27	000010	12.0	24.0	
28	000001	5.0	10.0	A_3 B_2
29	000001	7.0	16.0	
30	000001	7.0	20.0	
31	000001	7.0	14.0	
32	000001	8.0	14.0	
33	000001	10.0	14.0	
34	000001	12.0	24.0	

Note. The column labeled "dummy" contains values for six dummy variables that may be used in a least squares ANOVA or MANOVA of these two dependent variables.

less perplexing in these two cases because the multivariate results are consistent with the univariate results. [The reader should also note that the univariate U-statistic is simply the ratio of the unconstrained residual variance to the constrained residual variance. This means that $U = [1 - R(t)]/[1 - R(c)]$, where $R(t)$ is the R-square for the unconstrained model and $R(c)$ is the R-square for the constrained model.]

Table 8.3 Univariate and multivariate ANOVA results for the experimental data given in Table 8.2.

Constraints	Dep Var	R^2	$S(ee)$	F	df	Prob	U	df
None	Y_1	.445	3.79					
	Y_2	.554	15.15					
H_0: A	Y_1	.395	4.13	1.25	2,28	.294	.918	1,2,28
	Y_2	.507	16.77	1.49	2,28	.242	.904	1,2,28
	multivariate			1.40	4,54	.247	.821	2,2,28
H_0: B	Y_1	.137	5.89	15.51	1,28	<.001	.644	1,1,28
	Y_2	.183	27.79	23.36	1,28	<.001	.545	1,1,28
	multivariate			11.41	2,27	<.001	.542	2,1,28
H_0: $A \times B$	Y_1	.368	4.32	1.95	2,28	.163	.878	1,2,28
	Y_2	.427	19.47	3.99	2,28	.030	.778	1,2,28
	multivariate			2.20	4,54	.081	.739	2,2,28

Multivariate Computations: MANOVA *Least Squares*

In order to perform a MANOVA on the data in Table 8.2, it is necessary to fit the data to a multivariate regression model (like that in equation 8.4), and then to place appropriate constraints on that model to test classical null hypotheses for main effects and interactions. These procedures are handled automatically by most MANOVA programs. Thus, the following procedures include many more steps than would be necessary in actual practice.

The reader should recall that two approaches to univariate ANOVA were outlined in Chapter 7. The first made use of an ordinary least squares estimation procedure and included the usual intercept term. The OLS approach required the use of dummy variables coded as -1, 0, and $+1$ (see Section 7.3 and equation 7.12). The second approach to estimation was labeled the ANOVA *least squares* approach in Chapter 7 and does not include the intercept term in the model. This latter approach has the advantage of simplicity in that the dummy variables are more easily coded (one for each cell of the experiment, coded 0 or 1) and the model coefficients are cell means.

The application of the OLS approach to the multivariate case is exactly analogous to the procedures applied to the nonexperimental example in Section 8.4. When constraints corresponding to particular main effect or interaction null hypotheses are placed upon the model (by assuming that certain coefficients or contrasts are zero), these constraints are simply placed upon the coefficients for *both* dependent variables simultaneously. Thus, for the data in Table 8.2, the following outline of computations deals only with the more intuitively appealing MANOVA least squares approach. (The reader should use a multivariate regression program and the OLS method of dummy variable coding to analyze the data in Table 8.2 as an exercise. The resulting F ratios and U-statistics should be exactly the same as those given in Table 8.3.)

We begin the analysis by specifying a multivariate linear model for the data in Table 8.2:

$$Y_1 = b^*_{11}X_1 + b^*_{21}X_2 + b^*_{31}X_3 + b^*_{41}X_4 + b^*_{51}X_5 + b^*_{61}X_6 + e^*_1 \quad (8.38a)$$

$$Y_2 = b^*_{12}X_1 + b^*_{22}X_2 + b^*_{32}X_3 + b^*_{42}X_4 + b^*_{52}X_5 + b^*_{62}X_6 + e^*_2 \quad (8.38b)$$

where

$$X_j(i) = \begin{cases} 1 \text{ if the } i\text{th person is in cell } j \\ 0 \text{ if otherwise} \end{cases}$$

Note that this scheme of dummy variable coding renders the coefficients of the model as cell means. There are six cells in this 3×2 design ($j = 1, \ldots, 6$): A_1B_1 ($j = 1$), A_2B_1 ($j = 2$), A_3B_1 ($j = 3$), A_1B_2 ($j = 4$), A_2B_2 ($j = 5$), and A_3B_2 ($j = 6$).

Equations 8.38 may be written in matrix form as

$$\underline{Y} = \underline{X}\,\underline{B}^* + \underline{E}^*$$

where

$$\underline{B}^* = \begin{bmatrix} b^*_{11} & b^*_{12} \\ b^*_{21} & b^*_{22} \\ b^*_{31} & b^*_{32} \\ b^*_{41} & b^*_{42} \\ b^*_{51} & b^*_{52} \\ b^*_{61} & b^*_{62} \end{bmatrix}$$

Now, if $\underline{E}^*$ has the appropriate properties, then $\underline{B}^*$ may be estimated from sample data by

$$\underline{B} = \begin{bmatrix} b_{11} & b_{12} \\ b_{21} & b_{22} \\ b_{31} & b_{32} \\ b_{41} & b_{42} \\ b_{51} & b_{52} \\ b_{61} & b_{62} \end{bmatrix} = \begin{bmatrix} 10.0 & 20.0 \\ 13.0 & 26.0 \\ 12.0 & 28.0 \\ 9.2 & 18.4 \\ 9.0 & 18.0 \\ 8.0 & 16.0 \end{bmatrix} \quad (8.39)$$

where $\underline{B}$ minimizes the diagonal elements of $(\underline{E}'\underline{E})$ in

$$\underline{Y} = \underline{X}\,\underline{B} + \underline{E} \quad (8.40)$$

Note that the elements of $\underline{B}$ in equation 8.39 are the means of the cells in Table 8.2. The first column of $\underline{B}$ gives the means for Y_1 and the second column gives the means for Y_2. These values are the least squares solution to equation 8.40—the values of $\underline{B}$ that minimize the sum of squared residuals for each univariate equation.

Now that the experimental results have been fit to the multivariate regression model, we can proceed to place constraints upon the model and test multivariate null hypotheses. Three multivariate null hypotheses will be tested: one for the main effect of A, one for the main effect of B, and one for the interaction between A and B. These null hypotheses may be stated in terms of the model 8.38 as follows:

H_0: no main effect of A for Y_1 or Y_2

$$b^*_{11} + b^*_{41} = b^*_{21} + b^*_{51} = b^*_{31} + b^*_{61} \text{ (for } Y_1\text{)}$$

and

$$b^*_{12} + b^*_{42} = b^*_{22} + b^*_{52} = b^*_{32} + b^*_{62} \text{ (for } Y_2\text{)}$$

H_0: no main effect of B (knowledge) for Y_1 or Y_2

$$b^*_{11} + b^*_{21} + b^*_{31} = b^*_{41} + b^*_{51} + b^*_{61}$$

and

$$b^*_{12} + b^*_{22} + b^*_{32} = b^*_{42} + b^*_{52} + b^*_{62}$$

H_0: no interaction between A and B for Y_1 or Y_2

$$b^*_{11} - b^*_{41} = b^*_{21} - b^*_{51} = b^*_{31} - b^*_{61}$$

and

$$b^*_{12} - b^*_{42} = b^*_{22} - b^*_{52} = b^*_{32} - b^*_{62}$$

The task that now remains is construction of the hypothesis matrices $\underline{A}$, $\underline{C}$, and $\underline{D}$ that apply appropriate constraints to equation 8.40 to test each one of these null hypotheses.

Recall that the constraints placed upon a multivariate regression model may be stated in terms of the matrix equation

$$\underline{D} = \underline{A}\,\underline{B}\,\underline{C}$$

In the case of MANOVA least squares, the matrix $\underline{A}$ defines *contrasts* among cell means and $\underline{C}$ defines the number of dependent variables involved in the constraints. The matrix $\underline{D}$ specifies the values to which the contrasts are constrained (these values are usually zero in MANOVA, but may be not be zero in special circumstances). The hypothesis matrices that specify the constraints appropriate for testing the multivariate null hypotheses are as follows:

A: constraint to test null hypothesis for A

$$\begin{matrix} \underline{D} & = & \underline{A} & \underline{B} & \underline{C} \\ \begin{bmatrix} 0 & 0 \\ 0 & 0 \end{bmatrix} & = & \begin{bmatrix} 1 & -1 & 0 & 1 & -1 & 0 \\ 1 & 0 & -1 & 1 & 0 & -1 \end{bmatrix} & \underline{B} & \begin{bmatrix} 1 & 0 \\ 0 & 1 \end{bmatrix} \end{matrix}$$

B: constraint to test null hypothesis for B

$$[0 \;\; 0] = [1 \;\; 1 \;\; 1 \;\; -1 \;\; -1 \;\; -1]\underline{B}\begin{bmatrix} 1 & 0 \\ 0 & 1 \end{bmatrix}$$

$A \times B$: constraints to test null hypothesis for $A \times B$

$$\begin{bmatrix} 0 & 0 \\ 0 & 0 \end{bmatrix} = \begin{bmatrix} 1 & -1 & 0 & -1 & 1 & 0 \\ 1 & 0 & -1 & -1 & 0 & 1 \end{bmatrix} \underline{B} \begin{bmatrix} 1 & 0 \\ 0 & 1 \end{bmatrix}$$

If the data in Table 8.2 are now fit to equation 8.40 with these three sets of constraints, the covariance matrix of the residuals can be computed in each case and compared with the residual covariance matrix, for the unconstrained solution. The four solutions for $\underline{B}$ are as follows:

Unconstrained		A constraints		B constraints		$A \times B$ constraints	
10.0	20.0	10.6	15.3	8.38	16.02	11.12	22.93
13.0	26.0	12.1	25.0	11.4	22.0	12.4	25.7
12.0	28.0	12.3	27.0	10.4	22.4	11.4	25.4
9.2	18.4	9.8	20.4	10.8	22.4	8.1	15.5
9.0	18.0	8.4	17.3	10.2	20.8	9.4	18.2
8.0	16.0	8.2	15.3	9.2	18.8	8.4	17.9

[Note that the fit of each of the constrained models to the observed data can be assessed by comparing the "fitted cell means" (i.e., the constrained coefficients) to the observed cell means. The more the fitted means deviate from the observed means, the poorer is the fit of the constrained model. In fact, the deviations of the fitted means from the observed means could be used as a measure of relative fit and is related to the differences between the residual covariance matrices.] If these four solutions to equation 8.40 are labeled $\underline{B}$, $\underline{B}_1$, $\underline{B}_2$, and $\underline{B}_3$, then each of the residual matrices could be defined as

$$\underline{S}(EE) = (1/N)(\underline{E}'\underline{E}) = (1/N)[(\underline{Y}'\underline{Y}) - (\underline{Y}'\underline{X}\,\underline{B})]$$
$$\underline{S}(E_1E_1) = (1/N)(\underline{E}_1'\underline{E}_1) = (1/N)[(\underline{Y}'\underline{Y}) - (\underline{Y}'\underline{X}\,\underline{B}_1)]$$
$$\underline{S}(E_2E_2) = (1/N)(\underline{E}_2'\underline{E}_2) = (1/N)[(\underline{Y}'\underline{Y}) - (\underline{Y}'\underline{X}\,\underline{B}_2)]$$
$$\underline{S}(E_3E_3) = (1/N)(\underline{E}_3'\underline{E}_3) = (1/N)[(\underline{Y}'\underline{Y}) - (\underline{Y}'\underline{X}\,\underline{B}_3)]$$

The actual values obtained for the residual covariance matrices are

$$\underline{S}(EE) = \begin{bmatrix} 3.79 & 5.58 \\ 5.58 & 15.15 \end{bmatrix}$$
$$\underline{S}(E_1E_1) = \begin{bmatrix} 4.13 & 6.10 \\ 6.10 & 16.77 \end{bmatrix}$$
$$\underline{S}(E_2E_2) = \begin{bmatrix} 5.89 & 10.73 \\ 10.73 & 27.79 \end{bmatrix}$$
$$\underline{S}(E_3E_3) = \begin{bmatrix} 4.32 & 6.96 \\ 6.96 & 19.47 \end{bmatrix}$$

And these residual covariance matrices can now be used in equation 8.24 to obtain U-statistics for each multivariate null hypothesis:

$$\begin{aligned} U_1 &= \det[\underline{S}(EE)]/\det[\underline{S}(E_1E_1)] \\ &= 26.30/32.04 = .821 \qquad df = 2,\ 2,\ 28 \\ U_2 &= \det[\underline{S}(EE)]/\det[\underline{S}(E_2E_2)] \\ &= 26.30/48.54 = .542 \qquad df = 2,\ 1,\ 28 \\ U_3 &= \det[\underline{S}(EE)]/\det[\underline{S}(E_3E_3)] \\ &= 26.30/35.58 = .739 \qquad df = 2,\ 2,\ 28 \end{aligned}$$

where U_1, U_2, and U_3 are for the null hypothesis for A, B, and $A \times B$, respectively; and the degrees of freedom are equal to the number of *columns* in $\underline{C}$, the number of *rows* in $\underline{A}$, and $(N - 6)$, respectively. (Note that the number of columns in $\underline{C}$ is equal to the number of dependent variables involved in the constraints, whereas the number of rows in $\underline{A}$ is equal to the number of contrasts involved in the null hypothesis.) These values of the U-statistic are listed in Table 8.3 for the corresponding null hypothesis and constraints. Substituting these values of U into equation 8.18 (Rao's approximate F, which is exact in all these examples) yields the multivariate F ratios given in Table 8.3.

Analysis of Repeated Measures Designs

The advantage of repeated measures designs (crossing persons with independent variables rather than nesting persons within treatment combinations) is that variability in the dependent variable that is accounted for by stable individual differences can be extracted from the category of "noise" in the experiment. This reduction in the amount of noise may manifest itself in more powerful tests of null hypotheses. An intuitive sense of why that is the case can be gleaned from a simple example.

Suppose that we return to our standard example of assessing the effect of threat on F-scale scores. If a simple experiment involving a control group and an experimental group is conducted, then we might fit the data from this nested design to the linear model

$$Y = b_0 + b_1X_1 + e$$

where X_1 is zero for the control group and one for the experimental group, b_1 represents the effect of the manipulation on Y (F-scale scores), and e is the residual. Now, the standard error of b_1 is a function of the variance of the residual—$S(ee)$. And the variance of the residual will be determined by the extent of individual differences within the two groups in the experiment.

Now, for any realistic manipulation, we would not expect the manipulation of social threat to explain a very large portion of the total variability in Y. However, if the manipulation increased the F-scale score of almost everyone in the experimental group, then we would have a very important and reliable finding—even though the magnitude of b_1 relative to the standard deviation of Y may be small. But notice that the simple experiment does not enable a researcher to measure how much individuals may have changed due to the manipulation. Rather, the amount of change is inferred from the difference between the experimental and control groups—relying upon the assumption that randomization yielded equivalent groups prior to the manipulation.

Suppose, then, that a different experimental design is developed in which the F-scale scores of both groups are measured before *and* after the experimental manipulation. The observations from this study might be modeled with the equations

$$Y_1 = b_{11} + b_{21}X_2 + e_1 \tag{8.41a}$$

$$Y_2 = b_{12} + b_{22}X_2 + e_2 \tag{8.41b}$$

where Y_1 is the F-scale score before the manipulation and Y_2 is the F-scale score after the manipulation and X_2 is the dummy variable that is one for the experimental group and zero for the control group. (This model is depicted in Figure 8.2.) The focus of attention in this study is now upon the difference between the coefficients b_{22} and b_{21}, where b_{22} is the difference between the two groups after the manipulation and b_{21} is any pretreatment difference that may occur due to chance. The null hypothesis that $(b^*_{22} - b^*_{21})$ is equal to zero is then easily tested with the methods of multivariate regression. But the advantage of this analysis over the experiment with only a post-treatment measure is more apparent if equation 8.41a is subtracted from equation 8.41b to yield

$$Y_2 - Y_1 = (b_{12} - b_{11}) + (b_{22} - b_{21})X_2 + (e_2 - e_1) \tag{8.42}$$

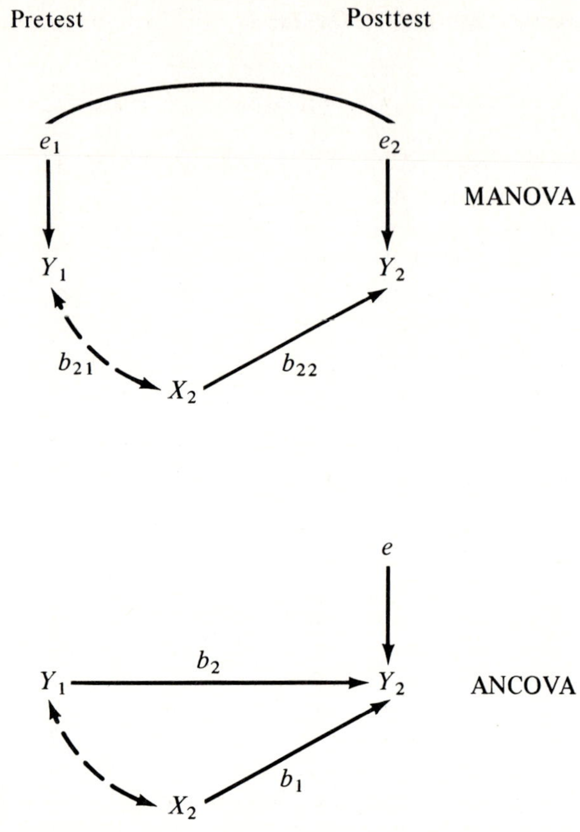

Figure 8.2 Path diagrams of the MANOVA (equation 8.41) and ANCOVA structural models for a randomized experiment with pre and post measures. The curved paths are dashed to indicate that the population covariance betweeen the pretest and the manipulation is zero when the manipulation is randomized.

where $(b_{22} - b_{21})$ is the slope of the change score $(Y_2 - Y_1)$ regressed on X_2 if $(e_2 - e_1)$ is uncorrelated with X_2. Now, if the assumption that $(e_2 - e_1)$ is uncorrelated with X_2 is valid, then equation 8.42 provides a much more powerful test of the research hypothesis than would the postest design when e_2 and e_1 are correlated. The reason for this is that the difference between two positively correlated variables can have a variance that is smaller than the variance of either variable. Thus, when b_{21} (pretreatment difference) is close to zero, we have $b_1 = b_{22}$; the standard error of b_1 is a function of the variance of e_2, while the standard error of $(b_{22} - b_{21})$ is a function of the variance of $(e_2 - e_1)$. And when the two dependent variables are highly correlated, the variance of $(e_2 - e_1)$ can be much smaller than the variance of e_1 or e_2. [The exact relationship is that

$$\text{Var}(e_2 - e_1) = \text{Var}(e_1) + \text{Var}(e_2) - 2\text{Cov}(e_1, e_2)$$

which reveals that the variance of the difference between the residuals can be much larger than the variance of e_2 if e_1 and e_2 are not strongly correlated. In fact, if e_1 and e_2 are *negatively* correlated, then the variance of the difference can have a variance that approaches twice the sum of the individual variances. However, in the case of a measure like the F-scale, the likelihood of substantial positive correlation over time is so great that a repeated measures design would almost certainly decrease the standard error of $(b_{22} - b_{21})$ relative to the standard error of b_1—and yet we would expect the magnitude of these two regression coefficients to be about the same (since b_{21} should differ from zero only by chance).]

For this simple design with two dependent variables (or, equivalently, two levels of the repeated measures factor) the F ratio for the null hypothesis that the manipulation has no effect will be the same for the repeated measures model in equation 8.42 and for the multivariate regression model in equation 8.41. That is, the standard error of $(b_{22} - b_{21})$ in equation 8.42 is equal to the standard error of the contrast that could be obtained from the multivariate F-test derived from the hypothesis matrices

$$0 = [0 \;\; 1] \begin{bmatrix} b_{11} & b_{12} \\ b_{21} & b_{22} \end{bmatrix} \begin{bmatrix} -1 \\ 1 \end{bmatrix} \tag{8.43}$$

Thus, for this simple design, the univariate approach to the analysis of repeated measures (equation 8.42) is equivalent, and considerably simpler, than the use of a multivariate model (equation 8.41). [This equivalence is the same as saying that the residual $(e_2 - e_1)$ is equal to the least squares solution when the difference score $(Y_2 - Y_1)$ is regressed on the dummy variable defining group membership. Rather than provide an algebraic proof of this fact, it will be more instructive for the reader to concoct a data set for this simple design and "see" that the difference between the two residuals (e_2 and e_1) is equal to the residual obtained from the change score regression.]

The more experienced reader may raise the question at this point of whether or not Y_1 should be treated as a covariate rather than a second dependent variable (or that the ANCOVA model may be more appropriate than the change score model). For example,

$$Y_2 = b_0 + b_1X_2 + b_2Y_1 + e$$

seems like a plausible model, since b_1 is statistically adjusted for differences in Y_1. Or, in other words, when persons are equated on Y_1, b_1 is the net difference between the experimental and control groups. Although this topic will be treated more completely when we discuss longitudinal designs (Chapter 11), some points relevant to experimentation will be outlined here.

The differences between the MANOVA and ANCOVA approaches to analysis of this simple experimental design may be summarized as follows. If the two groups of persons in the experiment do not differ on the pretest (i.e., the mean of Y_1 is the same for the two groups), then the ANCOVA model will provide the same estimate of the effect magnitude, but it can provide a *smaller* standard error than the MANOVA (or change score) model. On the other hand, as the pretreatment difference between groups becomes larger, the ANCOVA can give quite a different estimate of the effect of the manipulation on the dependent variable. The reason for this difference is that the

ANCOVA model assesses the change in the mean for the experimental group in the context of the pretreatment differences between the two groups. If the control group has a higher mean on Y_1 than the experimental group, and the experimental group increases after the manipulation, then the ANCOVA model assumes that some (or all) of that increase is due to "regression toward the mean." That is, the ANCOVA model assumes that both groups are selected from the same population and that the overall mean for the two groups is a better estimate of the population mean. Thus, it would be expected that the lower group would tend to move up over time and the higher group would tend to move down. (Just as a baseball team at the top of the league will be more likely to have a decline rather than an increase in its standing, the team at the bottom is more likely to improve than fall farther behind—since it is already at the bottom.)

Thus, for this simple design, the ANCOVA model is more powerful than the MANOVA or change score model in most instances where pretreatment differences are minimal. Also, when the pretreatment differences are sizable and group membership was determined randomly, it is reasonable to assume that the participants were drawn from the same population and that the overall mean on the pretest predicts the direction in which the individual means are most likely to move over time—without the aid of an experimental manipulation. However, as we shall see later, when the assumption that the two groups are samples from the same population is false, the ANCOVA model can lead to serious misinterpretations (see Cook and Campbell, 1979).

There is an even more important reason for using the pretest as a covariate rather than to perform a MANOVA or change score analysis. The reason is that the structural model in the ANCOVA case is more plausible. Reference to Figure 8-2 provides a hint of why the MANOVA model presents a problem. Notice that there is no direct path leading from Y_1 to Y_2. This specification means that stability of Y over time is presumed to arise completely because of the influence of other factors (X_2, e_1, and e_2). For most attributes of persons or social entities, this is an inappropriate assumption. Nevertheless, the testing of more plausible structural models requires a relaxation of the assumption that all residuals are uncorrelated with all right-hand variables. And the multivariate regression model is insufficient to this task. The LISREL model described in Chapter 10 is an improvement over the multivariate regression model in that respect, and the application of that model to the analysis of longitudinal designs is discussed in Chapter 11. In the meantime, the following section describes the analysis of repeated measures experimental designs in a more traditional manner. The shortcomings of these methods will become more apparent in Chapter 11.

Computations for a Multivariate Regression of Data from a Repeated Measures Design

As a detailed example of the use of a multivariate regression model to analyze data from a repeated measures design, suppose that the experiment in Chapter 7 was conducted with only two samples of persons. That is, suppose that persons were nested within the factor B (no knowledge/knowledge) and *crossed* with factor A (control/medium threat/high threat). This means that the same person is observed under the different levels of A, whereas different persons are observed under the two levels of B.

Table 8.4 Hypothetical data from an experiment based upon a two-way design with repeated measures on one factor (A).

Participant number	X_1	X_2	A_1 Y_1	A_2 Y_2	A_3 Y_3	
1	1	0	7	11	12	
2	1	0	9	9	16	
3	1	0	10(10)	12(12)	13(13)	B_1 (no knowledge)
4	1	0	11	15	10	
5	1	0	13	13	14	
6	0	1	7	10	7	
7	0	1	9	8	11	
8	0	1	10(10)	11(11)	10(10)	B_2 (knowledge)
9	0	1	11	14	9	
10	0	1	13	12	13	

Note. The numbers in parentheses are cell means.

The hypothetical results of such an experiment are given in Table 8.4. Notice that each participant in the study has three different F-scale scores: one in the control condition A_1 (Y_1), one in the medium threat condition A_2 (Y_2), and one in the high threat condition A_3 (Y_3). Now, the methodological problems created by such a design (e.g., the effect of repeatedly taking the same test) may be considerable, but we will proceed as if there were no such problems.

The data in Table 8.4 can be modeled with the multivariate linear equations

$$Y_1 = b^*_{11}X_1 + b^*_{21}X_2 + e^*_1 \tag{8.44a}$$

$$Y_2 = b^*_{12}X_1 + b^*_{22}X_2 + e^*_2 \tag{8.44b}$$

$$Y_3 = b^*_{13}X_1 + b^*_{23}X_2 + e^*_3 \tag{8.44c}$$

where

$$X_1(i) = \begin{cases} 1 \text{ if participant } i \text{ is in the group assigned to level } B_1 \\ 0 \text{ otherwise} \end{cases}$$

$$X_2(i) = \begin{cases} 1 \text{ if participant } i \text{ is in the group assigned to level } B_2 \\ 0 \text{ otherwise} \end{cases}$$

Or equations 8.44 may be written more compactly as

$$\underline{Y} = \underline{X}\,\underline{B}^* + \underline{E}^*$$

where $\underline{Y}$ is an $N \times 3$ matrix of scores on the three repeated measures, $\underline{X}$ is an $N \times 2$ matrix of scores on the dummy variables defining group membership, $\underline{B}^*$ is the 2×3 matrix of population regression coefficients, and $\underline{E}^*$ is an $N \times 3$ matrix of population disturbances (residuals). The distinguishing feature of equations 8.44 relative to the models for nested designs presented in Chapter 7 is that only the factor B is defined by dummy variables; the factor A is defined by the inclusion of three different dependent variables. [The reader should note that the more general least squares approach to dummy variable coding has been adopted here, since it is most easily adapted to the null hypotheses of classic procedures. The factor that has participants nested within it—factor B in this instance—could be coded to include an intercept term and would yield the same F ratios as the procedure followed here. However, when an intercept term is included in the model, it is necessary to use the $1/0/-1$ coding scheme outlined in Chapter 7 for OLS ANOVA models.]

Now that model 8.44 has been specified, it is a straightforward procedure to apply constraints to the sample estimates in order to test null hypotheses for main effects and interactions. We begin by estimating 8.44 with the data in Table 8.4 and note that the least squares solution for the six coefficients corresponds to the six cell means:

$$\underline{B} = \begin{bmatrix} b_{11} & b_{12} & b_{13} \\ b_{21} & b_{22} & b_{23} \end{bmatrix} = \begin{bmatrix} 10 & 12 & 13 \\ 10 & 11 & 10 \end{bmatrix} \tag{8.45}$$

where $\underline{B}$ minimizes the three sums of squared residuals in

$$\underline{Y} = \underline{X}\,\underline{B} + \underline{E}$$

(the sample estimate of equations 8.44).

The next step is to test the main effects and interaction null hypotheses. These null hypotheses may be stated as follows:

H_0 = no main effect of A in the population

$$b_{11}^* + b_{21}^* = b_{12}^* + b_{22}^* = b_{13}^* + b_{23}^*$$

or

$$[0 \quad 0] = [1 \quad 1]\underline{B}^* \begin{bmatrix} 1 & 1 \\ -1 & 0 \\ 0 & -1 \end{bmatrix}$$

H_0 = no main effect of $\underline{B}$ in the population

$$b_{11}^* + b_{12}^* + b_{13}^* = b_{21}^* + b_{22}^* + b_{23}^*$$

or

$$[0] = [1 \quad -1]\underline{B}^* \begin{bmatrix} 1 \\ 1 \\ 1 \end{bmatrix}$$

H_0 = no interaction $A \times B$ in the population

$$b_{11}^* - b_{21}^* = b_{12}^* - b_{22}^* = b_{13}^* - b_{23}^*$$

or

$$[0 \quad 0] = [1 \quad -1]\underline{B}^* \begin{bmatrix} 1 & 1 \\ -1 & 0 \\ 0 & -1 \end{bmatrix}$$

Now, each of these null hypotheses can be tested by placing constraints on the least squares estimates of $\underline{B}^*$. Those constraints are easily stated in terms of the hypothesis matrices used above to state the null hypotheses in matrix form. For example, to test the interaction null hypothesis, we solve

$$\underline{Y} = \underline{X}\,\underline{B}_3 + \underline{E}_3$$

for $\underline{B}_3$ by using the constraint that

$$[0 \quad 0] = [1 \quad -1]\underline{B}_3 \begin{bmatrix} 1 & 1 \\ -1 & 0 \\ 0 & -1 \end{bmatrix}$$

and the least squares criterion that the diagonal elements of $(\underline{E}_3'\underline{E}_3)$ are minimums. [The reader should note at this point that when constraints apply across different linear equations in a multivariate model—rather than within individual equations—then the residuals for individual equations may not sum to zero. This means that the cross product matrix $(\underline{E}_3'\underline{E}_3)$ is probably not equal to N times the residual covariance matrix because covariances are computed about means. Thus, when null hypotheses involve constraints between equations, we must use the residual cross product matrix rather than the residual covariance in the computation of the multivariate U-statistic. This means that analyses of repeated measures designs should be computed with programs that provide output in terms of cross products (e.g., BMD11V).] The resulting residual cross product matrices are

$$\underline{E}'\underline{E} = \begin{bmatrix} 40 & 24 & 16 \\ 24 & 40 & -16 \\ 16 & -16 & 40 \end{bmatrix}$$

and

$$\underline{C}'(\underline{E}'\underline{E})\underline{C} = \begin{bmatrix} 32 & -16 \\ -16 & 48 \end{bmatrix}$$

for the unconstrained model and

$$\underline{C}'(\underline{E}_3'\underline{E}_3)\underline{C} = \begin{bmatrix} 34.5 & -8.5 \\ -8.5 & 72.5 \end{bmatrix}$$

for the estimation with constraints. These results are then substituted into the equation for the U-statistic

$$U = \det[\underline{C}'(\underline{E}'\underline{E})\underline{C}]/\det[\underline{C}'(\underline{E}_3'\underline{E}_3)\underline{C}]$$

where $\underline{C}$ is the hypothesis matrix for the interaction null hypothesis. The numerical result in this instance is

$$U = 1280/2360 = .542 \qquad df = 2, 1, 8$$

(note that the degree of freedom parameters are equal to the number of columns in $\underline{C}$, the number of rows in $\underline{A}$ and $N - p$, respectively.) And this value of U may be transformed into Rao's F-approximation—which is exact in this instance—to give

$$F = 2.95 \qquad df = 2, 7$$
$$\text{Prob}(F > 2.95 \mid H_0) = .118$$

Table 8.5 Results of the multivariate regression analysis of the data from a repeated measures design (Table 8.4).

Source	*U*	*df*	*F*	*df*	Prob
A (threat)	.337	2,1,8	6.89	2,7	.022
B (knowledge)	.808	1,1,8	1.90	1,8	.205
$A \times B$	.542	2,1,8	2.95	2,7	.118

Note. The factor *A* is crossed with participants while participants are nested within factor *B*. Thus, the experimental design has repeated measures on only one of the two independent variables.

The U-statistics and F ratios for all three of the null hypotheses are given in Table 8.5.

Use of the Computer Program BMD11V for Multivariate Regression One of the most versatile computer programs for multivariate regression that is also widely available is the element BMD11V of the older BMD series of statistical programs. (The newer series is termed BMDP.) Although newer programs, such as the SPSS subprogram for MANOVA, may be more convenient, the user is advised to check the results of such routines with the well-tested BMD11V.

BMD11V is given the title "multivariate general linear hypothesis." The adjective *multivariate* indicates that multiple left-hand (dependent) variables may be included in a model, and the adjective *general* refers to the fact that the user is free to define a wide range of constraints upon the coefficients of right-hand (independent or covariate) variables. However, the program has no facility for automatic generation of dummy variables or hypothesis testing constraints. For learning purposes, the fact that the user must specify constraints in matrix form and input dummy variables to define independent variables is useful. One measure of its usefulness is the requirement that the user must understand more of what he is doing. More automatic programs have the dubious virtue of producing plausible appearing output based on the most erroneous of inputs.

The reader of this chapter should have little trouble deciphering the program description provided in Dixon (1968, pp. 739–743). The analysis of the data in Table 8.2 that is presented in Table 8.3 can be reproduced with the following control statements and BMD11V:

```
PROBLM 6 2 9   34     1MANOVAB
(15X,6F1.0,2F11.5)
          1          100000   7.000000     14.00000
          2          100000   9.000000     22.00000
          3          100000   10.00000     20.00000
          4          100000   11.00000     18.00000
          5          100000   13.00000     26.00000
          6          010000   10.00000     20.00000
          7          010000   12.00000     28.00000
          8          010000   13.00000     26.00000
          9          010000   14.00000     24.00000
```

```
        10      010000  16.00000    32.00000
        11      001000  9.000000    22.00000
        12      001000  11.00000    30.00000
        13      001000  12.00000    28.00000
        14      001000  13.00000    26.00000
        15      001000  15.00000    34.00000
        16      000100  7.000000    14.00000
        17      000100  8.000000    20.00000
        18      000100  9.000000    18.00000
        19      000100  10.00000    16.00000
        20      000100  12.00000    24.00000
        21      000010  6.000000    12.00000
        22      000010  8.000000    20.00000
        23      000010  8.000000    20.00000
        24      000010  9.000000    18.00000
        25      000010  10.00000    16.00000
        26      000010  10.00000    16.00000
        27      000010  12.00000    24.00000
        28      000001  5.000000    10.00000
        29      000001  7.000000    16.00000
        30      000001  7.000000    20.00000
        31      000001  7.000000    14.00000
        32      000001  8.000000    14.00000
        33      000001  10.00000    14.00000
        34      000001  12.00000    24.00000
(6F4.1)
HYPOTHTHREAT 2
0010 -10 000 010 -10 000
 010 000 -10 010 000 -10
HYPOTHKNOWLE 1
 010 010 010 -10 -10 -10
HYPOTHINTERA 2
 010 -10 000 -10 010 000
 010 000 -10 -10 000 010
HYPOTHTHR Y1 2 1
0010 -10 000 010 -10 000
 010 000 -10 010 000 -10
 010 000
HYPOTHKNO Y1 1 1
 010 010 010 -10 -10 -10
 010 000
HYPOTHAXB Y1 2 1
 010 -10 000 -10 010 000
 010 000 -10 -10 000 010
 010 000
HYPOTHTHR Y2 2 1
0010 -10 000 010 -10 000
 010 000 -10 010 000 -10
 000 010
HYPOTHKNO Y2 1 1
 010 010 010 -10 -10 -10
 000 010
HYPOTHAXB Y2 2 1
 010 -10 000 -10 010 000
 010 000 -10 -10 000 010
 000 010
FINISH
```

The reader will find that the first three hypotheses are multivariate, whereas the last six are univariate. It should also be noted that the ANOVA least squares method of model specification has been used in this example.

The inclusion of covariates in a MANOVA is accomplished by simply including the covariates as right-hand variables in addition to the dummy variables defining the independent variables. When constraints are placed upon the coefficients of the dummy variables to test various hypotheses, no constraints are placed upon the coefficients of the covariates. This approach yields significance tests of effects that are "net" of relationships and variance explained by the covariates.

This general program will handle unbalanced and incomplete experimental designs provided that complete data are available for each participant (or case). This means that *repeated measures* factors in ANOVA or MANOVA designs (factors crossed with cases rather than factors within which cases are nested) must be completely crossed with other factors in the design.

As a second example of the use of BMD11V, the following set of statements performs the analysis of the repeated measures design (Table 8.4) that is presented in Table 8.5:

```
PROBLM 2 3 5   10     1FACREP8
(2F2.0,6X,3F11.5)
 1 0   1    7.000000     11.00000     12.00000
 1 0   2    9.000000     9.000000     16.00000
 1 0   3    10.00000     12.00000     13.00000
 1 0   4    11.00000     15.00000     10.00000
 1 0   5    13.00000     13.00000     14.00000
 0 1   6    7.000000     10.00000     7.000000
 0 1   7    9.000000     8.000000     11.00000
 0 1   8    10.00000     11.00000     10.00000
 0 1   9    11.00000     14.00000     9.000000
 0 1  10    13.00000     12.00000     13.00000
(3F3.0)
HYPOTH AMAIN 1 2
   1  1
   1 -1  0
   1  0 -1
HYPOTH BMAIN 1 1
   1 -1
   1  1  1
HYPOTH AXBIN 1 2
   1 -1
   1 -1  0
   1  0 -1
HYPOTHPLAN 1 1
   1 -1
HYPOTHPLAN 2 1 1
   1 -1
  -1 .5 .5
FINISH
```

In addition to the constraints testing the two main effects and the two-way interaction, two additional hypotheses are also tested. These additional constraints are plausible planned or *a priori* comparisons that flow logically from the design of the experiment. Upon execution of this program, the reader will discover that the planned comparisons are highly significant—even though the overlapping effects in the standard ANOVA model (main effect of *B* and interaction between *A* and *B*) are not. But the reader should bear in mind that the probabilities associated with planned constraints are biased downwards if the planned comparisons are actually concocted *after* inspection of the pattern of results.

REFERENCES AND SUGGESTED READING

Anderson, T. W. *Introduction to multivariate analysis.* New York: Wiley, 1958. (The classic text for mathematical statisticians.)

Bock, R. D. *Multivariate statistical methods in behavioral research.* New York: McGraw-Hill, 1975. (A comprehensive treatment of multivariate analysis that is accessible to readers without formal training in mathematical statistics.)

Cook, T. D., and Campbell, D. T. *Quasi-experimentation: Design and analysis issues for field settings.* Chicago: Rand McNally, 1979.

Dixon, W. J. (ed.) *BMD biomedical computer programs.* Los Angeles: UCLA, 1968.

Finn, J. D. *MULTIVARIANCE: Univariance and multivariance analysis of variance, covariance and regression.* Chicago: National Educational Resources, 1972.

Harris, R. J. *A primer of multivariate statistics.* New York: Academic Press, 1975.

Kirk, R. E. *Experimental design: Procedures for the behavioral sciences.* Belmont, Calif.: Brooks/Cole, 1968.

Morrison, D. F. *Multivariate statistical methods.* San Francisco: McGraw-Hill, 1967.

Overall, J., and Klett, C. J. *Applied multivariate analysis.* San Francisco: McGraw-Hill, 1972.

Rao, C. R. Some problems involving linear hypotheses in multivariate analysis. *Biometrika,* 1959, *46,* 49–58.

Tatsuoka, M. M. *Multivariate analysis: Techniques for educational and psychological research.* New York: Wiley, 1971.

Van de Geer, J. P. *Introduction to multivariate analysis for the social sciences.* San Francisco: Freeman, 1971.

Wilks, S. S. Certain generalizations in the analysis of variance. *Biometrika,* 1932, *24,* 471–494.

9

Measurement Models

9.1 THE PROBLEM OF MEASUREMENT ERROR

Social and behavioral scientists are almost immune to pronouncements concerning the nexus of "the problem" that is delaying their research relative to that of the physical scientists. Nevertheless, I shall add to the long list of such pronouncements by claiming that measurement problems have been crucial to the apparent failure of most attempts to establish empirical generalizations about social and/or psychological phenomena. This claim is based upon the following arguments.

1. Information comes from movement. Physics blossomed with Newton's discovery of mental models of the movement of objects. Similarly, social scientists seek models that describe and elucidate the movement of social and psychological phenomena. But those phenomena are not objects so much as they are *organized patterns* of movement. We are not concerned with the movement of body parts or of groups of human bodies *per se*. Our concern is with the scenario or story line that such movements are part of. If observation of falling objects might be considered fundamental to physics, then the fundamental observation in the human sciences is the one act play—on stage or in real life.

2. If it is accepted that the essential unit of observation in the human sciences is the flow of organized activity, complete with the motives and purposes that characterize a story line, then the empirical task is to test models that propose to predict and explain regularities in that flow. And this means that characteristics of that flow must be measured so that the models may be tested. The only essential characteristic of that measurement process, aside from its "making sense" to its users, is that it be replicable by doubters. In this sense, then, measurement is synonymous with observation when scientific model testing is the task at hand. Model testing always involves counting events in some way to "see" whether or not humans act in the way that the model predicts: Does the flow of activity move in accordance with the scenario outlined in the model?

3. In order to carry out this measurement process, indeed in order to theorize about

human activity at all, it is necessary to specify the *dimensions* that are used to describe the human activity. For example, observation of a group of persons may lead to the hypothesis that there are stable individual differences in patterns of sexual interaction; the dimensions of these differences must then be specified. Similarly, observation of a group of persons may lead to the hypothesis that persons in different work settings manifest different patterns of behavior toward people from minority groups. In order to test such a hypothesis, the dimensions of behavior pattern differences must be specified in such a manner that any determined observer could decide whether or not the observations were consistent with the model.

4. Now, the measurement *problem* for the social scientist arises because dimensions of organized activity are often very difficult to count. Suppose that a model includes the dimension "attitude toward black women" as a variable that relates in some regular way with variability on other dimensions. Given unlimited resources, how would we go about measuring such a dimension? Or, more fundamentally, how would we determine whether or not such a dimension makes empirical sense? One approach would be to obtain a sample of persons from the population of interest and hire undercover agents to follow the people in the sample. The agents would then surreptitiously film and tape record all interactions between the persons and black women. The investigator might also arrange some meetings so that some aspects of the encounters could be standardized. The films could then be observed and rated by a number of different raters to see if the sampled people could be reliably ranked along a dimension negative at one pole and positive at the other pole. Alternatively, they may be ranked on several more specific dimensions that purport to summarize individual differences in behavior pattern. A mental attitude may then be hypothesized to be the remnant of past activity that underlies current regularities in overt behavior. Now, the reality is that dimensions in the human sciences are almost never measured with this kind of thoroughness. Instead, the resources available to most researchers in the social sciences require that measurement of such behavior patterns be based upon self-report questionnaires, brief interviews or—with very small samples—a few contrived encounters in a laboratory. [Meanwhile, physicists in the United States spend the bulk of their money and time on the most routine of measurements.]

5. Finally, then, this situation of poor measurement greatly limits the model testing capacity of the social scientist because poorly measured variables cannot supply good information concerning the magnitude of relationships. This means that only bivariate models with directional hypotheses can be tested with these measures. Models that involve three variables, in which it is hypothesized that change in the second variable mediates the influence of the first variable on the third—a rather simple model—is outside the realm of what is testable when measures are only approximate. We could determine whether or not the three variables in this model were positively or negatively correlated, but it would not be possible to determine how one variable related to another when the third variable was held constant, for example.

In terms of a stochastic linear model, this latter point means that the partial coefficients of multiple regression equations are of dubious interpretability when the right-hand variables are poorly measured. And this is an indictment of much of the empirical research that has been conducted in the social and behavioral sciences. But more importantly, this fact points to the crucial contribution that could be made by a methodology that gives an indication of which dimensions in a specified research area might deserve the kind of attention to measurement that was described above. And the goal of this chapter is the presentation of one such methodology: a methodology for using several poor measures to estimate the "poorness" of each measure and, thereby, to adjust the structural coefficients of linear models for the distorting effects of measurement error.

9.2 THE EFFECT OF MEASUREMENT ERROR ON ESTIMATION OF STRUCTURAL COEFFICIENTS AND SLOPES

Suppose that a simple linear model described the causal relationship between three variables:

$$Y = b_1^* X_1^* + b_2^* X_2^* + s \tag{9.1}$$

where the asterisk on the coefficients indicates a population parameter, the asterisk on the variable label indicates an accurate measure, and all variables have a mean of zero. The coefficients b_1^* and b_2^* have the usual meaning: that an increase of one unit in X_1^*, say, will cause a change of b_1^* units in Y if X_2^* is held constant and s is uncorrelated with X_1^* (and similarly for b_2^*).

Now, suppose that X_1 is a rough measure of X_1^* and that X_2 is a rough measure of X_2^*. More precisely, suppose that

$$X_1 = k_{11} + k_{12} X_1^* + d_1 \tag{9.2a}$$

$$X_2 = k_{21} + k_{22} X_2^* + d_2 \tag{9.2b}$$

where d_1 and d_2 are uncorrelated with X_1^*, X_2^* and one another. Equations 9.2 describe a measurement model in which the observed measure (say, X_1) is a linear function of both the "true" score (X_1^*) and an error component (d_1).

The question that concerns us at this point is how the measurement error in X_1 and X_2 would affect estimates b_1^* and b_2^* based upon observation of Y, X_1, and X_2. That is, how would the coefficients b_1 and b_2 in

$$Y = b_1 X_1 + b_2 X_2 + s_2$$

relate to b_1^* and b_2^*?

The Bivariate Case

In order to answer this question, we must first determine the effect of measurement error on the magnitude of covariances, variances, and correlation coefficients—since sample estimates of these parameters are the building blocks of estimates of partial slopes and structural coefficients. For example, we know that b_1 and b_2 are functions of the covariance between X_1 and X_2 [$\text{Cov}(X_1,X_2)$]. How, then, does the covariance between the observed variables relate to the covariance between the perfectly measured variables (henceforth referred to as *latent variables*)? We can answer this question by performing some simple algebraic procedures to equations 9.2; the procedure is analogous to the *method of moments* used to solve for regression coefficients in Chapters 4 and 6. We begin by recalling the definition of covariance:

$$\begin{aligned}\text{Cov}(X_1,X_2) &= E\{[X_1 - E(X_1)][X_2 - E(X_2)]\}\\ &= E[X_1\ X_2] - E[X_1]E[X_2]\end{aligned}$$

and

$$\text{Cov}(X_1^*,X_2^*) = E\{[X_1^* - E(X_1^*)][X_2 - E(X_2^*)]\}$$

for the observed and latent variables, respectively. This definition of $\text{Cov}(X_1,X_2)$ can now be applied to equations 9.2 to obtain

$$\begin{aligned}\text{Cov}(X_1,X_2) &= E\{[k_{11} + k_{12}X_1^* + d_1 - E(k_{11} + k_{12}\,X_1^* + d_1)]\\ &\quad \times [k_{21} + k_{22}X_2^* + d_2 - E(k_{21} + k_{22}X_2^* + d_2)]\}\\ &= k_{12}k_{22}\,E\,\{[X_1^* - E(X_1^*)][X_2^* - E(X_2^*)]\}\\ &= k_{12}k_{22}\,\text{Cov}(X_1^*,X_2^*)\end{aligned} \tag{9.3}$$

since we assume that

$$\begin{aligned}E(d_1) &= E(d_2) = 0\\ E(d_1d_2) &= \text{Cov}(d_1,d_2) = 0\\ E(d_1X_1^*) &= \text{Cov}(d_1,X_1^*) + E(d_1)E(X_1^*) = 0 + 0 = 0\\ E(d_1X_2^*) &= \text{Cov}(d_1,X_2^*) + E(d_1)E(X_2^*) = 0 + 0 = 0\\ E(d_2X_2^*) &= \text{Cov}(d_2,X_2^*) + E(d_2)E(X_2^*) = 0 + 0 = 0\end{aligned}$$

and

$$E(d_2X_1^*) = \text{Cov}(d_2,X_1^*) + E(d_2)E(X_1^*) = 0 + 0 = 0$$

Several aspects of the result in equation 9.3 are of particular importance. First, the variance in the observed variables due to the error components (d_1 and d_2) appears to have no distorting effect on $\text{Cov}(X_1,X_2)$ as an estimate of $\text{Cov}(X_1^*,X_2^*)$—only the slopes k_{12} and k_{22} enter into the relationship. This means that if $k_{12} = k_{22} = 1$, as is the case when measurement errors are "well behaved", then the errors of measurement do not distort the covariances among the observed variables relative to the covariances among the latent variables. Second, note that the assumptions concerning the lack of covariance between the measurement errors and between the errors and the latent variables are not to be made lightly. There are many instances in which such assumptions are quite inappropriate—for example, when X_1 and X_2 involve the same *method of measurement,* or when floor effects produce asymmetrically distributed errors at the lower end of a dimension. The assumptions incorporated into equation 9.3 will be discussed further in Sections 9.3 and 9.4 of this chapter, but it should be pointed out now that these assumptions are most appropriate when errors of measurement are completely *random.* When measurement errors are random, they are just as likely to be positive as negative (i.e., they are just as likely to give a score that is too high as to give too low a score).

We can now turn to the variance of the observed variables. Recalling that

$$\text{Var}(X_1) = E\{[X_1 - E(X_1)]^2\}$$

we have from equations 9.2 that

$$\text{Var}(X_1) = E\{[k_{12}(X_1^* - E(X_1^*)) + d_1]^2\} = (k_{12}^2)\ \text{Var}(X_1^*) + \text{Var}(d_1) \quad (9.4)$$

The result in equation 9.4 reveals that measurement error in an observed variable directly *inflates* its variance relative to the latent variable. Dividing both sides of equation 9.4 by the variance of the latent variable gives

$$\text{Var}(X_1)/\text{Var}(X_1^*) = (k_{12}^2) + \text{Var}(d_1)/\text{Var}(X_1^*) \quad (9.5)$$

which reveals that the ratio of observed to true variance is equal to the square of the regression of the observed variable on the latent variable *plus* the ratio of the error variance to the true variance. If the variance of d_1 is zero, then the ratio in equation 9.5 is simply the square of k_{12}—which raises the issue of the meaning of the magnitude of k_{12} (and k_{22}).

If k_{11} equals zero and k_{12} is equal to one, and the expected value of d_1 is zero for every value of X_1^*, then the *expected value* of X_1 is X_1^*:

$$E(X_1 | X_1^*) = E(X_1^*) + E(d_1) = X_1^*$$

This means that repeated independent measurements of a particular case would produce a series of scores with a mean that is the true score for that case. For example, suppose that a study was conducted in which the physical height of persons in some population was a variable of interest. Further, suppose that the investigator was unable to make direct measurements of height with the usual (highly accurate) methods; rather, the investigator had to rely upon the estimates of one or more observers who simply encountered the persons in some social situation. In such a circumstance, X_1^* would be the true height of each person in the population (defined as the height obtained from a "doctor's office measurement"), and X_1 would be the observed height of each individual (obtained from the estimates of an observer). Now, if the *observed* heights were regressed on the *true* heights (assuming that the true heights became available at some later point), then the regression slope would be equal to one if the observers' estimates of height involved *symmetrically distributed errors.* That is, k_{12} would be one when the observed height was randomly too high or too low relative to the true height of the individual—and repeated estimates of height would produce a mean that converged on the true height.

But now suppose that one observer of heights gives very accurate *relative* estimates of height, but all estimated heights are 10 percent too high (the estimated height X_1 of each individual in the sample is 1.1 times the true height of the individual, with only negligible additional error). In this circumstance, the variance of the error term d_1 would be essentially zero, the slope k_{12} would be 1.1, and the correlation between X_1 and X_1^* would be close to one. Thus, the error d_1 may be redefined as the measurement error in the observed score relative to a linear transformation of the true score (i.e., $k_{11} + k_{12}X_1^*$) when constant additive (k_{11}) or multiplicative (k_{12}) errors of measurement occur. (A constant additive error would occur in the height estimation task if the estimates tended to be $-k_{11}$ centimeters too short, or k_{11} centimeters too tall, for all individuals.) The errors of measurement represented by k_{11} differing from zero or k_{12} differing from one are constant and linear and therefore may not be considered errors at all in situations where the units and zero point of a scale are arbitrary. When the units are not arbitrary, then k_{12} reflects the extent to which the units of the observed scale are stretched ($k_{12} < 1$) or shrunken ($k_{12} > 1$) relative to the units of the latent variable; meanwhile, the magnitude of $\text{Var}(d_1)$ reflects the extent to which the observed scores deviate about the linearly transformed true scores.

Now that we have seen the impact of well-behaved measurement errors on covariance and variance, we are in a position to assess the impact of measurement error on a metric regression coefficient and a standardized regression coefficient (the latter being synonymous with the correlation coefficient when only one predictor variable is involved). Suppose that our interest is in the regression of X_1^* on X_2^*, where

$$X_1^* = c_0 + c_1 X_2^* + u \tag{9.6}$$

and

$$\text{Cov}(X_2^*, u) = 0$$

But suppose that we had only X_1 and X_2 available as rough measures of X_1^* and X_2^*, respectively. How would the slope of X_1 regressed on X_2 compare to c_1? We can make this comparison by recalling that f_1 in

$$X_1 = f_0 + f_1 X_2 + u_2$$

is given by

$$f_1 = \text{Cov}(X_1, X_2)/\text{Var}(X_2)$$

whereas c_1 in equation 9.6 is given by

$$c_1 = \text{Cov}(X_1^*, X_2^*)/\text{Var}(X_2^*)$$

And equations 9.3 and 9.4 can be used to show that

$$\begin{aligned} f_1 &= k_{12}k_{22}\,\text{Cov}(X_1^*, X_2^*)/[k_{22}^2\,\text{Var}(X_2^*) + \text{Var}(d_2)] \\ &= c_1\{[k_{12}k_{22}\,\text{Var}(X_2^*)]/[k_{22}^2\,\text{Var}(X_2^*) + \text{Var}(d_2)]\} \end{aligned} \tag{9.7}$$

This equation is most easily understood if we assume that $k_{12} = k_{22} = 1$; then

$$f_1 = c_1\{\text{Var}(X_2^*)/[\text{Var}(X_2^*) + \text{Var}(d_2)]\}$$

which means that the absolute value of f_1 will be less than c_1 if $\text{Var}(d_2)$ is greater than zero—since the term in braces will be less than one. This reduction in the magnitude of the regression coefficient is referred to as attenuation due to measurement error. Notice that if there is no measurement error in X_2 (meaning that $\text{Var}(d_2)$ is zero), then there is no attenuation due to the measurement error *in* X_1. This is an important fact to remember; one of the advantages of experimental manipulation is the plausibility of the assumption that measurement error is only present in the dependent variable. When this is the case, and where the slope of the observed dependent variable on the latent dependent variable is one, then the measurement error in the dependent variable does not affect the magnitude of the structural coefficients—although it does increase the standard errors of estimates of those coefficients.

When the units of variables are arbitrary, it is often convenient to standardize variables for interpretation of relative magnitudes. If X_1, X_2, X_1^*, and X_2^* are standardized with zero mean and standard deviations of one, then

$$k_{11} = k_{21} = c_0 = f_0 = 0$$
$$k_{12} = \text{Corr}(X_1,X_1^*)$$
$$k_{22} = \text{Corr}(X_2,X_2^*)$$

and

$$c_1 = \text{Corr}(X_1^*,X_2^*)$$
$$f_1 = \text{Corr}(X_1,X_2)$$

Furthermore, applying the relations developed in Chapter 5, we find that

$$\text{Var}(d_1) = 1 - k_{12}^2$$

and

$$\text{Var}(d_2) = 1 - k_{22}^2$$

These relations can now be substituted into equation 9.7 to obtain

$$\begin{aligned} f_1 &= c_1[k_{12}k_{22}/(k_{22}^2 + 1 - k_{22}^2)] \\ &= c_1(k_{12}k_{22}) \end{aligned}$$

or

$$\text{Corr}(X_1,X_2) = k_{12}k_{22}\,\text{Corr}(X_1^*,X_2^*) \tag{9.8}$$

The result in equation 9.8 is conceptually convenient. It states that when measurement errors in two variables are random (or otherwise well behaved), the correlation between the observed variables is attenuated relative to the correlation between the true variables to an extent that is equal to the product of the correlation between each observed variable and its corresponding latent variable (i.e., $k_{12}k_{22}$). This means that standardized regression coefficients are attenuated by measurement error in both the independent *and the dependent* variable. Thus, yet another disadvantage is associated with the use of standardized variables (recall that the other major disadvantage is the effect of the range of scores on the independent variables that is included in the study).

In this standardized format, the coefficients k_{12} and k_{22} will be referred to as *validity* coefficients, since they reflect how well the observed variable correlates with the corresponding latent variable. (The square of the validity coefficient of a measure is sometimes referred to as the reliability of the measure, but this usage will not be adopted in this context. Subsequently, reliability will be distinguished from validity as a related but distinct characteristic of a measure.) Thus, the validity of X_1 as a measure of X_1^* is k_{12}—and similarly for X_2, X_2^*, and k_{22}. Various schemes for estimating these validities when the true scores are not available may be familiar to the reader. For example, Cronbach's (1952) coefficient *alpha* has been used to correct correlations between psychological tests for attenuation due to measurement error. Cronbach used indices of internal consistency to estimate k_{12} and k_{22}, and then divided the observed correlation by these estimates to obtain an estimate of the correlation between the latent (true) variables. Although such procedures may be useful in some circumstances, they are not generally recommended or described in this context for two reasons. First, estimates of validity based upon internal consistency tend to confound method variance with latent variable variance—thereby inflating the estimates of validity. A better approach to the assessment of validity involves use of method divergent measures of latent variables as well as repeated use of the same (or highly similar) measures. The term *method divergent* refers to the use of distinct methods of measurement such that the systematic errors of measurement (as opposed to *random* errors of measurement) are unrelated across methods of measurement. This approach to validity estimation is similar to that proposed by Campbell and Fiske (1969)—multitrait multi-method approach. We shall return to a more detailed presentation of this topic in Section 9.4 of this chapter. The second objection to the use of such statistics as Cronbach's alpha is that the measurement model assumed by Cronbach cannot be rejected by the observations; in other words, there is no way to test the dimensionality hypothesis that is implicit in the use of most psychological scales. The procedures developed in Section 9.3 of this chapter provide an investigator with the opportunity to test whether or not his assumptions concerning the latent variables being measured are consistent with the data. (A third objection to such measures as Cronbach's alpha is based upon the focus of the measures on standardized variables. This focus means that estimates of validity will be dependent upon the range of the true scores included in a particular study. Placing an emphasis on *metric* validities avoids this problem with interstudy comparisons.)

The important conclusion to draw from this foray into the impact of measurement error on bivariate slopes and correlations is that the measurement error *attenuates* the observed slope or correlation relative to the true slope or correlation between the latent variables. In practical terms, then, the use of error-laden measures makes the observed measures of association conservative estimates of true associations or causal relationships. (Thus the usual comment in research reports that the rather modest observed relationship is important because the measures employed were so poor.) This state of affairs is tolerable in that the attenuation is always in the conservative direction, and random measurement error cannot be responsible for a true positive relationship appearing negative, or vice versa. Furthermore, the extent of attenuation is a direct function of the validity of each measure, so that efforts for achieving validities of .8 or .9 would assure that conclusions from samples about the direction of relationships would not be misleading. But now we move to the case in which more than the direction of a relationship (whether it is positive or negative) is of interest. That is, we move to a situation in which the magnitude of relationships becomes important information because we want to determine, for example, how a bivariate relationship is transformed by the controlling of other variables. Obviously, such statistical control is most important in studies that do not involve experimental control over important variables. Thus, in order to move to a methodology appropriate for quasi-experimentation, as well as randomized experimentation, it will be necessary to develop methods for assessing measurement error in a testable manner. Before describing that methodology in greater detail, the following discussion outlines the rather dramatic distortions in *partial* regression slopes and structural coefficients that may be produced by even moderate amounts of measurement error.

The Trivariate Case

Now that the effects of random or well-behaved measurement error on covariances, variances, and correlations between observed variables have been outlined, we are in a position to investigate the much more bothersome effects of measurement error on partial regression coefficients—whether interpreted as slopes or structural coefficients. The trivariate case is sufficient for this demonstration (i.e., one left-hand variable and two right-hand variables), since the addition of more variables simply compounds the problem. The goal here is to compare the magnitude and direction of the regression coefficient *when the latent* (true) scores are used with the regression coefficient *when observed* scores (flawed by measurement error) are used.

Recall that the three variable regression model was given in equation 9.1:

$$Y = b_1^* X_1^* + b_2^* X_2^* + s$$

where X_1^* and X_2^* are latent variables (true scores). If we then observed the variables X_1 and X_2 where (as in equations 9.2a and 9.2b)

$$X_1 = k_{11} + k_{12} X_1^* + d_1$$

$$X_2 = k_{21} + k_{22} X_2^* + d_2$$

then our concern here is with how the results of regressing Y on X_1 and X_2 would compare with the results of regressing Y on X_1^* and X_2^*. This question can be answered by solving for b_1 in the multiple regression equation

$$Y = b_1 X_1 + b_2 X_2 + s_2$$

and then using equations 9.2 (the measurement model) to relate this value to b_1^*:

$$\begin{aligned}
(1)\ b_1 &= \frac{\text{Cov}(X_2,X_2)\text{Cov}(Y,X_1) - \text{Cov}(Y,X_2)S(X_1,X_2)}{\text{Cov}(X_1,X_1)\text{Cov}(X_2,X_2) - \text{Cov}(X_1,X_2)^2} \\
(2)\ \text{Cov}(X_1,X_1) &= k_{12}^2\text{Cov}(X_1^*,X_1^*) + \text{Cov}(d_1,d_1) \\
(3)\ \text{Cov}(X_2,X_2) &= k_{22}^2\text{Cov}(X_2^*,X_2^*) + \text{Cov}(d_2,d_2) \\
(4)\ \text{Cov}(X_1,X_2) &= k_{12}k_{22}\text{Cov}(X_1^*,X_2^*) \\
(5)\ \text{Cov}(Y,X_1) &= k_{12}\text{Cov}(Y,X_1^*) \\
(6)\ \text{Cov}(Y,X_2) &= k_{22}\text{Cov}(Y,X_2^*)
\end{aligned} \qquad (9.9)$$

Notice that all the bivariate relations may be derived as above. The remaining task is to substitute the true score variances and covariances into equation 9.9 so that the relationship between b_1 and b_1^* can be assessed. The result of this substitution is equation 9.10:

$$b_1 = \frac{[k_{22}^2\,\mathrm{Var}(X_2^*) + \mathrm{Var}(d_2)][k_{12}\,\mathrm{Cov}(Y,X_1^*)] - [k_{22}\,\mathrm{Cov}(Y,X_2^*)][k_{12}k_{22}\,\mathrm{Cov}(X_1^*,X_2^*)]}{[k_{12}^2\,\mathrm{Var}(X_1^*) + \mathrm{Var}(d_1)][k_{22}^2\,\mathrm{Var}(X_2^*) + \mathrm{Var}(d_2)] - [k_{12}k_{22}\,\mathrm{Cov}(X_1^*,X_2^*)]^2} \tag{9.10}$$

Unfortunately, however, equation 9.10 is not easily interpreted in this form.

One of the best ways to explore a complex algebraic expression such as equation 9.10 is with a heuristic computer program that steps through a range of values for each parameter in the function, making the desired comparisons along the way. We can apply such a procedure to equation 9.10 and assess the relationship between b_1 and b_1^* along the way. The parameters that will be varied (or stepped) are the variances and covariances among the latent variables, the slopes k_{12} and k_{22}, and the variances of the errors d_1 and d_2. Understanding of this heuristic procedure is enhanced if we assume that X_1, X_2, X_1^*, X_2^*, and Y are standardized variables. Then b_1^* and b_2^* are beta weights, k_{12} and k_{22} are correlation coefficients relating the observed variables to the latent variables, and the variances of the measurement errors are equal to one minus the square of the respective validity. Table 9.1 presents some results of this heuristic exercise.

Table 9.1 gives the correlations among the latent variables, the value of b_1^*, and then the value of b_1 that would be obtained if the independent variables were measured with the accuracy indicated by the correlations k_{12} and k_{22}. Note that when the validities of the observed variables are quite high, $k_{12} = k_{22} = .9$, there can be extreme distortion in b_1 relative to b_1^*. Not only can b_1 be attenuated relative to b_1^*, but b_1 can be larger than b_1^*, or it can be a different sign. It should be pointed out that these distortions are fairly extreme and do not represent the distortions to be expected throughout the entire range of values of the correlations among the latent variables. Nevertheless, when only observed variables are available and validities are not estimated, then distortions such as these and worse can, and surely have, occurred.

Table 9.1 Some results from a heuristic assessment of the effects of measurement error on partial regression coefficients.

$\mathrm{Corr}(Y,X_1^*)$	$\mathrm{Corr}(Y,X_2^*)$	$\mathrm{Corr}(X_1^*,X_2^*)$	b_1^*	b_1	b_1/b_1^*
$k_{12} = k_{22} = .90$					
.6	.8	.3	.395	.388	.981
.2	.6	.6	−.250	−.108	.432
.4	.6	.6	.063	.128	2.044
.6	1.0	.6	.000	.134	∞
.2	.4	.7	−.157	−.036	.227
.6	.8	.7	.078	.194	2.476
.6	.8	.8	−.111	.127	−1.139
$k_{12} = .9$ and $k_{22} = .6$					
.2	.8	.2	.042	.103	2.452
.4	.8	.4	.095	.215	2.263
.4	.6	.5	.133	.231	1.737
.4	.8	.5	.000	.179	∞
.4	.8	.6	−.125	.142	−1.136
.2	.8	.7	−.706	−.108	.153
.6	.8	.9	−.632	.249	−.394

When the right-hand variables are correlated only slightly, then measurement error in the right-hand variables affects the regression coefficients in a manner analogous to the bivariate case. In other words, if k_{12} and k_{22} are both 0.6 and if Corr(X_1^*,X_2^*) is close to zero, then the regression coefficients for the observed variables will be attenuated by 0.4 (b_1/b_1^* would equal 0.6). Thus, quasi-experiments that deal with naturally correlated variables are subject to greater dangers of very misleading results due to measurement error than are experiments in which randomization reduces the correlation between independent variables and covariates, as well as the correlation among independent variables, to near zero levels.

The bottom one-half of Table 9.1 is particularly interesting, since it represents a situation that is not uncommon in the social sciences. Note that k_{12} is 0.9, indicating a substantial validity. Now suppose that X_1^* is the variable that is of primary interest, so we have taken great pains to measure it accurately. Then, in order to test some more elaborate hypotheses, additional (but less carefully measured) variables are added to the model. The message to be gleaned from those examples given in the bottom one-half of Table 9.1 is that such a practice is most dangerous; the coefficient of the carefully measured variable is grossly distorted by the measurement error in the less valid measure.

At this point the reader may be prepared to abandon quasi-experimentation. "If all our measures must be extremely valid before we can obtain reasonable estimates of partial regression coefficients, then there is little prospect of doing important quasi-experimental research in the near future," might be an appropriate reply to the above discussion. Especially in research involving large representative samples of populations and enough variables to even approach the apparent complexity of social systems, the prospect of using highly valid measures seems remote. But there is a solution to the dilemma. It is certainly not the only solution, and it may be a solution that will be surpassed in the future, but it is a solution that is workable with current resources and procedures of data collection. That solution is the use of multiple flawed indicators to triangulate on the true scores. This means that the structural coefficients relating latent variables can be estimated without distortion (bias) even though the true scores are never observed. The only requirement for achieving this bit of Merlin's work is that multiple measures with well-behaved errors must be developed. This requirement, of course, is no small task in most instances, but it is one that is within the reach of many research projects in the human sciences. Section 9.3 of this chapter provides a more detailed discussion of multiple indicator measurement models.

9.3 INTRODUCTION TO MULTIPLE INDICATOR MEASUREMENT MODELS

A Numerical Example

As a preliminary to a more mathematical presentation of multiple indicator models, a simple numerical example can make the idea less mysterious. Many people react to the claim that poor measures can be used to obtain results that would be obtained with good measures with a healthy disbelief. Or, the reaction is that if such poor measures can be used to estimate regression coefficients among latent variables, why cannot those poor measures be used to obtain the true scores? The following example should answer these questions in a concrete manner and allay any fears that some sleight of hand is involved.

For the sake of simplicity, suppose that the heights of persons in some population are the true scores of concern. Our goal is to regress the variable Y (e.g., true weight) on the heights of these persons to obtain the slope b_1^*. Let the variable Y be a measure of bodily weight that involves only minute amounts of random measurement error. Suppose that reality is depicted by the model

$$Y = b_0^* + b_1^* X^* + s \tag{9.11}$$

where X^* is the true height of persons, Cov(s,X^*) is zero, and $E(s|X^*) = 0$. The model in equation 9.11 states that weight increases linearly with height, and that each increase of one unit in height is associated, on the average, with an increase of b_1^* units of weight. (In fact, of course, the relationship between the height and weight of most populations would not be linear. This would mean that Y or X^* would have to be transformed prior to calculation. A logarithmic transformation would be a good candidate in this instance. However, we shall ignore this complexity for the moment.)

Now suppose that we must rely on some very rough measures of X^*—estimates of height by two observers who have seen the persons in the sample in some social situation. These two measures are X_1 and X_2. The question now is whether or not we could obtain an unbiased estimate of b_1^* based upon the measures X_1 and X_2 (assuming that we did not have the true heights X^*). The answer to this question depends upon how the rough measures (called *observed variables* or indicators) relate to the true scores. If the errors in the indicators are random, and the expected value of the indicator is the true score (i.e., repeated measurement would eventually produce a mean with a

Table 9.2 Contrived data from the observation of height (X) and weight (Y) for 16 persons.

Person	True weight Y	True height X^*	First height indicator X_1	Second height indicator X_2	Error d_1	Error d_2
1	30	150	145	142	−5	−8
2	70	150	155	158	+5	+8
3	40	155	150	163	−5	+8
4	80	155	160	147	+5	−8
5	86	158	153	150	−5	−8
6	46	158	163	166	+5	+8
7	90	160	155	168	−5	+8
8	50	160	165	152	+5	−8
9	50	160	155	152	−5	−8
10	90	160	165	168	+5	+8
11	54	162	157	170	−5	+8
12	94	162	167	154	+5	−8
13	100	165	160	157	−5	−8
14	60	165	170	173	+5	+8
15	110	170	165	178	−5	+8
16	70	170	175	162	+5	−8

value that would converge on the true score), then there is a way to use X_1 and X_2 to triangulate on b_1^*—even though triangulation on X^* is not possible. (We will develop less restrictive requirements in subsequent measurement models.)

The statements concerning the way the indicators relate to the true scores (latent variable) constitute a *measurement model.* In this contrived example we are going to have the true scores, and so we will be able to assess the relationship between the indicators and the latent variable. This will enable us to see that the two indicators could be used to obtain the same result that would be obtained with the true scores. However, in many instances, the true scores will simply be unavailable or even *unobservable.* In such instances, it will be necessary to develop methods to test our measurement model even though we do not have the true scores. The traditional notions of factor theory will be used for this purpose.

To return to our example, suppose that the data in Table 9.2 resulted from observation of the height and weight of a sample of 16 persons. The most striking aspect of these data is the symmetry of the measurement errors. Such an outcome would be extremely unlikely, but in a larger sample it is often the case that measurement errors have a symmetrical distribution about the true scores. The measurement errors (d_1 and d_2) in Table 9.2 were constructed so that they would not only be symmetrical about the true scores, but also so that d_1 and d_2 are uncorrelated with one another and with the residuals that will result from a regression of Y on X^*.

The correlations and covariances among the variables in Table 9.2 are given in Table 9.3. It should be noted in Table 9.3 that the measurement errors are correlated only with the indicator that each is a part of. The remaining correlations involving d_1 and d_2 are zero—as one would expect in a *large* sample when d_1 and d_2 are truly random. It should also be noted that the correlation between Y and the true variable (X^*) is larger than the correlation between Y and the error-laden indicators X_1 and

Table 9.3 Correlations and covariances among the variables given in the height/weight example (Table 9.2).

Correlation matrix

	Y	X^*	X_1	X_2	d_1	d_2
Y	1.00000					
X^*	.493818	1.00000				
X_1	.370633	.750546	1.00000			
X_2	.285846	.578848	.434452	1.00000		
d_1	.000000	.000000	.660819	.000000	1.00000	
d_2	.000000	.000000	.000000	.815435	.000000	1.00000

Covariance matrix

	Y	X^*	X_1	X_2	d_1	d_2
Y	529.000					
X^*	64.5000	32.2500				
X_1	64.5000	32.2500	57.2500			
X_2	64.5000	32.2500	32.2500	96.2500		
d_1	.000000	.000000	25.0000	.000000	25.0000	
d_2	.000000	.000000	.000000	64.5000	.000000	64.0000

X_2. The correlation involving the true scores is .494, whereas the other two indicators of height correlate .371 and .286 with the weight variable. In fact, these two lower correlations are attenuated relative to the true correlation by an amount derived from the correlation between the indicator and the true scores:

$$\begin{aligned} r(Y,X_1) &= r(Y,X^*)r(X_1,X^*) \\ &= (.494)(.751) \\ &= .371 \end{aligned}$$

and

$$\begin{aligned} r(Y,X_2) &= r(Y,X^*)r(X_2,X^*) \\ &= (.494)(.579) \\ &= .286 \end{aligned}$$

(see the derivation of equation 9.8). Clearly, then, if we did not know $r(Y,X^*)$, but we knew $r(X_1,X^*)$ or $r(X_2X^*)$, then we could solve for $r(Y,X^*)$ in terms of $r(Y,X_1)$, say, and $r(X_1,X^*)$. But without actually observing X^*, it is still unclear just how we might even obtain $r(X_1,X^*)$—we have not arrived at that point yet. Noting the variances of the variables in the covariance matrix shows that the measurement errors inflate the variance of the indicators relative to the true variable. The extent of this inflation is again a function of the validities of the indicators:

$$\begin{aligned} S(X_1,X_1) &= S(X^*,X^*) + S(d_1,d_1) \\ &= 32.25 + 25.00 \\ &= 57.25 \end{aligned}$$

and the estimate of the population variance is

$$\begin{aligned} NS(X_1,X_1)/(N-1) &= (16)(57.25)/(16-1) \\ &= 61.07 \end{aligned}$$

where

$$\begin{aligned} S(X_1) &= S(X^*)/r(X_1,X^*) \\ &= (5.68)/(.751) \\ &= 7.57 \end{aligned}$$

As we shall see, the reason that all these relations work out so simply is that X_1 and X_2 are *tau equivalent* measures of X^*.

The next piece in our puzzle is the computation of an estimate of b_1^* in equation 9.11. We obtain this estimate by regressing Y on X^* to obtain

$$\begin{aligned} Y &= -250 + 2.00X^* + s \\ R\text{-square} &= .244 \end{aligned} \qquad (9.12)$$

where

$$\begin{aligned} 2.00 &= S(Y,X^*)/S(X^*,X^*) \\ &= 64.50/32.25 \end{aligned}$$

Now, what if Y had been regressed on the *indicators* rather than the true scores? We would expect the resulting slopes to be attenuated relative to the value of 2.00, and this is the case

$$Y = -110 + 1.13X_1 + u_1 \qquad \text{R-square} = .137 \tag{9.13}$$

$$Y = -37 + .67X_2 + u_2 \qquad \text{R-square} = .082 \tag{9.14}$$

where

$$1.13 = S(Y,X_1)/S(X_1,X_1)$$
$$.67 = S(Y,X_2)/S(X_2,X_2)$$

Clearly, the slopes obtained in equations 9.13 and 9.14 would be rather misleading if included in some larger model. For example, if a researcher was attempting to control statistically for height when relating weight to other variables, the use of X_1 or X_2 in place of X^* would very much overstate the variance in Y that was linearly independent of X^*.

The reader may now be wondering if a better estimate of b_1^* would be obtained if Y was regressed on the mean of X_1 and X_2. The result of such a calculation is

$$Y = -119 + 1.18[(X_1 + X_2)/2] + u_3 \qquad \text{R-square} = .144 \tag{9.15}$$

which improves the situation only slightly over what was obtained with the better of the two indicators (see equation 9.13). And yet another idea may occur to the reader: include X_1 and X_2 as two variables in a multiple regression equation to see if the R-square is comparable to that of the true scores. The result of these calculations is

$$Y = -136 + .92X_1 + .36X_2 + u_4 \qquad \text{R-square} = .157 \tag{9.16}$$

And once again the result is essentially comparable to the regression that involved only the better of the two indicators. The reader may be getting the impression that there is little prospect of using X_1 and X_2 to obtain an estimate of b_1^* that really improves on equation 9.13. But this is not the case.

The secret to the riddle is to specify the measurement model and use it to obtain an estimate of b_1^*. The measurement model is specified as follows:

$$X_1 = k_{11} + k_{12}X^* + d_1 \tag{9.17a}$$
$$X_2 = k_{21} + k_{22}X^* + d_2 \tag{9.17b}$$

Now, if we incorporate the assumption that d_1 and d_2 are symmetrically distributed about the true scores, then equations 9.17 may be rewritten as

$$X_1 = X^* + d_1$$
$$X_2 = X^* + d_2$$

(And if the data in Tables 9.2 and 9.3 are actually used to compute the regression of X_1 and X_2 on X^*, we find that

$$X_1 = 0.0 + 1.00X^* + d_1$$
$$X_2 = 0.0 + 1.00X^* + d_2$$

because of the way in which the measurement errors are distributed about the true scores in the concocted data.) Thus, if we multiply equation 9.17a by equation 9.17b for each case and sum, we find that

$$\begin{aligned} E(X_1X_2) &= E(X^*X^*) + E(d_1X^*) + E(d_2X^*) + E(d_1d_2) \\ &= E(X^*X^*) \end{aligned}$$

if d_1 and d_2 are uncorrelated with one another and X^*. And

$$E(X_1) = E(X_2) = E(X^*) = X^*$$

implies that

$$\text{Cov}(X_1,X_2) = \text{Var}(X^*,X^*) \tag{9.18}$$

Now result 9.18 is of extreme importance, since it tells us that we can estimate the variance of X^* from the covariance of X_1 and X_2! Since this variance is the denominator of b_1^*, we only lack an estimate of Cov(Y,X^*)—the numerator of b_1^*—in our search for an estimate of b_1^* based on X_1 and X_2. As noted in Section 9.2, it turns out that Cov(Y,X_1) and Cov(Y,X_2) are both equal to Cov(Y,X^*), since the measurement errors are well behaved. This can be seen by multiplying equation 9.17a or 9.17b by Y and taking the expectation:

$$\begin{aligned} E(YX_1) &= E(YX^*) + E(Yd_1) \\ &= E(YX^*) \end{aligned}$$

and

$$\begin{aligned} E(YX_2) &= E(YX^*) + E(Yd_2) \\ &= E(YX^*) \end{aligned}$$

if d_1 and d_2 are uncorrelated with Y. And since $E(X_1) = E(X_2) = E(X^*)$, we have

$$\text{Cov}(Y,X_1) = \text{Cov}(Y,X^*) \tag{9.19a}$$

and

$$\text{Cov}(Y,X_2) = \text{Cov}(Y,X^*) \tag{9.19b}$$

Thus, we now have two ways to estimate the numerator of b_1^*:

$$b_1^* = \text{Cov}(Y,X_1)/\text{Cov}(X_1,X_2) \tag{9.20a}$$

and

$$b_1^* = \text{Cov}(Y,X_2)/\text{Cov}(X_1,X_2) \tag{9.20b}$$

If we now return to the data in Table 9.3 and substitute into equations 9.20, we find that

$$S(Y,X_1)/S(X_1,X_2) = 64.50/32.25 = 2.00$$

and

$$S(Y,X_2)/S(X_1,X_2) = 64.50/32.25 = 2.00$$

And this all means that we could have obtained an estimate of b_1^* comparable to that provided by the true scores, even though we had only observed the error-laden indicators.

In order to have accomplished this, however, we would have had to specify the measurement model without the benefit of being able to actually estimate k_{11}, k_{12}, k_{21}, and k_{22}. That is we would have had to *assume* that X_1 and X_2 were tau equivalent measures of X^* in order to obtain the equations 9.20. But note that the equations 9.20 would provide some opportunity to *test* these assumptions, since two ways of estimating b_1^* emerged. Thus, if the measurement model is true, then

$$\text{Cov}(Y,X_1) = \text{Cov}(Y,X_2)$$

and $S(Y,X_1)$ should be close to $S(Y,X_2)$—with differences between the two accounted for by sampling error. This equality of covariances is called an *overidentifying restriction* and provides a test of the measurement model even though the true scores were not available. Of course, these two covariances could be equal for some reason other than that the measurement model is true, but at least the model predicted the outcome. Note that in our contrived data these two covariances are exactly equal, since the measurement errors meet the requirements of tau equivalence exactly [i.e., $k_{12} = k_{22} = 1$, $\text{Cov}(d_1,d_2) = \text{Cov}(d_2,X^*) = \text{Cov}(d_2,X^*) = 0$].

Although we will not be developing methods for the estimation of standard errors of estimates of b_1^* until Chapter 10, it is instructive at this point to contrast the standard error of an estimate of b_1^* when the true scores are available and when they are only inferred from error-laden indicators. The slope of 2.00 for Y regressed on X^* has an estimated standard error of .94 if the standard procedures for computing such an estimate are followed. In contrast, when the as yet undescribed method of *maximum likelihood* is used to estimate b_1^* from observations of Y, X_1 and X_2, the resulting slope is 2.00 and has an estimated standard error of 1.40. The slope obtained using the error-laden indicators is equal to the result using the true scores, but note that the estimated standard error of this statistic is nearly 50 percent larger. This means that larger samples will be required to obtain accurate estimates from multiple indicator models than would be necessary if true scores were available.

Path Diagrams of Measurement Models

Linear structural models can be depicted as path diagrams as outlined in Chapter 6. The addition of a measurement model is straightforward. As with the structural portion of the model, the conventions are (1) straight arrows lead from right-hand variables to left-hand variables and (2) curved lines connect correlated right-hand variables. Thus, the structural model in equation 9.11 and the *measurement model* in equations 9.17 would be depicted as in Figure 9.1. Note that the arrows connecting indicators with the latent variable lead *from* the latent variable *to* the observed indicator. This arrangement reflects the presumption that the indicators *reflect* changes in the latent variable and not *vice versa.*

The assumptions concerning the lack of correlation between the measurement errors and other variables in the model are reflected in the absence of any paths connecting d_1 and d_2 with any variables other than X_1 and X_2, respectively. The tau equivalence of the two indicators of X^* is reflected in the diagram by the association of a path coefficient of one with both of the arrows leading from X^* to X_1 and X_2. In subsequent models, this constraint will be relaxed; but this constraint can be relaxed only when more than two indicators are available.

For most people, it is much easier to think in terms of path diagrams than equations.

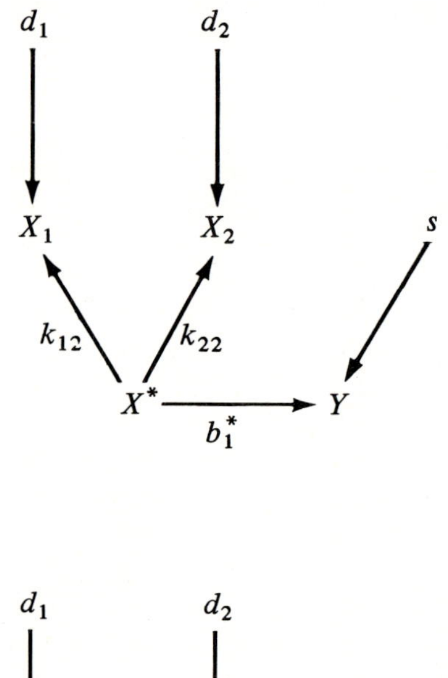

Figure 9.1 Path diagram of a structural model (equation 9.11) that incorporates a measurement model (equation 9.17). Y = weight, X^* = true height, X_1 = estimated height, X_2 = estimated height, d = measurement error.

Thus, it is useful to be able to generate the equations from the diagram. This is accomplished by writing a linear equation for each variable, whether observed or latent, in the diagram that is at the tip of an arrow. The variable at the tip of the arrow is treated as a left-hand variable in a linear equation that includes as right-hand variables all those variables at the tails of arrows that lead to the left-hand variable. The coefficients of the right-hand variables are then those on the arrows leading from the respective right-hand variable to the left-hand variable. (Since it is almost always the case that the coefficients of arrows leading *from* disturbances and measurement errors are given the value one, these values are often not included in path diagrams involving metric variables.) The path model does not provide for the inclusion of intercept terms, however. When the zero point of variables is arbitrary, this exclusion suggests treating all variables as deviations about their means, but there is certainly no necessity to do so. The linear equations may simply be written as outlined above with the addition of intercept terms.

If we apply this procedure for writing linear equations from path diagrams to Figure 9.1, we can easily reconstruct equations 9.11 and 9.17. First, we note that three variables in the model have arrows leading to them: Y, X_1, and X_2. Thus, there are going to be three equations in the full model. Second, we note which variables are at the tails of arrows leading to each of these three variables:

$$s \text{ and } X^* \text{ lead to } Y$$

$$d_1 \text{ and } X^* \text{ lead to } X_1$$

$$d_2 \text{ and } X^* \text{ lead to } X_2$$

Finally, we may write the three equations using the coefficients given on the paths and adding intercepts:

$$Y = b_0^* + b_1^* X^* + u$$

$$X_1 = k_{11} + X^* + d_1$$

$$X_2 = k_{21} + X^* + d_2$$

If X^* and Y are presumed to be causally related, then the first of these three equations is termed the *structural* portion of the model, while the latter two equations comprise the *measurement model.* Together with a statement concerning the pattern of correlations assumed to hold among the measurement errors and disturbances, the set of three equations are said to comprise a *linear structural equation model,* in which causal relations are presumed to hold among the perfectly measured and/or latent variables only.

A More Formal Approach

Up to this point in Section 9.3 the discussion has been organized to introduce the newcomer to the methods that allow estimation of linear relations among true variables even when those true scores are not, or cannot be, observed. The following is a more formal presentation of these ideas.

We begin by defining two types of unobserved variables that may be reflected in observed indicators. The first type is the *unobservable* variable. Unobservable variables are hypothetical constructs that cannot, with current technologies, be observed directly. As physics has matured as a science, it has relied more and more upon hypothetical constructs that are not directly observable. Niels Bohr never looked at an electron; and Murray Gell-Mann will never see a quark. Rather, these notions are hypothetical constructs that bring order to a constellation of direct observations—such as the deflections of a needle on a meter or spots on a photographic plate. Although only the thermometer reading is observed directly, theoretical progress derives from the presumption that a change in those readings reflects intermolecular processes that cannot be observed directly. As measurement technologies develop, certain hypothetical dimensions are so accurately measured that the procedure becomes synonymous with the hypothetical dimension. But this close relationship between indicator and hypothetical construct should not be allowed to obscure the importance of unobservable abstractions in the course of theory building.

The second type of unobserved variable is the *directly observable variable* that is nevertheless measured indirectly for practical or theoretical reasons. For example, the number of people that reside in a particular part of New York City at a particular time is a directly observable entity. And yet, most any study that included a variable such as this would rely upon indirect indicators of the true population—such as the count at the last census, interim estimates by the Planning Department, Tax Department information on the number of dwellings in the area, and so on. This kind of unobserved variable is clearly distinct from a hypothetical construct (such as the quality of life in the area, level of alienation, etc.) that population density may influence. Nevertheless, from a mathematical point of view, the measurement models for unobservable and unobserved observable variables have the same form and logic. Thus, both of these types of unobserved variables will be referred to as *latent variables* in the remainder of this book. (There is another category of variable that will be used in Chapters 10 and 11. This additional category is that of the *unmeasured* variable in a structural equation model. An unmeasured variable is one that is incorporated in a model even though there is no attempt to observe indicators of the variable. Obviously, the interpretation of such variables is very restricted, and the effects of such a variable can be estimated only if certain assumptions about the relationship between the unmeasured variables and the measured latent variables are accepted. The disturbance in a structural model is an unmeasured variable. Also, the environmental and genetic factors in models fit to data from twins are examples of commonly used unmeasured variables.)

The relationship between the observed variables and the latent variables constitutes the measurement model. The specification of this relationship that stems from the psychometric tradition is the *factor model.* If $\underline{X}$ is a vector of q observed variables,

$$\begin{aligned}\underline{X} &= [X_1 \quad X_2 \quad X_3 \quad \ldots \quad X_q]' \\ &= \begin{bmatrix} X_1 \\ X_2 \\ X_3 \\ \vdots \\ X_q \end{bmatrix}\end{aligned}$$

then we presume in a classic factor model that $\underline{X}$ is a function of n latent variables $(F_1, F_2, \ldots, F_n)$, q measurement errors, and, for simplicity at this point, q intercepts of zero:

$$
\begin{aligned}
X_1 &= k_{11}F_1 + k_{12}F_2 + \ldots + k_{1n}F_n + d_1 \\
X_2 &= k_{21}F_1 + k_{22}F_2 + \ldots + k_{2n}F_n + d_2 \\
&\vdots \\
X_q &= k_{q1}F_1 + k_{q2}F_2 + \ldots + k_{qn}F_n + d_q
\end{aligned}
$$

or

$$\underline{X} = (\underline{Lx})(\underline{F}) + \underline{d} \tag{9.21}$$

where

$$\underline{Lx} = \begin{bmatrix} k_{11} & k_{12} & \ldots & k_{1n} \\ k_{21} & k_{22} & \ldots & k_{2n} \\ \vdots & & & \\ k_{q1} & k_{q2} & \ldots & k_{qn} \end{bmatrix}$$

$$\underline{F} = [F_1 \quad F_2 \quad \ldots \quad F_n]'$$

and

$$\underline{d} = [d_1 \quad d_2 \quad \ldots \quad d_q]'$$

This model is depicted in path diagram form in Figure 9.2. Note the similarity to a multivariate regression model. Now equation 9.21 probably seems odd since the right-hand variables in the equations are all latent variables for which there are no scores

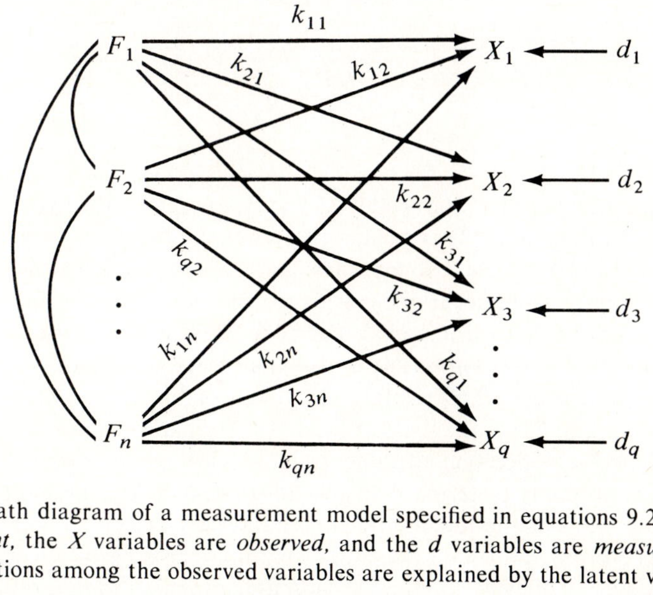

Figure 9.2 Path diagram of a measurement model specified in equations 9.21. The F variables are *latent*, the X variables are *observed*, and the d variables are *measurement errors*. Linear associations among the observed variables are explained by the latent variables.

on which to base computations. However, as demonstrated in the height and weight example above, it is possible to estimate the coefficients of these latent variables, and their covariance matrix, if certain constraints are placed upon the coefficients in $\underline{Lx}$, the covariances among the measurement errors [Cov($\underline{d}$)], and the covariances among the latent variables [Cov($\underline{F}$)]. In fact, with a sufficient number of constraints, the free parameters in the model become overidentified (usually only a few of them), and these overidentifications provide the basis for testing the fit of the model to observed data.

To facilitate specification of these constraints, the following definitions are made:

$$\begin{aligned}\underline{PH} &= \text{covariance matrix of the latent variables } (\underline{F}) \\ &= \text{Cov}(\underline{F}) \\ &= \begin{bmatrix} \text{Cov}(F_1,F_1) & \text{Cov}(F_1,F_2) & \ldots & \text{Cov}(F_1,F_n) \\ \text{Cov}(F_2,F_1) & \text{Cov}(F_2,F_2) & \ldots & \text{Cov}(F_2,F_n) \\ \vdots & & & \\ \text{Cov}(F_n,F_1) & \text{Cov}(F_n,F_2) & \ldots & \text{Cov}(F_n,F_n) \end{bmatrix} \\ &= \begin{bmatrix} ph_{11} & ph_{12} & \ldots & ph_{1n} \\ ph_{21} & ph_{22} & \ldots & ph_{2n} \\ \vdots & & & \\ ph_{n1} & ph_{n2} & \ldots & ph_{nn} \end{bmatrix}\end{aligned}$$

$$\begin{aligned}\underline{Td} &= \text{covariance matrix of measurement errors } (\underline{d}) \\ &= \text{Cov}(\underline{d}) \\ &= \begin{bmatrix} \text{Cov}(d_1,d_1) & \text{Cov}(d_1,d_2) & \ldots & \text{Cov}(d_1,d_q) \\ \text{Cov}(d_2,d_1) & \text{Cov}(d_2,d_2) & \ldots & \text{Cov}(d_2,d_q) \\ \vdots & & & \\ \text{Cov}(d_q,d_1) & \text{Cov}(d_qd_2) & \ldots & \text{Cov}(d_q,d_q) \end{bmatrix} \\ &= \begin{bmatrix} t_{11} & t_{12} & \ldots & t_{1q} \\ t_{21} & t_{22} & \ldots & t_{2q} \\ \vdots & & & \\ t_{q1} & t_{q2} & \ldots & t_{qq} \end{bmatrix}\end{aligned}$$

In order for the parameters in the model 9.21 to be *identified,* the matrix $\underline{Td}$ is usually constrained to be diagonal. (This means that only the diagonal elements of $\underline{Td}$ are nonzero; since the variances of the measurement errors are located in the diagonal of $\underline{Td}$, this constraint sets the covariances between errors of different indicators to zero. Thus, to constrain $\underline{Td}$ to be a diagonal matrix is to assume that measurement errors are uncorrelated with one another.) It is also customary to assume that the measurement errors are uncorrelated with latent variables. And in some instances it is appropriate to assume that $\underline{PH}$ is diagonal. When $\underline{PH}$ is diagonal, the latent variables are said to be *orthogonal,* since they are uncorrelated with one another.

The best way to come to terms with the matrices defining this factor model is to consider simple cases of the model and then build up to more complex cases. In the simple cases the problem of identification of parameters is easily seen. When models are large, however, it is wise to allow a computer program to determine whether or

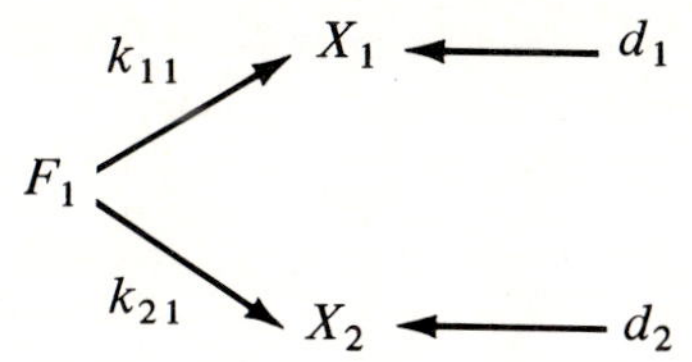

Figure 9.3 Path diagram of the measurement model specified in equations 9.22. The parameters of this model are identified only when very restrictive assumptions are made.

not the measurement model is identified. If a particular model is *underidentified,* it means that some of the parameters in the model have no unique solution and additional constraints must be placed upon the model. If no reasonable constraints can be added, then the model may have to be abandoned as untestable with the current array of indicators.

The Two-Indicator Case

The simplest multiple indicator measurement model is one in which two indicators of one latent variable are included. In this case the model 9.21 reduces to

$$X_1 = k_{11}F_1 + d_1 \qquad (9.22a)$$

$$X_2 = k_{21}F_1 + d_2 \qquad (9.22b)$$

where we assume that the means of all variables are zero (see Figure 9.3). Now the goal of such a model may be twofold. On the one hand, a researcher may want to know if observed data fit such a model; on the other hand, a researcher may be interested in estimating the validities of X_1 and X_2 as indicators of F_1. In order to determine whether or not this simple model could be helpful with data for only two indicators, it is necessary to generate the equivalent of the normal equations in multiple regression. That is, each of the observed variances and covariances are equated to a function of the parameters that are the object of investigation. In this case, the parameters of interest are k_{11}, k_{21}, Var(F_1), Var(d_1), and Var(d_2). And the covariances of the observed variables would be Var(X_1), Var(X_2), and Cov(X_1,X_2). Clearly, the latter of these parameters could be estimated directly from sample observation of X_1 and X_2, but the task of estimating the other five parameters appears beyond the data—and this is usually the case unless some very restrictive constraints are reasonable.

We begin the analysis of the model in equation 9.22 by noting that

$$\underline{Lx} = \begin{bmatrix} k_{11} \\ k_{21} \end{bmatrix}$$

$$\underline{PH} = \text{Cov}(\underline{F}) = \text{Var}(F_1)$$

and

$$\underline{Td} = \begin{bmatrix} \text{Var}(d_1) & \text{Cov}(d_1,d_2) \\ \text{Cov}(d_2,d_1) & \text{Var}(d_2) \end{bmatrix} = \begin{bmatrix} t_{11} & t_{12} \\ t_{21} & t_{22} \end{bmatrix}$$

These three matrices contain seven parameters in all. The covariance matrix of the indicators contains three *unique* elements:

$$\text{Cov}(\underline{X}) = \begin{bmatrix} \text{Var}(X_1) & \text{Cov}(X_1,X_2) \\ \text{Cov}(X_2,X_1) & \text{Var}(X_2) \end{bmatrix}$$

As stated previously, the task of estimating the elements of Cov($\underline{X}$) is straightforward, since the simple $[N/N - 1)]$ adjustment of the sample covariances provides unbiased estimates of the population covariances. In order to see how we might estimate some of the elements in $\underline{Lx}$, $\underline{PH}$, and $\underline{Td}$ from these observed covariances, we square the equations in the model 9.22, multiply the two equations together, and take the expectation across all cases:

squaring equation 9.22a gives

$$\begin{aligned} E(X_1X_1) &= \text{Var}(X_1) = E[(k_{11}F_1 + d_1)^2] \\ &= k_{11}^2 \text{Var}(F_1) + \text{Var}(d_1) + 2k_{11} \text{Cov}(F_1,d_1) \end{aligned}$$

Squaring 9.22b gives

$$\text{Var}(X_2) = k_{21}^2 \text{Var}(F_1) + \text{Var}(d_2) + 2k_{21} \text{Cov}(F_1,d_2)$$

Multiplying 9.22a by 9.22b gives

$$\text{Cov}(X_1,X_2) = k_{11}k_{21} \text{Var}(F_1) + \text{Cov}(d_1,d_2) + k_{11} \text{Cov}(F_1,d_2) + k_{21} \text{Cov}(F_1,d_1)$$

Clearly, these three equations are not soluble in terms of the covariances among the observed indicators. Thus, if we are to proceed, some constraints must be placed upon the elements of $\underline{Lx}$, $\underline{PH}$, and $\underline{Td}$ (which are also the right-hand terms in these three equations).

The most restrictive set of constraints to place upon such a model is to assume that the two indicators are *parallel*. In psychometric theory, for example, when two tests of some ability consist of the same number of items, and when both tests are comprised of items randomly selected from the same population of items, it may be reasonable to assume that the two tests are parallel. The assumption that X_1 and X_2 are parallel is constituted by the following set of constraints upon the elements of $\underline{Lx}$, $\underline{PH}$, and $\underline{Td}$:

$$\begin{aligned} k_{11} &= k_{21} \\ \text{Var}(d_1) &= \text{Var}(d_2) \\ \text{Cov}(d_1,d_2) &= 0 \end{aligned}$$

In addition the more general assumptions are also made

$$\text{Cov}(F_1,d_1) = \text{Cov}(F_1,d_2) = 0$$

These assumptions render our three expressions for the variances and covariances among the indicators as

$$\text{Var}(X_1) = k_{11}^2 \text{Var}(F_1) + \text{Var}(d_1) \tag{9.23a}$$

$$\text{Var}(X_2) = k_{11}^2 \text{Var}(F_1) + \text{Var}(d_2) \tag{9.23b}$$

$$\text{Cov}(X_1,X_2) = k_{11}^2 \text{Var}(F_1) \tag{9.23c}$$

which is a bit more manageable. In fact, we solve equations 9.23 for $\text{Var}(d_1)$ and $\text{Var}(d_2)$—which are presumed to be equal—directly:

$$\text{Var}(d_1) = \text{Var}(X_1) - \text{Cov}(X_1,X_2)$$
$$\text{Var}(d_2) = \text{Var}(X_2) - \text{Cov}(X_1,X_2)$$

which implies that $\text{Var}(X_1) = \text{Var}(X_2)$ if the assumption of parallel indicators is invoked. Thus, the parallel constraints were sufficient to *overidentify* the variances of the measurement errors (i.e., there are two solutions for $\text{Var}(d_1)$ or $\text{Var}(d_2)$ in terms of the variances and covariances of the observed indicators—$[\text{Var}(X_1) - \text{Cov}(X_1,X_2)]$ or $[\text{Var}(X_2) - \text{Cov}(X_1,X_2)]$).

However, there is no unique solution for the parameters k_{11} or $\text{Var}(F_1)$ in terms of the elements of $\text{Cov}(\underline{X})$. There are two approaches to the same solution to this dilemma. The first is to assume that the expected value of the observed indicators is the true score; this is equivalent to assuming that the measurement errors are symmetrically distributed around the true scores. This assumption is also equivalent to the constraint that $k_{11} = k_{21} = 1$. The second approach that leads to an analogous constraint is the assumption that the units of the latent variable are arbitrary, and thus any value of k_{11} and k_{21} is acceptable. And if any value is acceptable, the value one is convenient.

When we add the $k_{11} = k_{21} = 1$ constraint to our three expressions 9.23, they reduce to

$$\text{Var}(X_1) = \text{Var}(F_1) + \text{Var}(d_1)$$
$$\text{Var}(X_2) = \text{Var}(F_1) + \text{Var}(d_2)$$
$$\text{Cov}(X_1,X_2) = \text{Var}(F_1)$$

which gives

$$\text{Var}(F_1) = \text{Cov}(X_1,X_2)$$
$$\text{Var}(d_1) = \text{Var}(X_1) - \text{Cov}(X_1,X_2)$$
$$\text{Var}(d_2) = \text{Var}(X_2) - \text{Cov}(X_1,X_2)$$

Thus, the assumption that X_1 and X_2 are parallel indicators of F_1 leads to identification of all parameters in the model 9.22 and overidentification of $\text{Var}(d_1)$ and $\text{Var}(d_2)$; and this overidentification leads to one overidentifying restriction—that $\text{Var}(X_1) = \text{Var}(X_2)$.

If we relaxed the assumption that $\text{Var}(d_1) = \text{Var}(d_2)$, then the remaining set of constraints would be summarized by saying that X_1 and X_2 are presumed to be *tau equivalent* measures of F_1. The only distinction then, between parallel and tau equivalent indicators is that the latter are not assumed to have equal error variances. (Recall in Table 9.1 that the two indicators were concocted with different error variances. But both indicators had a slope of one when regressed on the true scores, so the two measures were tau equivalent rather than parallel.) But it should also be noted that when the equality constraint between $\text{Var}(d_1)$ and $\text{Var}(d_2)$ is relaxed, there is no longer an overidentification of any of the parameters in the model. Thus, the model 9.22 is *just identified* under the assumption that X_1 and X_2 are tau equivalent indicators of F_1. This means that any two variables that intercorrelate could be considered tau equivalent indicators of a common latent variable, and the parameters of the model could be estimated. Clearly, when only two indicators are available, only the assumptions of parallelism could lead to a useful exercise in testing the validity of hypothesized indicators. But the assumption of parallelism is quite unrealistic in most circumstances—especially when different methods of measurement are used as multiple indicators. Thus, we move on to models with more indicators where less restrictive constraints are necessary to identify and test a measurement model.

The Three-Indicator Case

Suppose now that the three indicators X_1, X_2, and X_3 are presumed to measure the same latent variable F_1. Then the general model in 9.21 reduces to

$$X_1 = k_{11}F_1 + d_1 \tag{9.24a}$$

$$X_2 = k_{21}F_1 + d_2 \tag{9.24b}$$

$$X_3 = k_{31}F_1 + d_3 \tag{9.24c}$$

where we assume that all variables have a mean of zero (see Figure 9.4). We now proceed to square the individual equations and multiply equations so that expectations can be taken. To simplify the algebra, we will incorporate the assumption that F_1 is uncorrelated with the measurement errors straightaway:

Squaring each equation gives

$$\text{Var}(X_1) = k_{11}^2 \text{ Var}(F_1) + \text{Var}(d_1) \tag{9.25a}$$

$$\text{Var}(X_2) = k_{21}^2 \text{ Var}(F_1) + \text{Var}(d_2) \tag{9.25b}$$

$$\text{Var}(X_3) = k_{31}^2 \text{ Var}(F_1) + \text{Var}(d_3) \tag{9.25c}$$

Multiplying equations together gives

$$\text{Cov}(X_1,X_2) = k_{11}k_{21} \text{ Var}(F_1) + \text{Cov}(d_1,d_2) \tag{9.25d}$$

$$\text{Cov}(X_1,X_3) = k_{11}k_{31} \text{ Var}(F_1) + \text{Cov}(d_1,d_3) \tag{9.25e}$$

$$\text{Cov}(X_2,X_3) = k_{21}k_{31} \text{ Var}(F_1) + \text{Cov}(d_2,d_3) \tag{9.25f}$$

Thus, the six unique elements of Cov($\underline{X}$) are expressed as functions of three elements of $\underline{Lx}$ (k_{11}, k_{21}, and k_{31},), one element of $\underline{PH}$ [Var(F_1)], and six elements of $\underline{Td}$ (the

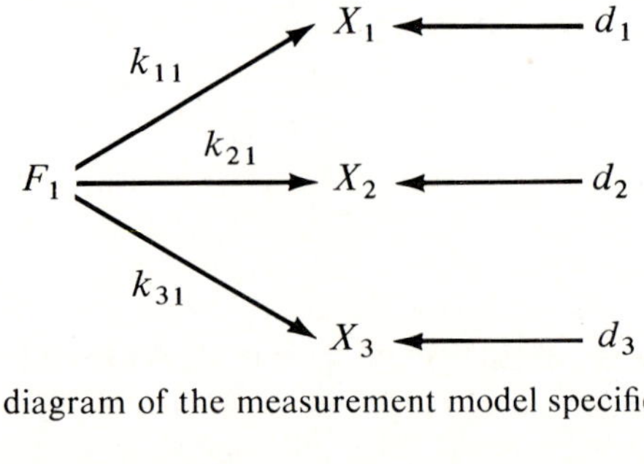

Figure 9.4 Path diagram of the measurement model specified in equations 9.24.

covariance matrix of the measurement errors). This means that there are ten unknowns on the right side of the six equations in 9.25. Clearly, these ten parameters cannot all be identified in terms of the six elements of Cov($\underline{X}$).

The task at hand is to specify constraints that will allow solution of equations 9.25 for the parameters on the right sides of the equations. We could, for example, assume that the three measures are tau equivalent. That is,

$$k_{11} = k_{21} = k_{31} = 1$$

and

$$\text{Cov}(d_1,d_2) = \text{Cov}(d_1,d_3) = \text{Cov}(d_2,d_3) = 0$$

These constraints on equations 9.24 lead to identification of all parameters in the model with two overidentifying restrictions:

$$\begin{aligned}
\text{Var}(F_1) &= \text{Cov}(X_1,X_2) \\
\text{Var}(F_1) &= \text{Cov}(X_1,X_3) \\
\text{Var}(F_1) &= \text{Cov}(X_2,X_3) \\
\text{Var}(d_1) &= \text{Var}(X_1) - \text{Var}(F_1) \\
\text{Var}(d_2) &= \text{Var}(X_2) - \text{Var}(F_1) \\
\text{Var}(d_3) &= \text{Var}(X_3) - \text{Var}(F_1)
\end{aligned}$$

Note that if we added the additional restriction of parallelism, we would be constraining the variances of the three measurement errors to be equal, and this would produce two more overidentifying restrictions. As we shall see subsequently, sample data can be used to test the model by measuring the extent to which the sample variances and covariances of the indicators fit the overidentifying restrictions. [For example, in this three-indicator case, the assumption of tau equivalence leads to the overidentification of Var(F_1) and the two overidentifying restrictions that

$$\text{Cov}(X_1,X_2) = \text{Cov}(X_1,X_3) = \text{Cov}(X_2,X_3)$$

which may be tested by gathering data and computing $S(X_1,X_2)$, $S(X_1,X_3)$, and $S(X_2,X_3)$ to see how similar they actually are.]

There is an additional set of constraints that can be placed upon the model in equations 9.24 that will probably be the most useful in actual research projects. This set of constraints is associated with the eminent Swedish statistician Karl Joreskog. The three indicators are said to be *congeneric* measures of F_1 if

$$k_{11} = 1 \text{ (or } k_{21} = 1 \text{ or } k_{31} = 1)$$

and

$$\text{Cov}(d_1,d_2) = \text{Cov}(d_1,d_3) = \text{Cov}(d_2,d_3) = 0$$

That is, if any one of the three validities can be set to one, and the measurement errors are uncorrelated with one another, then the three indicators are said to be congeneric. And in the three-indicator case, the congeneric model is *just* identified. This means that all the parameters in the model that are not constrained by assumptions are identified in only one way—there is only one solution for each free parameter in terms of the covariances of the observed indicators. The solutions are

$$\begin{aligned}
\mathrm{Var}(F_1) &= \mathrm{Cov}(X_1,X_2)\ \mathrm{Cov}(X_1,X_3)/\mathrm{Cov}(X_2,X_3) \\
k_{21} &= \mathrm{Cov}(X_1,X_2)/\mathrm{Var}(F_1) \\
k_{31} &= \mathrm{Cov}(X_1,X_3)/\mathrm{Var}(F_1) \\
\mathrm{Var}(d_1) &= \mathrm{Var}(X_1) - \mathrm{Var}(F_1) \\
\mathrm{Var}(d_2) &= \mathrm{Var}(X_2) - [\mathrm{Var}(F_1)k_{21}^2] \\
\mathrm{Var}(d_3) &= \mathrm{Var}(X_3) - [\mathrm{Var}(F_1)k_{31}^2]
\end{aligned}$$

with no overidentifying restrictions.

The advantage of the congeneric model over the parallel and tau equivalent models is that it does not require that the multiple indicators of a latent variable be measured in the same units. Thus, in a congeneric model with the latent variable of population size, the multiple indicators may be measured in kilowatts (average consumption of electricity), numbers of persons (census data), and number of dwelling units. In this instance, it would be most appropriate to constrain (to one) the path from the latent variable to the indicator that is actually measured in persons so that the units of the latent variable would be true population size if the slope of the indicator on true population was one.

The Four-Indicator Case

The congeneric model is particularly suited to instances in which four indicators of a latent variable are measured. The reason for this is that the four-indicator case provides two overidentifying restrictions when the assumption of uncorrelated measurement errors is made. This means, of course, that a parallel or tau equivalent model will have even larger numbers of overidentifying restrictions that must be consistent with the data to achieve a good fit. As will be demonstrated, the tau equivalent model has five such restrictions, whereas the parallel model has eight.

In order to specify the four-indicator/one-latent variable model, we reduce equation 9.21 to

$$X_1 = k_{11}F_1 + d_1 \tag{9.26a}$$

$$X_2 = k_{21}F_1 + d_2 \tag{9.26b}$$

$$X_3 = k_{31}F_1 + d_3 \tag{9.26c}$$

$$X_4 = k_{41}F_1 + d_4 \tag{9.26d}$$

or

$$\underline{X} = (\underline{Lx}\ \underline{F}) + \underline{d}$$

where

$$\underline{X} = [X_1 \quad X_2 \quad X_3 \quad X_4]' \qquad \text{(observed indicators)}$$

$$\underline{Lx} = [k_{11} \quad k_{21} \quad k_{31} \quad k_{41}]' \qquad \text{(factor loadings)}$$

$$\underline{F} = [F_1] \qquad \text{(latent variable)}$$

$$\underline{d} = [d_1 \quad d_2 \quad d_3 \quad d_4]' \qquad \text{(measurement errors)}$$

The "normal equations" for this model are then generated from equations 9.26 by multiplying equations and taking expectations over individuals in the population. If we assume that all variables have a mean of zero and that the measurement errors are uncorrelated with the latent variable, then these normal equations are as follows:

$$\begin{aligned}
\text{Var}(X_1) &= k_{11}^2 \text{ Var}(F_1) + \text{Var}(d_1) \\
\text{Var}(X_2) &= k_{21}^2 \text{ Var}(F_1) + \text{Var}(d_2) \\
\text{Var}(X_3) &= k_{31}^2 \text{ Var}(F_1) + \text{Var}(d_3) \\
\text{Var}(X_4) &= k_{41}^2 \text{ Var}(F_1) + \text{Var}(d_4) \\
\text{Cov}(X_1,X_2) &= k_{11}k_{21} \text{ Var}(F_1) + \text{Cov}(d_1,d_2) \\
\text{Cov}(X_1,X_3) &= k_{11}k_{31} \text{ Var}(F_1) + \text{Cov}(d_1,d_3) \\
\text{Cov}(X_1,X_4) &= k_{11}k_{41} \text{ Var}(F_1) + \text{Cov}(d_1,d_4) \\
\text{Cov}(X_2,X_3) &= k_{21}k_{31} \text{ Var}(F_1) + \text{Cov}(d_2,d_3) \\
\text{Cov}(X_2,X_4) &= k_{21}k_{41} \text{ Var}(F_1) + \text{Cov}(d_2,d_4) \\
\text{Cov}(X_3,X_4) &= k_{31}k_{41} \text{ Var}(F_1) + \text{Cov}(d_3,d_4)
\end{aligned}$$

If we now assume that the measurement errors are uncorrelated with one another (i.e., $\text{Cov}(d_1,d_2) = \text{Cov}(d_1,d_3) = \ldots = 0$), and if we assign the value 1.00 to the coefficient k_{11} (where the choice of coefficient to be assigned this value is arbitrary), then these normal equations may be solved for the parameters of the model:

$$\begin{aligned}
\text{Var}(F_1) &= [\text{Cov}(X_1,X_2)\ \text{Cov}(X_1,X_3)]/\text{Cov}(X_2,X_3) \\
\text{Var}(F_1) &= [\text{Cov}(X_1,X_2)\ \text{Cov}(X_1,X_4)]/\text{Cov}(X_2,X_4) \\
\text{Var}(F_1) &= [\text{Cov}(X_1,X_3)\ \text{Cov}(X_1,X_4)]/\text{Cov}(X_3,X_4) \\
k_{11} &= 1 \\
k_{21} &= \text{Cov}(X_1,X_2)/\text{Var}(F_1) \\
k_{31} &= \text{Cov}(X_1,X_3)/\text{Var}(F_1) \\
k_{41} &= \text{Cov}(X_1,X_4)/\text{Var}(F_1) \\
\text{Var}(d_1) &= \text{Var}(X_1) - \text{Var}(F_1) \\
\text{Var}(d_2) &= \text{Var}(X_2) - k_{21}^2\text{Var}(F_1) \\
\text{Var}(d_3) &= \text{Var}(X_3) - k_{31}^2\text{Var}(F_1) \\
\text{Var}(d_4) &= \text{Var}(X_4) - k_{41}^2\text{Var}(F_1)
\end{aligned}$$

Note that we may solve for the variance of the latent variable in terms of three distinct functions of observable covariances. This means that the parameter $\text{Var}(F_1)$ is overidentified and that the model 9.26 implies two overidentifying restrictions when the measurement errors are assumed to be mutually uncorrelated.

These two overidentifying restrictions may be obtained from the three expressions for $\text{Var}(F_1)$ above and written as

$$\begin{aligned}\text{Cov}(X_1,X_2)\ \text{Cov}(X_3,X_4) &= \text{Cov}(X_1,X_3)\ \text{Cov}(X_2,X_4)\\ &= \text{Cov}(X_1,X_4)\ \text{Cov}(X_2,X_3)\end{aligned}$$

Thus, if four observed variables are indeed flawed measures of the same latent variable, and if the flaws (errors) of measurement are well behaved, then the observed covariances among the four observed variables would be expected to have approximately this relationship with one another. In other words, for a sample of observations from a population in which equation 9.26 holds, we would expect the expressions

$$S(X_1,X_2)S(X_3,X_4)$$
$$S(X_1,X_3)S(X_2,X_4)$$
$$S(X_1,X_4)S(X_2,X_3)$$

to be approximately equal. If our sample covariances did not adhere to this prediction, then we may be motivated to reject the hypothesis that the model 9.26 actually holds for this population. The crucial question, of course, concerns how much of a deviation from the model's predictions constitutes a sufficient violation to warrant rejection of the model. Some answers to this question will be presented when methods for the estimation of overidentified models are presented in Section 9.5.

The two overidentifying restrictions associated with the congeneric model (in the four observed variable/one latent variable instance) may also be treated as sources of information concerning correlations among errors. For example, if t_{21} [$\text{Cov}(d_1,d_2)$] or t_{43} [$\text{Cov}(d_3,d_4)$] are treated as free parameters in the model 9.26, then the model is identified. Such a model is depicted in Figure 9.5. The reader should solve the normal equations generated by 9.26 for $\text{Cov}(d_1,d_2)$ or $\text{Cov}(d_3,d_4)$ to see that either solution is unique. Furthermore, the information required for these estimates is precisely that

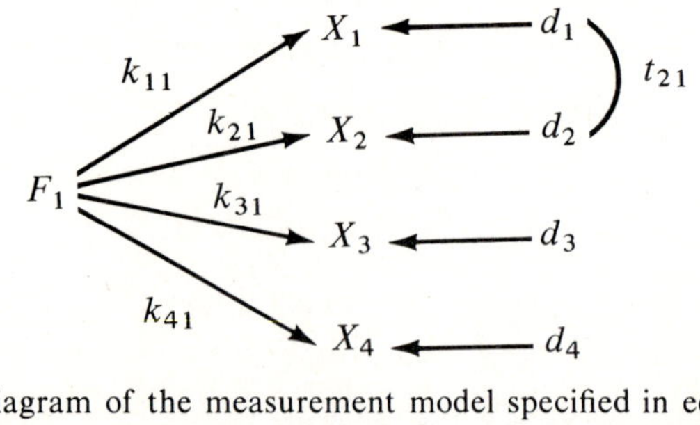

Figure 9.5 Path diagram of the measurement model specified in equations 9.26, but with correlated measurement errors. In the case of the congeneric constraints, a four-indicator model allows identification of one covariance among the measurement errors.

which provided for the overidentification in the case of uncorrelated measurement errors.

Up to this point we have placed only the constraints associated with congeneric measures upon equation 9.26. Suppose now that construction of the measurement instruments warranted the conclusion that each observed variable differed from the other only in the magnitude of symmetric errors—that is, in the variances of the measurement errors d_1, d_2, d_3, and d_4. Such a model would include the constraint that

$$k_{11} = k_{21} = k_{31} = k_{41} = 1$$

This assumption of tau equivalence thus adds three overidentifying restrictions to the model, and these three together with the two under the congeneric assumption give a total of five. The number of overidentifying restrictions added by one or more constraints upon parameters of the model is equal to the number of constraints. Thus, constraining k_{21}, k_{31}, and k_{41} to also be equal to one adds three additional overidentifying restrictions.

In a similar manner, the assumption that X_1, X_2, X_3, and X_4 are *parallel* measures of F_1 leads to additional overidentifying restrictions. The parallel specification adds to the tau equivalence constraints the constraints that

$$\mathrm{Var}(d_1) = \mathrm{Var}(d_2) = \mathrm{Var}(d_3) = \mathrm{Var}(d_4)$$

Since this constraint means that giving a value to any one of the error variances determines the other three, this constraint produces three additional overidentifying restrictions. Thus, the assumption that the observed variables are parallel measures of F_1 yields a total of eight overidentifying restrictions.

Failure to Reject a Model

A question that usually arises when models such as the above are presented is the following. Suppose that a sample from some population is observed and that four indicators of some hypothesized latent variable are recorded for each case. Now, suppose that the covariances and variances among these four indicators come very close to meeting the requirements of the overidentifying restrictions for the four-indicator/one latent variable congeneric model. What does this result mean? If the observed moments (variances and covariances) had obviously violated the predictions from the model, then the logic leading to rejection of the model is quite straightforward: the model predicted that certain relations between the moments would hold, and the observations do not conform to those predictions. Thus, we reject the model as a description of dynamics in the population, since the probability of our observations would be small if the model actually did hold in the population. But what if the moments *do* conform to the predictions of the model? Does this mean that the hypothesized measurement model is true?

The answer to this question, of course, is clearly "no." The observation that a set of data is consistent with the predictions of a model means that this data set is an *insufficient basis for rejecting the model.* In other words, the model has survived the current attempt to demonstrate its falsehood. Inherent in this conclusion is the possibility that alternative models would make the same predictions about the observed moments. This means that the real power of these procedures for the testing of measurement models comes from the ability to specify the empirical predictions that flow from *various* models that may be proposed to explain the interrelations of indicators, and then estimation techniques can be used to see if the predictions of one of the models is superior to the others. Procedures for comparing the "fit" of data to various models will be demonstrated in Section 9.5.

It is important to contrast this procedure for supporting a measurement model with the more familiar rejection of the null hypothesis in experimentation. If a researcher has hypothesized that an increase in X will cause an increase in Y, then a true experiment tests the null hypothesis that a randomized manipulation of X produces no increase in Y. If observations are then sufficiently unlikely under the assumption of the null hypothesis, then the null hypothesis is rejected and the logical alternative to the null hypothesis is the researcher's hypothesis that an increase in X causes an increase in Y. The statistical result in an instance favorable to the researcher's hypothesis is that the linear model specifying the null hypothesis evidences a "poor fit" with the observed moments, while the fit of the linear model corresponding to the researcher's hypothesis evidences "good fit." However, in most instances, the null hypothesis involves overidentifying restrictions (constraints), while the researcher's hypothesis does not. Thus, many researchers are unaccustomed to testing the "fit" of hypothesized models. In contrast, when measurement models are the object of inquiry, the successful model is the one with the set of overidentifying restrictions most consistent with the observations. (As argued in Chapter 10, more convincing experimental designs in the social and behavioral sciences will often include overidentifying restrictions.)

Although more will be made of this point subsequently, it is important for the reader to distinguish between significance testing in the more usual correlational sense and "model fitting." In the correlational context, the null hypothesis is that a regression coefficient is zero in a population. Any unlikely deviation from that condition motivates a rejection of the null hypothesis—even though the observed correlation may be rather small. On the other hand, an overidentified measurement model must reproduce the observed covariance matrix with errors no greater than what would be expected by chance variation. Thus, in the model fitting case, the "fit" must be almost perfect, whereas in the correlation problem the fit need only be different from zero. When the null hypothesis is rejected in the correlational case, the investigator is essentially rejecting an overidentified model (the one incorporating the null hypothesis) and accepting a just identified model. But only in the case of true experimentation is a just identified model very compelling evidence of a causal relation.

The testing of measurement models raises the philosophical and methodological issues that are described in the social sciences as concerns about *construct validation.* In other words, how do we evaluate the claim that a set of operations is generating numbers that measure a particular construct? In more technical terms, how do we evaluate the *validity* of an indicator? In the final analysis, there is no definite answer to this question, since the judgment that a construct and its indicators are scientifically useful stems from a complex relationship between results obtained with the indicators and other facets of work within the theoretical and methodological domain that contains the construct. This domain and the relations between the elements of the domain have been referred to as a *nomological net*. However, it is possible to specify some of the more usual elements of a nomological net and thus to outline some procedures for construct validation that can be applied in many instances. The set of such procedures to be outlined in this chapter is known as the multitrait-multimethod approach to construct validation.

The essential principle in the multitrait-multimethod approach to construct validation is one that emphasizes uniqueness in a complex cluster of constructs and their indicators. In order for a construct and its indicators to be judged valid, it is necessary to demonstrate that (1) multiple measures of the construct, each employing a distinct *method of measurement,* intercorrelate in a predicted direction and (2) the pattern of intercorrelations among indicators of the construct and those of closely related constructs demonstrates the uniqueness of the construct. In terms of the two-, three-, and four-indicator models that have been described above, this principle means that the validity of a construct cannot be sufficiently established if only one construct (latent variable) is measured in a study. The models described above and their overidentifying restrictions would thus be a first step in the process of validating a construct. A convincing case for the validity of a construct would require more complex models that involved several related latent variables and a variety of measurement procedures.

This discussion of the meaning of obtaining a good fit between the predictions of a measurement model and actual data may be summarized by the statement that greater confidence in the validity of a latent variable stems from the inclusion of more latent variables in the model. The more complex the model, the more specific are the overidentifying restrictions that must be satisfied. As the labyrinth of such restrictions becomes more complex, it becomes more difficult to concoct substantively distinct alternative models to the one proposed.

9.4 ESTIMATION OF MEASUREMENT MODELS AND MODEL "FITTING"

For each of the latent variable models presented in Section 9.3, a set of normal equations was generated. Each equation expressed an observable population variance or covariance as a function of parameters of the measurement model. That is, elements of Cov($\underline{X}$) are expressed in the normal equations as functions of elements of $\underline{Lx}$ (factor loadings), $\underline{PH}$ (variances and covariances of latent variables), and $\underline{Td}$ (variances and covariances of measurement errors). When the model is identified, it is possible to solve for each model parameter in terms of the population variances and covariances among the observable variables. When the model is *over*identified, it is possible to solve for some of the parameters in terms of more than one unique function of the elements of Cov($\underline{X}$). And it is this latter case that presents the greatest problems for estimation: when several estimates of a parameter are possible, and the several do not agree exactly, how can the multiple estimates be combined to obtain the best estimates, and how can the differences between the various possible estimates be used to assess the fit between the observed data and the predictions of the measurement model?

The solution to this problem to be presented here is that of maximum likelihood estimation (MLE). The algebra of MLE is rather complex, but the principle behind the procedure is quite simple. Suppose that we begin by observing $\underline{X}$ for a sample from some population. From these observations we compute $\underline{S}(\underline{X})$, the sample variance covariance matrix. If the elements of this matrix are multiplied by $N/(N-1)$, where N is the sample size, then the resulting matrix is an unbiased estimate of Cov($\underline{X}$). [Henceforth, the symbol $\underline{S}$ will be used to refer to this unbiased estimate of Cov($\underline{X}$).] The next step in our model testing exercise would be to specify the measurement model to be tested. And the crucial point to understand here is that if estimates of the parameters in the measurement model are assigned to the various parameters, then an estimate of Cov($\underline{X}$) can be obtained *from the model.* This fact is the basis for the application of MLE to the estimation of overidentified measurement models. When the choice of estimates is not a simple matter of substituting sample moments for population moments in the normal equations (i.e., the *just identified* case), the choice of estimates is based upon the agreement between the estimate of Cov($\underline{X}$) based upon the model and the estimate $\underline{S}$. The final choice of estimates is the set of such estimates that minimize the differences between $\underline{Z}$ and $\underline{S}$, where $\underline{Z}$ is an estimate of Cov($\underline{X}$) based upon the set of parameter estimates. A numerical example should clarify the meaning of these statements for the newcomer.

The complexity of MLE becomes apparent when the details of comparing $\underline{Z}$ and $\underline{S}$ are presented. The goal is to find a set of estimates of the elements of $\underline{Lx}$, $\underline{PH}$, and $\underline{Td}$ that minimize the differences between $\underline{Z}$ and $\underline{S}$. We could, for example, subtract $\underline{Z}$ from $\underline{S}$, square all the resulting differences, and sum. Our choice of parameter estimates would then be those that minimized this sum. Unfortunately, this intuitively appealing procedure has mathematical drawbacks that are avoided by the use of a somewhat more complex minimization function known as a (multivariate normal) *likelihood function.*

A transformation of the likelihood function that is often used in ML computer programs is written as

$$H = \log[\det(\underline{Z})] + \mathrm{tr}[\underline{S}\ \mathrm{inv}(\underline{Z})] - \log[\det(\underline{S})] - q \tag{9.27}$$

where

"log" means natural logarithm
"det" means determinant
"tr" means trace (sum of main diagonal)
"inv" means inverse
q = number of variables in $\underline{X}$
$\underline{S}$ = sample estimate of Cov($\underline{X}$)
$\underline{Z}$ = estimate of Cov($\underline{X}$) based on estimates of model parameters

Now, for different values of the parameter estimates, H takes on different values because $\underline{Z}$ takes on different values. The *maximum likelihood* estimates of the parameters are obtained when H is a minimum. And, in fact, it can be shown that where H is a minimum, the differences between $\underline{S}$ and $\underline{Z}$ tend to be minimized. Thus, for the less mathematically sophisticated, minimization of H can be visualized as a minimization of ($\underline{S} - \underline{Z}$). Furthermore, a specified model is identified if the minimum for H

is achieved for a unique set of estimates of the parameters of the model. If any one of the parameter estimates can be changed after H has been minimized, and the change does not increase H, then that parameter is not identified—and neither is the model. (As we shall see, however, the identified parameters in an underidentified model can be estimated, and the fit of the model to the data can be assessed. In such an instance, the parameters that are not identified may be assigned arbitrary values.)

An intuitive understanding of equation 9.27 is not easily attained, but some progress can be made by grouping the terms as follows:

$$\log[\det(\underline{Z})] - \log[\det(\underline{S})]$$

and

$$\mathrm{tr}[\underline{S}\ \mathrm{inv}(\underline{Z})] - q$$

The first of these two differences will clearly be close to zero when $\underline{Z}$ and $\underline{S}$ are similar—even though it is not clear why a minimization of this difference, when $\underline{Z}$ and $\underline{S}$ are not identical, results in some kind of more desirable estimates of population parameters. The reason, in fact, for choosing likelihood functions as the basis for estimation is twofold. First, it can be shown that the resulting estimators have desirable properties (sufficiency, consistency, and efficiency are the technical terms for these desirable properties) when the variables are multivariate normal in distribution. Second, in the case of linear measurement models, the likelihood functions satisfy "regularity conditions." And these regularity conditions mean that the ML estimators are obtained by partial differentiation of the likelihood function. This latter characteristic makes it possible for computer programs to be written that easily accomplish the otherwise insurmountable task of calculating ML estimates.

The second half of equation 9.27 presented above is less obviously zero when $\underline{Z}$ and $\underline{S}$ are equal. In order to see this, it is necessary to work out some elaborate algebra. If this is done, it becomes apparent that the diagonal of $[\underline{S}\ \mathrm{inv}(\underline{Z})]$ will be a series of ones when $\underline{Z}$ and $\underline{S}$ are equal—because the inverse of $\underline{Z}$ is then the same as the inverse of $\underline{S}$. The trace of this product, which is equal to the sum of the diagonal elements, would then be equal to the number of rows (or columns) in $\underline{S}$, and that number is q (the number of observed variables). Thus, although the reason that minimization of equation 9.27 is a desirable way of estimating model parameters may not be completely understood, the reader should be able to see that minimization of H tends to maximize the similarity of $\underline{Z}$ and $\underline{S}$. And this means that minimization of H produces sample estimates of model parameters that enable the model to reproduce the observed data with a minimum of distortion.

[*Technical note.* Equation 9.27 is described above as a *transformation* of a likelihood function. A conventional likelihood function is actually expressed in terms of the probability density of the observed variables. This means that the shape of the distribution of scores must be described mathematically, and clearly the multivariate normal distribution is a favorite among statisticians.

In the general case, suppose that the variables X_1, X_2, and X_3 are distributed according to the density function $f(X_1,X_2,X_3;a_1,a_2)$—a multivariate density that may be used to determine the probability of any three-dimensional region in the space defined by X_1, X_2, and X_3. The parameters a_1 and a_2 define the particular shape of the density (e.g., if f is a multivariate normal density, then a_1 and a_2 could be the mean and variance for each variable—presuming that each variable had the same mean and variance). Now suppose that a random sample of N cases is drawn from the population. The *likelihood function* for these N observations is then

$$\begin{aligned} L &= g[(X_{11},X_{21},X_{31}), \ldots (X_{1N},X_{2N},X_{3N});a_1,a_2] \\ &= f(X_{11},X_{21},X_{31};a_1,a_2)f(X_{12},X_{22},X_{32};a_1,a_2) \ldots f(X_{1N},X_{2N},X_{3N};a_1,a_2) \end{aligned}$$

This expression is thus the product of the density evaluated at each of the N sample points. And the values a_1' and a_2' that maximize L are the maximum likelihood estimates of a_1 and a_2.

When f is the multivariate normal distribution, finding the maximum for L is simplified, since L can be differentiated with respect to each parameter (a_1 and a_2), and each partial differential equation can be set to zero to obtain the maximum of L—if a unique maximum exists. For the case of just one variable X, recall that the normal density is

$$f(X;m,v) = (2pv)^{-.5}e^{-(X-m)^2/2v}$$

where

$$\begin{aligned} m &= \text{population mean of } X \\ v &= \text{population variance of } X \\ p &= pi = \pi = 3.141592 \ldots \\ e &= \text{base of natural logarithms} = 2.71828 \ldots \end{aligned}$$

The likelihood function is then the product of $f(X;a_1,a_2)$ for each observed value of X, and the ML estimates of m and v are the values a_1' and a_2' that maximize that product. The resulting density defined by $f(X;m',v')$ is then the normal density with the following characteristic: the *likelihood* of the observed sample is greater given $f(X;m',v')$ than given any other normal density. (The reader can use the following analogy to understand the maximum likelihood concept. Suppose that a roulette player knew that some roulette tables in a casino were "fixed" such that even numbers occurred with a probability of .55 (rather than .50), whereas other tables were "fair." If the player observed 100 spins of a particular wheel and found that the winning number was even on 53 of the trials, then the "maximum likelihood estimator" of the population probability of an even number for that roulette table would be .55. The reason for this choice is that 53 even numbers would be more likely to occur at the fixed table than at the fair table. The two types of roulette tables are analogous to the different normal densities defined by different means and variances; the 100 outcomes of the roulette wheel are analogous to a sample of observations of some variable that is presumed to be normally distributed. Finally, the choice of .55 as the estimator of the probability of an even number is analogous to choosing the sample mean and sample variance as the estimates of the population distribution from which the sample of scores was drawn.)

Although the algebra of ML estimation is quite complex when several variables are involved and the variables are presumed to be multivariate normal in distribution, estimation of mean and variance are quite straightforward in the univariate case. This simple case then provides an intuitive understanding of how maximum likelihood estimation works in more complex cases. Suppose that a sample of four cases is drawn from a normal density $f(X;m,v)$. The four observed values of X are $X_1, \ldots, X_4$. The likelihood function is then defined as follows:

$$L(m,v) = f(X_1;m,v)f(X_2;m,v)f(X_3;m,v)f(X_4;m,v)$$

The goal here is to find the values m' and v' of m and v, respectively, that maximize L. Note that maximizing L is then equivalent to finding the normal density that maximizes the product of the "heights" of the density function evaluated at X_1, X_2, X_3, and X_4. Returning to the equation defining normal densities, we can rewrite L as

$$L(m,v) = (2pv)^{-2}e^{-[(X_1-m)2+(X_2-m)2+(X_3-m)2+(X_4-m)2]/2v}$$

since multiplying constants raised to powers is accomplished by adding the powers. Our goal now is to find the estimators of m and v (labeled m' and v') that maximize L. From a knowledge of derivatives we know that the maximum for a continuous function occurs where the partial derivatives are zero. Thus, we must differentiate L with respect to m and with respect to v to find the maximum likelihood estimators of m and v—assuming that X is normally distributed.

It turns out that the natural logarithm of L is easier to handle in the normal case, and may be written as

$$\log(L[m,v]) = [-N\log(2p)/2] - [N\log(v)/2] - [NS/2v]$$

where

$$\begin{aligned}
p &= \text{pi} \\
N &= \text{sample size} = 4 \\
v &= \text{Var}(X) \\
S &= [(X_1 - m)^2 + (X_2 - m)^2 + (X_3 - m)^2 + (X_4 - m)^2]/N \\
&= \text{sample variance about population mean}
\end{aligned}$$

Now, if this expression is differentiated with respect to m and then v, then ML estimators of m and v are obtained:

Partial derivative of L with respect to m

$$= [(X_1 - m) + (X_2 - m) + (X_3 - m) + (X_4 - m)]/v$$

Partial derivative of L with respect to v

$$= -N/2v + [(X_1 - m)^2 + (X_2 - m)^2 + (X_3 - m)^2 + (X_4 - m)^2]/(2v^2)$$

Setting the first differential equation equal to zero and solving for m' gives

$$\begin{aligned}
0 &= [(X_1 - m') + (X_2 - m') + (X_3 - m') + (X_4 - m')] \\
&= (X_1 + X_2 + X_3 + X_4) - 4m'
\end{aligned}$$

or

$$\begin{aligned}
ML(m) &= m' \\
&= (X_1 + X_2 + X_3 + X_4)/4 \\
&= \text{sample mean of } X
\end{aligned}$$

Setting the second differential equation equal to zero and solving for v' gives

$$0 = -N/2v' + [(X_1 - m)^2 + (X_2 - m)^2 + (X_3 - m)^2 + (X_4 - m)^2]/(2v'^2)$$

or

$$v' = [(X_1 - m)^2 + (X_2 - m)^2 + (X_3 - m)^2 + (X_4 - m)^2]/N$$

And we now substitute the maximum likelihood estimate of m into the equation and find that

$$ML(v) = v' = \text{sample variance of } X$$

The result is that the sample mean and the sample variance are the *ML* estimators, and the latter is biased in small samples—because the expected value of v' is $(N - 1)\ \mathrm{Var}(X)/N$. Fortunately, the bias becomes inconsequential as sample size increases.

In the case of measurement models, the parameters to be estimated are functions of the variances and covariances among a set of multivariate normal variables. This population covariance matrix is a matrix of parameters in the multivariate normal distribution in the same way that the variance is a parameter of the family of univariate normal distributions. If the parameters of the model are unique functions of the variances and covariances, then the likelihood function can be minimized in terms of partial differential equations with respect to the *model parameters* rather than the variances and covariances. (Computer programs such as LISREL presume that higher-order moments provide no additional information about dependencies among the variables.) For example, if a covariance matrix for three indicators was fit to a measurement model in which the three indicators were presumed to be parallel indicators of a single latent variable, then a_1 would be ph_{11} and a_2 would be t_{11}. Since estimates of ph_{11} and t_{11} can be used to define estimates of $\mathrm{Var}(\underline{X})$, the estimates that maximize L could be found.

The reader should note that the transformation of the likelihood function that is described in equation 9.27 resembles a reciprocal of a likelihood function, since it is evaluated at its minimum rather than its maximum. This inversion is done for the sake of computing convenience.]

Estimation of the Three-Indicator/One Latent Variable Model

In order to clarify these mathematical developments, consider the three-indicator/one latent variable model described in equations 9.24 and the corresponding "normal equations" in 9.25. For the sake of simplicity, suppose that the measurement model to be tested includes the specification that the three observed variables are parallel indicators of the one latent variable. In this instance, then, equations 9.24 may be rewritten as

$$X_1 = F_1 + d_1$$
$$X_2 = F_1 + d_2$$
$$X_3 = F_1 + d_3$$

or

$$\underline{X} = \underline{Lx}\,\underline{F} + \underline{d} \tag{9.28}$$

where

$$\underline{X} = [X_1 \quad X_2 \quad X_3]'$$
$$\underline{Lx} = [1 \quad 1 \quad 1]'$$
$$\underline{F} = [F_1]$$
$$\underline{d} = [d_1 \quad d_2 \quad d_3]'.$$

And the normal equations may be rewritten as

$$\begin{aligned}
\mathrm{Var}(X_1) &= \mathrm{Var}(F_1) + \mathrm{Var}(d_1) = ph_{11} + t_{11} \\
\mathrm{Var}(X_2) &= \mathrm{Var}(F_1) + \mathrm{Var}(d_2) = ph_{11} + t_{22} \\
\mathrm{Var}(X_3) &= \mathrm{Var}(F_1) + \mathrm{Var}(d_3) = ph_{11} + t_{33} \\
\mathrm{Cov}(X_1,X_2) &= \mathrm{Var}(F_1) = ph_{11} \\
\mathrm{Cov}(X_1,X_3) &= \mathrm{Var}(F_1) = ph_{11} \\
\mathrm{Cov}(X_2,X_3) &= \mathrm{Var}(F_1) = ph_{11}
\end{aligned}$$

where

$$\mathrm{Var}(d_1) = \mathrm{Var}(d_2) = \mathrm{Var}(d_3)$$

or

$$t_{11} = t_{22} = t_{33}$$

Notice that since the three measurement error variances are constrained to be equal, there are only two parameters in the model: ph_{11} (the variance of the latent variable) and t_{11} (the variance of the measurement errors). Each of these parameters is over-identified, since there are three solutions for each:

$$\begin{aligned}
ph_{11} &= \mathrm{Cov}(X_1,X_2) \\
&= \mathrm{Cov}(X_1,X_3) \\
&= \mathrm{Cov}(X_2,X_3)
\end{aligned}$$

and

$$\begin{aligned}
t_{11} &= \mathrm{Var}(X_1) - ph_{11} \\
&= \mathrm{Var}(X_2) - ph_{11} \\
&= \mathrm{Var}(X_3) - ph_{11}
\end{aligned}$$

There are two problems facing us if we observe $\underline{X}$ and wish to test this model. The first is to obtain estimates of ph_{11} and t_{11}, and the second is to assess the fit of the model to the observed variances and covariances.

An intuitively appealing approach to the estimation problem would be to replace the population variances and covariances in the normal equations with the unbiased sample estimates. Then the model parameters could be replaced by estimators of those parameters, the estimators could be solved for in terms of the sample moments, and the problem of multiple solutions could be resolved by some sort of averaging. In fact, this is precisely what the ML estimate procedure would do in this instance, and, for this simple model, the ML estimate is equivalent to a kind of *least squares* solution that should be easy for the reader of this book to understand.

The least squares approach to estimation of ph_{11} and t_{11} consists of constructing a multiple regression of the observed moments (rather than scores) on the estimators of the model parameters. This multiple regression may be specified as follows:

$$\begin{bmatrix} S_{11} \\ S_{22} \\ S_{33} \\ S_{12} \\ S_{13} \\ S_{23} \end{bmatrix} = \begin{bmatrix} 1 \\ 1 \\ 1 \\ 1 \\ 1 \\ 1 \end{bmatrix} \text{est}(ph_{11}) + \begin{bmatrix} 1 \\ 1 \\ 1 \\ 0 \\ 0 \\ 0 \end{bmatrix} \text{est}(t_{11}) + \begin{bmatrix} r_1 \\ r_2 \\ r_3 \\ r_4 \\ r_5 \\ r_6 \end{bmatrix}$$

where r_i $(i = 1, \ldots, 6)$ is a residual. Note that in this multiple regression the sample moments serve as cases for the dependent variable, while estimators of model parameters are treated as slopes or intercepts. The solution to this multiple regression problem is that est(ph_{11}) is the mean of the three sample covariances and est(t_{11}) is the mean of the three differences between sample variances and est(ph_{11}). As noted above, this is also the ML estimate solution:

$$\text{est}(ph_{11}) = (S_{12} + S_{13} + S_{23})/3$$

$$\text{est}(t_{11}) = (S_{11} + S_{22} + S_{33})/3 - \text{est}(ph_{11})$$

The multiple regression results would also give us a measure of the fit of the data to the predictions of the model—the usual R-square. (It is instructive at this point to observe that the null hypothesis in this model fitting exercise is that the R-square between the population covariances and the predictions of the model is one. This situation is in contrast to the correlational problem in which the null hypothesis is that the population correlation is zero.) Unfortunately, however, the multiple regression solution to the estimation and model fit problems is only applicable to *very* simple models. As soon as we progress to the congeneric model, for example, we must resort to some alternative such as the ML estimate. But the important point here is that the logic of the ML solution is analogous to the multiple regression solution and may be understood in terms of these familiar concepts.

To summarize, ML estimation is a procedure that may be used to fit observed moments to estimators of model parameters. It provides both numerical values for the estimators and measures of overall fit between the observed covariance matrix and the covariance matrix implied by the parameter estimates ($\underline{Z}$). In the case of the three parallel indicator/one latent variable model,

$$\underline{S}(\underline{X}) = \begin{bmatrix} S_{11} & S_{12} & S_{13} \\ S_{21} & S_{22} & S_{23} \\ S_{31} & S_{32} & S_{33} \end{bmatrix}$$

and

$$\underline{Z} = \begin{bmatrix} \hat{ph}_{11} + \hat{t}_{11} & \hat{ph}_{11} & \hat{ph}_{11} \\ \hat{ph}_{11} & \hat{ph}_{11} + \hat{t}_{11} & \hat{ph}_{11} \\ \hat{ph}_{11} & \hat{ph}_{11} & \hat{ph}_{11} + \hat{t}_{11} \end{bmatrix}$$

where the symbol ˆ indicates an estimate based on a sample. Computer programs that compute ML estimates use the likelihood function (equation 9.27) to minimize the differences between $\underline{S}$ and $\underline{Z}$. (Note that the elements of $S(x)$ are, in this instance, the unbiased estimates rather than the ML estimates.)

On the Efficiency of Maximum Likelihood Estimates

The reader will recall that least squares estimators of slopes and structural coefficients had some very desirable properties. In general, if the residuals were well behaved, the OLS estimators were the best linear unbiased estimators of population parameters. The obvious question that should occur to the reader concerning ML estimates applied to measurement models is how well these estimators estimate population parameters. Joreskog (1973) has demonstrated that when the observed variables are distributed as the multivariate normal distribution, the estimates of parameters from linear models that result from minimization of H are maximum likelihood estimates that are *efficient in large samples.* The term "efficient" is used in this context to mean "squared-error asymptotically efficient," a technical classification used by mathematical statisticians. When an estimator qualifies for this classification it means that there is no other *consistent* estimator of the population parameter that will be closer to the population parameter as the sample size becomes very large. And an estimator is *consistent* if it becomes more accurate as sample size increases (where accuracy is measured in terms of squared error or squared residuals). One important point here, however, is that an efficient estimator is not necessarily unbiased. For example, if the mean of a sample of scores is m' and the sample size is N, then $(Nm')/(N + 1)$ is an efficient estimator of the population mean even though it is biased. Similarly, when X is distributed normally, the ML estimator of Var(X) is $S(XX)$ rather than $[NS(XX)]/(N - 1)$, and the former is efficient even though it is a biased estimator. These examples point to the importance of using large samples for the estimation of overidentified measurement models [i.e., the bias in $S(XX)$, for example, becomes inconsequential as N becomes large].

The fact that an assumption about the distribution of the observed variables, such as the assumption that they are multivariate normal, is required to demonstrate estimator efficiency raises the problem of how trustworthy the estimates are when the indicators deviate from multivariate normality; and it also raises the problem of how to deal with such deviations.

The problem of estimation efficiency when variables deviate considerably from multivariate normality is a sticky wicket at the moment, since simulation studies concerning this question are few in number and narrow in scope (see Bentler, 1980, for relevant references). This gap in our knowledge should be filled in within the next few years, and it is my expectation that findings will corroborate the rule of thumb that violation of the normality assumption is disastrous only in the most extreme cases.

When a set of indicators clearly violates the normality assumption, there are procedures available for "robustifying" covariance matrices. These procedures are outlined by Gnanadesikan and Kettenring (1972, cited in Joreskog and Sorbom, 1978). A less elegant approach to this problem would be to use the SPSS routine for normalizing individual variables. This latter approach does not really address the *multivariate* normality problem, but it is probably an adequate corrective for most situations. (To be truly multivariate normal, the conditional distribution of each variable should also be normal. Normalizing the variables individually does not guarantee this outcome.)

Assessing the "Fit" of the Model to the Data

It was suggested above that an analog to the R-square for the regression of the observed moments on parameter estimators could be used to assess the fit between $\underline{S}$ and $\underline{Z}$. That very rough analog is the *likelihood ratio chi-square.* This chi-square statistic ranges from 0 to N, where N is the sample size. The chi-square statistic is zero when a model is *just identified;* that is, when there are no overidentifying restrictions, and each parameter in the model is a *unique* function of moments. When the observed variables are multivariate normal in distribution, *and the model is true in the population,* then this likelihood ratio is distributed as a chi-square with degrees of freedom equal to the number of unique moments in $\underline{S}$ minus the number of estimated parameters in the model. This means that the specified model is treated as the null hypothesis. If the observed covariances would be unlikely to occur if the specified model were true in the population, then the chi-square will be large relative to its degrees of freedom. However, if the sample observations are closer to what would be expected if the model held in the population, then the chi-square will have a value closer to its degrees of freedom.

The likelihood ratio chi-square is computed from the value of H at its minimum. If $\min(H)$ is the minimum value of H, and f is the number of free parameters that are to be estimated, then

$$\text{chi-square} = \min(H)(N/2) \tag{9.29}$$

with

$$df = [q(q-1)/2 + q] - f$$

where

$$\begin{aligned} N &= \text{sample size} \\ q &= \text{number of observed variables} \\ f &= \text{number of free parameters estimated} \end{aligned}$$

Thus, the parameter estimates that minimize H also produce the smallest possible chi-square statistic.

Confusion concerning the interpretation of this chi-square statistic is common. This confusion stems from the fact that the value of the statistic itself is *smaller* when the hypothesized model evidences a *better* fit with the data. In contrast, the F ratio for an R-square in multiple regression is *larger* when the right-hand variables produce a "tighter" fitting predictor of the left-hand variable. Furthermore, to make matters worse, when it makes sense to think of the solution of the normal equations for a measurement model as a multiple regression of sample moments on dummy variables, then the resulting R-square would become larger as the chi-square became smaller. But the real source of confusion here is due to the the confounding of two distinct notions of model "fitting."

(1) The R-square in multiple regression is a measure of the extent to which the right-hand variables are *related* to the left-hand variable. In other words, the R-square is a measure of the magnitude of the relationship between the left- and right-hand variables (keeping the limitations of R-square in mind). In this sense, the left-hand variable is *fit* to the right-hand variables, and when the fit is better the R-square is larger. This type of "fitting" exercise will henceforth be referred to as an assessment of *explained variation.* (2) The second way in which the term "fit" is used is in the context of an overidentified measurement (or structural) model. In this latter context, whether relationships among variables (paths in causal diagrams) are large or small may be irrelevant to the fit of the model with the observed data. The important parameter for what we will henceforth call "model fit" is the extent to which an hypothesized model can reproduce the observed *relations among variables.* In other words, how well does the covariance matrix predicted by the model match with the observed covariance matrix? There is little concern with prediction of scores on a particular variable in this context. However large or small the relationships among observed variables may be, the crucial question in model fit is the extent to which the observed associations are consistent with the overidentifying restrictions that are predicted by the hypothesized model.

This crucial distinction between variance explained and model fit is nicely demonstrated by the fitting of covariances among four observed variables to a single latent variable model (see equations 9.26). Suppose that such a matrix has been fit to a congeneric model involving one latent variable and zero correlation among the measurement errors. We might then ask, "How well do the observed indicators fit the latent variable?" In other words, how closely does the variation in each indicator reflect variation in the latent variable? This question concerns *explained variation* and would be addressed by the magnitude of the correlation (or its square) between the latent variable and a particular indicator. Thus, we might find that 80 percent of the variation in one of the indicators is explained by the latent variable—indicating a "good fit." On the other hand, before we would even want to consider an estimate of such a measure of explained variance, we would want to know if the overidentifying restrictions implied by the model were consistent with the observations. That is, do the observed covariances conform with the specification that

$$\begin{aligned}\mathrm{Cov}(X_1,X_2)\ \mathrm{Cov}(X_3,X_4) &= \mathrm{Cov}(X_1,X_3)\ \mathrm{Cov}(X_2,X_4)\\ &= \mathrm{Cov}(X_1,X_4)\ \mathrm{Cov}(X_2,X_3)\end{aligned}$$

Note that these constraints are almost independent of the absolute magnitudes of the covariances: they focus upon the *relative* magnitudes of the covariances. And these relativistic constraints must be satisfied before we would entertain the hypothesis that the four indicators are indeed measures of one latent variable. This exercise in model fitting would be a prerequisite to estimation of variance explained in one latent variable by another or of variance in an indicator explained by a latent variable.

Another example of the contrast between explained variance and model fitting stems from the generation of significance tests for individual coefficients in a structural model. Recall that the procedure for such was based upon the observed decline in the R-square when the coefficient of the variable in question is deleted from the model. We referred to the linear model with the variable deleted as the *constrained model,* since the coefficient of the deleted variable had been constrained to zero. In fact, such a constrained model is simply an *overidentified model.* And the differential between the R-square for the full and constrained models is a test of the fit of the overidentified model. If the constraint (overidentifying restriction) produces a marked decline in the R-square, then the associated F ratio will be significant, and the constrained model will be rejected. The F ratio in this instance is analogous to the likelihood ratio chi-square; as constraints are added, the chi-square increases. But the crucial question is whether or not the increase is greater than what one would expect by chance. Notice that the absolute value of the R-square for the full (or the constrained) model is not a major concern here. The major concern is with how much the addition of certain constraints *decreases* the R-square and *increases* the F ratio. More generally, the central concern is the extent to which the constrained model can reproduce the observed covariance matrix. (Recall that the normal equations of multiple regression usually have a unique solution. This means that the full linear model reproduces the observed covariance matrix exactly—and this is what we would expect from a model that implies no overidentifying restrictions.)

Hierarchical Comparisons of Model Fit

The likelihood ratio chi-square given in 9.29 is particularly useful when two or more hierarchically related models are of interest. For example, in the case of one latent variable and several indicators, the parallel model is hierarchically related to the tau equivalent model. This is the case because the parallel model includes all the overidentifying restrictions implied by the tau equivalent model *plus* some additional constraints. Thus, in general, model M_1 is hierarchically related to model M_2, and M_2 is above M_1, if the constraints placed by M_2 (on the covariance matrix of observed variables) include all those implied by M_1 *plus* one or more additional constraints.

Suppose now that M_1 and M_2 are hierarchically related models where M_2 is above M_1. A sample covariance matrix could then be used to perform the following sort of model testing. Suppose that the fit of M_2 to the data yields a chi-square of 16 with 10 degrees of freedom. Such a fit is insufficient to reject M_2 at the 5 percent level, since the probability of observing a chi-square of 16 or greater with ten degrees of freedom is .10. This means that the observed covariance matrix is not an unlikely event if the specified model holds in the population from which the sample was randomly drawn. Our concern is now with model M_1: is the fit of M_1 with the observed covariances sufficiently better than the fit of M_2, such that M_2 can be rejected? This question is answered by computing the chi-square associated with the fit of M_1 to the observed covariance matrix and comparing the decrease to the number of overidentifying restrictions that are relaxed. Suppose, for example, that M_1 deletes 3 degrees of freedom (because of relaxing three overidentifying restrictions) and yields a chi-square of 8 with 7 degrees of freedom. If we now treat M_2 as the null hypothesis, then the difference between the two chi-squares ($16 - 8 = 8$) is also distributed as a chi-square statistic with degrees of freedom equal to the difference between the degrees of freedom for the two models (10 − 7 = 3). The probability of this chi-square then tells us the probability of the observed deviations from the constraints peculiar to M_2 if M_2 held in the population. In the above case, the probability of an observed chi-square greater than 8, with $df = 3$, is less than .02. Thus, we would be motivated to reject M_2 in favor of M_1, since the additional constraints implied by M_2 do not conform well with observed covariances. If the difference between the chi-squares for M_1 and M_2 was smaller, say closer to 3 (the difference in degrees of freedom), then M_2 would not be rejected.

This entire procedure is analogous to treating a constrained linear model as the null hypothesis in an ANOVA or other structural model. If the R-square for the unconstrained model is sufficiently larger (as indicated by the F ratio) than the R-square for the constrained model (null hypothesis), then the constrained model is rejected in favor of the unconstrained model, which includes the independent variable being tested. As in the case of measurement models, recall also that under some circumstances it is desirable in ANOVA to use a constrained model that is simply *less* constrained than the null hypothesis (rather than *un*constrained) for the calculation of F ratios (e.g., when numerous independent variables are involved and some of the higher-order interactions are presumed to be zero *a priori*). This is the norm in the testing of measurement models.

In Section 9.5, numerical examples of the procedures outlined in this section are presented. It is unlikely that readers will achieve a very thorough understanding of these methods without paying rather close attention to the examples and to the computation of their own example problems.

9.5 SOME NUMERICAL EXAMPLES

The Three-Indicator/One Latent Variable Case

The simplest measurement model that is likely to be of practical interest is that involving three parallel or tau equivalent measures of a single latent variable. (Recall that the congeneric model is not overidentified when only three indicators are measured.) For example, Dwyer, Boyer, Roberts, and Stanley (1981) administered a questionnaire concerning racial attitudes and authoritarianism to 109 white undergraduates. The questionnaire included the following three items:

Q1: Some candidates for Congress are taking the position that there is no need for any kind of affirmative action programs promoting job opportunities for blacks. How would such a position influence your vote?

Q2: Would you vote for a state bond issue to allocate funds for the implementation of job training programs for blacks?

Q3: Do you favor or oppose laws to give blacks better job opportunities?

For each of these items, the respondent was asked to give a number from 0 to 9 indicating his or her reaction to the item.

The question that concerns us here is whether or not the responses to these three

Table 9.4 The covariance and correlation matrices for three questionnaire items concerning support for expansion of job opportunities for blacks ($N = 109$).

	X_1	X_2	X_3	Mean
X_1	6.09	3.46	3.02	3.79
X_2	.517	7.37	4.36	4.83
X_3	.479	.627	6.55	3.81

Note. The correlations are given in the lower triangle and the variance/covariance matrix is given in the upper triangle and main diagonal.

items are indicative of a *single* latent dimension. All three items refer to job opportunities for blacks, but each item is distinct in its focus upon voting for candidates, voting for bonds, and favoring or opposing the passage of legislation. Furthermore, one item refers to affirmative action, whereas another refers to job training programs. It is certainly conceivable that these items reflect distinct psychological dimensions. On the other hand, it is also plausible that responses to all three of these items reflect the same psychological dimension; that is, if the random error could be eliminated from responses to each item, then responses to item Q1 would order the sample in exactly the same sequence as Q2 or Q3. Furthermore, we might hypothesize that each item reflects the same "true score" with the same amount of measurement error. This means that we are hypothesizing a parallel model to explain observed correlations among responses to the three items.

The parallel model is testable with the covariance matrix derived from the 109 respondents. That is, we can determine how likely the observed covariance matrix would be if the parallel model held in the population. If this probability is small, then the parallel model can be rejected. Unfortunately, the more elaborate model, in which each item reflects a distinct dimension, cannot be tested with this small number of variables. For example, a model involving three correlated latent variables with one observed indicator for each is hopelessly *under*identified, and this means that virtually any pattern of observed covariances could be construed as consistent with the model (covariances with a sign opposite to that predicted would be necessary for rejection).

The covariance and correlation matrices for these three questionnaire items are given in Table 9.4. (In fact, items Q2 and Q3 were "reflected" to avoid negative covariances; this was done for simplicity, however, since the "raw" covariance matrix could have been used.) Note that the three covariances among the variables range from 3.02 to 4.36, and the variances range from 6.09 to 7.37. Since the assumption that these three items are parallel measures implies that the population covariances are equal and that the population variances are equal, our analysis involves the probability that the deviations from equality in Table 9.4 would occur by chance if a parallel model held in the population.

The specification of the parallel model for the population was given in equations 9.24

$$X_1 = k_{11}F_1 + d_1$$
$$X_2 = k_{21}F_1 + d_2$$
$$X_3 = k_{31}F_1 + d_3$$

where

$$k_{11} = k_{21} = k_{31} = 1$$
$$\text{Var}(k_1) = \text{Var}(d_2) = \text{Var}(d_3)$$

and

$$\text{Cov}(d_1,d_2) = \text{Cov}(d_1,d_3) = \text{Cov}(d_2,d_3) = 0$$

In matrix notation, we would specify the measurement model as

$$\underline{X} = \underline{Lx}\,\underline{F} + \underline{d}$$

where

$$\underline{Lx} = [1 \quad 1 \quad 1]'$$
$$t_{11} = t_{22} = t_{33}$$

and

$$\underline{Td} \text{ is diagonal (off-diagonal elements are zero)}$$

The next step is to estimate the covariance matrix of the latent variables ($\underline{PH}$), the matrix of factor loadings ($\underline{Lx}$), and the covariance matrix of measurement errors ($\underline{Td}$) with the observed covariances and variances. This could be accomplished by solving equations 9.25, given the parallel constraints, for the *free* elements of these three matrices. The free elements are those that have not been constrained to any particular population value. This means that there are no free parameters in $\underline{Lx}$ (since all loadings have been constrained to equal one), one free parameter in $\underline{PH}$ (ph_{11}, the variance of the latent variable F_1), and one free parameter in $\underline{Td}$ (t_{11}, since t_{22} and t_{33} are constrained to be equal to t_{11} and since $\underline{Td}$ has been constrained to be diagonal). Thus, we must estimate a total of two free parameters (ph_{11} and t_{11}).

Equations 9.25 provide three solutions for ph_{11} and three solutions for t_{11} in terms of covariances among observable variables:

$$ph_{11} = \text{Cov}(X_1,X_2)$$
$$ph_{11} = \text{Cov}(X_1,X_3)$$
$$ph_{11} = \text{Cov}(X_2,X_3)$$
$$t_{11} = \text{Var}(X_1) - ph_{11}$$
$$t_{11} = \text{Var}(X_2) - ph_{11}$$
$$t_{11} = \text{Var}(X_3) - ph_{11}$$

Now, if we simply replace the population covariances in these equations with the sample values given in Table 9.4, we obtain three estimates of ph_{11}:

$$\text{first estimate of } ph_{11} = 3.46$$
$$\text{second estimate of } ph_{11} = 3.02$$
$$\text{third estimate of } ph_{11} = 4.36$$

Before we proceed to estimate t_{11}, it would be convenient to settle upon a single estimate of ph_{11}. The most straightforward approach to combining these three estimates would be to use their mean, which is 3.61. In fact, for this simple model, the mean is also the maximum likelihood estimate of ph_{11}. And we can then compute our three estimates of t_{11}:

$$\text{first estimate of } t_{11} = 6.09 - 3.61 = 2.48$$
$$\text{second estimate of } t_{11} = 7.37 - 3.61 = 3.76$$
$$\text{third estimate of } t_{11} = 6.55 - 3.61 = 2.94$$

If we again use the simple mean of these three estimates as our estimate of t_{11}, then we have

$$\text{est}(ph_{11}) = 3.61$$
$$\text{est}(t_{11}) = 3.06$$

(once again, it should be noted, the mean of the three estimates of t_{11} is also the maximum likelihood estimate of t_{11}).

Our concern is now focused upon how well this parallel model fits these observed variances and covariances. The first step in this procedure is to return to the normal equations in 9.25, in which population variances and covariances are expressed as functions of the parameters of the model:

$$\text{Var}(X_1) = \text{Var}(X_2) = \text{Var}(X_3) = ph_{11} + t_{11}$$
$$\text{Cov}(X_1,X_2) = \text{Cov}(X_1,X_3) = \text{Cov}(X_2,X_3) = ph_{11}$$

Substituting the estimates of ph_{11} and t_{11} obtained above into these equations gives us the covariance matrix of $\underline{X}$ that is *predicted by the model* (i.e., the matrix $\underline{Z}$):

$$\underline{Z} = \begin{bmatrix} 6.67 & 3.61 & 3.61 \\ 3.61 & 6.67 & 3.61 \\ 3.61 & 3.61 & 6.67 \end{bmatrix}$$

Now, since the elements of $\underline{Z}$ are based upon the ML estimates of the free parameters in the model, then if the values of $\underline{S}$ given in Table 9.4 and the values of $\underline{Z}$ given above were substituted into equation 9.27 (the transformed likelihood function) we would have a minimum value of H. That minimum value is actually equal to .0949 for these data. If we now recall equation 9.29 for the likelihood ratio chi-square, we can compute

$$\begin{aligned}\text{chi-square} &= \min(H)(N/2)\\ &= (.0949)(109/2)\\ &= 5.1705\end{aligned}$$

where

$$\begin{aligned}df &= [q(q-1)/2 + q] - f\\ &= [3(3-1)/2 + 3] - 2\\ &= 4\end{aligned}$$

(recalling that q = the number of observed indicators and f = the number of free parameters estimated).

The next step in this procedure is to determine how often a chi-square this large would occur if the hypothesized parallel model held in the population from which the sample was drawn. This question is answered by using a table or computer program to compute

$$\text{Prob(chi-square} > 5.17 \mid \text{model holds)} = .27$$

which is simply the probability associated with the chi-square distribution for 4 degrees of freedom. If the assumption of multivariate normality of the observable variables holds, then this probability is not sufficiently small to motivate a rejection of the parallel model. This means that the data are consistent with the model, that the deviations of the observed moments from those predicted by the model were well within the range of what would be expected from sampling error, and that the parallel model has survived an attempt to reject it. This does not mean, it is important to remember, that the hypothesized model is "true," or that it is the only model that fits these data. Rather, these data had the opportunity to lead to a rejection of the parallel model, but failed to do so.

At this point, the skeptical reader may be thinking, "All right, the probability of observed deviations from the model's predictions may be .27 if the parallel model holds, but what is the probability of these observations if the parallel model *does not hold?*" It may be the case, for example, that the "true" model is quite distinct from the parallel model, but that the overidentifying restrictions of the true model are sufficiently similar to those of the hypothesized model that a good fit is obtained. These are important concerns, and proper interpretation of model fitting requires that these pitfalls be recognized. The fact that a model evidences a good fit with a set of data means that the data are insufficient to reject that model, but such a failure is not very compelling evidence if the likelihood of rejection is not very great *even when the hypothesized model is not true.*

This discussion is focusing upon the issue of statistical *power*. That is, how often will the hypothesis test lead to rejection of the null hypothesis when the null hypothesis is indeed false (and false to a particular extent)? The only way to answer this question is to conduct monte carlo type simulations for specific measurement models. But since this is rather impractical for most researchers, and the generality of results would be limited to those specific models used, the following recommendations are advanced:

(1) the fact that a measurement model is consistent with an observed covariance matrix should be interpreted conservatively, and (2) researchers should strive to construct their research problems in terms of hierarchically related models so that a specific model can be rejected *relative to another specific model.* (3) Furthermore, with the specification of measurement models, including "method factors" and multiple latent variables, a researcher can increase her confidence in the validity of a model through its mesh with an *a priori* nomological net. (4) The most compelling result of a model testing venture is *rejection* of a model; in the tradition of Karl Popper, this point of view stems from the fact that a "good fit" means that the hypothesized model is one of many plausible models, whereas rejection is definite.

For the sake of the reader who suspects that most any three correlated variables would fit a parallel or tau equivalent measurement model, consider the data in Table 9.5. This covariance matrix is based upon three other items from the questionnaire on race relations that was described above. In this case, the three items are versions of an item from the California F-scale:

Q4: A sane, normal, decent person might have to hurt a close friend or relative at some point in their life.
(disagree 0 1 2 3 4 5 6 7 8 9 agree)

Q5: A sane, normal, decent person might have to hurt a close friend or relative at some point in their life.
(disagree -------------------------- agree)

Q6: (A) No sane, normal decent person could ever think of hurting a close friend or relative. (B) A sane, normal, decent person might have to hurt a close friend or relative at some point in their life.
(A 0 1 2 3 4 5 6 7 8 9 B)

The first two of these F-scale items differ only in that they involve different methods of response (numbers versus lines), whereas the third is distinguished by the "forced choice" format (although the inclusion of a fairly continuous range of numbers for response moderates the extent to which a choice between the two statements is "forced").

Except for the fact that the correlations among the three F-scale items are stronger than those for the three racial attitude items in Table 9.4, the matrix in Table 9.5 would appear to suggest that these three items can be construed as indicators of the

Table 9.5 The covariance and correlation matrices for three versions of an F-scale item (N = 109).

	X_4	X_5	X_6	Mean
X_4	6.13	6.12	4.78	6.95
X_5	.859	8.29	5.85	6.35
X_6	.712	.749	7.35	6.33

Note. The correlations are given in the lower triangle and the variance/covariance matrix is given in the upper triangle and main diagonal.

one latent variable. However, when these data are actually fit to such models, the fit is rather poor.

Fitting a measurement model of the form

$$X_4 = k_{11}F_1 + d_1$$
$$X_5 = k_{21}F_1 + d_2$$
$$X_6 = k_{31}F_1 + d_3$$

to the data in Table 9.5 yields the results given in Table 9.6. Note that the fit of the parallel and tau equivalent models produces rather large chi-square statistics relative to their respective degrees of freedom. Furthermore, the probabilities associated with these observed chi-squares are quite small, indicating that these observations would be an unusual occurrence if the hypothesized measurement model actually held in the population from which the sample was drawn.

The three measurement models fitted in Table 9.6 are *hierarchically* related, since, as we move from right to left in the table (i.e., from the congeneric to the tau equivalent to the parallel model), all previous constraints are included in the new model, and the new model is distinguished by the addition of constraints. For example, the constraint that $k_{11} = 1.00$ is part of all three models, whereas the constraint that

$$k_{21} = k_{31} = 1$$

both distinguishes the tau equivalent model from the congeneric model and is incorporated in the still more restrictive parallel model. This hierarchical relationship among these three models means that differences between chi-squares can be used to test the hypothesis that the additional constraints of the more restrictive model hold in the population. For example, we may treat the parallel model as our null hypothesis

Table 9.6 Results of fitting the covariance matrix in Table 9.5 to three hierarchically related measurement models.

Matrix estimated	Parameter estimated	Measurement model		
		Parallel	Tau equiv	Congeneric
Lx	k_{11}	1.00	1.00	1.00
	k_{21}	1.00	1.00	1.22(.08)
	k_{31}	1.00	1.00	.96(.07)
PH	ph_{11}	5.58(.84)	5.66(.83)	5.00(.79)
Td	t_{11}	1.67(.16)	.80(.24)	1.13(.27)
	t_{22}	1.67(.16)	1.45(.30)	.84(.35)
	t_{33}	1.67(.16)	2.80(.45)	2.75(.43)
Chi-square		24.35	10.77	0.00
Degrees of freedom		4	2	0
Probability		.0001	.0046	—

Note. The values in parentheses are approximate standard errors of the parameter estimates. If the model holds in the population and the observed variables are multivariate normal in distribution, then the probability of the observed chi-square (or one greater) is the probability of the observed deviation (or one greater) from the predictions of the model.

and ask the following question: "How likely is the reduction of the observed chi-square from 24.35 to 10.77 if the parallel model holds in the population?" If the indicators are multivariate normal, then the answer to this question is obtained by subtracting the smaller chi-square from the larger. The degrees of freedom for this difference between chi-squares is the difference between their respective degrees of freedom. Thus,

$$\text{chi-square} = 24.35 - 10.77 = 13.58$$

where

$$df = 4 - 2 = 2$$

And reference to the chi-square distribution in the appendix reveals that

$$\text{Prob}(\text{chi-square} > 13.58 \mid df = 2) < .0001$$

This result means that the better fit of the tau equivalent model to the observed data would occur very infrequently if the parallel model held in the population.

This result does not, of course, mean that the tau equivalent model is the appropriate model; rather, this result means that the fit of the tau equivalent model is *significantly* better than the fit of the parallel model at the .0001 level. Having made this discovery, the next step is to see if some other model provides a significantly better fit than the tau equivalent model. In fact, if we test the tau equivalent model against the congeneric model, we find that the constraints peculiar to the tau equivalent model can be rejected. The difference between the chi-squares for these two models is 10.77 with 2 degrees of freedom. The probability of this improvement in fit is very small (.0046) if the tau equivalent model holds in the population.

The finding that the tau equivalent model can be rejected relative to the congeneric model is a bit unsatisfying as a final result, however. The reason for this is that the congeneric model does not place any constraints upon the observed covariances and variances. The lack of testable constraints means that the congeneric model cannot be rejected by the observed moments among three indicators. For this reason, then, the congeneric model is most useful when four or more indicators of a single latent variable have been measured.

{*Technical note.* In a strict sense, the congeneric model does place a constraint upon the observed moments, but it is a "weak" constraint. The constraint placed by the congeneric model is that the diagonal elements of $\underline{PH}$ and $\underline{Td}$ must be positive—since they are all variances. For example, if we solve the congeneric model for estimates of t_{11}, we find that

$$\text{est}(t_{11}) = S(X_1,X_1) - [S(X_1,X_2)S(X_1,X_3)/S(X_2,X_3)]$$

which implies that

$$S(X_1,X_1)S(X_2,X_3) > S(X_1,X_2)S(X_1,X_3)$$

or

$$r(X_2,X_3) > r(X_1,X_2)\ r(X_1,X_3)$$

if

$$\text{est}(t_{11}) > 0.00$$

Thus, the congeneric model for three indicators predicts that the product of any two correlations will be less than the third correlation coefficient. The reason that this constraint is considered "weak" is that it is in the form of an *inequality* that could be satisfied by a range of results. In contrast, the parallel model with three indicators, or the congeneric model with four indicators, implies *equality* constraints. Nevertheless, there may be circumstances in which the just identified congeneric model is a useful one, since it may be rejected when the inequality constraint is significantly violated. This would be accomplished by constraining to zero those elements of $\underline{PH}$ and $\underline{Td}$ that had been estimated to be less than zero. If this model could not be rejected relative to the unconstrained congeneric model, then the negative estimates could be treated as arising from sampling variation. However, if the restricted model could be rejected, then the negative estimates must be accepted as an indication that the unconstrained congeneric model places weak constraints upon the moments, and the observed moments significantly violate those constraints. As an exercise, the reader should fit a congeneric model to the following covariance matrix for three items from the above-mentioned questionnaire (three versions of the F-scale item "Nobody ever learned anything really important except through suffering"):

	Q7	Q8	Q9
Q7	8.31		
Q8	6.38	7.42	
Q9	2.67	3.55	6.23

Using the LISREL computer program to fit these data to a congeneric model will yield an estimate of t_{22} equaling -1.08 [i.e., a negative estimate of $\mathrm{Var}(d_2)$]. However, this negative coefficient is not significantly different from zero, so the constraint that $t_{22} = 0$ cannot be rejected as a null hypothesis. But to presume that t_{22} is zero is equivalent to the presumption that responses to Q8 include no error component—which is a dubious conclusion on *a priori* grounds.}

Where do these analyses leave us? They leave us with the conclusion that the three questionnaire items (Q4 to Q6) are not parallel or tau equivalent indicators of a single latent variable. They may be congeneric indicators, but the lack of overidentifying restrictions for the congeneric model does not provide us an opportunity to test the congeneric model against a less restricted model.

Maximum Likelihood Standard Errors

Note that Table 9.6 includes standard errors of parameters that are estimated by the maximum likelihood procedure. These standard errors are interpreted in the same manner as standard errors of slopes in ordinary least squares regression. When the assumption of multivariate normality holds for the observed variables, then the standard errors may be interpreted probabilistically. That is, the ML estimate plus and minus two standard errors defines a 95 percent confidence interval. This means that if the model holds in the population, then 95 percent of such confidence intervals from samples will contain the population value of the parameter.

The actual mathematics of standard error estimation are quite complex in the case of ML procedures. However, an understandable method of computing standard errors with programs such as LISREL can be outlined—even though this method need not be used, since LISREL provides the estimated standard errors as optional output. This method of estimating standard errors should be understandable to the reader of this book, since it is analogous to the procedures outlined for the estimation of standard errors of slopes in OLS regression. Recall that the F ratio for the null hypothesis that a population slope was zero flowed from excluding the relevant right-hand variable from the regression equation (i.e., constraining it to zero) and utilizing the resultant reduction in the R-square.

An analogous procedure applies to the testing of a null hypothesis that a parameter in a measurement model is zero. After the model in which the parameter is estimated has been fit to the data (yielding a goodness of fit chi-square), then the parameter of interest is constrained to zero and the new model is fit to the data. The difference between these two goodness of fit chi-squares is then distributed as a chi-square with one degree of freedom (since one restriction has been added to the model by the constraint of one parameter to a value of zero) if the model including the null hypothesis holds in the population. And the square root of a chi-square statistic with one degree of freedom is distributed as the normal distribution. Thus, if b^* is the parameter being estimated, b is the ML estimate, then

$$\text{approx std error } (b) = b/[(c_0 - c_1)^{.5}]$$

where

c_0 = goodness of fit chi-square for model including the constraint that $b^* = 0$

c_1 = goodness of fit chi-square for model with b^* a free parameter

As has been emphasized in this text many times, estimated standard errors are an invaluable tool, but they should be taken with a grain of salt. They are an attempt to predict what would happen if the study was replicated, but they usually accomplish this task only approximately—and should be interpreted accordingly.

Transformation of Variables to Standardized Form

Once the parameters of a measurement model have been estimated, it is often convenient to standardize the variables in the model so that covariances become correlations and slopes become standardized regression coefficients or "beta weights." The LISREL computer program includes standardized output as an option, but only the latent variables are standardized. The observed variables, measurement errors, and disturbances are left in their unstandardized metric. In the case of congeneric indicators, this type of standardization amounts to an alternative specification of the units of the latent variable. Recall that the congeneric model adopts the units of one of the indicators as the units of the latent variable. An alternative specification is that the variance of the latent variable is equal to one (i.e., standardized). This specification is generally less useful than the congeneric specification, since it can obscure important metric information that can be used when comparing different samples and different populations. Nevertheless, standardization of a model can be helpful in communicating the relative magnitude of relationships among variables.

Consider, for example, the results of fitting the data in Table 9.4 to a parallel measurement model. For these three questionnaire items dealing with support for efforts to increase job opportunities among blacks in the United States, we found that the ML estimate of the variance of the latent variable (ph_{11}) was 3.61, whereas the ML estimate of the variance of the measurement errors (t_{11}, t_{22}, and t_{33}) was 3.06. Now, we return to the equations that specify the population measurement model:

$$X_1 = F_1 + d_1$$
$$X_2 = F_1 + d_2$$
$$X_3 = F_1 + d_3$$

where $t_{11} = t_{22} = t_{33}$. If we then standardized each variable in these equations, we would have

$$Z(X_1) = a_{11}Z(F_1) + a_{12}Z(d_1)$$
$$Z(X_2) = a_{21}Z(F_1) + a_{22}Z(d_2)$$
$$Z(X_3) = a_{31}Z(F_1) + a_{32}Z(d_3)$$

where Z indicates a standardization transformation and the coefficients $a_{11}, \ldots, a_{32}$ are standardized coefficients. The specification that the latent variable is uncorrelated with the measurement errors leads to the conclusion that

$$a_{11} = \text{Corr}(X_1,F_1)$$
$$a_{21} = \text{Corr}(X_2,F_1)$$
$$a_{31} = \text{Corr}(X_3,F_1)$$

and

$$1 = a_{11}^2 + a_{12}^2$$
$$1 = a_{21}^2 + a_{22}^2$$
$$1 = a_{31}^2 + a_{32}^2$$

In order to estimate these coefficients in terms of the estimates of $\underline{Lx}$, $\underline{PH}$, and $\underline{Td}$, we recall that

$$\text{Corr}(X_1,F_1) = \text{Cov}(X_1,F_1)/[\text{Var}(X_1)\ \text{Var}(F_1)]^{.5}$$

and that

$$\text{Var}(F_1) = ph_{11}$$
$$\text{Cov}(X_1,F_1) = ph_{11}$$
$$\text{Var}(X_1) = ph_{11} + t_{11}$$

Thus, we have

$$a_{11} = [ph_{11}/(ph_{11} + t_{11})]^{.5}$$

$$a_{21} = [ph_{11}/(ph_{11} + t_{22})]^{.5}$$
$$a_{31} = [ph_{11}/(ph_{11} + t_{33})]^{.5}$$

Now, since all the parameters on the right-hand side of these equations have been estimated, we can simply substitute those estimates into these equations:

$$\text{est}(a_{11}) = [3.61/(3.61 + 3.06)]^{.5}$$
$$= .54^{.5} = .74$$
$$\text{est}(a_{21}) = .74$$
$$\text{est}(a_{31}) = .74$$

And these results then give the following:

$$\text{est}(a_{12}) = (1 - .54)^{.5} = .68$$
$$\text{est}(a_{22}) = .68$$
$$\text{est}(a_{32}) = .68$$

These estimates of the standardized coefficients may now be portrayed in a path diagram (see Figure 9.6). They indicate that 54 percent of the variance in the observed indicators reflects variation in the latent variable ("true variance"), whereas 46 percent of the observed variance is attributed to measurement error. The value .74 is termed the *standardized validity* of each indicator in this parallel model. The problem with such standardized measures is that they depend heavily upon the range of scores included in a particular population. That is, the correlation between X_1 and F_1 clearly depends upon the variance of F_1 (ph_{11}). In contrast, if d_1 is truly independent of F_1, then the variance of d_1 (t_{11}) would be independent of the range of F_1 included in a study. Thus, the estimate of t_{11} is a more useful piece of information for an investigator who plans to utilize an indicator in a number of contexts.

The standardized coefficients are most informative for comparisons within a particular model or when exhaustive sampling has guaranteed inclusion of the full range of true scores. For example, the congeneric model fit to the data in Table 9.5 (see Table 9.6) yields quite different estimates for the three loadings and the variances of the three measurement errors. Transforming these results into a standardized form gives

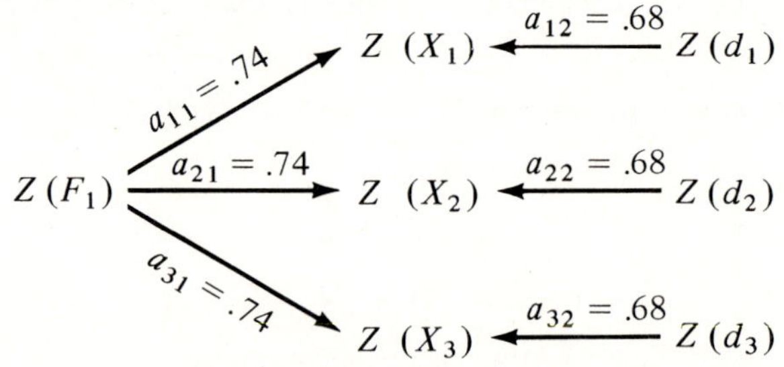

Figure 9.6 Path diagram of a three-indicator measurement model in standardized form. All variables have a mean of zero and variance of one. The sum of the squares of all paths leading to observed variables (X_1, X_2, X_3) are equal to one.

$$\begin{aligned}
\text{est}(a_{11}) &= [5.00/(5.00 + 1.13)]^{.5} \\
&= .82^{.5} = .90 \\
\text{est}(a_{21}) &= [7.44/(7.44 + .84)]^{.5} \\
&= .90^{.5} = .95 \\
\text{est}(a_{31}) &= [4.61/(4.61 + 2.75)]^{.5} \\
&= .63^{.5} = .79
\end{aligned}$$

Thus, Q4, Q5, and Q6 are estimated to be correlated .90, .95, and .79 with their common latent variable, respectively. This result indicates that the use of a line for responses to Q7 involved the smallest error variance—*if the congeneric model holds in the population.*

To restate the disadvantage of the standardization, the correlations between a latent variable and an indicator is an aggregation of information that can be more useful in the disaggregated form. For example, consider the results of fitting the F-scale data to a congeneric model (Table 9.6). Noting that k_{21} is estimated as 1.22, whereas k_{11} is constrained to 1.00, suggests that the "true variance" in the line response format was greater than in the number response format (scores were scaled on a zero to nine metric for the line format item). This means that respondents used more of the 0 to 9 range when making their response on a line than when the response was given in terms of a number from 0 to 9. The extent to which this is the case is estimated by the square of k_{21}:

$$\begin{aligned}
\text{true variance in } X_4 &= \text{est}(ph_{11}) = 5.00 \\
\text{true variance in } X_5 &= (1.22^2) \times \text{est}(ph_{11}) \\
&= 7.44
\end{aligned}$$

[In order to see why the above calculations are appropriate, take the equation

$$X_5 = k_{21}F_1 + d_2$$

square both sides, and take expectations over the population. The result is

$$\begin{aligned}
\text{Var}(X_5) &= k_{21}^2 \text{ Var}(F_1) + \text{Var}(d_2) \\
&= k_{21}^2 ph_{11} + t_{22}
\end{aligned}$$

which reveals $(k_{21}^2 ph_{11})$ to be the "true variance" in X_5 (responses to Q5).] Thus, although the estimated variance of measurement error is less for Q5 than Q4 (.84 versus 1.13), this difference is not statistically significant and is not nearly so important a factor in determining the larger standardized validity of Q5 as is the greater "true variance" of Q5 (7.44 versus 5.00).

The Four-Indicator/One Latent Variable Case

The use of four indicators of a single latent variable crosses an important threshold, since the congeneric model is also overidentified. This means that a goodness of fit chi-square can be obtained for the congeneric model, and this means that observed data can reject the congeneric model—which was not the case with three indicators. However, the reader should keep in mind that failure to reject a model must be interpreted conservatively when the alternative is an undefined model (i.e., one with no overidentifying restrictions for the data in hand).

For our first example, we will use data of a rather different sort from the psychological scaling data used thus far. They are taken from a study of socioeconomic variables relevant to reading achievement in the New York City school system (Modoono, 1981). In an attempt to explain inter-school differences in reading achievement, Modoono gathered scores on a number of indicators. Observations relevant to the communities surrounding 160 junior high schools in New York City included the following:

X_1 = percent of housing units owner occupied in 1970

X_2 = percent of the school population enrolled in the "Free Lunch" program for low income families (1974)

X_3 = median income of families living in the school district in 1970 (in thousands of dollars)

X_4 = "Litter Index," based upon visible litter in the streets included in the school district (1974)

These four indicators are clearly measures of distinct differences among the school districts of New York City. However, our concern here is with the hypothesis that the covariance among these indicators (across school districts) arises because of the effects of a single latent variable. If such a hypothesis is true, then these indicators could be used as multiple indicators of that latent variable. Furthermore, if the variability in these indicators that is not explained by that latent variable is uncorrelated with other theoretically relevant variables, then relationships between this latent variable and other variables can be assessed without distortion due to measurement error.

We proceed to test whether or not the observations of the four indicators are consistent with a single factor model by fitting the following covariance matrix to a congeneric model:

	X_1	X_2	X_3	X_4	Mean	SD
X_1	372				30.4	19.3
X_2	−200	530			55.1	23.0
X_3	32.2	−31.1	6.31		9.32	2.51
X_4	−27.5	20.9	−3.66	2.87	14.9	1.69

Since each of these indicators is measured in a unique metric, it would make no sense to presume that they were parallel or tau equivalent measures of a common factor. Rather, the congeneric model is the most strict that could be applied.

The congeneric model for the four variable case is specified with the equations

$$X_1 = k_{11}F_1 + d_1$$

$$X_2 = k_{21}F_1 + d_2$$

$$X_3 = k_{31}F_1 + d_3$$

$$X_4 = k_{41}F_1 + d_4$$

Table 9.7 Results of fitting the covariances among four indicators of socioeconomic differences among 130 school districts in New York City to models with a single latent variable.

Matrix estimated	Parameter estimated	Model 1	Model 2	Standardized
Lx	k_{11}	1.00	1.00	.810
	k_{21}	−.74(.12)	−.75(.11)	−.509
	k_{31}	.13(.01)	.13(0.01)	.827
	k_{41}	−.11(.01)	−.11(.01)	*
PH	ph_{11}	255(45)	244(44)	1.00
Td	t_{11}	117(16)	128(18)	.344
	t_{22}	391(48)	393(48)	.741
	t_{33}	1.82(.26)	1.99(.29)	.316
	t_{44}	−.12(.09)	−.24(.11)	*
	t_{32}	0.00	−6.74(2.50)	−.241
Chi-square		8.40	0.78	
Degrees of freedom		2	1	
Probability		.015	.377	

Note. The numbers in parentheses are approximate standard errors. Coefficients without standard errors are constrained or "fixed" parameters. The asterisk indicates that the standardized coefficient is undefined due to the negative estimate of t_{44} (the variance of d_4).

where k_{11}, for example, is constrained to equal one, and the measurement errors $d_1, \ldots, d_4$ are uncorrelated with F_1 and one another. The fit of this model with the covariance matrix is given in the first column of Table 9.7. Model 1 in Table 9.7 is the usual congeneric model. Note that the fit of the model is unsatisfactory for two reasons. First, the goodness of fit chi-square is large relative to the degrees of freedom and would occur less than two times out of 100 if the model held in some population from which the observations are a random sample. (This sampling assumption is hypothetical in this case, since all the schools with complete data were included in the analysis. Nevertheless, it is still useful in such instances to know these "hypothetical probabilities," since we would still like to know how likely these results would be *if* the observations studied were a random sample.) Second, the ML estimate of t_{44} (the variance of d_4) is negative. This result is referred to as the Heywood Case in the factor analytic literature. The combination of these two counterindications is strongly suggestive that variability in Modoono's four indicators is *not* consistent with the predictions of a congeneric model.

Suppose, however, that an investigator expected the measurement errors in X_2 and X_3 to be correlated, since each is an income measure. Then model 2 in Table 9.7 could have been tested against the congeneric model (see Figure 9.7). Model 2 does not constrain the covariance between d_2 and d_3 (t_{32}) to zero, thus adding one unconstrained ("free") parameter to the model and reducing the degrees of freedom from two to one. The difference between the chi-squares for the two models can then be used to estimate the probability of the observed difference in fit of the two models if the more restricted model held in some population. The difference between the two chi-squares is 7.62 (8.40 − .78) with 1 degree of freedom, which is clearly an unlikely event if model 1 holds in the population ($p < .001$). This result means that the investigator's hypothesis would have been retained and the congeneric model rejected.

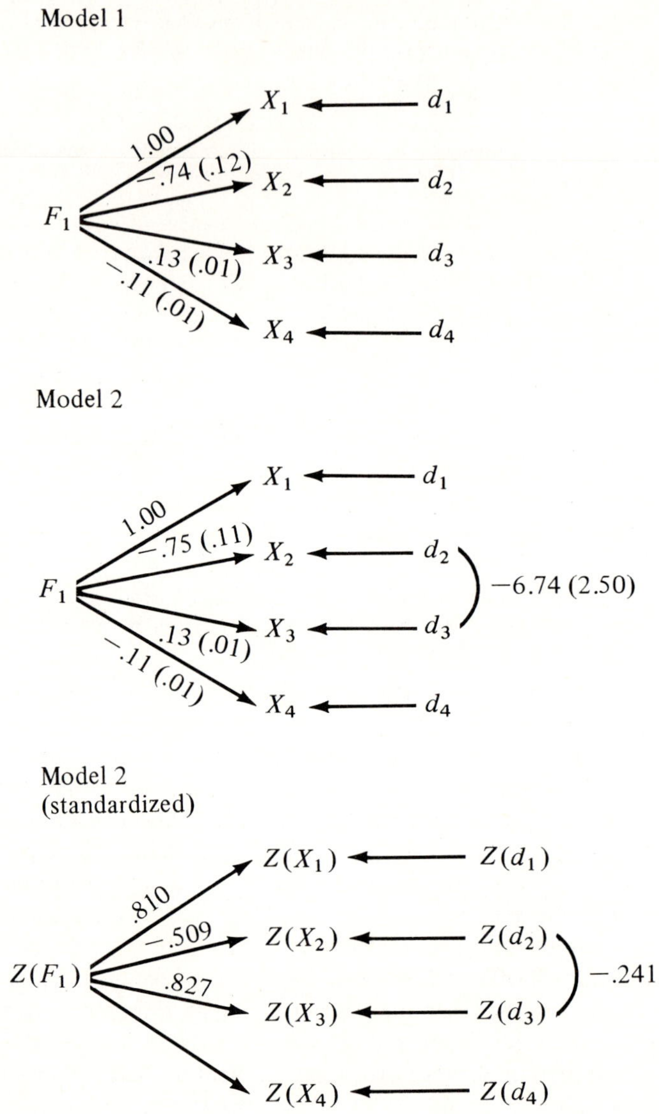

Figure 9.7 Path diagram of models 1 and 2 from Table 9.7 (data from Modoono, 1981). X_1 = houses owner occupied, X_2 = children on free lunch, X_3 = median family income, and X_4 = litter index.

But further inspection of the results for model 2 indicates that the Heywood Case is still with us. In fact, the negative variance is now more than twice its standard error in absolute value. At this point, the conservative conclusion to draw from the analysis is that the single latent variable models are not consistent with these data. The question of why these models do not fit is best framed in terms of which other models *would* be consistent with these data. And the fact of the matter is that with only four indicators, very few alternative models are testable (i.e., they do not generate overidentifying restrictions for these data). One may speculate, of course, that there are several correlations among the measurement errors, that there are two latent variables responsible for the covariances among the indicators, and so on. The rather sizable variance of d_2, together with the substantial covariance between d_2 and d_3, is particularly suggestive that these two income-related indicators may reflect one or more latent variables that are distinct from those reflected in X_1 (owner-occupied housing) and X_4 (street litter). But more elaborate data sets will be necessary to explore such alternatives.

This example raises the issue of correlated measurement errors and their meaning. If we accepted model 2 in Table 9.7, how would we interpret the estimate -6.74 of t_{32}? To begin with, we would note that 74 percent of the variation in X_2 (free lunch) and 32 percent of the variation in X_3 (median income) is estimated to be uncorrelated with the latent variable. This means that if we chose school districts that were equal on the latent variable, there would still be substantial variability in X_2 and some variability in X_3. Furthermore, the covariance between d_2 and d_3 of -6.74 means that if F_1 were held constant, there would still be a sizable correlation between X_2 and X_3—although this would not be the case for any other pair of these four indicators. Why might such a situation arise? If we set aside the conclusion that the model is simply erroneous, then one plausible explanation of the result would be that the latent variable is "social status" (an intentionally nebulous dimension at this point). And the two income-related indicators *share* the flaw of reflecting an income dimension. The income dimension is presumed to be correlated with the social status dimension, but to be distinct from it. Thus, if a community has a median income that is high relative to the social status of its members, then the free lunch indicator would tend to understate social status whereas the median income indicator would overstate the value of the latent variable for that community (school district)—and *vice versa.* In contrast, we might expect that the litter indicator would overstate and understate F_1 in a manner that was unrelated to the overstatements and understatements of the income indicators.

This line of reasoning reflects the principles that will be developed under the rubric of the mutitrait multimethod matrix. Indicators X_2 and X_3 would be treated as insufficiently "method divergent" as indicators of the social status latent variable. From this perspective, then, the covariance t_{32} would reflect the existence of a *method factor* or *method latent variable* that inflates the covariance and correlation beyond that which is explicable in terms of the theoretically relevant latent variable. A more systematic approach to the inclusion of method factors in measurement models can provide very specific knowledge of the magnitude and associations of these method factors. And this more systematic approach will be discussed in Section 9.7 of this chapter.

For our last example of a measurement model with a single latent variable, we shall return to the questionnaire concerning racial attitudes. To items Q1, Q2, and Q3 (see above), three additional items will be added: Q1′, Q2′, and Q3′. Q1′ is essentially the same as Q1, but a slightly different response format was used. Whereas agreement/disagreement was expressed with a number from 0 to 9 in Q1, a mark on a line (which was subsequently divided into 10 units for scoring) was used for Q1′. Thus, Q1′, Q2′,

Table 9.8 Covariance matrix for six items from a questionnaire concerning racial attitudes ($N = 109$).

Item	Variable	X_1	X_2	X_3	X_4	X_5	X_6
Q1	X_1	6.09					
Q2	X_2	3.46	7.37				
Q3	X_3	3.02	4.36	6.55			
Q1′	X_4	3.31	3.19	2.82	5.80		
Q2′	X_5	2.95	4.17	3.26	3.42	7.55	
Q3′	X_6	1.95	3.07	3.70	2.41	3.61	5.72

Note. The primes are used to indicate that the same item was repeated with a different response format (line versus number). Although some items were reversed, the scales were reflected so that all covariances are positive.

and Q3′ replicated the wording of Q1, Q2, and Q3, respectively. The covariances among these six items are given in Table 9.8. Our goal is to contrast the fit of various models with these observed data (covariances and variances). The models to be tested are generated from constraints placed upon the following "single factor" model:

$$X_1 = k_{11}F_1 + d_1$$
$$X_2 = k_{21}F_1 + d_2$$
$$X_3 = k_{31}F_1 + d_3$$
$$X_4 = k_{41}F_1 + d_4$$
$$X_5 = k_{51}F_1 + d_5$$
$$X_6 = k_{61}F_1 + d_6$$

The specific models defined by particular constraints could be the following:

(1) Parallel model. This model presumes that all observed variables have the same slope when regressed on the true scores

$$\underline{Lx} = [k_{11} \quad k_{21} \quad k_{31} \quad k_{41} \quad k_{51} \quad k_{61}]'$$
$$= [1 \quad 1 \quad 1 \quad 1 \quad 1 \quad 1]'$$

that the measurement errors for each indicator are equal in variance

$$t_{11} = t_{22} = t_{33} = t_{44} = t_{55} = t_{66}$$

and that the measurement errors are uncorrelated across observed indicators

$$\underline{Td} = \text{a diagonal matrix}$$

This measurement model involves only two free parameters, ph_{11} and t_{11}. Since there are 21 covariances and variances in $\underline{S}(\underline{X}\ \underline{X})$, the estimation of two free parameters leaves 19 degrees of freedom for the goodness of fit chi-square.

(2) Tau equivalent model. This model relaxes the constraint that the measurement errors have equal variance. In all other respects it is equivalent to the parallel model. Five free parameters (t_{22}, t_{33}, t_{44}, t_{55}, and t_{66}) are added to the parallel model for a total of 7—leaving $(21 - 7) = 14$ degrees of freedom for the goodness of fit chi-square.

(3) Congeneric model. This model relaxes the constraint that all indicators have the same slope when regressed on the true scores. In all other respects, it is equivalent to the tau equivalent model. Five free parameters (k_{21}, k_{31}, k_{41}, k_{51}, and k_{61}) are added to the tau equivalent model for a total of 12. Thus, the goodness of fit chi-square for the congeneric model has $(21 - 12) = 9$ degrees of freedom. Note that the constraint $k_{11} = 1$ is retained.

(4) Item specific model. This model is identical with the parallel model, except that it allows nonzero (but equal) covariances between errors in items with the same wording

$$t_{41} = t_{52} = t_{63} = \text{free parameter}$$

This model derives from the hypothesis that identical items may elicit similar errors of measurement ("item specific errors") that are uncorrelated with the latent variable of interest. The addition of one free parameter to the parallel model yields a total of 3. The goodness of fit chi-square then has $(21 - 3) = 18$ degrees of freedom.

Using a ML procedure such as LISREL to fit these four models to the data in Table 9.8 gives the results in Table 9.9.

With 19 degrees of freedom, the parallel model (model 1 in Table 9.9) yields a

Table 9.9 Results of maximum likelihood fitting of five measurement models to the covariances among six questionnaire items concerning job opportunities for blacks (Table 9.8).

	Measurement model				
Parameter	(1)	(2)	(3)	(4)	(5)
k_{11}	1.00	1.00	1.00	1.00	1.00
k_{21}	1.00	1.00	1.32(.20)	1.00	1.00
k_{31}	1.00	1.00	1.22(.19)	1.00	1.00
k_{41}	1.00	1.00	1.02(.17)	1.00	1.00
k_{51}	1.00	1.00	1.22(.20)	1.00	1.00
k_{61}	1.00	1.00	1.02(.17)	1.00	1.00
ph_{11}	3.25(.52)	3.23(.51)	2.53(.72)	3.13(.52)	3.05(.51)
t_{11}	3.27(.20)	3.46(.54)	3.56(.55)	3.39(.22)	3.42(.23)
t_{22}	3.27(.20)	3.32(.53)	2.95(.53)	3.39(.22)	3.42(.23)
t_{33}	3.27(.20)	2.90(.47)	2.75(.48)	3.39(.22)	3.42(.23)
t_{44}	3.27(.20)	3.04(.49)	3.20(.50)	3.39(.22)	3.42(.23)
t_{55}	3.27(.20)	3.84(.59)	3.77(.61)	3.39(.22)	3.42(.23)
t_{66}	3.27(.20)	3.06(.49)	3.09(.49)	3.39(.22)	3.42(.23)
t_{41}	.00	.00	.00	.60(.22)	.64(.22)
t_{52}	.00	.00	.00	.60(.22)	.64(.22)
t_{63}	.00	.00	.00	.60(.22)	.64(.22)
t_{61}	.00	.00	.00	.00	−.90(.38)
t_{32}	.00	.00	.00	.00	.94(.32)
Chi-square	33.44	31.36	24.93	25.05	11.80
df	19	14	9	18	16
Probability	.0214	.0049	.0031	.1236	.7577

Note. The numbers in parentheses are approximate standard errors. Coefficients without standard errors are fixed at the given values. Series of equal parameter estimates are constrained to be equal.

goodness of fit chi-square of 33.44 (Prob = .0214). The parallel model can then be treated as a null hypothesis against which the fit of the tau equivalent model can be tested. The chi-square for the tau equivalent model is 31.36 with 14 degrees of freedom. This means that the parallel model adds 5 degrees of freedom, but increases the chi-square by only 2.08 (33.44 − 31.36). Thus, the parallel model cannot be rejected relative to the tau equivalent model.

Next, we contrast the fit of the parallel model with the fit of the congeneric model. The difference between the degrees of freedom for the two models is 10 (19 − 9), and the difference between the chi-squares for the two models is only 8.51 (33.44 − 24.93). Thus, the parallel model cannot be rejected relative to the congeneric model either. The probability of the observed difference between the two chi-squares is quite large if the parallel model held in the population [Prob(chi-square > 8.51) = .58 with df = 10].

Model (4) in Table 9.9, the "item specific model," includes all the constraints of the parallel model except that $\underline{Td}$ (the covariance matrix of the measurement errors) is not constrained to be diagonal. Rather, the errors in Q1 are allowed to correlate with the errors in Q1′ and so on. Since these three covariances among errors are constrained to be equal, they subtract only one degree of freedom. Nevertheless, the reduction in the chi square is 8.39 for the item specific model relative to the parallel model. This difference would be extremely unlikely if the parallel model held in the population [Prob(chi-square > 8.39) = .0038 with df = 1]. Thus, the parallel model can be rejected in favor of a model that includes item specific factors (see Figure 9.8).

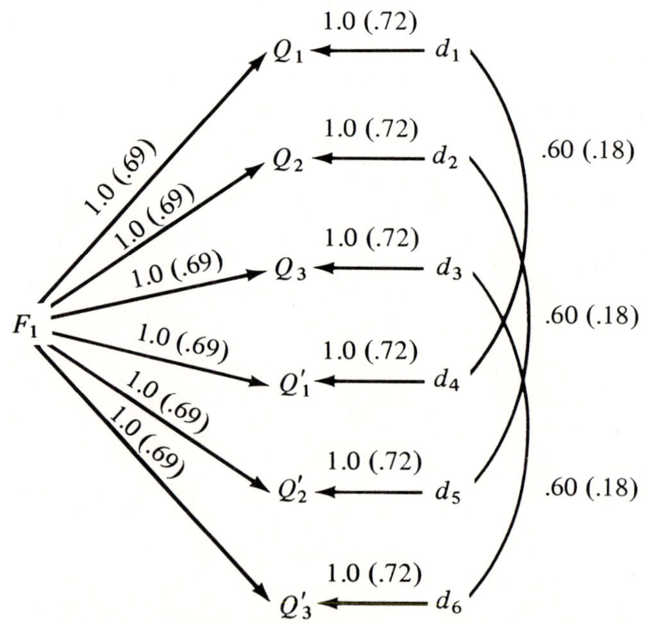

Figure 9.8 Path diagram of model 4 in Table 9.9, the "item specific model." $Q_1, \ldots, Q_3'$ are responses to six questionnaire items concerning racial prejudice. The numbers in parentheses are standardized forms of the coefficients.

The above series of model tests was based upon *a priori* hypotheses concerning measurement properties of the six questionnaire items. The final model presented in Table 9.9 (model 5) deviates from this pattern of hypothesis testing. In this final model, two additional measurement error covariances are estimated, and the relaxation of these two constraints produces a dramatic reduction in the goodness of fit chi-square (from 25.05 to 11.80). These constraints were not relaxed for any theoretical reason, however. Rather, they were freed because an inspection of the residual matrix ($\underline{Z} - \underline{S}$) revealed that the covariances $S(X_6, X_1)$ and $S(X_3, X_2)$ were poorly predicted by the item specific model. And constraints that flow from inspection of the data have a different logical status; one that is analogous to the *a posteriori* comparison of means in the analysis of variance. The probabilities associated with such constraints are not accurately assessed by the standard significance test (based upon the difference in goodness of fit chi-squares) because inspection of the data may involve a selection from dozens of possible constraints.

Whether a researcher decided to settle for model 4, or preferred to make constraints based upon inspection of the data, conclusions about the structures that may be measured by these six items are similar. Model 4 attributes 48.0 percent of indicator variance to the "true scores" [3.13/(3.13 + 3.39) = .480], whereas model 5 attributes 47.1 percent [3.05/(3.05 + 3.42) = .471]. It is clear that a model including item specific latent variables (implied by the covariances t_{41}, t_{52}, and t_{63}) provides a better fit than one that presumes no such factors. The ML estimate of model 4 indicates that the covariances between items using the same wording were inflated 9.2 percent by the item specific factors [.60/(3.13 + 3.39) = .092]. Inflated, that is, beyond that which would be expected on the basis of the relationship between each observed item and the latent variable that is common to all of them. The negative and positive covariances in model 5 (t_{61} and t_{32}, respectively) could be given similar *numerical* interpretations—in terms of factors that attenuated and inflated covariances—but any theoretical interpretations would be *post hoc*. (It should be pointed out that the negative covariance t_{61} suggests that positive errors in responses to Q1 are associated with negative errors in responses to Q3′. This negative covariance is at least consistent with the fact that these two items were presented in reversed formats: a high score was associated with a pro-black position for item Q1, whereas a low score was pro-black for item Q3′. The covariances in Table 9.8 are all positive because reverse format items were "reflected" prior to computations.)

The scores used in these analyses were not transformed in any way. Thus, it is possible that some transformation of the observed variables would produce a better fit between some of the models and the data—although any improvement on the fit of model 5 would raise suspicion of capitalizing upon chance. Psychophysicists, for example, usually model the relationships between estimates of magnitude and "true magnitude" in terms of power functions. This is accomplished in measurement models of the sort described in this chapter by logarithmically transforming the observed scores prior to computation of the covariance matrix. When this is done, the elements of $\underline{Lx}$ (factor loadings) are interpreted as powers to which the latent variable must be raised to best predict the observed score. For example, suppose that

$$X = uF^a$$

describes the relationship between an observed variable X and a latent variable F. The term $\underline{a}$ is the power to which F must be raised in order to minimize the variability of the multiplicative error term u. (Note that if $u > 1$, then X overestimates F; if $u < 1$, X underestimates F.) In order to translate this power function into a linear form that is congenial with the estimation procedures outlined above, we simply take the natural logarithm (ln) of both sides of the equation

$$\ln(X) = \ln(uF^a) = a \ln(F) + \ln(u)$$

Thus, the power a becomes the loading in a linearized version of the power function, the natural log of the latent variable becomes the latent variable, and the natural log of the multiplicative error term becomes the additive error term.

It should also be noted that economists are prone to multiplicative models. They also tend to prefer logarithmic transformations of money income when the income of individuals is included as a variable in a causal model.

9.6 THE CASE OF MULTIPLE LATENT VARIABLES

The main reason for hypothesizing and testing measurement models is to relate variables in linear models without the distorting effects of measurement errors. This means that actual research situations utilizing measurement models will involve more than one latent variable. In terms of the general measurement model specified in equation 9.21, the inclusion of multiple latent variables means that $\underline{PH}$ has n rows and columns (where n is the number of latent variables in the model). Some of these latent variables may not actually be latent, however, since some variables are presumed to be measured without error.

Before delving into the intricacies of estimating measurement models with multiple indicators of multiple latent variables, the next section discusses some problems with a widely used technique that is related, but distinct from, the methods to be outlined. That technique is *exploratory factor analysis.*

Can Measurement Models "Emerge from" Correlational Data? or, "On the Proper Interpretation of Exploratory Factor Analysis"

Those readers with some knowledge of factor analysis are accustomed to a context in which numerous observed variables are presumed to reflect a more parsimonious set of latent variables. For example, Cattell's well-known personality inventory, the 16PF, was derived from a factor analysis of hundreds of questionnaire items. The resulting 16 latent variables that presumably "emerged" from this analysis were treated as the fundamental dimensions of inter-individual differences in personality.

There are at least two tacks that one may take in criticizing such efforts in *exploratory factor analysis.* The first is that the domain measured by responses to questionnaires is too restricted to yield useful statements about individual differences in "personality." We shall not pursue this line of criticism. The second criticism of exploratory factor studies is more statistical in focus. The heart of this criticism is the following: (1) most any covariance matrix among a set of observed indicators that is consistent with a multiple latent variable measurement model will also be consistent with alternative measurement models that provide radically different interpretations of the dimensions latent in the observations; (2) canned computer programs for the computation of exploratory factor analyses mask the arbitrary nature of their solution through the use of mystifying criteria, such as the number of eigenvalues greater than one, Kaiser normalization, and rotation for simple structure. Thus, exploratory factor analysis is a useful technique when an investigator wants to identify clusters of observed variables that are more closely correlated within clusters than between clusters. But interpretation of the results of such analyses should not be confused with interpretations given to tests of measurement hypotheses through the use of *confirmatory factor analysis* (an alternative label for the fitting of linear measurement models to covariance or correlation matrices).

The kind of model testing outlined in this chapter requires that an investigator hypothesize the relationship among latent variables and observed indicators. The results of the analysis will communicate the degree to which the predictions of the model are consistent with the observations. The model can always be rejected by the data. In the case of exploratory factor analysis, however, a solution model is always found *on the basis of criteria that are built into the program to deal with problems of underidentification.* Thus, if an exploratory factor analysis program is used in the hope that the latent variable structure will "emerge from the data," the investigator is presuming that the rather complex set of assumptions built into the program will choose the appropriate solution from among those that are consistent with the data. Such a presumption is at least risky.

The arbitrary nature of factor solutions is nicely demonstrated by data from Lord (1956, cited and analyzed in Saris, 1980). Lord administered four vocabulary tests to 649 students. Two of the tests (X_1 and X_2) included 15 items and a liberal allotment of time; the other two tests (X_3 and X_4) included 75 items and a strict time limit. The covariance matrix for these four vocabulary tests was as follows:

$$
\begin{array}{lc}
\text{15 items} & X_1 \\
 & X_2 \\
\text{75 items} & X_3 \\
 & X_4
\end{array}
\begin{bmatrix}
86.3937 & & & \\
57.7751 & 86.2632 & & \\
56.8651 & 59.3177 & 97.2850 & \\
58.8986 & 59.6683 & 73.8201 & 97.8192
\end{bmatrix}
$$

The model fit to these data by Saris was the following:

$$
\begin{aligned}
X_1 &= F_1 + d_1 \\
X_2 &= F_1 + d_2 \\
X_3 &= F_2 + d_3 \\
X_4 &= F_2 + d_4
\end{aligned}
$$

where

$$t_{11} = t_{22} \quad \text{and} \quad t_{33} = t_{44}$$

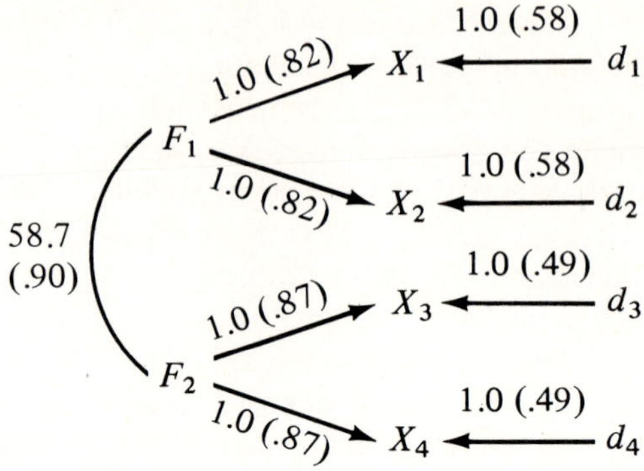

Figure 9.9 Path diagram of the "two factor" measurement model fit to scores on four vocabulary tests (data from Lord, 1956, cited in Saris, 1980). The numbers in parentheses are standardized forms of the coefficients.

This measurement model specifies that the two 15-item tests are parallel measures of the latent variable F_1, the two 75-item tests are parallel measures of F_2, and F_1 may be correlated with F_2 (see Figure 9.9). The matrix form of the model is

$$\begin{bmatrix} X_1 \\ X_2 \\ X_3 \\ X_4 \end{bmatrix} = \begin{bmatrix} k_{11} & 0 \\ k_{21} & 0 \\ 0 & k_{32} \\ 0 & k_{42} \end{bmatrix} \begin{bmatrix} F_1 \\ F_2 \end{bmatrix} + \begin{bmatrix} d_1 \\ d_2 \\ d_3 \\ d_4 \end{bmatrix}$$

where

$$k_{11} = k_{21} = k_{32} = k_{42} = 1$$

$$t_{11} = t_{22}$$

and

$$t_{33} = t_{44}$$

The covariance matrix for the latent variables is

$$\underline{PH} = \begin{bmatrix} ph_{11} & \\ ph_{21} & ph_{22} \end{bmatrix}$$

where ph_{11} and ph_{22} are the variances of F_1 and F_2, respectively, and ph_{21} is the covariance between F_1 and F_2. ML estimation of the parameters of this model yields

$$\text{est}(\underline{PH}) = \begin{bmatrix} 57.79(4.08) & \\ 58.70(3.85) & 73.84(4.81) \end{bmatrix}$$

$$\text{est}(\underline{Td}) = \begin{bmatrix} 28.56(1.59) & & & \\ 0.00 & 28.56(1.59) & & \\ 0.00 & 0.00 & 23.73(1.32) & \\ 0.00 & 0.00 & 0.00 & 23.73(1.32) \end{bmatrix}$$

where approximate standard errors are given in parentheses. The fit of this model with the data is excellent. The goodness of fit chi-square is 1.9335 with 5 degrees of freedom (Prob = .8583).

Now, the overidentifying restrictions that provide for the test of fit may be obtained from solution of the equations for the various moments:

$$\mathrm{Cov}(X_1,X_3) = \mathrm{Cov}(X_1,X_4) = \mathrm{Cov}(X_2,X_3) = \mathrm{Cov}(X_2,X_4)$$

$$\mathrm{Var}(X_1) = \mathrm{Var}(X_2)$$

and

$$\mathrm{Var}(X_3) = \mathrm{Var}(X_4)$$

Inspection of the covariance matrix makes it quite clear that Lord's data are consistent with these overidentifying restrictions.

However, and this is the crux of the current argument, there is a rather distinct alternative measurement model that places the same restrictions upon the second moments. The alternative model presumes that all four tests are measuring a general latent variable (vocabulary knowledge); and in addition, each pair of similar tests is measuring a "test specific" latent variable that is irrelevant to the vocabulary latent variable (see Figure 9.10). This new model is specified by placing the constraints

$$ph_{11} = ph_{22} = ph_{21}$$

on $\underline{PH}$ (this equates F_1 and F_2); freeing k_{32} and k_{42} to be different from one, but equal to each other; and freeing t_{21} and t_{43} to be nonzero. Although this new model is underidentified, it is nevertheless straightforward to show that it implies exactly the same overidentifying restrictions as the model proposed by Saris. Thus, the goodness of fit chi-square for the single latent variable model is 1.9335—the same value obtained for the two latent variable model.

Single factor model

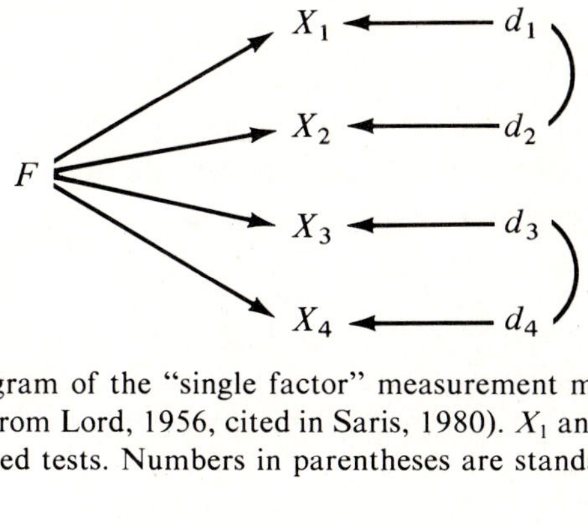

Figure 9.10 Path diagram of the "single factor" measurement model fit to scores on four vocabulary tests (data from Lord, 1956, cited in Saris, 1980). X_1 and X_2 are 15-item tests, X_3 and X_4 are 75-item timed tests. Numbers in parentheses are standardized versions of coefficients.

The substantive difference between these two models is demonstrated nicely by the differing estimates of "true variance" that is latent in, for example, X_1. The two latent variable model partitions the variance of X_1 into a component explained by the latent variable (ph_{11} or 57.79) and an error component (t_{11} or 28.56). According to this model, then, the validity of X_1 is .669 [estimate of $ph_{11}/(ph_{11} + t_{11}) = 57.79/(57.79 + 28.56)$]. In contrast, a single latent variable model estimates ph_{11} as 48.97 and estimates t_{11} as 37.36. These new estimates yield an estimate of the standardized validity of X_1 equal .567. The difference between the .669 and .567 is the proportion of variance in X_1 that is attributed to the method of measurement (type of vocabulary test) by the model with a single latent variable. (The reader should keep in mind that these particular values are somewhat arbitrary in this instance, since the model is underidentified. There are certain limits on the parameter estimates, but the range of "reasonable" values is quite large.) Whether or not test specific variance should or should not be distinguished from true variance is a theoretical question. However, if one is interested in the relationship between knowledge of vocabulary and some other variable, then it seems likely that theoretical considerations would exclude the test specific variance as an irrelevancy.

So the reader may be wondering, which model is correct? The one that depicts performance on the two types of tests as distinct dimensions or the one that posits test specific factors that are uncorrelated with the latent variable of interest (vocabulary knowledge)? There is, of course, no way to choose between these two models on the basis of these data. The important point here is that application of an exploratory factor analysis to these data could come up with either model as the solution. The assumptions built into the program would determine which of the solutions was given (which might not even be one of the two outlined above). [In order to demonstrate that one of these models is superior to the other, it would be necessary to find some experimental manipulation, or network of correlated variables, that provided differential predictions in the two cases. For example, if the single factor model is true, then estimates of the two factor model confound true variance and test specific variance. In equation form, if F is the latent variable in the single factor model, u_1 is the test specific factor for the 15-item tests, and u_2 is the test specific factor for the 75-item tests, then the two latent variables estimated in the two-factor model are actually

$$F_1 = F + u_1$$
$$F_2 = (F + u_2)a$$

where $\text{Cov}(u_1,u_2)$, $\text{Cov}(F,u_1)$, and $\text{Cov}(F,u_2)$ are zero. This model predicts that any training in vocabulary that does not affect the test specific factors will have the same effect on F_1 as on F_2 (except for the proportionality constant $\underline{a}$). Thus, if a training program in vocabulary could be concocted that, for example, influenced F_1 and not F_2, then the single-factor model could be rejected. Such a result would mean that F_1 and F_2 were distinct, although correlated dimensions of knowledge or ability. However, if every manipulation produced comparable effects upon F_1 and F_2 (except those aimed at "test styles"), then the single-factor model would be indicated.]

Thus, to return to the question that heads this section, it is my view that social and behavioral scientists are not likely to produce much that is scientifically valuable if they rely upon exploratory factor analysis programs to "discover" their theoretical dimensions. The methods described in this book are in sharp contrast to the practice of measuring all relevant variables and throwing them into the factor analysis pot. The methods advocated here focus upon the careful investigation of each hypothesized dimension in a theoretical domain. Once the characteristics of a set of indicators are fairly well established, then relationships among the established latent variable and other latent variables can be studied. For the reader who may despair over the amount of time that must be invested in such an approach to social science, I can only reply that the bulk of all scientific work has been devoted to measurement techniques and devices. The measurement aspect of scientific work is seldom the most glamorous, but it is the basis for the tests of hypotheses—and it is the outcome of those tests that makes the scientific project exciting.

In concluding this section, I summarize my views on exploratory factor analysis by arguing that it is appropriate when an investigator is "searching for structure." In other words, when no explicit theoretical constructs are available to the investigator then exploratory factor analysis may be a useful technique for data reduction. However, when investigation is directed by theory, then the investigator can test the validity of hypothesized structures. The latter enterprise is the subject of this book. It is clear to me that theory testing has been a useful procedure in science generally. Exploratory procedures, however, are more likely to manifest implicit orientations in the investigator and hidden assumptions in the statistical procedure. For this reason, I recommend the simplest and least opaque procedures for exploratory analyses.

Problem To demonstrate the futility of searching for the "correct" model with exploratory factor analysis procedures, consider the following. Suppose that two latent variables, F_1 and F_2, are each measured by three indicators. The measurement model is depicted in Figure 9.11 and may be specified in equation form as follows:

$$
\begin{aligned}
X_1 &= .5F_1 + d_1 \\
X_2 &= .7F_1 + d_2 \\
X_3 &= .9F_1 + d_3 \\
X_4 &= .4F_2 + d_4 \\
X_5 &= .6F_2 + d_5 \\
X_6 &= .8F_2 + d_6
\end{aligned}
$$

where the observed and latent variables are standardized, but the measurement errors are not.

The above equations or the path diagram (Figure 9.11) can be used to produce the correlations among the six observable variables. For example, the correlation between X_1 and X_4 is obtained by multiplying the paths connecting these two variables

$$\text{Corr}(X_1, X_4) = (.5)(.7)(.4) = .14$$

Follow these rules to construct the correlation matrix among the observable variables.

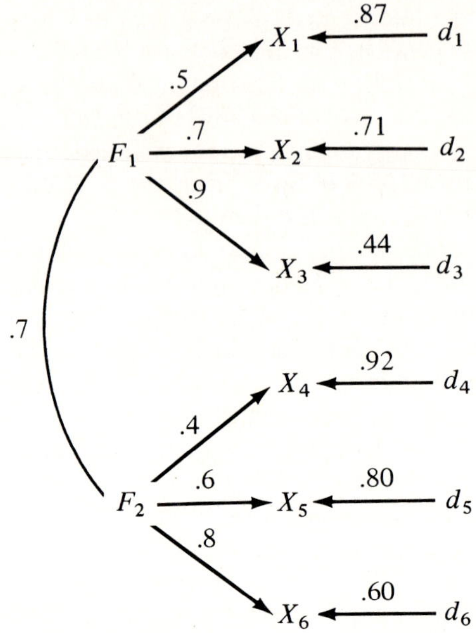

Figure 9.11 Path diagram of a three indicator measurement model. All variables are standardized.

Now that the correlation matrix has been obtained, use a packaged exploratory factor analysis program such as the SPSS subprogram FACTOR to see what kind of factor structure it "discovers" for you. The reader will find that the SPSS program below yields a solution with only one latent variable instead of two. The structure that produced the intercorrelations is uncovered by the SPSS subprogram only if the user specifies that two latent variables should be extracted—and this is precisely the information that the exploratory technique was supposed to provide.

```
RUN NAME          EXPLORATORY FACTOR ANALYSIS WITH SPSS/FACTOR
VARIABLE LIST     X1,X2,X3,X4,X5,X6
N OF CASES        200
INPUT MEDIUM      CARD
FACTOR            VARIABLES=X1,X2,X3,X4,X5,X6/
                  TYPE=PA2/
                  ROTATE=OBLIQUE/
                  VARIABLES=X1,X2,X3,X4,X5,X6/
                  TYPE=PA2/
                  VARIABLES=X1,X2,X3,X4,X5,X6/
                  TYPE=PA2/
                  ROTATE=OBLIQUE/
                  NFACTORS=2/
OPTIONS           3
STATISTICS        2,3,4,5,6,7,8
```

```
READ MATRIX
 1.0       .35       .45       .14       .21       .28
 .35       1.0       .63       .196      .294      .392
 .45       .63       1.0       .252      .378      .504
 .14       .196      .252      1.0       .24       .32
 .21       .294      .378      .24       1.0       .48
 .28       .392      .504      .32       .48       1.0
FINISH
```

In contrast to the output from an exploratory factor analysis program, a confirmatory program such as Joreskog's LISREL (the use of which will be demonstrated in Chapter 11) would indicate that the single latent variable model was inconsistent with this correlation matrix, although it would be (perfectly) consistent with the true measurement model. If a model as simple as this one produces such problems for an exploratory program, imagine the difficulties presented by a more complex model with correlated errors.

Numerical Example: Control over the Pace of Work

As was outlined in the previous section, the transition from measurement models with a single latent variable to multiple latent variable models raises some problems when the actual number of latent variables in a model is not hypothesized. In the following examples, however, the number of latent variables in a model follows quite directly from the design of the study. That is, studies are planned with multiple indicators of hypothesized latent variables. This means that the data are used to see if the observations are consistent with the predictions of the hypothesized measurement model. There may be several models to chose from, or rejection of a model may lead to some exploration of alternatives, but these will occur in the context of a very structured search.

The first multiple latent variable model to be fit to data is taken from a doctoral dissertation by Wehrle (1982). Wehrle set out to measure several theoretical constructs relevant to a study of alienation among a sample of white female secretaries working in the United States. Each hypothetical construct was measured by three different methods so that multiple indicators could be used to test the measurement hypotheses. Furthermore, the study employed a longitudinal design with a lag time of three months. This lag was included so that certain causal models could be tested, but this feature of the design also means that the same latent variable was measured at two points in time—if the hypothesized measurement model holds.

One latent variable of interest was the amount of control that the woman had over the pace of her work. In the absence of direct observation of the workplace, this latent variable was measured by three methods: (1) response to a Likert scale composed of five statements that expressed progressively greater control over the pace of work; (2) a rating by the respondent from 1 to 10 where 1 was no control and 10 was complete control; and (3) a rating from 1 to 5 of the respondent's control over the pace of work based upon the respondent's narrative description of her work. The ratings that comprised the third indicator were made by a trained rater who was unaware of other

Table 9.10 Covariance matrix for six indicators of the latent variable "control over pace of work" at two points of time (from Wehrle, 1982).

		X_1	X_2	X_3	X_4	X_5	X_6
Time 1	Likert	.976					
	Number	1.056	3.480				
	Rating	.339	.699	.517			
Time 2	Likert	.475	.793	.213	1.025		
	Number	.949	1.660	.508	1.694	4.625	
	Rating	.248	.494	.212	.438	.918	.537

Note. These covariances are based upon the 119 cases with complete data for all six variables.

responses by the respondents. Since these measurements were gathered at two points in time, we have a total of six observed variables that are hypothesized to measure two latent variables—control over pace at time one (F_1) and control over pace at time two (F_2). The covariances among these six indicators are given in Table 9.10.

The following model can now be tested with these data

$$\begin{bmatrix} X_1 \\ X_2 \\ X_3 \\ X_4 \\ X_5 \\ X_6 \end{bmatrix} = \begin{bmatrix} k_{11} & 0 \\ k_{21} & 0 \\ k_{31} & 0 \\ 0 & k_{42} \\ 0 & k_{52} \\ 0 & k_{62} \end{bmatrix} \begin{bmatrix} F_1 \\ F_2 \end{bmatrix} + \begin{bmatrix} d_1 \\ d_2 \\ d_3 \\ d_4 \\ d_5 \\ d_6 \end{bmatrix}$$

where F_1 and F_2 are expected to covary.

Since X_1 and X_4 involve the same method of measuring the same latent variable (at different points in time), we might hypothesize that $k_{11} = k_{42}$. This hypothesis presumes that the slope of the indicator regressed on the true scores is the same at the two points in time. Thus, we can hypothesize that

$$k_{11} = k_{42},\ k_{21} = k_{52}, \text{ and } k_{31} = k_{62}$$

In order to identify the model, we can make the further assumption that $k_{11} = 1$.

There are two additional hypotheses that may be tested with these data. One is that the measurement errors are *autocorrelated*—correlated over time for reasons that are uncorrelated with the latent variable. This hypothesis would be tested with the constraint

$$t_{41} = t_{52} = t_{63} = 0$$

Another hypothesis that can be tested is that the variance of the measurement errors is smaller in the second interview than in the first interview. This expectation would derive from the rationale that the respondents would be thinking about the questions asked in the first interview during the lag time and would more accurately report their circumstances in the second interview. The constraint that must be rejected to provide support for this hypothesis is

$$t_{11} = t_{44}, t_{22} = t_{55}, \text{ and } t_{33} = t_{66}$$

We begin the model testing exercise by fitting the most constrained measurement model to the data. This model includes the constraints

$$k_{11} = k_{42} = 1$$
$$k_{21} = k_{52}$$
$$k_{31} = k_{62}$$
$$t_{11} = t_{44}, t_{22} = t_{55}, t_{33} = t_{66}$$

and

$$t_{41} = t_{52} = t_{63} = 0$$

The 15 covariances and six variances among these six variables yield a potential of 21 degrees of freedom. This most constrained model includes eight parameters (k_{21}, k_{31}, ph_{11}, ph_{22}, ph_{21}, t_{11}, t_{22}, and t_{33}), leaving 13 degrees of freedom for the goodness of fit chi-square. The resulting chi-square is 18.14 (Prob = .1521). Clearly, this is a reasonable fit. The question now before us is whether or not any of the alternative models provides a significantly better fit. Table 9.11 presents these results. Note that relaxation of two of the sets of constraints produces substantial reductions in the chi-square statistic. If both of these relaxations are incorporated into one model (see Figure 9.12), the resulting chi-square is 2.62 with 7 degrees of freedom (Prob = .918). The only problem with this model is that the estimate of t_{52} is negative (−.156); however, this result is not a serious problem, since the absolute value of this estimate is not even as large as its approximate standard error of .170. Thus, the final model for these data includes the constraint that t_{52} is zero, but this constraint, being *a posteriori*, is not allowed to add a degree of freedom for the goodness of fit chi-square. The estimates of the parameters of the model are given in Table 9.12, and a path diagram of the model is presented in Figure 9.12. Note that the ML fit of this model to the data esti-

Table 9.11 Results of fitting several measurement models to the longitudinal data presented in Table 9.10.

Constraints relaxed	Chi-square	*df*	Prob	Chi-square difference	*df*	Probability
None	18.14	13	.152	0	0	
$k_{21} = k_{52}, k_{31} = k_{62}$	16.62	11	.120	1.52	2	.468
$t_{11} = t_{44}, t_{22} = t_{55}, t_{33} = t_{66}$	11.05	10	.353	7.09	3	.069
$t_{41} = t_{52} = t_{63} = 0$	9.94	10	.446	8.20	3	.042

Note. The column Chi-square difference gives the difference between the goodness of fit chi-square for the model with all three sets of constraints (no constraints relaxed), and the models specified by the relaxation of certain constraints.

Table 9.12 Maximum likelihood estimate of a measurement model fit to the data in Table 9.10.

Matrix estimated	Estimates
Lx	$\begin{bmatrix} 1.000 & 0.000 \\ 2.055(.172) & 0.000 \\ .576(.061) & 0.000 \\ 0.000 & 1.000 \\ 0.000 & 2.055(.172) \\ 0.000 & .576(.061) \end{bmatrix}$
PH	$\begin{bmatrix} .543(.107) & \\ .411(.092) & .805(.137) \end{bmatrix}$ $\{\text{est}[\text{Corr}(F_1,F_2)] = .621\}$
Td	$\begin{bmatrix} .43(.08) & & & & & \\ .00 & 1.30(.29) & & & & \\ .00 & .00 & .31(.05) & & & \\ .05(.05) & .00 & .00 & .22(.06) & & \\ .00 & .00 & .00 & .00 & 1.13(.27) & \\ .00 & .00 & .08(.03) & .00 & .00 & .29(.04) \end{bmatrix}$

Chi-square with 7 degrees of freedom = 3.235; Prob = .862

Note. The numbers in parentheses are approximate standard errors.

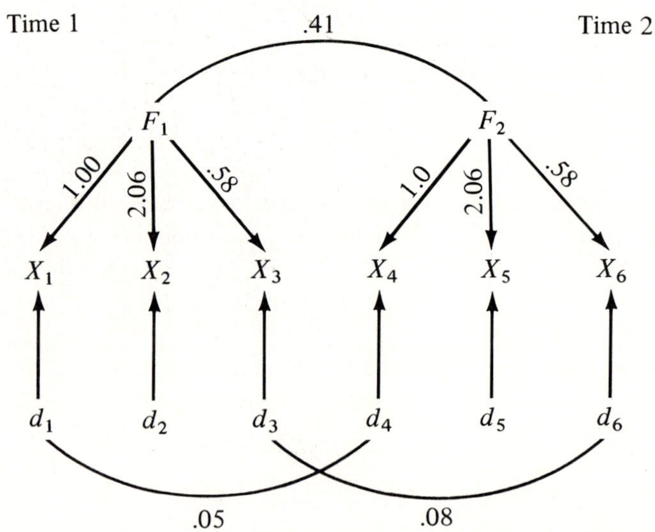

Figure 9.12 Path diagram of the measurement model estimated in Table 9.12 where the observed variables are indicators of "perceived control over the pace of work" (from Dwyer, Wehrle, and Phelan, 1981). The autocorrelation between F_1 and F_2 is $.411/[(.543)(.805)]^{.5} = .621$.

mates the autocorrelation of the latent variable at .621. Furthermore, the standardized validities of the indicators increased substantially from time 1 to time 2:

$$\text{estimated validity of } X_1 = .543/(.543 + .433) = .556$$

$$\text{estimated validity of } X_4 = .805/(.805 + .216) = .788$$

for example. Note that the increase in the validity is due to both an *increase* in the variance of the latent variable (from .543 to .805) and a *decrease* in the variance of the measurement errors (from .433 to .216). The reduced variance of the errors is not unexpected in a longitudinal design. Perhaps relying upon a worker's reports of control over pace transforms the latent variable into some product of that environment and the worker's understanding of it. Thus, the greater variance of F_2 may reflect the effect of the first interview upon the respondent's conception of her work environment—that is, a greater differentiation of work environment characteristics.

Numerical Example: General Feelings of Control in the Workplace

The example of the latent variable "control over pace" was discussed in some detail to demonstrate the procedure of model fitting when more than one latent variable has been measured. The next example is taken from the same study and concerns the measurement of a related hypothetical construct. This construct is derived from the Marxian concept of powerlessness that has been used so widely in sociology and social psychology. Werhle's goal, in this instance, was to assess the participating secretaries' subjective feelings of control in their work situation.

Once again, three methods of measurement were used to measure this latent variable at time 1 (F_3) and time 2 (F_4). The three measures were as follows:

1. Likert item

   How often do you feel that you can change things about your work when you think they need to be improved?

   a. Nearly all the time

   b. Very often

   c. Some of the time

   d. Rarely or never

2. Number response

   How much influence do you feel you have over your work situation? Give a number from 1 to 10 where 1 is "no influence" and 10 is "complete control."

3. Rating of narrative response to the following question (trained, and "blind," raters used a four-level Likert scale)

   If you made a suggestion at your job about how your work could be organized in a more efficient way, do you feel that your suggestion would have a chance of being implemented?

Each respondent was asked each of these questions in both the first and second interview—yielding six response variables. The covariance matrix for these items is given in Table 9.13.

Table 9.13 Covariance matrix for items measuring general feelings of control over the work process for a sample of secretaries (from Wehrle, 1980).

			X_7	X_8	X_9	X_{10}	X_{11}	X_{12}
Time 1	Likert	X_7	.903					
	Number	X_8	1.522	7.126				
	Rating	X_9	.508	1.577	.933			
Time 2	Likert	X_{10}	.387	1.081	.527	1.117		
	Number	X_{11}	1.036	3.499	1.363	1.715	6.065	
	Rating	X_{12}	.441	1.404	.674	.649	1.742	.974

Note. Covariances and variances are based upon a sample size of 119 respondents with complete data.

With our experience from the previous example, we could hypothesize a similar measurement model:

$$\begin{bmatrix} X_7 \\ X_8 \\ X_9 \\ X_{10} \\ X_{11} \\ X_{12} \end{bmatrix} = \begin{bmatrix} 1.00 & 0.00 \\ k_{83} & 0.00 \\ k_{93} & 0.00 \\ 0.00 & 1.00 \\ 0.00 & k_{11,4} \\ 0.00 & k_{12,4} \end{bmatrix} \begin{bmatrix} F_3 \\ F_4 \end{bmatrix} + \begin{bmatrix} d_7 \\ d_8 \\ d_9 \\ d_{10} \\ d_{11} \\ d_{12} \end{bmatrix}$$

where

$$k_{83} = k_{11,4}$$
$$k_{93} = k_{12,4}$$

and

$$\text{Cov}(d_{12},d_9) = t_{12,9} = \text{free parameter}$$

This model depicts (X_7,X_8,X_9) as three congeneric indicators of F_3 and (X_{10},X_{11},X_{12}) as three congeneric indicators of F_4. The latent variable is expected to have considerable stability over the three-month time span, since the work situation and subjective orientation of the respondents is expected to be fairly stable.

Using a ML program such as LISREL to estimate this model with the data in Table 9.13 gives the results in Table 9.14. Note that the fit of this model with the data is quite good. The goodness of fit chi-square is approximately equal to the degrees of freedom (which is equal to the number of second moments among the observed variables minus the number of free parameters estimated). Thus, this model cannot be rejected, since the observations are consistent with our expectations under the assumption that the model holds in the population from which the sample was drawn. Furthermore, the fact that this model fits without *a posteriori* "fiddling" increases our confidence in the interpretation of the probability associated with the observed goodness of fit chi-square.

The successful fitting of this longitudinal measurement model to the observed scores

Table 9.14 Maximum likelihood estimates of a measurement model fit to the data in Table 9.13.

Matrix	Maximum likelihood estimate					
Lx	1.000	0.000				
	2.872(.253)	0.000				
	1.086(.103)	0.000				
	0.000	1.000				
	0.000	2.872(.253)				
	0.000	1.086(.103)				
PH	.493(.097)					
	.427(.088)	.564(.111)				
	[Corr(F_3,F_4) = .809]					
Td	.444(.072)					
	0.000	2.870(.498)				
	0.000	0.000	.337(.064)			
	0.000	0.000	0.000	.515(.079)		
	0.000	0.000	0.000	0.000	1.514(.349)	
	0.000	0.000	.135(.045)	.000	0.000	.294(.058)

(.045) Chi-square = 8.390, with df = 9, Prob = .495

Note. Numbers in parentheses are approximate standard errors. Estimates without standard errors are fixed by hypothesis. Results are presented in matrix format.

from indicators of two separate latent variables (control over the pace of work and general feelings of powerlessness in the workplace) leads to the notion of a larger measurement model that includes both of these latent variables. In fact, the *construct validity* of these two latent variables can be explored by seeing if these two latent variables are truly distinct from one another; Can they withstand a test of *divergent validity?* The fitting of such a model comprises our next numerical example.

Numerical Example: A Model with Four Latent Variables

We shall now specify a measurement model that incorporates all 12 of the observed indicators used in the previous two examples (X_1 to X_{12}). As has been noted many times in this chapter, the fact that a data set is consistent with a measurement model does not mean that another model may not be "more correct." Thus, the good fit between the longitudinal measurement model specified and the data for control of pace of work and feelings of general control may arise even though the model is false. One way to further test these measurement models is to incorporate them into larger models. This incorporation into larger models places additional restrictions upon the observations and may reveal inconsistencies that were not apparent in the more isolated models. For example, F_1 and F_2 may represent a composite of two or more latent variables that relate to one another in such a way that the single factor model fits the data. This could occur if the two-component factors were not causally related and if the relations among the indicators and latent variables abided by some strict criteria of stationarity (see Kenny, 1980). As the complexity of the model grows, we would become more and more convinced that the "good fit" was not due to such fortuitous accidents—or we would conclude that the model should be rejected.

In order to specify a measurement model incorporating all 12 of the indicators used in the previous two examples, we specify a model as follows:

$$\begin{bmatrix} X_1 \\ X_2 \\ X_3 \\ X_4 \\ X_5 \\ X_6 \\ X_7 \\ X_8 \\ X_9 \\ X_{10} \\ X_{11} \\ X_{12} \end{bmatrix} = \begin{bmatrix} 1.0 & 0.0 & 0.0 & 0.0 \\ k_{21} & 0.0 & 0.0 & 0.0 \\ k_{31} & 0.0 & 0.0 & 0.0 \\ 0.0 & 1.0 & 0.0 & 0.0 \\ 0.0 & k_{52} & 0.0 & 0.0 \\ 0.0 & k_{62} & 0.0 & 0.0 \\ 0.0 & 0.0 & 1.0 & 0.0 \\ 0.0 & 0.0 & k_{83} & 0.0 \\ 0.0 & 0.0 & k_{93} & 0.0 \\ 0.0 & 0.0 & 0.0 & 1.0 \\ 0.0 & 0.0 & 0.0 & k_{11,4} \\ 0.0 & 0.0 & 0.0 & k_{12,4} \end{bmatrix} \begin{bmatrix} F_1 \\ F_2 \\ F_3 \\ F_4 \end{bmatrix} + \begin{bmatrix} d_1 \\ d_2 \\ d_3 \\ d_4 \\ d_5 \\ d_6 \\ d_7 \\ d_8 \\ d_9 \\ d_{10} \\ d_{11} \\ d_{12} \end{bmatrix}$$

with the constraints that

$$k_{21} = k_{52};\ k_{31} = k_{62}$$

$$k_{83} = k_{11,4};\ k_{93} = k_{12,4}$$

$$\underline{Td} \text{ diagonal except for } t_{63} \text{ and } t_{12,9}$$

This model essentially combines the previous two into one model such that the relations among the latent variables can be assessed. This means that the covariances among all 12 variables must be consistent with the restrictions of the model.

One-half of the covariances among these 12 variables were given in Tables 9.10 and 9.13. The remaining covariances necessary to estimate the full model are given in Table 9.15. The result of fitting these data to the specified measurement model is given

Table 9.15 Observed covariances between the indicators of "control over pace of work" (X_1 to X_6) and indicators of "general feelings of control over work" (X_7 to X_{12}).

	X_1	X_2	X_3	X_4	X_5	X_6
X_7	.272	.277	.202	.180	.194	.125
X_8	.756	1.797	.879	.542	1.075	.363
X_9	.303	.440	.239	.264	.520	.223
X_{10}	.270	.432	.234	.388	.785	.311
X_{11}	.586	1.131	.604	.849	2.087	.794
X_{12}	.340	.603	.238	.315	.690	.341

Note. These data are taken from Wehrle (1982). The remainder of the covariance matrix among the 12 variables is contained in Tables 9.10 and 9.13.

Table 9.16 Maximum likelihood estimate of a measurement model including four latent variables.

Matrix	Maximum likelihood estimates			
Lx	1.000	0.000	0.000	0.000
	2.038(.157)	0.000	0.000	0.000
	.599(.060)	0.000	0.000	0.000
	0.000	1.000	0.000	0.000
	0.000	2.038(.157)	0.000	0.000
	0.000	.599(.060)	0.000	0.000
	0.000	0.000	1.000	0.000
	0.000	0.000	2.896(.248)	0.000
	0.000	0.000	1.062(.100)	0.000
	0.000	0.000	0.000	1.000
	0.000	0.000	0.000	2.896(.248)
	0.000	0.000	0.000	1.062(.100)
PH	.539(.103)			
	.419(.089)	.802(.134)		
	.267(.067)	.240(.072)	.498(.098)	
	.258(.069)	.341(.082)	.429(.088)	.567(.111)

Td											
.421(.076)											
	1.442(.283)										
		.298(.045)									
			.231(.056)								
				1.133(.248)							
		.074 (.031)			.281(.042)						
						.445(.071)					
							2.660(.471)				
								.355(.064)			
									.506(.077)		
										1.474(.334)	
								.130 (.044)			.304(.056)

Chi-square = 65.636 df = 50 Prob = .068

Note. Numbers in parentheses are approximate standard errors. Coefficients without standard errors are fixed by hypothesis.

in Table 9.16 and Figure 9.13. Note that the fit of this model is acceptable by the usual 5 percent criterion, but that the fit of this combined model is not as good as each of the two separate models. For the separate models, the goodness of fit chi-square was no greater than the degrees of freedom. But for this combined model, the chi-square is almost one-third again as large as the degrees of freedom. This result suggests that there may be some misspecification of this measurement model. However, the resulting fit is not so poor that additional analyses of the covariances among the latent variables are unwarranted. In fact, we shall return to these data in Chapter 11 when causal inference from longitudinal data is discussed.

One reason for the lack of a better fit may be the presence of an "occasion factor" that inflates covariances *within* time 1 and *within* time 2, but not *between* time 1 and

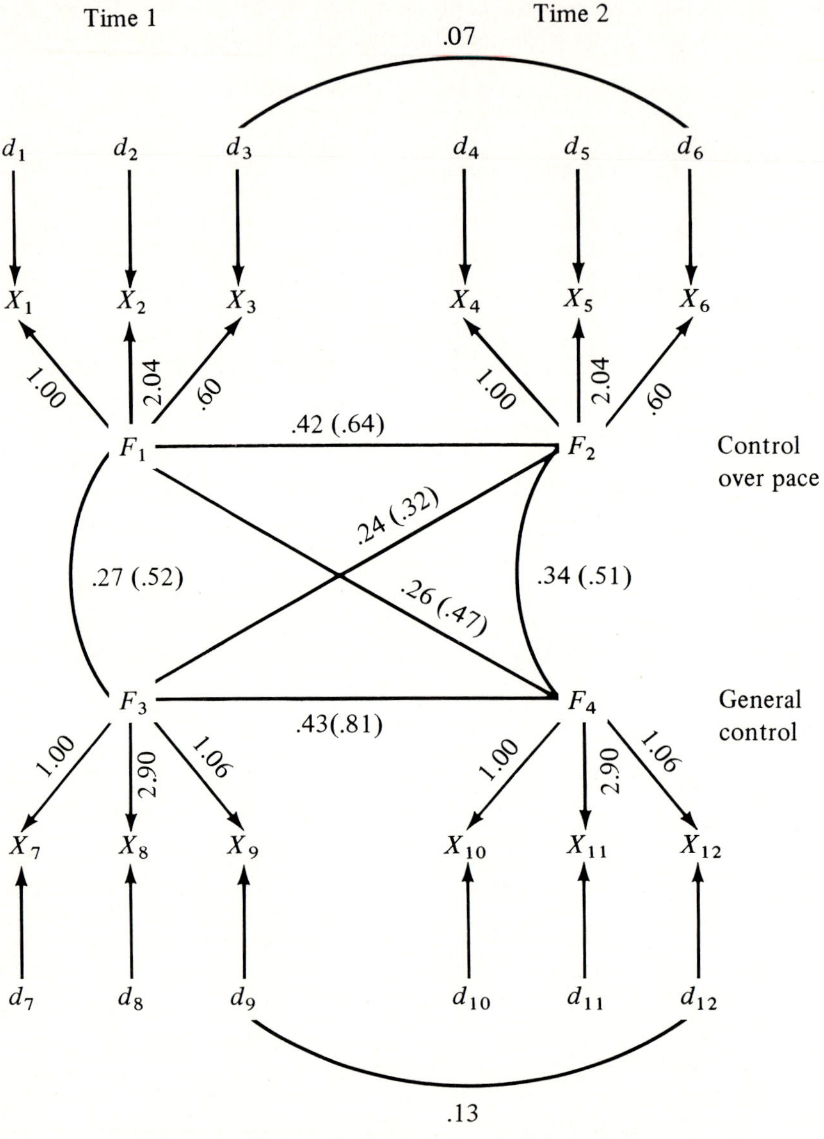

Figure 9.13 Path diagram of the measurement model estimated in Table 9.16 where three methods of measuring two latent variables are observed at two points in time (data from Wehrle, 1982; model from Dwyer, Wehrle, and Phelan, 1981). Numbers in parentheses are estimated correlations.

time 2. With the anticipation of such a factor and the inclusion of additional indicators, it would be possible to include such factors in the model. The extent to which such factors improved the fit of the model could then serve as a test of this hypothesis.

If the latent variables in the measurement model are standardized, we obtain the correlation matrix in Table 9.17. Given that the model holds, these coefficients are estimates of the correlations among the "true scores" that underlie the observed indi-

Table 9.17 Estimates of correlations among four latent variables (standardized version of the matrix $\underline{PH}$ in Table 9.16).

	F_1	F_2	F_3	F_4
F_1	1.000			
F_2	.638	1.000		
F_3	.515	.323	1.000	
F_4	.466	.505	.807	1.000

cators. Note, for example, that the correlation between F_1 and F_3 (control over pace and general control at time 1) is .515. In contrast, the correlations among the indicators of these two latent variables (X_1, X_2, X_3 and X_7, X_8, X_9) range from .156 to .458 (these correlations can be computed from the variances and covariances in Table 9.15). Clearly, if this measurement model is correct, the use of any single pair of indicators of these two latent variables would have given very misleading information concerning the *magnitude* of the relationship between these two latent variables.

One final bit of information that can be gained from the testing of this combined measurement model is a test of the null hypothesis that all six of the observed indicators at each time point are actually measuring the same latent dimension. It is at least possible that people do not differentiate between control over the pace of work and general control over the work process. This hypothesis would be tested by constraining F_1 to be equivalent with F_3 and F_2 to be equivalent with F_4. These constraints can be conceptualized in several ways, including the constraint that Corr(F_1,F_3) and Corr(F_2,F_4) are equal to one. In terms of the measurement model specified above, these constraints are specified as follows:

1. k_{73} and $k_{10,4}$ are freed, but constrained to equal one another.
2. $ph_{11} = ph_{33} = ph_{31}$.
3. $ph_{22} = ph_{44} = ph_{42}$.
4. $ph_{21} = ph_{43} = ph_{32} = ph_{41}$.

These constraints make the six indicators at each time point congeneric indicators of a single latent variable. These constraints add 6 degrees of freedom to the model and produce a chi-square of 241.506 (Prob = .0000). If we allow for all possible autocorrelations of errors, the resulting chi-square is 210.055 with 52 degrees of freedom. Clearly, the hypothesis that these indicators are measures of a single dimension can be rejected on the basis of these observations.

In Section 9.7 we demonstrate the specification of a measurement model that includes explicit hypothesis about systematic measurement errors.

9.7 THE MULTITRAIT-MULTIMETHOD APPROACH TO CONSTRUCT VALIDATION

Up to this point, we have treated *systematic* measurement errors in a rather haphazard manner. That is, the possibility of correlated measurement errors has been suggested in a number of instances—especially in the case of autocorrelation. The goal of this section is to outline a procedure for inclusion of explicit measurement error factors that may be construed as additional latent variables in a measurement model.

The inspiration for this line of thinking comes from the formulation of the multitrait multimethod approach to construct validity that was advanced by Don Campbell and Donald Fiske (1959). Empirically oriented social and behavioral scientists have always puzzled over the appropriate procedures for demonstrating that a hypothetical construct is more than an arbitrary or random organization of observations. For example, if we used the judgments of clinical psychologists to rank people in terms of the dimension "anxious," or to classify persons in terms of types of "anxiety," then the use of these measurements in a scientific investigation would motivate several questions. First, "How do we know that these rankings are other than random?" Second, "How can we determine if the basis for these rankings is independent of the particular person doing the ranking?" And third, "How do we demonstrate that the rankings are not based on some other dimension of individual differences that is being confused with anxiety?"

The first of these questions is rather easy to answer. If the rankings are correlated with any other variables, then the rankings cannot be random. The second question raises the issue of measurement method. Most scientists are suspicious of parameters that can be measured in only one way. Suppose, for example, that a physicist claimed to have discovered the universal particles that give rise to gravitation; but this physicist held that only his particular instrument could detect these particles. For better or worse, the canons of science would not allow us to accept this evidence because the scientific method includes the assumption that our measurements reflect something material in the world that is independent, to some extent, of our measurement of it. If this is not the case, then the procedure of measurement to test hypotheses becomes a circular one that serves no real purpose. Thus, a scientist must demonstrate that his hypothetical constructs can be measured by several different methods in order to convince others that his results are not artifacts of the measurement procedure.

The third question deals with the problem known as *discriminant validation.* That is, can the rank ordering of patients along the dimension of anxiety be distinguished from rank orderings along other dimensions? Suppose, for example, that rankings of anxiety are correlated with rankings of "social awkwardness." How can we be convinced that the behavioral patterns and judgment processes that give rise to these two rankings are distinct? Perhaps each is a reaction to the same differences in behavioral patterns, but the addition of random error to the measurement process attenuates the observed correlation to less than one.

The methodological procedure advocated by Campbell and Fiske for dealing with these issues is quite straightforward in concept, but quite complex in practice. The principle is that validation of a construct must involve the measurement of that construct (dimension in the linear case) with several methods of measurement and the measurement of other constructs that are expected to be closely related. The more *divergent* the methods of measurement, the more convincing the demonstration—if the methodologically divergent indicators agree in their rank orderings. For example, if the temperature of a liquid is measured via the expansion of mercury, electrical conductivity, and vapor pressure, *and* these three measures provide similar estimates of individual differences, then we would be hard pressed to resist the conclusion that differences in the hypothetical construct "temperature" were being measured. Furthermore, we would be strongly convinced that the differences in temperature were characteristics of the liquids and not of the measurement procedures.

Similarly, to return to the clinical psychology example, if ratings of anxiety by two clinicians correlated, we would begin to suspect that the ratings were not completely idiosyncratic. If the rankings of those clinicians were in turn correlated in the predicted direction with some paper and pencil test of anxiety, and all three of these were correlated with ratings of anxiety by a friend of each patient, then our confidence in the validity of the construct would increase.

If we now added additional constructs to our study, and measured each of them with the same number of methods, then we might be in a position to determine whether or not the anxiety construct could be discriminated from other related dimensions. For example, we might use four different methods of measuring anxiety, social awkwardness, knowledge of vocabulary, and introversion/extroversion. The more difficult statistical task is now to provide a procedure for determining whether or not this set of observations supports the measurement hypothesis or contradicts it. In our hypotheses about the dimensionality of these measurements are correct, we might expect to find that different measures of the same construct should correlate more strongly with one another than with indicators of other constructs. However, "between construct" covariances (i.e., those between indicators of different latent variables) may be inflated by covariance due to common methods of measurement. (For example, paper and pencil tests may evoke certain response patterns in participants that induce irrelevant correlations among paper and pencil measures of traits, states, cognitions, or behaviors.) Furthermore, relations among variables may be distorted by random measurement error in such a way that the magnitude of correlations is very misleading. Consider the following example.

Suppose that X_1 and X_2 are indicators of the latent variable "social status" (labeled F_1), and that Y_1 is an indicator of "knowledge of vocabulary" (F_2). Now suppose that

$$\mathrm{Corr}(X_1,F_1) = \mathrm{Corr}(X_2,F_1) = .5$$
$$\mathrm{Corr}(Y_1,F_2) = .9$$

and

$$\mathrm{Corr}(F_1,F_2) = .8$$

where all measurement errors are random. Under this assumption of random errors, it is straightforward to demonstrate that this model implies the following correlations among the observed indicators:

$$\mathrm{Corr}(X_1,X_2) = (.5)(.5) = .25$$
$$\mathrm{Corr}(X_1,Y_1) = (.5)(.8)(.9) = .36$$
$$\mathrm{Corr}(X_2,Y_1) = (.5)(.8)(.9) = .36$$

If attention was only paid to the relative magnitudes of these observed correlations, we would be misled about the common variable latent in X_1 and X_2. The fact that these two indicators correlated more strongly with Y_1 than with one another would obscure their connection. However, if a hypothesized latent variable structure were measured with multiple indicators employing divergent methods of measurement, then the procedures outlined in Section 9.6 of this chapter can be used to test the hypothesis. This procedure would not be subject to the misleading patterns of correlation that can arise from complex error structures in observable indicators.

A Numerical Example: Ratings of Personality Traits

The application of linear measurement models to multitrait multimethod matrix analysis will be demonstrated with data from a study of Fiske (1949). [These data have been fit to a measurement model by Saris (1980); the results reported here are analogous to those obtained by Saris.] A sample of 128 trainees in clinical psychology were rated on five personality dimensions. Each participant was measured on each of these five dimensions with three different methods of measurement. This design thus resulted in 15 scores for each participant.

The correlations among these 5 ratings are given in Table 9.18. Note that the five personality or temperament traits measured are assertive, cheerful, serious, unshakable poise, and broad interests. Each of these "linguistic dimensions" is then estimated via three different measurement operations: ratings by a clinical supervisor, ratings by teammates, and self-ratings.

If it is now reasonable to presume that the relations among the latent variables measured by these indicators are linear, then a linear measurement model can be tested that takes into account both the "substantive latent variables" (the personality dimensions) and the "method latent variables." In other words, it can be hypothesized that the linear relations among the observed indicators are explicable in terms of relations among a set of eight latent variables, where the three method factors are presumed to be uncorrelated with the substantive latent variables.

Such a measurement model may be specified as follows:

$$\begin{bmatrix} X_1 \\ X_2 \\ X_3 \\ X_4 \\ X_5 \\ X_6 \\ X_7 \\ X_8 \\ X_9 \\ X_{10} \\ X_{11} \\ X_{12} \\ X_{13} \\ X_{14} \\ X_{15} \end{bmatrix} = \begin{bmatrix} k_{11} & 0 & 0 & 0 & 0 & k_{16} & 0 & 0 \\ 0 & k_{22} & 0 & 0 & 0 & k_{26} & 0 & 0 \\ 0 & 0 & k_{33} & 0 & 0 & k_{36} & 0 & 0 \\ 0 & 0 & 0 & k_{44} & 0 & k_{46} & 0 & 0 \\ 0 & 0 & 0 & 0 & k_{55} & k_{56} & 0 & 0 \\ k_{61} & 0 & 0 & 0 & 0 & 0 & k_{67} & 0 \\ 0 & k_{72} & 0 & 0 & 0 & 0 & k_{77} & 0 \\ 0 & 0 & k_{83} & 0 & 0 & 0 & k_{87} & 0 \\ 0 & 0 & 0 & k_{94} & 0 & 0 & k_{97} & 0 \\ 0 & 0 & 0 & 0 & k_{10,5} & 0 & k_{10,7} & 0 \\ k_{11,1} & 0 & 0 & 0 & 0 & 0 & 0 & k_{11,8} \\ 0 & k_{12,2} & 0 & 0 & 0 & 0 & 0 & k_{12,8} \\ 0 & 0 & k_{13,3} & 0 & 0 & 0 & 0 & k_{13,8} \\ 0 & 0 & 0 & k_{14,4} & 0 & 0 & 0 & k_{14,8} \\ 0 & 0 & 0 & 0 & k_{15,5} & 0 & 0 & k_{15,8} \end{bmatrix} \begin{bmatrix} F_1 \\ F_2 \\ F_3 \\ F_4 \\ F_5 \\ F_6 \\ F_7 \\ F_8 \end{bmatrix} + \begin{bmatrix} d_1 \\ d_2 \\ d_3 \\ d_4 \\ d_5 \\ d_6 \\ d_7 \\ d_8 \\ d_9 \\ d_{10} \\ d_{11} \\ d_{12} \\ d_{13} \\ d_{14} \\ d_{15} \end{bmatrix}$$

where $\underline{PH}$ has ones in the diagonal, the substantive latent variables are correlated with one another, but not with the method latent variables, and $\underline{Td}$ is diagonal. Note that

Table 9.18 Correlations among 15 ratings of 128 trainees in clinical psychology. Each participant was rated on five dimensions with three different methods of measurement.

		X_1	X_2	X_3	X_4	X_5	X_6	X_7	X_8
Ratings by staff									
Assertive	X_1	1.00							
Cheerful	X_2	0.37	1.00						
Serious	X_3	−.24	−.14	1.00					
Unshakable poise	X_4	0.25	0.46	0.08	1.00				
Broad interest	X_5	0.35	0.19	0.09	0.31	1.00			
Ratings by teammates									
Assertive	X_6	0.71	0.35	−.18	0.26	0.41	1.00		
Cheerful	X_7	0.39	0.53	−.15	0.38	0.29	0.37	1.00	
Serious	X_8	−.27	−.31	0.43	−.06	0.03	−.15	−.19	1.00
Unshakable poise	X_9	0.03	−.05	0.03	0.20	0.07	0.11	0.23	0.19
Broad interest	X_{10}	0.19	0.05	0.04	0.29	0.47	0.33	0.22	0.19
Self-ratings									
Assertive	X_{11}	0.48	0.31	−.22	0.19	0.12	0.46	0.36	−.15
Cheerful	X_{12}	0.17	0.42	−.10	0.10	−.03	0.09	0.24	−.25
Serious	X_{13}	−.04	−.13	0.22	−.13	−.05	−.04	−.11	0.31
Unshakable poise	X_{14}	0.13	0.27	−.03	0.22	−.04	0.10	0.15	0.00
Broad interest	X_{15}	0.37	0.15	−.22	0.09	0.26	0.27	0.12	−.07

		X_9	X_{10}	X_{11}	X_{12}	X_{13}	X_{14}	X_{15}
Ratings by teammates								
Unshakable poise	X_9	1.00						
Broad interest	X_{10}	0.29	1.00					
Self-ratings								
Assertive	X_{11}	0.12	0.23	1.00				
Cheerful	X_{12}	−.11	−.03	0.23	1.00			
Serious	X_{13}	0.06	0.06	−.05	−.12	1.00		
Unshakable poise	X_{14}	0.14	−.03	0.16	0.26	0.11	1.	
Broad interest	X_{15}	0.05	0.35	0.21	0.15	0.17	0.31	1.00

Note. Data from Kelly and Fiske (1951), cited in Saris (1980).

there is an important difference in the way that this model has been specified relative to the previous measurement models. Since the units of measure are of little consequence for the testing of this model, the model has been specified such that the latent variables have unit variance and the data matrix is a correlation matrix rather than a covariance matrix. Although this method for specifying measurement models is not generally recommended, it is a simple approach in this particular case. (If the results of this model fitting were going to be used in a structural model, then analysis of the covariance matrix would be the more appropriate procedure—especially if studies involving different populations were to be compared.) The matrix $\underline{Lx}$ is comprised of standardized regression coefficients (beta weights) that result from the regression of observed variables on the latent variables; $\underline{PH}$ is a correlation matrix for the latent variables; and the diagonal elements of $\underline{Td}$ are the *proportion* of the variance in each indicator that is attributed to random measurement error.

We can now proceed to test several variants of this measurement model. First, for example, we can test to see if a model that does not include the method latent variables is consistent with the data. This is accomplished by the constraints

$$ph_{66} = ph_{77} = ph_{88} = ph_{76} = ph_{86} = ph_{87} = 0$$

$$k_{i6} = k_{i7} = k_{i8} = 0 \text{ for } i = 1, \ldots, 15$$

These constraints delete 18 free parameters from the model (leaving 40) and leave 80 degrees of freedom for the goodness of fit chi-square (there are 120 elements in the observed correlation matrix). The resulting chi-square is 140.46 (Prob = .0000). This is clearly an inadequate fit, and this more constrained model can serve as a null hypothesis relative to the one specified above that does include the method latent variables.

If the three method latent variables are allowed to intercorrelate, then the model involves 58 free parameters (15 "substantive" loadings in $\underline{Lx}$, 15 "method" loadings in $\underline{Lx}$, 13 correlations in $\underline{PH}$, and 15 error variances in $\underline{Td}$). The chi-square for this model is 59.331 with 62 degrees of freedom (Prob = .5726), which represents an acceptable fit and a sound basis for rejection of the model that did not include the method latent variables. Clearly, as Campbell and Fiske (1959) had argued, the method of measurement is an important confounding factor in correlations among measures of different dimensions.

Upon inspection of the ML estimates of the parameters of this model (which are not shown here), however, there is an obvious problem. The correlation between F_6 ("staff-rating" method of measurement) and F_8 ("self-rating" method of measurement) is estimated at -2.344 with a standard error of 2.708. Since a correlation coefficient cannot be smaller than -1, this result suggests that F_6 and F_8 may be almost perfectly correlated. This possibility can be treated as a null hypothesis relative to the current model by specifying the constraints

$$ph_{86} = -1$$

$$ph_{87} = ph_{76}$$

where the latter constraint is implied by the first one (if F_8 and F_6 are perfectly correlated, then any third variable will have the same correlation with either). Note that the sign of the correlations is arbitrary, since the signs of the loadings will be adjusted to produce the appropriate sign for the contribution of the method latent variables to the observed correlations. The addition of these two constraints increases the chi-square to 63.700 with 64 degrees of freedom. This increase of 4.369 (63.700 − 59.331) with 2 degrees of freedom has a probability of .113 if the constraints hold in the population. Thus, the null hypothesis can be accepted, and the equation of the staff and self-rating factors can be incorporated into the model. The ML estimate of this model is given in Table 9.19 and Figure 9.14.

Now, what do these results mean? First, the fact that the observed data fit the overidentifying restrictions of the model that incorporates the "method" factors means that the model is a plausible hypothesis that cannot be rejected by these data. Second, if this model is accepted, it means that the raters operate as if they hold a five-dimensional model of individual differences in their cognitive apparatus. And individuals,

Table 9.19 Maximum likelihood estimates of a measurement model for a multitrait multimethod matrix given in Table 9.18 (from Kelly and Fiske, 1951).

Matrix	Estimate							
Lx	.87	.00	.00	.00	.00	.00	.00	.00
	.00	.84	.00	.00	.00	−.08	.00	.00
	.00	.00	.58	.00	.00	.25	.00	.00
	.00	.00	.00	.82	.00	.21	.00	.00
	.00	.00	.00	.00	.72	.28	.00	.00
	.83	.00	.00	.00	.00	.00	.14	.00
	.00	.69	.00	.00	.00	.00	.29	.00
	.00	.00	.71	.00	.00	.00	.33	.00
	.00	.00	.00	.19	.00	.00	.60	.00
	.00	.00	.00	.00	.59	.00	.47	.00
	.55	.00	.00	.00	.00	.00	.00	.18
	.00	.45	.00	.00	.00	.00	.00	.24
	.00	.00	.43	.00	.00	.00	.00	.28
	.00	.00	.00	.40	.00	.00	.00	.43
	.00	.00	.00	.00	.63	.00	.00	.63
PH	1.00							
	.55	1.00						
	−.36	−.42	1.00					
	.38	.65	−.09	1.00				
	.57	.31	−.03	.42	1.00			
	.00	.00	.00	.00	.00	1.00		
	.00	.00	.00	.00	.00	.18	1.00	
	.00	.00	.00	.00	.00	−1.00	.18	1.00
Td	.24							
	.29							
	.59							
	.27							
	.37							
	.29							
	.47							
	.38							
	.60							
	.45							
	.68							
	.76							
	.75							
	.67							
	.27							

Chi-square = 63.700; *df* = 64; Prob = .487

Note. Coefficients with values +/− 1.00 or .00 are constrained by hypothesis.

including themselves, can be located in this five-dimensional space—although only roughly by any one type of rater. This "linguistic space" is sufficiently normative (i.e., shared by the members of the linguistic community included in the study) that ratings from the various participants in the study are consistent with a model that depicts each rating as an estimate of a latent variable that involves manageable errors. These

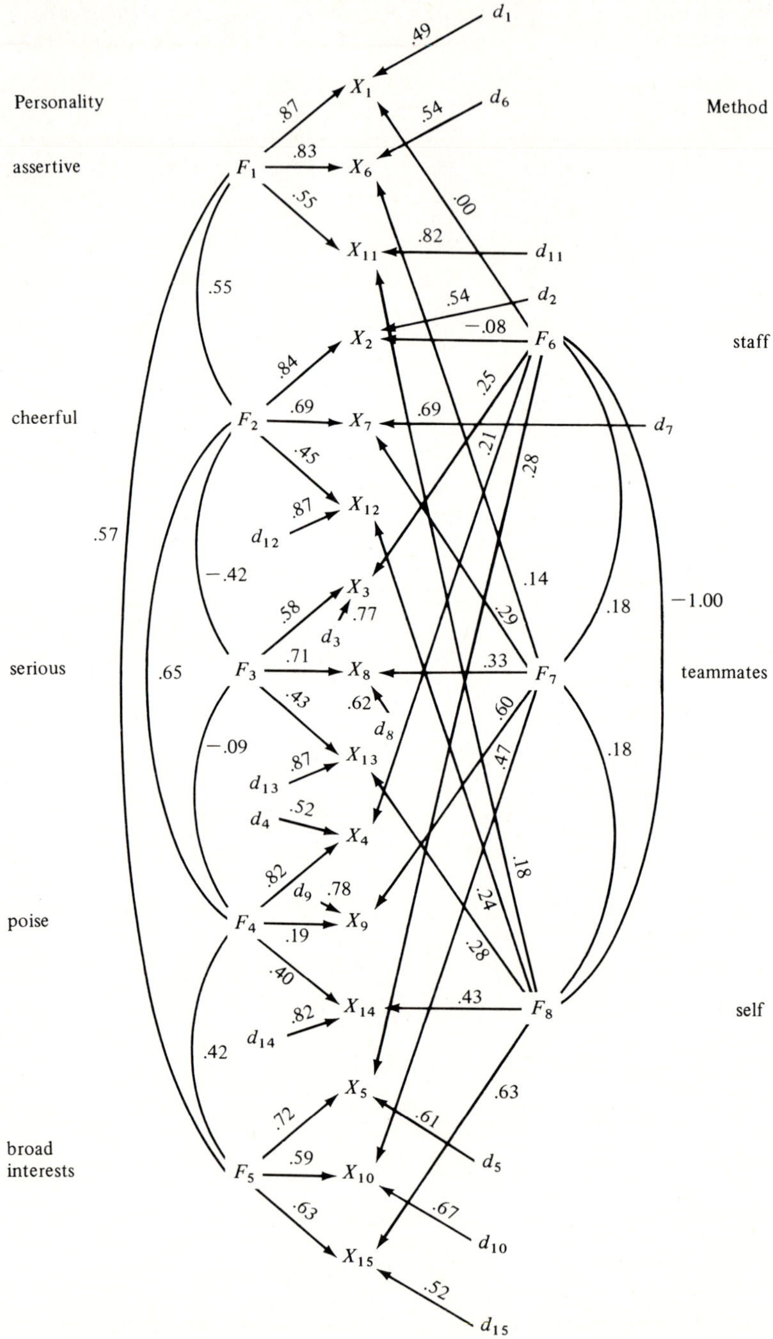

Figure 9.14 Path diagram of a measurement model fit to a multitrait multimethod matrix (from Table 9.18). In order to simplify the diagram, some correlations among latent personality traits are not shown. See Table 9.19 for the complete set of correlations.

errors are "manageable" in that they are either random relative to the true scores (d_1, $d_2, \ldots, d_{15}$) or systematically related to the method of measurement factor (F_6, F_7, F_8) that is also uncorrelated with the substantive latent variables.

The results obtained from the ML estimation are quite interesting in this case. Note that the assertiveness dimension (F_1) is better measured by the staff and teammates than the person being rated: the staff-rating (X_1) correlates .87 with the true scores, and the teammate rating (X_6) correlates .83, but the self-rating is only correlated .55 with F_1. A similar pattern holds for cheerfulness (F_2) and seriousness (F_3)—except that in the latter case the teammate validity is greater than that of the staff. In contrast, the teammates appear to be almost random in their ratings of "unshakable poise" (F_4), whereas the staff-ratings are quite valid and the self-ratings are only moderately valid.

The correlations between the ratings and the method factors are also of interest. Note that the largest loadings on these latent variables are for ratings of poise and breadth of interest by teammates and selves. This result suggests that ratings of these dimensions by these methods are very likely to produce distorted correlations or covariances if a model such as this one is not used to extract the method factors.

The most important research implication of these results is the sheer magnitude of the random and systematic errors in these ratings. Self-ratings, for example, are the most frequently employed method of measurement in social psychological research. The correlations between these ratings and their particular latent variable range from .40 to .63, indicating that between 61 and 84 percent of the variance in the observed indicators is attributable to measurement error. This means that correlations between these ratings and other variables would be so attenuated in most research settings that only a veiled hint of very strong relationships would be apparent in observed correlations. Two indicators that correlate .6 with their respective latent variables can correlate only .36—even if their latent variables are perfectly correlated—unless their errors of measurement are systematically related. Furthermore, if a number of items like X_{15} (self-ratings of "breadth of interests") were added together to form a measurement scale, the problem would not disappear.

Suppose, for example, that four items like X_{15} were added together to form a scale and that each had the same latent variable structure as X_{15}. This sum would then be

$$Y = Y_1 + Y_2 + Y_3 + Y_4$$

where Y_i ($i = 1,2,3,4$) is a self-report item measuring breadth of interests. From the model fit to the multitrait multimethod matrix we now have some basis to presume that

$$Y_i = .63F_5 + .63F_8 + d_{15}$$

where

$$\text{Var}(Y_i) = \text{Var}(.63F_5) + \text{Var}(.63F_8) + \text{Var}(d_{15})$$

The sum of four such items would then have the structure

$$Y = 4(.63F_5 + .63F_8) + (d_{15,1} + d_{15,2} + d_{15,3} + d_{15,4})$$

where $d_{15,i}$ is the random error component for the ith item in the "scale." Now, the variance of the sum of the four random components will, on the average, be equal to *four times* the variance of any one of the $d_{15,i}$. This is the case because

$$\mathrm{Cov}[(d_{15,i})(d_{15,i'})] = 0$$

when d_{15} is a *random* error and i is not equal to i'. On the other hand, the variance of the sum of the systematic components would be *greater than* four times the variance of the individual components:

$$\begin{aligned}\mathrm{Var}[4(.63F_5 + .63F_8)] &= E[16(.63F_5 + .63F_8)^2] \\ &= 16[\mathrm{Var}(.63F_5) + \mathrm{Var}(.63F_8)]\end{aligned}$$

since $\mathrm{Cov}(F_5, F_8) = 0$. Thus, as more items are added together, the variance of the *random* errors diminishes relative to the systematic component. But the gain here is illusory since one-half of the systematic component is a *method* latent variable that is uncorrelated with the latent variable of interest, but that may well be correlated with other variables. This "method variance" in the item is not attenuated by the repetition of additional items. Only random error is attenuated by repetitive measurement. The variance of a sum of random errors increases linearly with the number of items summed (presuming equal error variances for each item), while the variance of systematic components increases with the square of the number of items. Thus, the variance of X_{15} and the variance of a *sum* of four such items are partitioned as follows:

$$\mathrm{Var}(X_{15}) = \underset{\text{trait}}{.40} + \underset{\text{method}}{.40} + \underset{\text{random error}}{.27}$$

$$\mathrm{Var}(X_{15,1} + X_{15,2} + X_{15,3} + X_{15,4}) = \underset{\text{trait}}{6.35} + \underset{\text{method}}{6.35} + \underset{\text{random}}{1.08}$$

(Note that the variance of Y_{15} is not exactly one. This arises because of less than perfect fit between the model and the data.) The rather substantial systematic measurement error can be extracted from the assessment of relationships only through the use of measurement strategies that explicitly measure these errors.

The linear models presented in this chapter are one strategy for such isolation of theoretically relevant (valid) variance from theoretically irrelevant (error) variance in measures.

REFERENCES AND SUGGESTED READING

Aigner, D. J., and Goldberger, A. S. (eds.) *Latent variables in socioeconomic models.* Amsterdam: North Holland, 1977. (A useful collection of papers with numerous examples of latent variable models.)

Alwin, D. F. Approaches to the interpretation of relationships in the multitrait-multimethod matrix. In H. L. Costner (ed.), *Sociological Methodology 1973–1974.* San Francisco: Jossey-Bass, 1974.

Bentler, P. Multivariate analysis with latent variables. *Annual Review of Psychology,* 1980, *31,* 419–456.

Bohrnstedt, G. W., and Borgatta, E. F. (eds.), *Social measurement: Current issues.* Beverly Hills, Calif.: Sage, 1981.

Campbell, D. T., and Fiske, D. W. Convergent and discriminant validation by the multitrait-multimethod matrix. *Psychological Bulletin,* 1959, *56,* 81–105.

Costner, H. L., and Schoenberg, R. Diagnosing indicator ills in multiple indicator models. In A. S. Goldberger and O. D. Duncan (eds.), *Structural equation models in the social sciences.* New York: Seminar Press, 1973.

Cronbach, L. J. Coefficient alpha and the internal structure of tests. *Psychometrika,* 1951, *16,* 297–334.

Duncan, O. D. *Introduction to structural equation models.* New York: Academic Press, 1975.

Dwyer, J. H., Boyer, C., Roberts, R., and Stanley, C. On the latent variable structure of racial prejudice and authoritarianism. State University of New York at Stony Brook, mimeograph, 1981.

Dwyer, J. H., Wehrle, G., and Phelan, T. A latent variable model for causal inference from longitudinal designs: The case of powerlessness among female workers. State University of New York at Stony Brook, mimeograph, 1981.

Goldberger, A. S. Econometrics and psychometrics: A survey of communalities. *Psychometrika,* 1971, *36,* 83–107.

Goldberger, A. S., and Duncan, O. D. (eds.) *Structural equation models in the social sciences.* New York: Seminar Press, 1973. (A useful collection of papers with numerous examples of latent variable models.)

Harmon, H. H. *Modern factor analysis.* Chicago: University of Chicago Press, 1967 (2nd ed.). (This is a comprehensive treatment of factor analysis.)

Joreskog, K. G. Statistical analysis of sets of congeneric tests. *Psychometrika,* 1971, *36,* 109–133.

Joreskog, K. G. A general method for estimating a linear structural equation system. In Goldberger and Duncan (1973) (*q.v.*).

Joreskog, K. G. Analyzing psychological data by structural analysis of covariance matrices. In R. C. Atkinson, D. H. Krantz, and P. D. Suppes (eds.), *Contemporary developments in mathematical psychology* (Vol. 2). San Francisco: W. H. Freeman, 1974.

Joreskog, K. G., and Sorbom, D. *LISREL IV: Analysis of linear structural relationships by the method of maximum likelihood: A user's guide.* Chicago: International Educational Services, 1978.

Kenny, D. *Correlation and causality.* New York: Wiley, 1980.

Lawley, D. N., and Maxwell, A. E. *Factor analysis as a statistical method.* London: Butterworths, 1963. [The classic reference on the application of maximum likelihood techniques to the estimation of factor (latent variable) models.]

Lord, F. M., and Novick, M. R. *Statistical theories of mental test scores.* Reading, Mass.: Addison-Wesley, 1968.

Modoono, S. An investigation into the macro-determinants of scholastic achievement. State University of New York, doctoral dissertation, 1981.

Olsson, U. On the robustness of factor analysis against crude classification of observations. *Multivariate Behavioral Research,* 1977, *12,* 221–241.

Saris, W. Measurement models. Free University of Amsterdam, mimeograph, 1980. (This unpublished manuscript is one of the best sources of empirical examples available.)

Spearman, C., and Holzinger, K. The sampling error in the theory of two factors. *British Journal of Psychology,* 1924, *15,* 17–19.

Wald, A. The fitting of straight lines if both variables are subject to error. *Annals of Mathematical Statistics*, 1940, *11*, 284–300.

Wehrle, G. Alienation in work and leisure among secretaries. State University of New York at Stony Brook, doctoral dissertation, 1982.

Werts, C. E., Linn, R. L., and Joreskog, K. G. A congeneric model for platonic true scores. *Educational and Psychological Measurement*, 1973, *33*, 531–544.

Wilks, S. S. The large sample distribution of the likelihood ratio for testing composite hypotheses. *Annals of Mathematical Statistics*, 1938, *9*, 60–62.

10

Estimating Structural Equation Models

10.1 INTRODUCTION TO HYPOTHESIZING A CAUSAL MODEL

It is appropriate at this point in our journey through linear stochastic models to recall how far we have come—and how far we have to go. We began in Chapter 5 with a simple linear model relating one dependent variable to one independent variable. Such bivariate relations may be models of causal processes or simply descriptions of associations. If it is hypothesized that a one-unit change in the independent variable *causes* a change in the dependent variable that is constant across the levels of the independent variable, then the model is termed a structural model, and the ubiquitous linear coefficient is termed a structural coefficient.

As theory develops, hypotheses concerning the relations of numerous variables emerge. The first step in accommodating this greater complexity is presented in Chapter 6—where the multiple regression model is developed. When disturbances can be specified as uncorrelated with the multiple independent variables, estimation by ordinary least squares (OLS) allows assessment of partial relationships and interactions. A partial linear relationship, like a partial derivative or partial differential equation, is the relation that remains between two variables after other variables have been held constant. Interactions appear in a model when an hypothesis includes a specification that a partial relationship is dependent upon the level of other variables. In other words, a causal relationship between two variables is of greater or lesser magnitude depending upon the level of other variables in a system.

Chapter 7 details the application of a multiple regression model to the research design that motivates the least equivocal causal inference—the true experiment. Traditional methods of analysis of variance are recast in terms of the unifying least squares estimation of multiple regression models. The OLS procedure is found to be a bit cumbersome and is replaced by a least squares procedure that involves no "intercept term" in the structural model.

The constraint of having only one dependent variable is then relaxed in Chapter 8 with the introduction of a model for multiple equation regression. This general linear model allows the inclusion of several dependent variables in a single hypothesis, but it does not allow the dependent variables to be causally related to one another. Thus, the general linear model is only of interest when the computation of standard errors is the focus.

Finally, we are now prepared to introduce a general linear structural equation model that will accommodate multiple independent variables like multiple regression, multiple dependent variables like multivariate regression, *and* causal relations among the dependent variables. This general model will allow the specification that a variable dependent on other variables can also be a cause of other variables, or even of some of those variables that it depended upon. The measurement models described in Chapter 9 make it possible for social and behavioral scientists to estimate causal models of this complexity. Without the models for dealing with measurement error, that error would so distort the bulk of substantive estimates that the estimation of such models would obfuscate rather than inform. Of course, very little of the kind of work on measurement that would be necessary to launch immediately into the testing of causal models with the methods to be described in this chapter has been done. In this sense, then, the actual utilization of the general linear structural equation model will evolve as the work on measurement is advanced. The application of these structural models is actually bound by developments in measurement for most substantive areas of research. In fact, the empirical examples in Chapter 11 that involve multiple indicators of latent variables are mostly taken from very recent and unpublished research conducted by myself and several of my students.

10.2 A GENERAL LINEAR STRUCTURAL EQUATION MODEL

The general linear structural equation model will henceforth be referred to as the structural model. The structural model described here is analogous to the one known as the LISREL model that is associated with the Swedish statistician Karl Joreskog. The notation used here is congenial with that used by Joreskog in his maximum likelihood computer program for the estimation of model parameters. (LISREL is an acronym for LInear Structural RELations.)

The reader will recall from Chapter 8 that matrix notation is most convenient for the presentation of multiple dependent variable models. This is also certainly the case with more general structural models.

The mathematical complexity of structural models is nicely tamed by the correspondence between intuitively straightforward path diagrams and the actual equations that specify the model. One can even think of the equations as rather cumbersome representations of the diagram. Currently, the main reason for developing the equations is that computers are instructed in terms of "equation-like" commands. However, it is not unlikely that future developments in computer graphics will enable researchers to interact with their data in terms of diagrams rather than equations. In the meantime, we must learn how to translate the diagrams into algebraic equations.

The discussion in this chapter is primarily mathematical in that numerical examples are not described in detail. The reason for this is that Chapter 11 covers some particularly promising applications of the general linear model developed in this chapter. Some readers may prefer to review those examples prior to reading this more general, and therefore more mathematical and abstract, discussion.

We begin by supposing that an investigator conceives of the relations among numerous variables in terms of a causal model. This causal model is presumed to consist of linear structural relations such that change in any one of the causal variables has linear effects on other variables in the system. For the moment, we shall presume that all the variables in the system are measured without error.

The variables in the causal system are divided into two categories: *exogenous* (or independent) and *endogenous* (or dependent) variables. Change in the exogenous variables may produce change in other variables in the system, but they are not affected by changes in other variables in the system. Endogenous variables are all those variables that can, according to the hypothesized model, be changed by a change in other variables in the system. In other words, the variability in the exogenous variables is assumed to be due to factors not incorporated into the causal model. The model does not attempt to explain variability in the exogenous variables in terms of the influences of other variables within the system. In contrast, variability in the endogenous variables is explained to *some* extent by changes in other variables within the causal model. This distinction between exogenous and endogenous variables is not of crucial importance to an understanding of structural models. However, the distinction is ubiquitous in the literature, and an understanding of the distinction is necessary for the use of such computer programs as LISREL.

Variability in the endogenous variables is seldom completely explained by the other variables in a causal model—at least not in the social and behavioral sciences. The variability in the endogenous variables that is unexplained by the other variables in the model is thus attributed to *disturbances*. (The exogenous variables can also be depicted as influenced by disturbances—they are *completely* determined by disturbances.) Thus, the complete linear structural equation model involves an hypothesis that change in the endogenous variables is determined by change in (1) the exogenous variables, (2) change in certain other endogenous variables, and (3) change in factors that have not been included in the model (the disturbances). All these causal relations are presumed to be linear. Recall that the meaning of "linear" in this context is that the amount of change produced in the affected variable is a linear function of the amount of change that occurred in the causal variable. This means that if the amount of change in the cause doubles, then the amount of change produced also doubles, and so on.

A path diagram of such a causal model is presented in Figure 10.1. Note that paths lead only *from* the exogenous variables (labeled F_1, F_2, and so on) to other variables, but may lead both from *and* to the endogenous variables (E_1, E_2, . . .). Whether or not this pattern of causal asymmetry between exogenous and endogenous variables is valid is an empirical or methodological question. If the exogenous variables are randomized manipulations, then this causal asymmetry is clearly warranted; in other situations, the exogenous variables might be "background" variables that temporally precede changes in the endogenous variables. In still other situations, the classification of a variable as exogenous may not involve any convincing methodological arrangement; in such instances, the exogenous/endogenous distinction is simply an untested assumption.

Another important aspect of the causal model in Figure 10.1 is that one of the causal arrows violates a unidirectional causal sequence. That is, there is a path leading from one variable *back to* another variable when the other variable is a direct or indi-

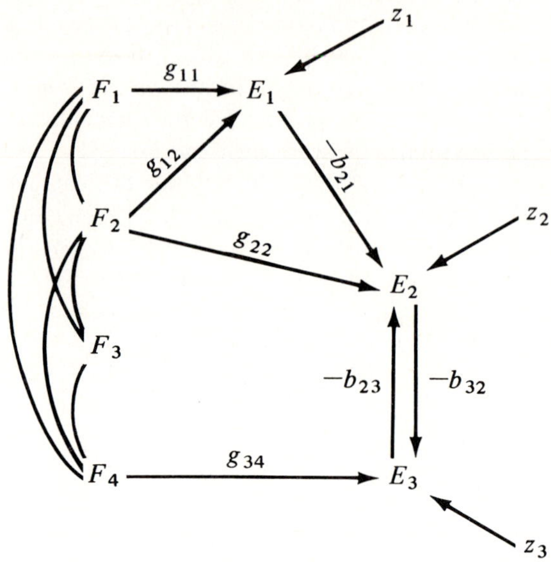

Figure 10.1 Path diagram of a structural equation model with exogenous (F) and endogenous (E) variables. Several logically plausible paths of causal influence have been deleted; these deletions correspond to hypotheses of structural coefficients with a value of zero. See the text for an explanation of the negative signs of the coefficients b.

rect determinant of the one variable. Change in E_3 affects E_2 and change in E_2 affects E_3. Such a model is referred to as a *nonhierarchical* model. The model in Figure 10.1 would be hierarchical if there was no such reciprocal causation (e.g., $b_{23} = 0$).

Finally, notice that the endogenous variables are completely determined by the exogenous variables, other endogenous variables, and the disturbances (z_1, z_2, . . .). In contrast, no arrows lead to the exogenous variables or the disturbances, indicating that variability in these variables is not determined by any of the variables within the causal system. Change in these variables is taken as a given; the variability in the exogenous variables and disturbances was determined by factors outside of the causal system. (Also, as in previous chapters, the curved arrows connecting variables or disturbances indicate an association that is not explained by causal processes within the causal model.)

The fact that the endogenous variables are completely determined by specified (although unmeasured, in the case of the disturbances) variables within the model suggests the linear equation form of such a path model. It is one in which the endogenous variables are expressed as functions of their determinants, and their determinants are all those variables with paths leading to them. (One need not think of the path diagram in strictly linear terms, of course. However, when actual empirical testing of the model is involved, the statistical procedures are greatly simplified for the linear case. This is particularly true when multiple indicator measurement models are added to the structural model.) Thus, we may translate the causal model in Figure 10.1 into a linear structural equation model as follows:

$$E_1 = g_{11}F_1 + g_{12}F_2 + z_1 \qquad (10.1a)$$
$$E_2 = g_{22}F_2 - b_{21}E_1 - b_{23}E_3 + z_2 \qquad (10.1b)$$
$$E_3 = g_{34}F_4 - b_{32}E_2 + z_3 \qquad (10.1c)$$

where all variables are measured as deviations about their respective means. (This simplifying assumption of zero expectations need not be made; see the analysis of nonequivalent control group designs in Chapter 11 for an instance in which the means of variables are a meaningful part of a structural model and "intercept" terms are included.)

The crucial question now before us is the following: If the causal model given in equations 10.1 describes a "real" process in some population, under what conditions could we use observations of $\underline{E}$ and $\underline{F}$ to estimate the coefficients of the model? (Note that

$$\underline{E} = [E_1 \quad E_2 \quad E_3]'$$

and

$$\underline{F} = [F_1 \quad F_2 \quad F_3 \quad F_4]')$$

One such set of conditions is the following: (1) $\underline{E}$ and $\underline{F}$ are measured without error; (2) the unknown parameters of the model (linear coefficients) are *identified* by the observed variances and covariances; and (3) the observed variables are jointly normal in distribution. Note that these conditions are sufficient to enable efficient maximum likelihood estimation of the model parameters *if the model holds in the population.* To say this is to say, for example, that an increase of one unit in F_1 produces an expected increase in E_1 of g_{11} units (if g_{11} is negative, then the increase in F_1 causes a decrease in E_1) if all other variables that both correlate with F_1 and influence E_1 are held constant. Furthermore, all the paths and covariances that are presumed zero in the model—such as the path from F_3 to E_3 or the covariance between z_1 and z_2—are not contradicted by the "real world."

Suppose now, however, that the state of the real world is not a given. Rather, we want to test this model against some sampled observations of $\underline{E}$ and $\underline{F}$. Can the observations be used to tell us whether or not the model holds in the population? The answer to this question is "no." But it is possible to answer the following question: "If this model were true, then how likely are the current observations?" The answer to this question is derived from the overidentifying restrictions that the model places on the observed variances and covariances. This situation is exactly analogous to that in Chapter 9 in which measurement models were tested. It is important to keep in mind, however, that the fit of a structural model to a set of data is distinct from the methodological procedures used to generate those data. The use of randomization, manipulation, and longitudinal or time series designs can greatly enhance a researcher's confidence that an estimate of a parameter in a linear equation is indeed an estimate of a structural coefficient (causal effect). But the fit of the model to the data and the design of the study are somewhat distinct factors that must be considered when assessing the results.

Matrix Notation for Structural Models

It is convenient to reformulate equations 10.1 into a general matrix format that is amenable to computer analysis. This is accomplished as follows:

$$\underline{B}\,\underline{E} = \underline{G}\,\underline{F} + \underline{z} \tag{10.2}$$

where

$\underline{B}$ = $(m \times m)$ matrix of structural coefficients b

m = the number of endogenous variables E

$\underline{E} = [E_1 \quad E_2 \quad \ldots \quad E_m]' = (m \times 1)$ vector of endogenous variables

$\underline{G}$ = $(m \times n)$ matrix of structural coefficients g

n = the number of exogenous variables F

$\underline{F} = [F_1 \quad F_2 \quad \ldots \quad F_n]' = (n \times 1)$ vector of exogenous variables

$\underline{z} = [z_1 \quad z_2 \quad \ldots \quad z_m]' = (m \times 1)$ vector of disturbances

In order to clarify this matrix form, we can now translate equations 10.1 into the form of equation 10.2:

$$\underline{B} = \begin{bmatrix} 1 & 0 & 0 \\ b_{21} & 1 & b_{23} \\ 0 & b_{32} & 1 \end{bmatrix}$$

$$\underline{E} = [E_1 \quad E_2 \quad E_3]'$$

$$\underline{G} = \begin{bmatrix} g_{11} & g_{12} & 0 & 0 \\ 0 & g_{22} & 0 & 0 \\ 0 & 0 & 0 & g_{34} \end{bmatrix}$$

$$\underline{F} = [F_1 \quad F_2 \quad F_3 \quad F_4]'$$

$$\underline{z} = [z_1 \quad z_2 \quad z_3]'$$

or

$$\begin{bmatrix} 1 & 0 & 0 \\ b_{21} & 1 & 0 \\ 0 & b_{32} & 1 \end{bmatrix} \begin{bmatrix} E_1 \\ E_2 \\ E_3 \end{bmatrix} = \begin{bmatrix} g_{11} & g_{12} & 0 & 0 \\ 0 & g_{22} & 0 & 0 \\ 0 & 0 & 0 & g_{34} \end{bmatrix} \begin{bmatrix} F_1 \\ F_2 \\ F_3 \\ F_4 \end{bmatrix} + \begin{bmatrix} z_1 \\ z_2 \\ z_3 \end{bmatrix} \tag{10.3}$$

The form of equation 10.3 should now make clear the reason for the negative signs for the structural coefficients of the endogenous variables in equations 10.1—it is simply a result of the conventions concerning matrix representation of an equation system.

Equation 10.2 is now a very general form of a linear structural equation model. The flexibility of 10.2 to accommodate a wide variety of path models derives from the generality of the matrices $\underline{B}$ and $\underline{G}$. The reader should practice the procedure of translating path diagrams into linear equations, and then transforming the linear equations into the matrix format of 10.2. Notice that the matrix $\underline{B}$ can even be specified to include *reciprocal causation* among the endogenous variables. The specification that

$$\underline{B} = \begin{bmatrix} 1 & 0 & 0 \\ b_{21} & 1 & b_{23} \\ 0 & b_{32} & 0 \end{bmatrix}$$

indicates that E_3 is a function of E_2 (since b_{32} is not constrained to zero), whereas E_2 is a function of E_3 (since b_{23} is not constrained to zero). However, whether or not the free parameters in $\underline{B}$ and $\underline{G}$ are identified is another matter. It is quite easy to specify models with equation 10.2 that are hopelessly *under*identified.

In order to address the identification problem more directly, it is convenient to first define the covariance matrices of the variables:

$$\begin{aligned} \underline{PH} = \text{Cov}(\underline{F}) &= (n \times n) \text{ covariance matrix of } \underline{F} \\ &= \text{expectation } (\underline{F}\,\underline{F}') \\ \underline{PS} = \text{Cov}(\underline{z}) &= (m \times m) \text{ covariance matrix of } \underline{z} \\ &= \text{expectation } (\underline{z}\,\underline{z}') \end{aligned}$$

To clarify the meaning of these covariance matrices, the following apply to the equation model specified in Figure 10.1 or equations 10.1:

$$\underline{PH} = \begin{bmatrix} \text{Var}(F_1) & \text{Cov}(F_1,F_2) & \text{Cov}(F_1,F_3) & \text{Cov}(F_1,F_4) \\ \text{Cov}(F_2,F_1) & \text{Var}(F_2) & \text{Cov}(F_2,F_3) & \text{Cov}(F_2,F_4) \\ \text{Cov}(F_3,F_1) & \text{Cov}(F_3,F_2) & \text{Var}(F_3) & \text{Cov}(F_3,F_4) \\ \text{Cov}(F_4,F_1) & \text{Cov}(F_4,F_2) & \text{Cov}(F_4,F_3) & \text{Var}(F_4) \end{bmatrix}$$

$$\text{Cov}(\underline{E}) = \begin{bmatrix} \text{Var}(E_1) & \text{Cov}(E_1,E_2) & \text{Cov}(E_1,E_3) \\ \text{Cov}(E_2,E_1) & \text{Var}(E_2) & \text{Cov}(E_2,E_3) \\ \text{Cov}(E_3,E_1) & \text{Cov}(E_3,E_2) & \text{Var}(E_3) \end{bmatrix}$$

$$\begin{aligned} \text{Cov}(\underline{E},\underline{F}) &= \text{expectation } (\underline{E}\,\underline{F}') \\ &= \begin{bmatrix} \text{Cov}(E_1,F_1) & \text{Cov}(E_1,F_2) & \text{Cov}(E_1,F_3) & \text{Cov}(E_1,F_4) \\ \text{Cov}(E_2,F_1) & \text{Cov}(E_2,F_2) & \text{Cov}(E_2,F_3) & \text{Cov}(E_2,F_4) \\ \text{Cov}(E_3,F_1) & \text{Cov}(E_3,F_2) & \text{Cov}(E_3,F_3) & \text{Cov}(E_3,F_4) \end{bmatrix} \end{aligned}$$

and

$$\underline{PS} = \begin{bmatrix} \text{Var}(z_1) & \text{Cov}(z_1,z_2) & \text{Cov}(z_1,z_3) \\ \text{Cov}(z_2,z_1) & \text{Var}(z_2) & \text{Cov}(z_2,z_3) \\ \text{Cov}(z_3,z_1) & \text{Cov}(z_3,z_2) & \text{Var}(z_3) \end{bmatrix}$$

The Identification Problem

The problem of identification can now be construed in terms of the matrices defined above. The parameters to be identified are elements of the matrices $\underline{B}$ and $\underline{G}$—as well as $\underline{PS}$. The elements of $\underline{PH}$, Cov($\underline{E}$), and Cov($\underline{E}$,$\underline{F}$) are observable (or identified from

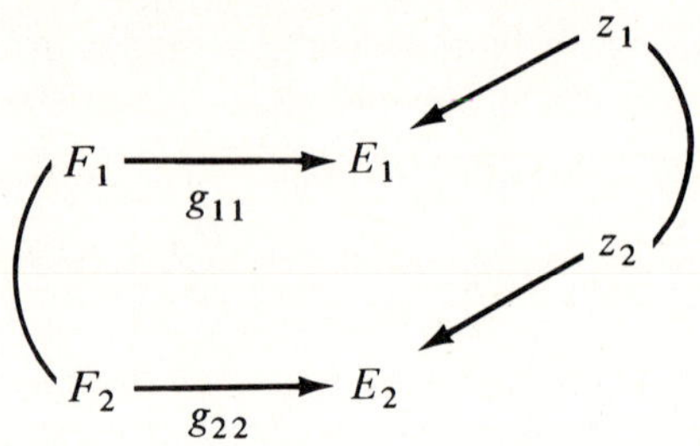

Figure 10.2 Path diagram of a simple structural model specified in equation 10.4.

a measurement model). Thus, an identified parameter is one that may be expressed in terms of at least one function of the elements of the observable matrices. The set of equations that express the elements of Cov($\underline{E}$) and Cov($\underline{E}$,$\underline{F}$) as functions of $\underline{B}$, $\underline{G}$, $\underline{PS}$, and $\underline{PH}$ are analogous to the "normal equations" of ordinary least squares. In fact, equation 10.2 can be solved with OLS methods when $\underline{PS}$ is diagonal (the disturbances are uncorrelated with one another), the model is hierarchical, and all the possible paths are included in the model. (When a model is both hierarchical and the disturbances are uncorrelated, then the label *recursive* is sometimes used.) The solution would be obtained by applying OLS procedures to each equation in 10.2. The model would be just identified. Clearly, however, OLS may be applied to only a very limited class of models.

There are two practical approaches to the determination of whether or not a structural model is identified. For relatively simple models, the algebra of expectation can be applied and then an algebraic solution can be sought. In more complex cases it is quite impractical to tabulate these algebraic solutions. Fortunately, computer programs like LISREL include a procedure for determining whether or not a model is identified. The procedure appears to be quite reliable. There may be a few odd instances in which the program indicates that an identified model is underidentified, or *vice versa.*

In order to demonstrate the algebraic identification procedure, suppose that a causal model involves two exogenous (F_1, F_2) and two endogenous (E_1, E_2) variables. Further suppose that the causal relations portrayed in Figure 10.2 are hypothesized to hold in the population from which the sample was drawn. This model translates into equation form as

$$E_1 = g_{11}F_1 + z_1 \tag{10.4a}$$

$$E_2 = g_{22}F_2 + z_2 \tag{10.4b}$$

The first step in the investigation of identification is to multiply the equations in 10.4 by the exogenous variables to obtain the elements of Cov($\underline{E}$,$\underline{F}$) as functions of the unknown parameters and estimable covariances. The second step will be to multiply equation 10.4a by equation 10.4b to express Cov($\underline{E}$) as a function of the same parameters. Finally, each equation in 10.4 will be squared to solve for the variances of the disturbances. Prior to making any assumptions about the disturbances, except that their means are zero, these equations are as follows:

1. Multiplying 10.4a and 10.4b by F_1 and F_2 and taking the expectation over the population gives

$$\begin{aligned}
\mathrm{Cov}(E_1,F_1) &= g_{11}\ \mathrm{Var}(F_1) + \mathrm{Cov}(F_1,z_1) \\
\mathrm{Cov}(E_1,F_2) &= g_{11}\ \mathrm{Cov}(F_1,F_2) + \mathrm{Cov}(F_2,z_1) \\
\mathrm{Cov}(E_2,F_1) &= g_{22}\ \mathrm{Cov}(F_1,F_2) + \mathrm{Cov}(F_1,z_2) \\
\mathrm{Cov}(E_2,F_2) &= g_{22}\ \mathrm{Var}(F_2) + \mathrm{Cov}(F_2,z_2)
\end{aligned}$$

2. Multiplying 10.4a by 10.4b and taking the expectation gives

$$\begin{aligned}
\mathrm{Cov}(E_1,E_2) = g_{11}g_{22}\ \mathrm{Cov}(F_1,F_2) &+ g_{11}\ \mathrm{Cov}(F_1,z_2) \\
&+ g_{22}\ \mathrm{Cov}(F_2,z_1) + \mathrm{Cov}(z_1,z_2)
\end{aligned}$$

3. Squaring 10.4a and 10.4b and taking the expectations gives

$$\begin{aligned}
\mathrm{Var}(E_1) &= g_{11}^2\ \mathrm{Var}(F_1) + 2g_{11}\ \mathrm{Cov}(F_1,z_1) + \mathrm{Var}(z_1) \\
\mathrm{Var}(E_2) &= g_{22}^2\ \mathrm{Var}(F_2) + 2g_{22}\ \mathrm{Cov}(F_2,z_2) + \mathrm{Var}(z_2)
\end{aligned}$$

As these equations stand, none of the parameters of the model are identified. There are two structural parameters (g_{11} and g_{22}), two disturbance variances and a covariance [$\mathrm{Var}(z_1)$, $\mathrm{Var}(z_2)$, and $\mathrm{Cov}(z_1,z_2)$], and four covariances between disturbances and exogenous variables [$\mathrm{Cov}(F_1,z_1)$, $\mathrm{Cov}(F_1,z_2)$, $\mathrm{Cov}(F_2,z_1)$, and $\mathrm{Cov}(F_2,z_2)$] that are unknown in these equations. Since the number of unknown parameters exceeds the number of equations (nine versus seven), we definitely have a case of underidentification. It can happen that some of the unknowns are identified in such a situation, but in this particular case none of the unknowns can be estimated uniquely from the observed data—i.e., these equations cannot be solved for the unknowns in terms of the knowns.

However, if we add some constraints, such as the ones implicit in Figure 10.2 concerning the disturbance, then some or all of the remaining unknowns can become identified. These constraints are essentially assumptions about some of the unknowns. For example, Figure 10.2 indicates that the disturbances may be correlated with one another, but they are uncorrelated with the exogenous variables (since there are no causal paths or correlational curves connecting them). If we add these constraints to the normal equations above, we have

$$\begin{aligned}
\mathrm{Cov}(E_1,F_1) &= g_{11}\ \mathrm{Var}(F_1) \\
\mathrm{Cov}(E_1,F_2) &= g_{11}\ \mathrm{Cov}(F_1,F_2) \\
\mathrm{Cov}(E_2,F_1) &= g_{22}\ \mathrm{Cov}(F_1,F_2) \\
\mathrm{Cov}(E_2,F_2) &= g_{22}\ \mathrm{Var}(F_2) \\
\mathrm{Cov}(E_1,E_2) &= g_{11}g_{22}\ \mathrm{Cov}(F_1,F_2) + \mathrm{Cov}(z_1,z_2) \\
\mathrm{Var}(E_1) &= g_{11}^2\ \mathrm{Var}(F_1) + \mathrm{Var}(z_1) \\
\mathrm{Var}(E_2) &= g_{22}^2\ \mathrm{Var}(F_2) + \mathrm{Var}(z_2)
\end{aligned}$$

The number of unknowns has now been reduced to five, and the two structural coefficients are overidentified:

$$g_{11} = \mathrm{Cov}(E_1,F_1)/\mathrm{Var}(F_1)$$
$$g_{11} = \mathrm{Cov}(E_1,F_2)/\mathrm{Cov}(F_1,F_2)$$
$$g_{22} = \mathrm{Cov}(E_2,F_2)/\mathrm{Var}(F_2)$$
$$g_{22} = \mathrm{Cov}(E_2,F_1)/\mathrm{Cov}(F_1,F_2)$$

whereas the elements of $\underline{PS}$ are just identified:

$$\mathrm{Cov}(z_1,z_2) = \mathrm{Cov}(E_1,E_2) - g_{11}g_{22}\,\mathrm{Cov}(F_1,F_2)$$
$$\mathrm{Var}(z_1) = \mathrm{Var}(E_1) - g_{11}^2\,\mathrm{Var}(F_1)$$
$$\mathrm{Var}(z_2) = \mathrm{Var}(E_2) - g_{22}^2\,\mathrm{Var}(F_2)$$

Thus, the entire model is identified with two overidentifying restrictions. These restrictions may be obtained by equating the two solutions for g_{11} and the two solutions for g_{22}. Note that the first solution for g_{11} and the first solution for g_{22} above are the OLS solutions to equations 10.4a and 10.4b, respectively. The maximum likelihood solution would deviate from the OLS solution to the extent that the overidentifying restrictions were violated by sample data.

Identification and Study Design

The parameters in any structural model become identified because of assumptions about some other parameters. Generally, these assumptions take the form of a specification that certain covariances or causal paths are zero. The most usual of such assumptions is that the disturbances are uncorrelated with certain other variables in the causal model; another common assumption is that the direction of causal influence is from one variable to another and not *vice versa.*

It is clarifying at this juncture to distinguish between two classes of untestable constraints (assumptions) that are invoked to achieve identification of certain parameters from a particular data set. The first class of assumptions is termed the "plausible" constraint category, whereas the second is termed the "by design" constraint category. The plausible constraint category includes those constraints that seem reasonable to one or more investigators, but that are not made plausible by explicit features of the research design. Rather, the constraint is deemed plausible for broader theoretical (or ideological) reasons, but not for methodological reasons. On the other hand, the design constraint class includes those constraints that follow from design features that were incorporated for the explicit purpose of making certain constraints plausible. Some concrete examples of each of these types of constraints are as follows:

Design Constraints The archtypical design constraint is the assumption that the disturbance is uncorrelated with the independent (exogenous) variable in a randomized experiment. The plausibility of this constraint does not follow from any theoretical considerations relevant to a particular study. Rather, the plausibility of this constraint flows from the adherence of all scientists to the presumption that randomized manipulations will tend to be uncorrelated with all variables except those that are influenced by the manipulation. This presumption is "extra-theoretical" relative to the theory that may be guiding an investigation in a particular substantive area. It is a presumption that would be acceptable to scientists from most any perspective within a particular field.

Another design constraint of particular interest to social and behavioral scientists is that of non-reciprocal causality in longitudinal designs. For example, when family characteristics of a person at age 15 are related to characteristics of that person at age 40, scientists also would tend to agree that the direction of causation is quite clear. There may be some rare instances in which stable characteristics of a person, such as a health problem, shape their family as well as their own future. However, where such variables as socioeconomic status are concerned, there would be virtual consensus that the direction of causal influence, if there be any, must be from the family to the individual—and not *vice versa.* Thus, by design, certain causal arrows (those leading from characteristics of the person at age 40 to family characteristics at age 15) can be presumed to be zero—because of the consensus concerning causation and the flow of time. (It should be noted, however, that the longitudinal design does not allow the assumption that the disturbance is uncorrelated with the variables that are presumed to be causal. There may be some "third variable" that is highly correlated with the temporally precedent variable that is actually responsible for the observed association. This issue will be discussed in more detail in Chapter 11.)

A third example of a design constraint comes from the study of trait heritability in human genetics. In an effort to sort out the relative contributions of genes and environment on IQ test performance, researchers have used twin designs for many years. The simplest linear model that is fit to these data is the following:

$$Y_1 = a_1 E_1 + a_2 G_1$$

$$Y_2 = a_1 E_2 + a_2 G_2$$

where Y_1 and Y_2 are the IQ scores of the first and second persons in a pair of same-sexed twins, respectively; E_1 and E_2 are the environmental factors that influence each of the twins; and G_1 and G_2 are the genetic factors that influence the twins' performance. All variables are standardized. Now, since E and G are not measured in these studies, a_1 and a_2 are clearly not identified, and the model is completely underidentified. In order to deal with this situation, researchers have utilized the widely accepted assumption that G_1 and G_2 should be almost perfectly correlated (across twins) for monozygotic (MZ) twins and correlated .50 to .60 for dizygotic (DZ) twins. (The particular value of the dizygotic correlation used depends upon assumptions about the degree to which assortative mating produces a correlation between the genetic characteristics of parents.) However, as can be easily demonstrated, this assumption is insufficient to identify a_1 or a_2. In order to achieve identification of a_2, researchers have thus resorted to constraints that seemed plausible to them for less universally acceptable reasons.

Plausible Constraints This latter example from IQ research leads us to the category of constraints based on plausibility, since it is these that many genetic researchers have had to concoct in order to identify the unknown parameters of the model. One set of such constraints that is sufficient to identify a_2 is the following:

1. There are no measurement errors in Y_1 and Y_2.
2. The level of correlation between E and G is the same for MZ and DZ twins.
3. The level of correlation between E_1 and E_2 is the same for MZ and DZ twins.
4. The effects of E and G on IQ are linear and additive (no interaction).

Now, whether or not any or all of these constraints are plausible does not depend upon widely held views within the scientific community. Consider, for example, assumption (3). This constraint means that the social environments of MZ twins are similar to the same degree as the social environments of DZ twins. Clearly, reality could be otherwise. Since MZ twins are much more similar in appearance, the social environment of the MZ twins may be more similar. This would be especially true in a society where appearance is an important basis for social evaluation.

Another more common example of a plausibility constraint is that invoked in cross-sectional designs. A researcher may, for example, want to estimate structural coefficients with cross-sectional data even though she knows that the assumption of causal direction is highly suspect. There is certainly nothing "improper" about analyses based upon a number of "what ifs" as long as it is clear to all involved that the constraints necessary for identification have been conjectured and not demonstrated in any way that would convince a skeptic. It is not uncommon for researchers to believe that the mere use of a structural equation model extracts causal information from what would otherwise be correlational data. It cannot be stressed sufficiently that design considerations are primary in this realm—not the statistical procedures used.

Plausible constraints are therefore of a different sort than design constraints in that they involve more uncertainty for the scientific community. For this reason, researchers should try alternatives to their plausible constraints to see how crucial their assumptions are to their conclusions. Arthur Goldberger, for example, has demonstrated that even small deviations from the usual constraints deemed plausible by behavioral geneticists result in substantively different conclusions concerning the heritability of performance on IQ tests. Such mixing of observed data with conjectured values of parameters borders on the procedures known as *simulation*. Simulations can be very illuminating in that they reveal consequences of assumptions that are virtually impossible to deduce verbally. But researchers should maintain a clear distinction between empirical tests of hypotheses and exercises exploring the logical consequences of various sets of assumptions. (It should be pointed out that the use of plausible constraints offers a special opportunity for the intrusion of the investigator's ideology. The twin studies are a good case in point. A researcher can consistently favor assumptions that solve the underidentification problem such that genes are given a more prominent role, or *vice versa*. The more objective approach would be to use such models to clarify which assumptions give more weight to genes and which assumptions give more weight to environment. In this regard, I support Goldberger's argument that the range of plausible conclusions that can be drawn from twin studies of IQ is usually too great to be of any real value in determining the direct effect of genes. The range of conclusions is so great because the range of plausible assumptions is so great. For example, suppose that persons in a society are differentially valued and trained on the basis of heritable darkness of skin color. In such a society, a pair of monozygotic twins would be treated more similarly than a pair of dizygotic twins—because, on the average, the monozygotic twins would be more similar in skin darkness.)

The structural equation approach to model testing has the virtue of making the assumptions that are necessary for identification quite explicit. And once these assumptions have been revealed they point toward the type of research that must be conducted to replace plausible constraints by design constraints. For example, the underidentification of the behavioral genetic model with twins indicates that studies that explicitly measure environmental similarity, with testable measurement models, are needed. Alternatively, a design involving persons related in some way that precludes an environmental correlation (such as separated twins reared in very divergent environments) would also be indicated.

There are, of course, many substantive areas of research in which models will be hopelessly underidentified without the inclusion of plausible constraints. This does not mean that work in these areas is less valuable than work in other areas. A little bit of tentative knowledge in an area of enormous practical importance may be of more value than solid evidence concerning an inconsequential relationship. What is incumbent upon the user of underidentified models is an explicit investigation of the impact of deviations from the constraints that were made to achieve identification of other parameters. This is achieved quite easily with computer programs such as LISREL; those parameters that have been fixed at some value—usually zero—can be fixed at other values to see whether or not deviations from the constraint have a major impact on important relationships.

10.3 MODELS WITH RECIPROCAL CAUSATION

Although social and behavioral scientists have often theorized in terms of reciprocal causal relationships, most have lacked empirical methods for the estimation of such processes. Linear structural models provide for the estimation of such causal effects when certain assumptions are acceptable. Longitudinal designs provide data that are amenable to the testing of such models; this topic will be discussed in Chapter 11.

As a clarifying example of what is meant by "reciprocal" in a causal context, consider the relationship between the values of a young person in high school and the values of that person's closest friend. Clearly, we would expect that the values of the one bring about changes in the other, and *vice versa*. Under what conditions could we estimate such effects? The answer to this question is the same as the answer to the question of how to identify the parameters of a structural equation model including reciprocal effects. And it is important to keep in mind a point that will be reiterated in the discussion of causal inference from longitudinal data: there is no universal model that will be applicable to each and every substantive area. Rather, models such as these must be developed in the context of theory and empirical experience that is directly relevant to the substantive research area.

Nevertheless, it is possible to outline the essentials of parameter identification in models involving reciprocal effects. We begin with a model similar to that specified in equation 10.4, except that E_1 and E_2 are hypothesized to influence one another reciprocally. This new model is depicted in Figure 10.3 and is specified as follows:

$$E_1 = g_{11}F_1 - b_{12}E_2 + z_1 \tag{10.5a}$$
$$E_2 = g_{22}F_2 - b_{21}E_1 + z_2 \tag{10.5b}$$

or

$$\begin{bmatrix} 1 & b_{12} \\ b_{21} & 1 \end{bmatrix} \begin{bmatrix} E_1 \\ E_2 \end{bmatrix} = \begin{bmatrix} g_{11} & 0 \\ 0 & g_{22} \end{bmatrix} \begin{bmatrix} F_1 \\ F_2 \end{bmatrix} + \begin{bmatrix} z_1 \\ z_2 \end{bmatrix} \tag{10.6}$$

where $\underline{E}$, $\underline{F}$, and $\underline{z}$ have means of zero; $\underline{F}$ and $\underline{z}$ are uncorrelated; $\underline{PS}$ is diagonal $[\text{Cov}(z_1,z_2) = 0]$. The normal equations for this model are then:

$$\text{Cov}(E_1,F_1) = g_{11}\ \text{Var}(F_1) - b_{12}\ \text{Cov}(F_1,E_2) \tag{10.7a}$$
$$\text{Cov}(E_1,F_2) = g_{11}\ \text{Cov}(F_1,F_2) - b_{12}\ \text{Cov}(F_2,E_2) \tag{10.7b}$$
$$\text{Cov}(E_2,F_1) = g_{22}\ \text{Cov}(F_1,F_2) - b_{21}\ \text{Cov}(F_1,E_1) \tag{10.7c}$$
$$\text{Cov}(E_2,F_2) = g_{22}\ \text{Var}(F_2) - b_{21}\ \text{Cov}(F_2,E_1) \tag{10.7d}$$

and

$$\text{Cov}(E_1,E_2) = g_{11}g_{22}\ \text{Cov}(F_1,F_2) + b_{12}b_{21}\ \text{Cov}(E_1,E_2) - g_{11}b_{21}\ \text{Cov}(F_1,E_1) - g_{22}b_{12}\ \text{Cov}(F_2,E_2) \tag{10.7e}$$

and

$$\text{Var}(E_1) = g_{11}\ \text{Cov}(F_1,E_1) - b_{12}\ \text{Cov}(E_1,E_2) + \text{Cov}(E_1,z_1) \tag{10.7f}$$
$$\text{Var}(E_2) = g_{22}\ \text{Cov}(F_2,E_2) - b_{21}\ \text{Cov}(E_1,E_2) + \text{Cov}(E_2,z_2) \tag{10.7g}$$

The reader can verify that equations 10.7a and 10.7b identify g_{11} and b_{12}, and equations 10.7c and 10.7d identify g_{22} and b_{21}. Equations 10.7f and 10.7g then identify the variances of the residuals [since $\text{Var}(z_1) = \text{Cov}(E_1,z_1)$ and $\text{Var}(z_2) = \text{Cov}(E_2,z_2)$; note that equations 10.5 were multiplied by E_1 and E_2 rather than squaring as above. This solution is equivalent.] Equation 10.7e then provides an overidentifying restriction since all the parameters in that equation are identified by other equations. This overidentifying restriction provides a test of the null hypothesis that $\text{Cov}(z_1,z_2) = 0$

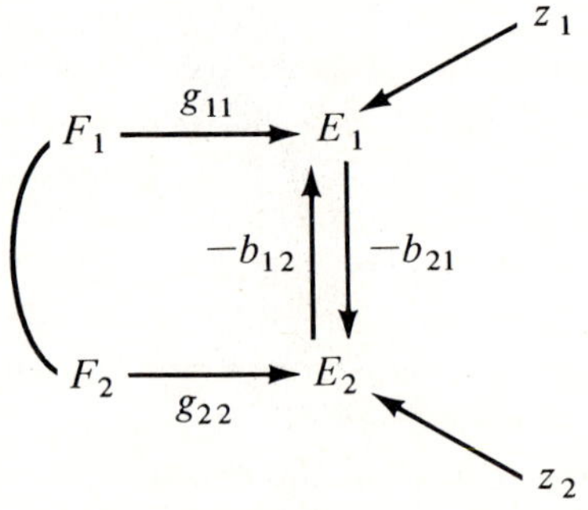

Figure 10.3 Path diagram of a structural model with reciprocal causation (see equations 10.5).

against the alternative hypothesis that the model holds in the population with a non-zero covariance between the two disturbances. If the disturbances are uncorrelated, it means that E_1 and E_2 are correlated *only* because F_1 and F_2 are correlated and because of the reciprocal effects between E_1 and E_2. If the disturbances are correlated, then it is presumed that factors independent of F_1 and F_2 contribute to the correlation between E_1 and E_2—either negatively or positively.

It is important to note at this point that the solution for the structural coefficients that is obtained from the above procedure is quite different than would be obtained if OLS procedures were applied. That is, if E_1 were regressed on F_1 and E_2, the regression coefficients obtained would be quite different from the estimates of g_{11} and $-b_{21}$. The reason for this difference is that OLS assumes a zero correlation between the residual and all the right-hand variables. In the model specified in equation 10.5a, in contrast, the disturbance z_1 is assumed to be correlated with E_2. This can be seen in the path diagram (Figure 10.3), in which a causal arrow leads from z_1 to E_1 and then from E_1 to E_2. This means that z_1 and E_2 are uncorrelated only if b_{21} is zero. Thus, when b_{21} is very different from zero, the application of OLS procedures to the estimation of g_{11} and b_{12} would lead to very biased and misleading estimates. For example, suppose that E_1 were regressed on F_1 and E_2:

$$E_1 = a_1F_1 + a_2E_2 + e$$

The OLS solution for a_2 is the now familiar formula

$$a_2 = \frac{\text{Cov}(E_1,E_2)\ \text{Var}(F_1) - \text{Cov}(F_1,E_1)\ \text{Cov}(F_1,E_2)}{\text{Var}(F_1)\ \text{Var}(E_2) - \text{Cov}(F_1,E_2)^2}$$

In contrast, the appropriate solution for $-b_{12}$ when the reciprocal effects model holds in the population is obtained from equations 10.7a and 10.7b as

$$-b_{12} = \frac{\text{Cov}(F_2,E_1)\ \text{Var}(F_1) - \text{Cov}(F_1,E_1)\ \text{Cov}(F_1,F_2)}{\text{Cov}(F_2,E_2)\ \text{Var}(F_1) - \text{Cov}(F_1,E_2)\ \text{Cov}(F_1,F_2)}$$

Although there is some similarity between the equations for a_2 and $-b_{12}$, the differences between the two are quite obvious. This little exercise demonstrates the importance of the assumption in OLS that the residuals are uncorrelated with all right-hand variables. If this assumption does not hold for the disturbances in a linear causal model, then the application of OLS can produce very misleading estimates. The more general structural equation model allows a researcher to construct a causal model in a form that is most realistic relative to the particular research tradition. Only experience with the models of a particular area will give the reader an adequate basis for judging whether or not a model is identified and whether or not the particular constraints necessary to attain identification are appropriate. For example, the reciprocal effects model described above includes the specification that F_1 does not affect E_2 and F_2 does not affect E_1. These assumptions may be reasonable in some situations, but not in others. Duncan's often cited example of a model with reciprocal effects is an especially good one (Duncan, Haller, and Portes, 1968). The reason is that E_1 and E_2 are values of two friends in high school, whereas F_1 and F_2 are characteristics of the friends' *respective* parents. Thus, it is quite plausible that the values of the one person are not affected by the income of the other's parents. In other situations, it may be necessary to align design and model characteristics in some alternative way to achieve identification of reciprocal effects.

Now that we have discussed the essential details of specifying a structural model, we can move to incorporate the type of measurement model developed in Chapter 9 with a structural model to specify a complete linear structural relations model.

10.4 MERGING MEASUREMENT AND STRUCTURAL MODELS

This section of Chapter 10 is the formal culmination of this book. The statistical model to be described in this section includes a measurement model for the exogenous variables and a measurement model for the endogenous variables of a structural model. The model is amenable to analysis with readily available computer programs, such as Joreskog's LISREL, and serves as a very general basis for causal modeling in the social and behavioral sciences.

The general model is comprised of three sets of equations, all of which have been introduced previously. The first set of equations defines the structural model:

$$\underline{B}\,\underline{E} = \underline{G}\,\underline{F} + \underline{z} \tag{10.8}$$

where $\underline{E}$ is a vector of m endogenous latent variables, $\underline{F}$ is a vector of n exogenous latent variables, and $\underline{z}$ is a vector of m disturbances that are uncorrelated with $\underline{F}$. The matrices $\underline{B}$ and $\underline{G}$ include the structural coefficients that define the pattern of causal relations in the model. The second set of equations specifies the measurement model for the observed indicators of the endogenous latent variables:

$$\underline{Y} = \underline{Ly}\,\underline{E} + \underline{e} \tag{10.9}$$

where $\underline{Y}$ is a vector of p observed indicators of $\underline{E}$, $\underline{Ly}$ is a $(p \times m)$ matrix of factor loadings, and $\underline{e}$ is a vector of p measurement errors. The third set of equations specifies the measurement model for the observed indicators of the exogenous latent variables:

$$\underline{X} = \underline{Lx}\,\underline{F} + \underline{d} \tag{10.10}$$

where $\underline{X}$ is a vector of q observed indicators of $\underline{F}$, $\underline{Lx}$ is a $(q \times n)$ matrix of factor loadings, and $\underline{d}$ is a vector of q measurement errors.

Now, in order to fully specify a structural model, we need to supply a bit more information than is supplied in equations 10.8 to 10.10. For example, we may want to specify certain patterns of covariance among the measurement errors or among the disturbances. The LISREL convention is to specify the model in terms of eight matrices, and we will follow that convention here. Four of these matrices are the coefficient matrices defined in the above equations: $\underline{B}$, $\underline{G}$, $\underline{Ly}$, and $\underline{Lx}$. The other four matrices used to specify a model are covariance matrices:

$$\text{Cov}(\underline{F}) = \underline{PH} \text{ (exogenous latent variables)}$$

$$\text{Cov}(\underline{z}) = \underline{PS} \text{ (disturbances)}$$

$$\text{Cov}(\underline{e}) = \underline{Te} \text{ (errors in endogenous indicators)}$$

$$\text{Cov}(\underline{d}) = \underline{Td} \text{ (errors in exogenous indicators)}$$

These eight matrices thus provide a great deal of flexibility in the specification of a causal model.

Table 10.1 provides a summary of the matrices and notation that will be used to specify causal models in Chapter 11. The table includes a listing of the Greek labels used in the LISREL program. For the most part, the notation used here parallels the notation used in that program. As far as the eight matrices crucial to specification of the model are concerned, this is always the case. Any reader who masters the notation used here will have little trouble using the LISREL program—especially version IV and its successors.

The variety of equation models that may be specified via these matrices is expanded by the capacity to place various constraints upon the parameters. The identified parameters of the model may be estimated as free parameters. Those parameters that are not identified can be fixed to some *a priori* value (e.g., zero or one) or involved in a constraint that allows identification. Another use of constraints and the "fixing" of parameters is their application to identified parameters for the purpose of testing a null hypothesis.

Version IV of LISREL only allows equality constraints (e.g., $t_{11} = t_{22} = t_{33}$). However, greater flexibility in the form of these constraints will probably be incorporated in subsequent versions of the program. In any case, it is important to distinguish between three categories of parameters in a structural equation model: (1) *free* parameters that are identified and estimated without constraint; (2) *fixed* parameters that are given an *a priori* value; and (3) *constrained* parameters that are estimated with the provision that the estimate satisfies a specified constraint.

10.5 MAXIMUM LIKELIHOOD ESTIMATION OF STRUCTURAL EQUATION MODELS

The procedure for fitting a matrix of observed variances and covariances to an hypothesized model is analogous to the procedure described in Chapter 9 for fitting a measurement model. The only difference is that constraints are placed upon the causal relations among the latent variables with the matrices $\underline{B}$ and $\underline{G}$. The same transformation of the likelihood function used to obtain maximum liklelihood estimates of measurement models is used to obtain estimates of parameters in a full structural equation model. In fact, the procedures outlined in Chapter 9 may be treated as a special case of the full model—the special case when only the exogenous measurement model was involved.

The transformed likelihood function given in equation 9.27 is repeated here:

$$H = \log[\det(\underline{Z})] + \mathrm{tr}[\underline{S}\ \mathrm{inv}(\underline{Z})] - \log[\det(\underline{S})] - u \qquad (10.11)$$

where

"log" means natural logarithm
"det" means determinant
"tr" means trace (sum of main diagonal)
"inv" means inverse

Table 10.1 Summary of notation for a complete LISREL model.

LISREL matrix	Matrix label	Matrix dimension	Element label	Definition
Variable vectors				
eta	$\underline{E}$	$(m \times 1)$	E_i	Latent endogenous
xi	$\underline{F}$	$(n \times 1)$	F_i	Latent exogenous
zeta	$\underline{z}$	$(m \times 1)$	z_i	Disturbance
epsilon	$\underline{e}$	$(p \times 1)$	e_i	Error in $\underline{Y}$
delta	$\underline{d}$	$(q \times 1)$	d_i	Error in $\underline{X}$
$\underline{Y}$	$\underline{Y}$	$(p \times 1)$	Y_i	Observed endogenous
$\underline{X}$	$\underline{X}$	$(q \times 1)$	X_i	Observed exogenous
Coefficient matrices				
lambda Y	$\underline{Ly}$	$(p \times m)$	j	$\underline{Y}$ loadings
lambda X	$\underline{Lx}$	$(q \times n)$	k	$\underline{X}$ loadings
beta	$\underline{B}$	$(m \times m)$	b	Endogenous paths
gamma	$\underline{G}$	$(m \times n)$	g	Exogenous paths
Covariance matrices				
phi	$\underline{PH}$	$(n \times n)$	ph	Cov($\underline{F}$)
psi	$\underline{PS}$	$(m \times m)$	ps	Cov($\underline{z}$)
theta epsilon	$\underline{Te}$	$(p \times p)$	s	Cov($\underline{e}$)
theta delta	$\underline{Td}$	$(q \times q)$	t	Cov($\underline{d}$)

Equation model:

$$\underline{B}\underline{E} = \underline{G}\underline{F} + \underline{z}$$
$$\underline{Y} = \underline{Ly}\underline{E} + \underline{e}$$
$$\underline{X} = \underline{Lx}\underline{F} + \underline{d}$$

where
$$\underline{B} = \begin{bmatrix} b_{11} & b_{12} & \cdots & b_{1m} \\ b_{21} & b_{22} & \cdots & b_{2m} \\ \vdots & & & \\ bm_{1} & b_{m2} & \cdots & b_{mm} \end{bmatrix}$$

$$\underline{E} = \begin{bmatrix} E_1 \\ E_2 \\ \vdots \\ E_m \end{bmatrix} = [E_1 \; E_2 \; \ldots \; E_m]'$$

$$\underline{G} = \begin{bmatrix} g_{11} & g_{12} & \cdots & g_{1n} \\ g_{21} & g_{22} & \cdots & g_{2n} \\ \vdots & & & \\ g_{m1} & g_{m2} & \cdots & g_{mn} \end{bmatrix}$$

$$\underline{F} = \begin{bmatrix} F_1 \\ F_2 \\ \vdots \\ F_n \end{bmatrix} = [F_1 \; F_2 \; \ldots \; F_n]'$$

$$\underline{z} = \begin{bmatrix} z_1 \\ z_2 \\ \vdots \\ z_m \end{bmatrix} = [z_1 \; z_2 \; \ldots \; z_m]'$$

$$\underline{Y} = [Y_1 \; Y_2 \; \ldots \; Y_p]'$$

$$\underline{Ly} = \begin{bmatrix} j_{11} & j_{12} & \cdots & j_{1m} \\ j_{21} & j_{22} & \cdots & j_{2m} \\ \vdots & & & \\ j_{p1} & j_{p2} & \cdots & j_{pm} \end{bmatrix}$$

LISREL matrix	Matrix label	Matrix dimension	Element label	Definition
$\underline{e} = [e_1 \ e_2 \ \ldots \ e_p]'$				
$\underline{X} = [X_1 \ X_2 \ \ldots \ X_q]'$				
$\underline{Lx} = \begin{bmatrix} k_{11} & k_{12} & \ldots & k_{1n} \\ k_{21} & k_{22} & \ldots & k_{2n} \\ \vdots & & & \\ k_{q1} & k_{q2} & \ldots & k_{qn} \end{bmatrix}$				
$\underline{d} = [d_1 \ d_2 \ \ldots \ d_q]'$				
$\underline{PH} = \text{Cov}(\underline{F}) = \text{expectation}(\underline{F}\underline{F}')$ $= \begin{bmatrix} ph_{11} & ph_{12} & \ldots & ph_{1n} \\ ph_{21} & ph_{22} & \ldots & ph_{2n} \\ \vdots & & & \\ ph_{n1} & ph_{n2} & \ldots & ph_{nn} \end{bmatrix}$				
$\underline{PS} = \text{Cov}(\underline{z}) = \text{expectation}(\underline{z}\underline{z}')$ $= \begin{bmatrix} ps_{11} & ps_{12} & \ldots & ps_{1m} \\ ps_{21} & ps_{22} & \ldots & ps_{2m} \\ \vdots & & & \\ ps_{m1} & ps_{m2} & \ldots & ps_{mm} \end{bmatrix}$				
$\underline{Te} = \text{Cov}(\underline{e}) = \text{expectation}(\underline{e}\underline{e}')$ $= \begin{bmatrix} s_{11} & s_{12} & \ldots & s_{1p} \\ s_{21} & s_{22} & \ldots & s_{2p} \\ \vdots & & & \\ s_{p1} & s_{p2} & \ldots & s_{pp} \end{bmatrix}$				
$\underline{Td} = \text{Cov}(\underline{d}) = \text{expectation}(\underline{d}\underline{d}')$ $= \begin{bmatrix} t_{11} & t_{12} & \ldots & t_{1q} \\ t_{21} & t_{22} & \ldots & t_{2q} \\ \vdots & & & \\ t_{q1} & t_{q2} & \ldots & t_{qq} \end{bmatrix}$				

$u = p + q$ = total number of observed indicators

$\underline{S}$ = unbiased sample estimate of covariance matrix of observed indicators

$\underline{Z}$ = estimate of covariance matrix of observed indicators based on estimates of model parameters

Minimization of this function is used to obtain estimates of model parameters from sample data. If the observed indicators are multivariate normal in distribution, then these estimates are efficient maximum likelihood estimates of the population values. This means that as the sample size increases, the sample estimates become closer to the population values in an almost ideal way.

Furthermore, the minimum value of H can be used to compute a likelihood ratio chi-square statistic that reflects the probability of the sample observations if the model holds in the population. This chi-square statistic is given by the relation

$$\text{chi-square} = \min(H)(N/2) \tag{10.12}$$

with

$$df = [(p + q)(p + q - 1)/2 + (p + q)] - f$$

where

$$\begin{aligned} N &= \text{sample size} \\ p + q &= \text{number of observed indicators} \\ f &= \text{number of free parameters estimated} \end{aligned}$$

Note that the degrees of freedom for the chi-square are equal to the difference between the number of observed variances/covariances and the number of free parameters estimated. This chi-square statistic can then be used to compute standard errors and contrast the fit of alternative models, as outlined in Chapter 9 with measurement models.

10.6 LIMITATIONS OF LINEAR MODELS

It is important that users of structural equation models understand the limitations of these models—especially the requirement of linearity. The assumption of linearity is misunderstood in two directions. On the one hand, the generality of the linear model is often underestimated by researchers because the use of "linearizing transformations" is overlooked. On the other hand, other researchers are rather cavalier about the assumption of linearity and fail to investigate scatterplots of bivariate relationships as an indispensable prerequisite to the use of linear models.

The introduction of multiple indicator models for research areas that necessarily measure variables with error makes the linearity assumption a bit more troublesome. This is the case because researchers are not free to investigate scatterplots for latent variables, nor are they able to explore various transformations of latent variables. And they are unable to do these things because scores on the latent variables cannot be computed (factor scores can be estimated, but they always involve measurement error). This means that only a very restricted category of nonlinearities can be accommodated within the most general structural equation model—i.e., one that involves measurement models as well. In order to understand these restrictions, consider the following example.

Suppose that variability in the latent variable E is determined by changes in the latent variable G and a disturbance z that is uncorrelated with G. Now, suppose that changes in E are a logarithmic function of changes in G such that

$$E = b_0 + b_1 \ln(G) + z$$

where $\ln(G)$ is the natural log of G and $\mathrm{Cov}[\ln(G),z]$ is zero. If E and G were observed without error, then this linearized model would present no problems for analysis by the usual methods. However, now suppose that E and G are measured with error by three indicators each (Y_1, Y_2, Y_3 and X_1, X_2, X_3). And further suppose that the relations between the latent variables and their indicators are linear (without any transformation). In this instance, there is no straightforward procedure for estimating any relation between E and G other than a linear one. And, to the extent that the true relationship is nonlinear, the linear model will give misleading results.

The second problem that arises with the use of multiple indicator models is the modeling of *interactions* that involve three latent variables. Suppose, for example, that the causal relationship between E, G_1, and G_2 is described by the following model:

$$\begin{aligned} E &= b_1G_1 + b_2G_2 + b_3(G_1G_2) + z \\ &= b_1G_1 + (b_2 + b_3G_1)G_2 + z \end{aligned}$$

Note that this model depicts the effect of changes in G_2 on E as a linear function of G_1—i.e., the effect of a unit increase in G_2 on E is $(b_2 + b_3G_1)$. If G_1 and G_2 are both measured indirectly via multiple indicators, then the estimation of this model presents serious problems. However, if, for example, levels of G_2 define separate populations, and population membership is measured without error, then the interaction between G_1 and G_2 can be estimated even though G_1 and E are measured with multiple indicators. An example of such an analysis is presented in Chapter 11 in the section dealing with nonequivalent control group designs.

In closing, then, it is important to emphasize that the use of latent variable structural equation models places certain limitations on a researcher. These limitations arise primarily because of the assumptions concerning linearity that are built into the measurement models. If researchers want to explore nonlinear relations among variables, they have two choices. First, study variables that have well-behaved nonlinearities—no interactions, latent variables that relate to one another and their indicators via power functions, and so on. Second, devote sufficient attention to the development of measurement instruments that measurement error can be reduced to less than a few percent of the variance of the measure. The distastefulness of each of these choices is, I believe, a realistic indication of the difficulties that confront the researcher who seriously plans to do quantitative work in the social and behavioral sciences.

REFERENCES AND SUGGESTED READING

Bentler, P. M. Multivariate analysis with latent variables. *Annual Review of Psychology*, 1980, *31*, 419–456.

Bielby, W. T., and Hauser, R. M. Structural equation models. *Annual Review of Sociology*, 1977, *3*, 137–161.

Duncan, O. D. *Introduction to structural equation models*. New York: Academic Press, 1975.

Duncan, O. D., Haller, A. O., and Portes, A. Peer influences on aspirations: A reinterpretation. *American Journal of Sociology*, 1968, *74*, 119–137.

Goldberger, A. S. Structural equation methods in the social sciences. *Econometrica*, 1972, *40*, 979–1000.

Heise, D. R. *Causal analysis*. New York: Wiley, 1975.

Joreskog, K. G. A general method for estimating a linear structural equation system. In A. S.

Goldberger and O. D. Duncan (eds.), *Structural equation models in the social sciences.* New York: Seminar Press, 1973.

Joreskog, K. G. Structural analysis of covariance and correlation matrices. *Psychometrika,* 1978, *43,* 443–477.

Joreskog, K. G., and Sorbom, D. *LISREL IV: Analysis of linear structural relationships by the method of maximum likelihood: A user's guide.* Chicago: International Educational Services, 1978.

Kenny, D. *Correlation and causality.* New York: Wiley, 1980.

Saris, W. Structural models. Free University of Amsterdam, mimeograph, 1980.

11

Application of Structural Equation Models to Causal Inference from Longitudinal Data

11.1 INTRODUCTION TO THE LONGITUDINAL DESIGN

There are two types of research designs that involve the observation of change over time. One is termed *longitudinal,* the other *time series.* Longitudinal (or panel) designs are characterized by observation of a sample of cases, such as persons, at two to ten or fifteen points in time. In contrast, the time series design usually involves observations on a single case at as many time points as possible (see McCleary and Hay, 1980).

The most common type of study that utilizes a longitudinal design is the panel survey. For example, the U.S. Department of Labor conducts a large survey of labor market activity known as the National Longitudinal Survey (NLS). Each participant in the survey is interviewed every two years (roughly). This means that information concerning employment, wages, and attitudes has been gathered from the same sample of persons in 1967, 1969, 1971, and so on. The use of a longitudinal design is motivated by the greater confidence gained from observed change rather than inferred change.

The need for nonexperimental methods of causal inference in the social and behavioral sciences arises for two kinds of reasons. The first kind of reason is that random manipulation of an independent variable is often impractical, unethical, or both. Investigation of the psychotropic effects of various compounds in humans is a good example of a relationship that may be unethical to pursue experimentally. Similarly, experimental assessment of the effect of rearing the children of middle-class parents in poverty families is both impractical and unethical. The second problem with true experiments in the social and behavioral sciences is more subtle—and is probably the more important. This second reason for seeking nonexperimental procedures for causal inference is that manipulation of an independent variable in a complex system can raise more serious obstacles to causal inference than are raised by the inherent weaknesses of nonexperiments. More specifically, as a social scientist draws closer and closer to the processes that are of interest, it is often the case that manipulation of a specific independent variable becomes more and more difficult. In fact, there are many independent variables that are of crucial importance for which the notion of manipulation does not even make good sense. One begins to get the feeling that the requirement of randomized manipulation is being imposed in a context that is not appropriate. The very idea of manipulating *only one* variable begins to sound absurd in the context of a social system in which all parts are interdependent.

Research on the effects of television violence on aggression in children is a good example of a substantive area in which the notion of manipulation has run aground. At first thought, it may seem quite straightforward to manipulate the amount of violent television that a number of groups of randomly selected children view. However, then consider that television viewing is an activity that usually involves choice on the part of the child. And that choice is influenced by advertising, comments of peers, family values, and so on. Now, if the television programs watched are manipulated by an experimenter, what is being manipulated? Are observed changes in the child due to a reduction in the amount of violent television viewing or to resentment over being restricted from viewing the programs that peers are viewing? Is there any way to manipulate the child's television viewing that does not confound the effect of program content with the effect of disrupting choice based on the child's interrelation with numerous other social actors? There is a way, but it would require manipulation of program content on a major social level—such that social subsystems, such as cities, became the unit of analysis. Although such experiments are not impossible, it would be unrealistic for social scientists to make the manipulation of entire cities the routine basis for empirical tests of their hypotheses. Manipulation on such a grand scale requires enormous political power and is likely to remain the perquisite of government and large corporations—not social scientists. Furthermore, should social scientists achieve the opportunity to manipulate at the level of social systems, there would, one hopes, be a strong basis for pursuing such an experiment, and that basis would have to be provided by nonexperimental research.

It is my hope that the reader of this chapter will achieve a balanced appreciation for both the strengths and the weaknesses of passive longitudinal research designs. By balanced I mean the avoidance of two extreme positions: on the one hand, the belief that the application of structural models to longitudinal data provide a basis for causal inference that rivals the true experiment; on the other hand, the belief that passive longitudinal data are of no greater value than cross-sectional correlations. The middle ground urged here includes the realization that no passive design rivals the true experiment in causal attributional confidence. But it also includes the argument that longitudinal designs have several advantages over cross-sectional designs that place them in a kind of "middle ground" between the passive cross-sectional and true experimental designs. These advantages include the measurement of change for each case rather than the inference of change from differences between cases, the greater plausibility of assuming that certain structural coefficients are zero, and the opportunity to test the fit of "restricted spuriousness" models before testing models that assume causal relations. These advantages lead to the possibility of gaining some insight into the *directionality* of observed correlations. That is, *given a particular model of change over time,* it is possible to assess the relative extent to which X shapes Y and Y shapes X—even though, as we shall see, acceptance of a given model of change includes many assumptions that must be investigated with some thoroughness in any particular substantive research area. In the end, the statement that "X shapes Y to a certain extent" must be tempered to "*X or some factor closely correlated with X* shapes Y to a certain extent." These advantages over cross-sectional designs are important, but they do not rival the experiment, since some model that incorporates untested assumptions must be assumed in the case in which randomization is not incorporated. In the final analysis, these advantages of passive longitudinal designs over passive cross-sectional designs simply consist of predictions involving *patterns* of intercorrelations. It is argued that prediction of a complex and specific pattern of intercorrelations is a more powerful test of an hypothesis than prediction of a simple association. However, when the most convincing evidence of an hypothesized causal relation in a social system is sought, and randomized manipulation is possible, there is no alternative to actually producing change in the hypothesized cause. (The reader should also recall the argument presented briefly in the preface concerning the capacity of experiments to demonstrate that an effect *can* occur, whereas passive designs are more likely to capture causal relations that *are* occurring.)

Although longitudinal designs allow rejection of what are called "restricted" spuriousness (or unmeasured third variable) models, there is always an additional third variable alternative that would require additional data to reject. If a randomized manipulation is sensible and possible, then a wholesale rejection of all spurious alternatives is logically—although not practically—achievable.

An example of a research area in which the contribution of true experimental and quasi-experimental studies is neatly contrasted is that concerning the efficacy of mastectomies in the treatment of breast cancer in women. A review of the studies comparing the outcomes of radical and more conservative surgical procedures was published by *Consumer Reports* in 1980. The authors were careful to point out that the studies in which women were randomly assigned to the radical or conservative treatment groups provided the strongest evidence, since patient and surgeon characteristics could not be correlated with group assignment. However, prior to such studies being ethically possible, it was necessary for patients to voluntarily opt for the less traditional treatment (the less radical surgery). The results of these nonequivalent control group studies (since patients choosing the less traditional treatment may differ from other patients in some way that affected their subsequent health) indicated that those patients choosing the less radical surgery fared as well as those treated with the more radical surgery. Interestingly, the subsequent randomized studies found the same result.

Another therapeutic example that has a different ending concerns the evaluation of psychotropic drugs. Here the introduction of "double blind" randomized designs—in which placebos were compared with the drug—revealed that apparent effects in quasi-experiments were due to expectations on the part of physicians and the patients.

The remainder of this chapter progresses as follows: After reviewing the concepts of *entropic decay* and *temporal inertia* in the context of time lagged observation of individual and social characteristics, a general model for longitudinal designs is presented. This model is then applied to analysis of (1) randomized experiments, (2) the nonequivalent control group design, and (3) the passive observational design. In each case, the same basic model is applicable with slight adaptations to particular situations. The role of measurement error is crucial in all these contexts and is dealt with using the measurement models presented in Chapter 9. (Such measurement models are often implicit rather than explicit, so the reader should be familiar with Chapter 9 before proceeding.) All models are then fit to data using the maximum likelihood procedures and the LISREL model described in Chapter 10.

Distinguishing Types of Longitudinal Designs: The Randomized Experiment, the Nonequivalent Control Group Design, and the Passive Observational Design

It is of some interest to list the assumptions that are incorporated in a causal inference from data derived in studies that proceed according to various designs. Such an exercise provides the most realistic picture of the advantages and disadvantages of passive longitudinal designs relative to other research designs—in which confidence in causal inferences are concerned.

We begin by identifying some assumptions that are necessary to interpret a linear coefficient as a structural parameter when a passive cross-sectional design generates the observations. Suppose, for example, that Y and $(X_1, \ldots, X_q)$ are observed at time 1. Now, under what circumstances would a linear model of the form

$$Y = b_0 + b_1X_1 + b_2X_2 + \ldots + b_qX_q + z$$

be both efficiently estimable from sample data *and* descriptive of a structural (causal) process—in which manipulated changes in X_i would actually produce the change in Y that is indicated by b_i? One such set of assumptions is the following:

1. There is no causal effect of changes in Y on X_i.
2. There are no third variables that have been "left out" of the model that influence Y and that are correlated with X_i (or, more strictly, no third variables influencing Y that have a nonzero partial correlation with X_i—partialling for the other variables included as causal variables in the model).
3. The interdependencies among the variables in the model are satisfactorily described by linear equations—with appropriate "linearizing" transformations when necessary.
4. The right-hand variables are measured without error.

Note that assumptions (1) to (3) are equivalent to saying that z is independent (uncorrelated, linearly or otherwise) of the variables X_i.

Note also that these assumptions include the assumption that differences between cases on the left- and right-hand variables are equivalent to *changes* in those variables. In this sense, change is unobserved and is inferred from differences between cases. This presumption usually includes the additional assumption that the observed differences between cases developed over time and that this temporal process has some time invariant properties (e.g., the magnitude of certain causal effects).

Passive *longitudinal* designs have several characteristics that can give them an advantaged status relative to passive cross-sectional designs. These advantages are as follows:

1. In the use of time lags, the assumption can be made that causal effects are unidirectional (i.e., that a change in Y at a later time cannot affect X at an earlier time) more plausible.
2. Some "third variable" models of spuriousness can be tested without actually measuring or identifying the third variable (we shall refer to such models as restricted null hypotheses).
3. Change in variables can actually be measured for each case rather than inferring change from differences among cases. This eliminates some uncertainty (as long as the intrusion of multiple measurement points does not disrupt the system being studied).

However, it is important to point out that *not all* third variable alternatives can be rejected with data from passive longitudinal designs, that time lags do not necessarily guarantee causal unidirectionality, and that assumptions of linearity and measurement accuracy are still required. Furthermore, all causal inferences from passive longitudinal designs require certain assumptions about the way in which relationships change over time. These assumptions must be explored in specific research areas and are thus much less categorical than those required for causal inference from randomized experiments.

The longitudinal quasi-experiment (i.e., the non-equivalent control group design with pre and post measures) gives to the passive longitudinal design the advantage of actually inducing the change in one or more independent variables. Here the independent variable(s) is manipulated, but not randomized. The evaluations of Head Start programs often involved such designs. Although it is quite clear in these instances that changes in the dependent variable do not influence the manipulation, we are nevertheless left with uncertainty about "third variables" that may be correlated with the manipulation and causal relative to the dependent variable. See Figure 11.1 for path diagrams of structural equation models applied to various research designs.

Finally, the true experiment overcomes all these threats to the internal validity of a causal inference from a data collection venture. The true experiment need not even be longitudinal, since association between a randomized independent variable and a dependent variable can arise only by chance or because of a causal effect. (The reader should keep in mind, of course, that actual experiments are subject to any number of misinterpretations, inaccurate executions, reactivity of measures, and so on. The above discussion focuses on the more formal categories of threats to internal validity.)

The Problem with Causal Inference from Cross-Lagged Correlations

Our primary concern in this chapter is with analysis of passive longitudinal and non-equivalent control group designs. The most prevalent misunderstanding about these designs is the extent to which a certain finding eliminates alternative explanations to the favored causal hypothesis. For example, the method of comparing cross-lagged correlations that was advanced by Donald Campbell and then formalized by David Kenny led many people to conclude that an observed difference between cross-lagged correlations indicated a lagged causal relation (see Figure 11.2). In other words, this

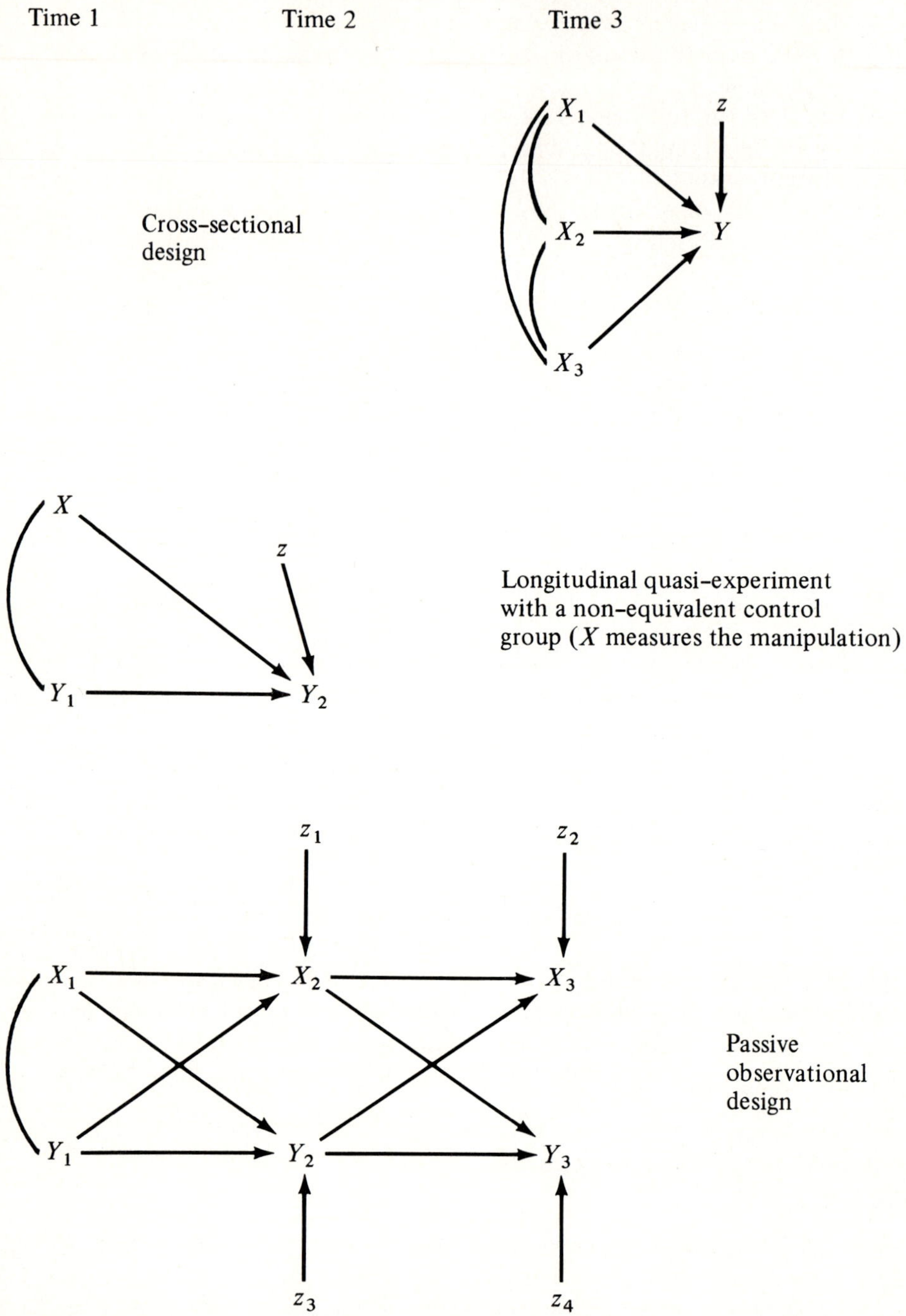

Figure 11.1 Path diagrams of typical structural models tested with data from three types of research designs.

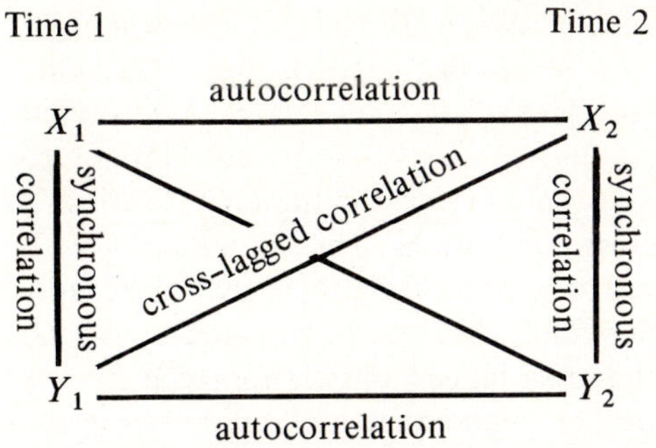

Figure 11.2 Diagram of intercorrelations among variables observed for an analysis of cross-lagged correlations. The main object of interest in the analysis of cross-lagged panel correlations is the difference between Corr(Y_1, X_2) and Corr(X_1, Y_2). The variable X might be social threat, the variable Y might be authoritarianism.

finding was believed to eliminate all (or most) alternative explanations. This is *far* from the case. (In fact, Kenny was very careful to point out that a difference between cross-lagged correlations was sufficient only to reject a *certain* spuriousness model—the "synchronous common factor model." However, his discussions of the method and certainly its application by researchers are characterized by a lack of appreciation for the numerous alternative explanations that remain quite plausible in spite of even substantial differences between cross-lagged correlations. As we shall see in Section 11.6, Kenny's important contribution on structural equation models for longitudinal processes was the insight that certain spuriousness models could be rejected with longitudinal data.) And the fact that this is far from the case is made apparent if a structural model is actually specified. The failure to specify such models has produced an exaggerated sense of the attributional power of the design.

Much work is needed on actual structural models with lagged causal effects. Such effects are clearly more realistic for analysis of social psychological systems (the real world). However, there is no empirical finding in such a situation that provides the confidence of causal inference that stems from the true experiment. Only specific models can be rejected, and any model that is consistent with the data is only one of many. Once a model has survived numerous tests with passive data, researchers should move on to manipulative designs if this is at all possible (and often it is not). When this is not possible, structural models must be elaborated and measurement procedures improved. And one of the best antidotes against overinterpretation is explicit specification of a structural model, as outlined in this chapter. The structural model makes the assumptions necessary for a causal inference painfully obvious to the researcher.

It is my impression that much of the confusion concerning causal inference from passive longitudinal designs stems from lack of clarity concerning the role of randomization in experimentation. As has been argued many times in this book, a correlation between a randomized independent variable and any other variable is potent evidence for causality. However, in the passive observational case, interpretation of any correlation must be based upon assumptions (which are part of a model) that are always more risky than the assumption that a randomized variable is uncorrelated with other factors influencing the dependent variable of interest. For example, a difference in "cross-lagged correlations" *could* be indicative of a lagged causal effect if a particular model actually holds (say Kenny's lagged factor model). However, if another model holds, then a difference between cross-lagged correlations could be consistent with causation in the opposite direction or a completely spurious explanation of the synchronous and lagged cross correlations. The beginning of wisdom in causal inference from passive longitudinal data is the realization that an accepted model is only as good as the generality of the models that can be rejected by the same data. (The primary error of the users of the cross-lagged correlation technique was to overestimate the generality of the spuriousness model they had sought to reject. They tended to confuse this spuriousness model with the null hypothesis in models fit to randomized experiments.)

For an excellent example of a published study making careful use of structural equation models in the context of a passive longitudinal design, the reader is referred to Bentler and Speckart (1979).

11.2 TEMPORAL INERTIA AND ENTROPIC DECAY

A crucial concept in the construction of longitudinal causal models is the tendency of most variables to have some stability over time—but not perfect stability. Borrowing from physics, the term *temporal inertia* is used to refer to the tendency of a characteristic of an individual or a social entity to remain stable over time. This temporal inertia is said to be less than complete to the extent that *entropic decay* occurs—borrowing once again from physics. The tendency of a set of cases ordered on some characteristic to undergo a random change in that ordering that increases with time manifests entropic decay, whereas a relation between an earlier and a later ordering manifests temporal inertia. Some longitudinal models include a component process in which certain types of correlations are expected to *decay* over time, whereas other models do not.

The choice of when to incorporate such a component in a model—and when *not* to—is a problem that has generated a considerable amount of confusion among methodologists. In fact, the apparent ambiguities in this area have been given a special label: Lord's paradox. This label has nothing to do with religion, however; it has acquired this auspicious title due to an article authored by Frederick Lord (1967). Lord was puzzled by the differing conclusions that could be drawn from longitudinal data analyzed with a linear model that did, or did not, incorporate an entropic decay component. Lord's paradox was one of semantics, but it is a common confusion that raises the most fundamental questions about causal inference from data generated by longitudinal designs in which randomized manipulation is not included or in which the randomization process has failed to produce equivalent groups (for reasons that are usually unknown). Lord's paradox is stated most often in terms of the choice between analysis of unconditional change scores and the analysis of "residualized" change scores. (In the latter case, scores at an earlier time are treated as a covariate.) As we shall see, the unconditional change score procedure is appropriate for testing the hypothesis that a certain type of equilibrium has been disturbed, whereas the residualized (conditional) change score procedure is appropriate for the estimation of actual structural models of longitudinal processes.

Temporal Inertia of Individual Characteristics

The best way to understand the longitudinal process of temporal inertia is via a concrete example in which the process can be "seen." For our example we will use data from the National Longitudinal Survey (NLS) of older men that was conducted by the University of Ohio for the U.S. Department of Labor. An 11-item version of Julian Rotter's measure of "locus of control" was administered to a sample of middle-aged men in 1969. The same scale was administered in 1971 to the same sample of middle-aged men. The scale consisted of 11 forced choice items: one choice was of the external type (indicating that the course of life is externally determined) and one was of the internal type (indicating that the individual determines the course of his life). In addition, there were two levels of preference for the chosen statement: the statement is *somewhat* closer to the respondent's belief than the other statement or *much* closer. For example, the first item on the scale was the following:

Statement 1: The misfortunes in people's lives are the result of mistakes that they make.
Statement 2: The unhappy things in people's lives are mostly due to bad luck.

The first of these statements is assumed to be in the internal direction, whereas the second one is in the external direction. Strong agreement with the internal statement is scored 1, and strong agreement with the external statement is scored 4; moderate agreement with the internal statement is scored 2, and moderate agreement with the external statement is scored 3.

This scoring scheme means that the scale ranges from 11 to 44 for the sum of 11 items. A score of 11 is the most extreme internal score, whereas 44 is the most extreme external score.

The means and standard deviations for the scale are given in Table 11.1. These statistics are given for 1969 and 1971 for black and white respondents.

Now, if the individual characteristic measured by the Rotter scale (presumably locus of control) behaves in a manner consistent with the assumption that this characteristic has temporal inertia, then we would expect that there would be some relation

Table 11.1 Means and standard deviations for Rotter's locus of control scale (from the NLS of older men) for 1969 and 1971.

	1969		1971	
	Mean	SD	Mean	SD
White men (N = 1992)	22.049	5.649	22.485	5.843
Black men (N = 678)	24.802	5.601	25.118	5.820

Note. Sample sizes are determined by the number of respondents with complete data on four variables (two of which have not yet been introduced).

between scores in 1969 and scores in 1971. More specifically, we might expect those persons with a high score in 1969 to be more likely to have a high score in 1971 than respondents with a low score in 1969. The fact that scores on the locus of control scale are consistent with such a mechanism may be seen in Table 11.2. Note, for example, that the 37 respondents with a score of 11 in 1969 (the lowest and most extreme internal score) have a mean score of 15.14 in 1971. Similarly, respondents with a 1969 score of 33 (a highly external score) averaged 28.88 in 1971. Clearly, these scores appear to evidence some kind of inertial tendency to remain stable over time, although it is also clear that some kind of disturbing process is also taking place, since the scores are changing.

Table 11.2 Regression to the mean for white respondents to the locus of control scale in the NLS of older men.

1969 score	*N*	Mean 1971 score	SD
11	37	15.14	3.90
12	25	17.12	4.95
13	37	17.54	4.60
14	77	18.70	5.02
15	75	18.44	4.61
16	78	17.95	3.52
17	129	19.95	4.64
18	125	19.59	4.03
19	126	20.03	4.35
20	125	21.84	5.15
21	130	21.34	3.99
22	131	22.67	4.66
23	125	23.20	5.01
24	116	23.88	4.78
25	113	24.55	5.31
26	123	25.02	4.67
27	86	25.86	5.92
28	72	25.13	5.11
29	64	27.95	5.36
30	49	26.69	5.46
31	31	28.52	5.43
32	34	27.24	6.70
33	33	28.88	5.21
34	18	27.44	5.26
35	9	30.89	6.03
36	4	31.75	6.70
37	5	32.00	3.24
38	6	24.83	4.36
39	1	35.00	0.00
40	3	35.00	3.61
41	4	31.25	4.50
42	1	21.00	0.00
43	0	...	...
44	0	...	...
Total	1992		

There is a rather interesting pattern to the stability and change (i.e., inertia and entropy) that we see in Table 11.2. As we select groups in 1969 with scores closer and closer to the 1969 mean, the amount of average change over the two-year period decreases in magnitude—but note that for almost every 1969 score with a sizable number of respondents, the subsequent 1971 mean is closer to the 1971 mean than the 1969 score was to the 1969 mean.

The next step is to ask ourselves what kind of a causal model would be consistent with this phenomena of time-lagged inertia and entropy. One of the simplest of such models is a linear model in which an entropic disturbance produces "regression to the mean" between time 1 and time 2, whereas an inertial term produces some stability over time. For example,

$$Y_2 = b_0 + b_1 X + b_2 Y_1 + z \tag{11.1}$$

where

Y_2 = Rotter scale score in 1971

Y_1 = Rotter scale score in 1969

X = unmeasured factors that affect Y over the two-year time lag

z = a disturbance that is independent of X and Y_1

is a linear model that could be estimated with OLS *if* X were measured without error. This model depicts Y at time 2 (Y_2) as a function of some unspecified factors X, Y at time 1 (Y_1), and a random disturbance (z). The model is depicted in Figure 11.3. The coefficient b_2 represents the inertia of the Y variable over the two-year period, whereas the variance of the disturbance [Var(z)] depicts the amount of entropic change that would occur between time 1 and time 2—even if the determinants X did not change, or if b_1 (the effect of changes in X) were zero.

The reader should not underestimate the importance of including a stochastic disturbance in this causal model. Such a component in the model means that change in Y will occur in the absence of systematic effects. Furthermore, if the linear form of the model holds, then entropic decay of Y will occur over time. This means that the

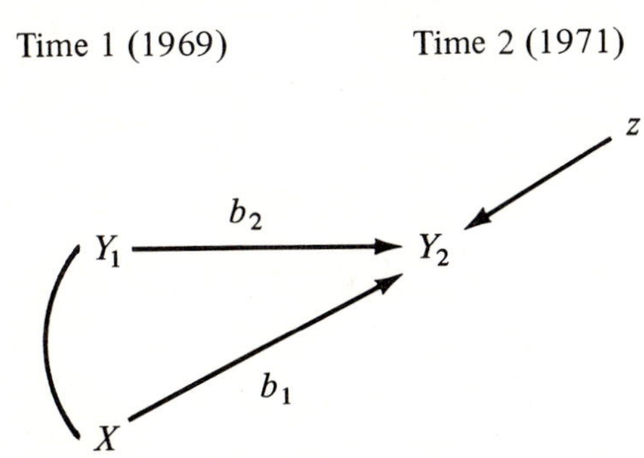

Figure 11.3 Path diagram of the longitudinal model specified in equation 11.1. The variable Y is a personality characteristic, X is a set of factors that influence Y over the observation time lag.

absence of change over time can be evidence of the *presence* of a force producing some kind of stability or equilibrium over time. (As we shall see, an understanding of this fact is the key to Lord's paradox.)

In order to provide a better understanding of how this linear model is suggested by the data in Table 11.2, it is useful to rewrite equation 11.1 in a different form. This different form is obtained by subtracting Y_1 from both sides of equation 11.1 and then applying some simple algebra:

$$\begin{aligned} Y_2 - Y_1 &= b_0 + b_1 X + b_2 Y_1 - Y_1 + z \\ &= b_0 + b_1 X + (b_2 - 1) Y_1 + z \end{aligned} \tag{11.2}$$

This form of the linear model reveals that *change* in Y (i.e., $Y_2 - Y_1$) is a linear function of X and Y_1. In fact, since we would expect the inertial coefficient b_2 to be positive and less than one (if the variance of Y is approximately the same at time 1 and time 2), equation 11.2 depicts change in Y to be a *negative* function of Y_1—since $(b_2 - 1)$ would be a negative number. In fact, this is precisely the pattern that is apparent in Table 11.2. Notice that the average change in each row of the table *decreases* steadily as the time 1 score *increases*. For example, respondents with a 1969 score of 12 changed an average of +5.12 points (17.12 − 12), while respondents with a 1969 score of 33 changed an average of −4.22 points (28.88 − 33).

Now, how well does this linear model "fit" the observed changes in the Rotter scale scores given in Table 11.2? In order to assess this fit, we first use OLS to estimate the coefficients in the linear equation and the R-square. [For the sake of simplicity, we will presume at this point that b_1 is zero or that X and Y_1 are uncorrelated. In the latter case z will include $(b_1 X)$ and b_0 absorbs the shift in intercept.]

$$Y_2 = 9.787 + .576 Y_1 + z \tag{11.3}$$

where

$$\text{R-square} = .310 \qquad (R = .557)$$

or

$$\begin{aligned} Y_2 - Y_1 &= 9.787 + (.576 - 1) Y_1 + z \\ &= 9.787 - .424 Y_1 + z \end{aligned}$$

In words, this result means that change in Y is described by this model as *averaging* $(9.787 - .424 Y_1)$. This means that when Y_1 is 11, the predicted change in Y is

$$9.787 - (.424)(11) = 5.123$$

Thus, the expected 1971 score for persons with a score of 11 in 1969 would be (11 + 5.123) or 16.123 according to this *linear* model of change in Y. This expected score could also be obtained from the equation

$$Y_2 = 9.787 + .576 Y_1 + z$$

where

$$\begin{aligned} (\text{predicted } Y_2 \mid Y_1 = 11) &= 9.787 + .576(11) \\ &= 9.787 + 6.363 \\ &= 16.123 \end{aligned}$$

The reader will note that the predicted average change in Y is not exactly the same as that actually observed in Table 11.2 (e.g., those respondents with a score of 11 in 1969 had an observed mean score of 15.14 rather than the mean of 16.12 predicted by the linear model). To the extent that these deviations from the linear change model's predictions are greater than we might expect by chance, the credibility of the model suffers. Thus, we need a nonlinear model against which to test this linear model.

A very general nonlinear model can be specified by simply defining a dummy variable for each level of Y_1:

$$X_{ij} = \begin{cases} 1 & \text{if } Y_1 = j \text{ for person } i \ (i = 1, \ldots, 1992;\ j = 11, \ldots, 44) \\ 0 & \text{otherwise} \end{cases}$$

This means that there would be one dummy variable for each of the 34 possible scores between 11 and 44. (In fact, only 32 dummy variables would be needed, since scores on Y_1 did not exceed 42 in actual magnitude.) The resulting regression is analogous to an analysis of variance in which the independent variable is Y_1 with 34 levels. The least squares solution for this problem yields an R-square of 0.3286. Note that this R-square is slightly larger than that obtained with the linear model. Inspection of the scattergram for the two variables may suggest a less general nonlinear model than the one used above (e.g., logarithmic, polynomial, etc.). In order to test the null hypothesis that this observed increase in the R-square is simply due to the fact that the nonlinear model capitalizes on chance, we can use the usual method for computing an F ratio with least squares estimates:

$$F = \frac{(.3286 - .3100)/(32 - 2)}{(1 - .3286)/(1992 - 32)}$$
$$= 1.808 \qquad (df = 30,\ 1960)$$
$$\text{Prob}(F > 1.808 \mid H_0) = .0047$$

where 32 is the number of parameters estimated in the nonlinear model (the mean of each group), 2 is the number of parameters estimated in the linear model (the slope and the intercept), and 1992 is the sample size. If we applied the usual rule of rejecting null hypotheses that generate probabilities less than 0.05, then we would reject the linear model in favor of the nonlinear model.

However, the significance test must be interpreted in light of the magnitude of the deviations from linearity. The standard error of estimate about the line fit to these data is 4.854, whereas the standard error of estimate for the nonlinear model (the square root of a weighted average of variances within groups) is 4.825. The linear model clearly fits the data about as well as the nonlinear model, even though the fit of the nonlinear model is enough better that the difference is unlikely to have occurred by chance. Whether or not this result would motivate an investigation of the rather minute nonlinearities would depend upon an investigator's judgment. (In fact, a logarithmic transformation of Y_1 and Y_2 reduces the difference in fit between the linear and nonlinear models such that the difference is no longer significant at the 5 percent level.) If the hypothesis of small nonlinearities stemmed from some substantive theory (such as an hypothesis about a "floor effect" arising from the positive skew of the scores), then they could be pursued. However, if the test for nonlinearity was computed simply as a check for violations of the assumption of linearity, then this result could be construed as support for the linearity assumption. (Recall that large samples render even the most trivial of differences as statistically significant.)

The reader should bear in mind the fact that the OLS estimate of b_2 obtained in this fashion is likely to be biased due to the assumptions that b_1 is zero and that Y_1 is measured without error. In other words, the inertia of the latent characteristic "locus of control" may be quite different from the value 0.577, since there are probably excluded factors correlated with Y_1 that increase the stability of Y over this time period—increase, that is, relative to pure inertia—and it is probably the case that there is a considerable amount of measurement error in Y_1. (Common sense suggests that the stability of personality characteristics is generally greater than the coefficient 0.577 indicates. Thus, measurement error in Y_1 has probably biased this estimate of stability downward, while the exclusion of causal factors correlated with Y_1 has biased it upwards.) The effect of excluding factors from the model can be seen by assuming that equation 11.1 is the true model (where b_1 is not zero) while we estimate the following model:

$$Y_2 = c_0 + c_1 Y_1 + e$$

Solving for c_1 we have

$$c_1 = \text{Cov}(Y_2, Y_1)/\text{Var}(Y_1)$$

If we solve equation 11.1 for the covariance and the variance, however, we find that

$$\text{Cov}(Y_2, Y_1) = b_1 \text{ Cov}(X, Y_1) + b_2 \text{ Var}(Y_1)$$

which implies that

$$\text{Cov}(Y_2, Y_1)/\text{Var}(Y_1) = b_1 \text{ Cov}(X, Y_1)/\text{Var}(Y_1) + b_2$$

or

$$c_1 = b_2 + b_1 \text{ Cov}(X, Y_1)/\text{Var}(Y_1)$$

Clearly, then, c_1 is equal to the inertia of the scores (b_2) plus a biasing amount that is a function of the effect of X on Y (b_1) and the covariance of X and Y_1.

The finding that changes in a locus of control scale over a two-year time lag fit an equation model in which that change is depicted as a linear function of the score at time 1 and a disturbance that is uncorrelated with Y_1 motivates a question about what other kinds of causal processes might generate this pattern of changes. And similarly, what kinds of causal processes would generate patterns of change that do not fit the linear regression change model?

One type of causal model that would generate data consistent with the linear regression model is a linear structural model in which the scores at the two points in time are depicted as flawed indicators of an unchanging latent variable. For example, suppose that

$$Y_1 = E + e_1$$

and

$$Y_2 = E + e_2$$

where e_1 and e_2 are measurement errors that are independent of one another and of the latent variable E—and the mean of the measurement errors is zero. This model presumes that Y_1 and Y_2 are parallel measures of the latent variable E. Now, it is not immediately apparent that such a model would generate change scores that fit the pattern seen in Table 11.2 (i.e., the negative correlation between change and Y_1, and the pattern of regression to the mean). In order to make this more clear, we begin by subtracting the first equation from the second to obtain

$$\begin{aligned} Y_2 - Y_1 &= (E - E) + (e_2 - e_1) \\ &= e_2 - e_1 \end{aligned}$$

Next, transform Y_1 and Y_2 into deviation scores by subtracting their respective means, Y_1' and Y_2':

$$(Y_2 - Y_2') - (Y_1 - Y_1') = e_2 - e_1 + Y_1' - Y_2'$$

We can now multiply by $(Y_1 - Y_1')$ and take expectations to obtain the covariance between Y_1 and $(Y_2 - Y_1)$:

$$\text{Cov}(Y_1, Y_2 - Y_1) = \text{Cov}(e_2, Y_1) - \text{Cov}(e_1, Y_1)$$

If we now recall that e_2 is uncorrelated with e_1 and E (and therefore with Y_1), then we have

$$\begin{aligned} \text{Cov}(Y_1, Y_2 - Y_1) &= -\text{Cov}(e_1, Y_1) \\ &= -\text{Var}(e_1) \end{aligned}$$

Thus, the model implies that change in Y will be negatively correlated with Y_1 when the variance of e_1 is positive. Furthermore, Y_2 is clearly a linear function of Y_1 from the rearrangement of the structural model as follows:

$$Y_2 = (Y_1 - e_1) + e_2$$

This latter form of the equation suggests another way of understanding how such a structural model could lead to the pattern of change seen in Table 11.2. If we write

$$Y_2 = Y_1 + (e_2 - e_1)$$

and take the conditional expectation

$$\text{Exp}[Y_2 | Y_1 = a] = a + \text{Exp}(e_2 | Y_1 = a) - \text{Exp}(e_1 | Y_1 = a)$$

we see that the expected value of Y_2 given that $Y_1 = a$ is equal to a plus the conditional expectation of e_2 and minus the conditional expectation of e_1. Now, in order for this equation to agree with the pattern of regression to the mean, the right-hand side must be less than a when Y_1 is greater than its mean and greater than a when Y_1 is less than its mean. This may seem odd since we are used to the assumption that residual terms like e_1 and e_2 have expectations of zero. In fact, however, the conditional nature of these particular expectations means that this is not the case. The usual assumption about e_1, for example, is that it is independent of the latent variable E and that the conditional expectation

$$\text{Exp}[e_1 | E] = 0$$

for all levels of E. However, the expectation of e_1 *conditional on* Y_l is another matter. It is straightforward to show, for example, that

$$\text{Cov}(Y_1,e_1) = \text{Var}(e_1)$$

which implies that Y_1 and e_1 are positively correlated. This means that e_1 is more likely to be positive when Y_1 is high, and e_1 is more likely to be negative when Y_1 is low. In other words, the measurement error is more likely to be in the direction of extremity for extreme scores, and thus the expected score at time 2 (the true score E) is closer to the mean than Y_1. (The reader can see how all of this works with the following set of data for Y_1, E, and e_1:

Y_1:	0	1	2	1	2	3	1	2	3	2	3	4
E:	1	1	1	2	2	2	2	2	2	3	3	3
e_1:	−1	0	+1	−1	0	+1	−1	0	+1	−1	0	+1

Note that E and e_1 are independent, but

$$\text{Exp}[e_1 \mid Y_1]$$

is negative when Y_1 is low and positive when Y_1 is high.)

As the reader may suspect, there are numerous structural models that would predict the "linear regression to the mean" pattern of change in a variable. It is always useful in such instances to also concoct a causal model in which the characteristic pattern would not be observed. Suppose, for example, that

$$Y_1 = b_0 + b_1X + z$$
$$Y_2 = c_0 + c_1X^2 + z_2$$

was the structural model for Y. Clearly, in this instance, Y_2 would not be a linear function of Y_1, and $(Y_2 - Y_1)$ would be a quadratic function of X rather than a linear function of Y_1. Another example of a structural model that would not produce the linear regression pattern of change is one in which the inertia of scores was related to the level of the score. Suppose, for example, that higher scores had more inertia than lower scores. The following structural model would have such a characteristic:

$$Y_2 = b_0 + b_1X + (b_2 + b_3Y_1)\, Y_1 + z$$

Note that the value of the inertial coefficient is a function of Y_1 itself. Thus, if b_3 were positive, the inertia would increase linearly with increases in Y_1. Such a model would produce patterns of change that would vary distinctly from the linear regression pattern; in fact, changes in Y would be a quadratic function of Y_1 rather than a linear function, and "regression" patterns would be asymmetric about the mean. This can be seen by rewriting the interactive model as

$$Y_2 = b_0 + b_1X + b_2Y_1 + b_3Y_1^2 + z$$

Fortunately, many psychological and sociological variables manifest the pattern of change noted in the locus of control scale. This means that opportunities for the application of linear structural models to longitudinal designs are plentiful. Furthermore, a sound understanding of the linear models of change provide a starting point for the development of more complex nonlinear models in substantive areas requiring such models.

Entropic Decay of Covariances

The causal model with a linear inertia component and an entropic disturbance component was concocted to explain how the pattern of change in the locus of control scores might come about. Our concern now is with patterns of change in covariances, slopes and correlations over time. For instance, suppose that the variables X and Y_1 are correlated in 1969. What then happens when we correlate the same X with Y_2 (1971)? Would $\text{Cov}(Y_2,X)$ tend to be larger, equal to, or smaller than $\text{Cov}(Y_1,X)$? Before discussing various models that would predict various patterns of change in relationships, we shall turn once again to the NLS data for white males to see what kind of longitudinal pattern is typical.

The Y variable for our exploration will be the same locus of control scale used above. The X variable will be another attitudinal measure: attitude toward current job in 1969. This variable was measured on a four-point scale where

1 = like it very much

2 = like it somewhat

3 = dislike it somewhat

4 = dislike it very much

Responses to this item were even more skewed than those for the IE scale: 53.5 percent of respondents claimed that they liked their jobs "very much," whereas only 1.8 percent claimed that they *dis*liked their jobs very much.

Our primary concern here is with the relationship between Y ("locus of control") and X (attitude toward job) and how that relationship changes over time. To begin with, we need to look at the synchronous relationship between X and Y_1—since both were measured in 1969 we refer to the relationship as *synchronous* rather than as longitudinal or lagged. The mean value of Y_1 for white respondents giving each of the four possible X scores are as follows:

$$X = 1,\ \text{mean of } Y_1 = 20.99\ (5.41)$$

$$X = 2,\ \text{mean of } Y_1 = 23.13\ (5.59)$$

$$X = 3,\ \text{mean of } Y_1 = 25.03\ (6.24)$$

$$X = 4,\ \text{mean of } Y_1 = 25.55\ (6.87)$$

(The number in parentheses is the standard deviation for each mean.) Clearly, there is an increase in the expected value of Y_1 as X increases; in fact, the relationship appears to be approximately linear. A nonlinear model using dummy variables for levels of X yields an R-square of .046, and the regression of Y_1 on X gives the result:

$$Y_1 = 19.28 + \underset{(.19)}{1.80}X + e \qquad (11.4\text{a})$$

$$\text{R-square} = .044$$

Table 11.3 Locus of control means for groups defined by attitude toward job.

Attitude toward job	Mean (SD) on IE scale Y_1 (1969)	Y_2 (1971)
Like very much	20.99(5.40)	22.00(5.69)
Like somewhat	23.13(5.59)	22.85(5.68)
Dislike somewhat	25.03(6.24)	24.48(6.44)
Dislike very much	25.55(6.87)	24.75(6.89)

The F ratio for the difference in fit between the nonlinear and linear models is 1.44 (df = 2,1988), which has a probability greater than 0.2 if the null hypothesis holds. The null hypothesis is, of course, the linear model; thus, we are on solid ground assuming that the relationship between these two variables is linear.

We now consider the relationship between X and Y at time 2. Table 11.3 gives the means of Y_1 and Y_2 for each of the four groups defined by scores on X. Note how the differences between the means of the four groups "shrink" between 1969 and 1971: the range from 21 to 25.5 in 1969 becomes 22 to 24.8 in 1971. This pattern of shrinkage is reminiscent of the pattern of regression to the mean in Table 11.2, and it suggests that relationships between variables may often decay in a manner analogous to the reordering of a single variable over time. This shrinkage is also apparent in the slope of Y_2 regressed on X:

$$Y_1 = 19.28 + 1.80X + e_1$$

whereas

$$Y_2 = 20.84 + 1.07X + e_2 \tag{11.4b}$$

The next question that presents itself is whether or not this decay in the magnitude of the slope of Y regressed on X is related to the inertia of Y.

The persistence (or inertia) of the Y on X slope over the two-year lag between 1969 and 1971 can be expressed as a ratio:

$$\text{lagged slope/synchronous slope} = 1.07/1.80 = .596$$

where the lagged slope is roughly 60 percent of the synchronous slope. If we compare this value of .596 to the inertia of Y (see equation 11.3), we find a suggestive similarity: the inertia of Y was .576, whereas the inertia of the *slope* of Y on X is .596. It is also of interest to look at the ratios of lagged to synchronous correlations and covariances:

$$\text{lagged Corr}(Y_2,X)/\text{synch Corr}(Y_1,X) = .121/.211 = .573$$

$$\text{lagged Cov}(Y_2,X)/\text{synch Cov}(Y_1,X) = .468/.789 = .593$$

Now, is this a fortuitous finding, or might we find that lagged slopes and covariances decay at the same rate as the dependent variable generally?

One way of addressing this question is to induce spuriously an even larger correlation between Y_1 and X by selecting cases such that the synchronous correlation is substantial. This procedure was followed with data for the 1992 white men on the IE scale (Y) and attitudes toward their job (X). A sample of 655 cases was selected that met the following criteria:

$$Z_1 - Z_2 < 0.5$$

and

$$Z_1 - Z_2 > -0.5$$

where

$$Z_1 = \text{standardized value of } Y_1$$

$$Z_2 = \text{standardized value of } X$$

The relationships among the three variables X, Y_1, and Y_2 are given in Table 11.4 for the entire sample of 1992 cases and the "spuriously" selected sample of 655 cases. Note that the synchronous correlation between Y_1 and X was spuriously increased to a value close to one: from .211 for the complete sample to .941 in the spurious subsample. Similarly, the slope of Y_1 on X was increased from 1.799 to 8.117. The question that now presents itself is whether or not this spuriously inflated synchronous relationship will evidence the same rate of decay (or inertia) as did the observed relationship for the entire sample. The inertia of the spurious relationship is measured by the ratio of lagged to synchronous coefficients:

$$\text{lagged slope/synch slope} = 4.542/8.117 = .560$$

$$\text{lagged Corr}(Y_2,X)/\text{synch Corr}(Y_1,X) = .453/.941 = .481$$

The rather surprising finding is that the inertia of the spuriously boosted slope is very close to that of the slope observed in the complete sample (.596), and it is almost exactly equal to the stability of the Y scores for the spurious subsample (.564; see slope of Y_2 on Y_1 in Table 11.4). The ratio of the lagged and synchronous correlations,

Table 11.4 Relations among three variables for a complete sample of 1992 cases and a subsample selected to increase the synchronous correlation between X and Y_1.

Complete sample (N = 1992)				Subsample (N = 655)		
Correlations						
	Y_1	Y_2	X	Y_1	Y_2	X
Y_1	1.000	.557	.211	1.000	.485	.941
Y_2		1.000	.121		1.000	.453
X			1.000			1.000
Slopes						
Y_1 on X: 1.799				Y_1 on X: 8.117		
Y_2 on X: 1.072				Y_2 on X: 4.542		
Y_2 on Y_1: .576				Y_2 on Y_1: .564		

on the other hand, seems to have strayed away from the stability of Y; but note that this ratio is very close to the correlation between Y_2 and Y_1 (called the "autocorrelation of Y") in the spurious subsample. This latter result demonstrates the greater invariance of inertial coefficients when they are expressed in an *un*standardized form. However, it also suggests that the autocorrelation of a dependent variable reflects its "standardized inertia."

These brief explorations of data from the NLS for white males suggest that synchronous slopes may have the same *lagged inertia* as the dependent variables in these systems. As we shall see, however, not all synchronous slopes decay at this rate. In fact, some do not decay at all, some increase in magnitude, others decay but have more inertia than the dependent variable, and so on. (As we shall see, deviations from the expected pattern of entropic decay can be used to indicate that a causal relation may be involved.) Nevertheless, the pattern of decay and inertia that was apparent with the locus of control and job satisfaction variables is quite a common one, and it is of interest to construct a model that would predict such an outcome.

A Causal Model that Produces Entropic Decay of Slopes

A simple and straightforward model that predicts the outcomes obtained with the NLS data above (i.e., both the pattern of regression to the mean for Y and the pattern of entropic decay of slopes that involves the same inertia as Y) is an adaptation of that presented in equation 11.1:

$$Y_2 = b_0 + b_1X + b_2Y_1 + z$$

where

$$b_1 = 0$$

and z (all other factors influencing Y_2) is presumed to be independent of Y_1 and X. According to this model, we have

$$\text{Cov}(Y_2,X) = b_1 \text{ Var}(X) + b_2 \text{ Cov}(Y_1,X)$$

or

$$\begin{aligned} b_1 &= [\text{Cov}(Y_2,X) - b_2 \text{ Cov}(Y_1,X)]/\text{Var}(X) \\ &= \text{slope of } Y_2 \text{ on } X - b_2 \text{ (slope of } Y_1 \text{ on } X) \end{aligned}$$

Thus, when b_1 is zero, we have the prediction that the lagged slope (Y_2 on X) will be equal to the synchronous slope (Y_1 on X) times the inertia of Y (b_2)—precisely our finding with the NLS data for locus of control and job satisfaction. Furthermore, this model implies that if X did have a positive lagged effect on Y, then the lagged slope should exceed the product of the synchronous slope and the inertia of Y. Thus, we would conclude from these data that job satisfaction in 1969 *did not* relate to Y_1 and Y_2 in a manner consistent with what we would expect if position on X influenced Y in the linear fashion indicated by equation 11.1 with $b_1 > 0$.

A crucial aspect of the argument up to this point should be reiterated. The fact that a linear causal model predicted the behavior of the NLS data may seem quite remarkable. Indeed, the power of the linear model in the real world is consistently astounding. Nevertheless, if the NLS data for these variables had been consistent with a linear model in which there was a lagged effect of job satisfaction (i.e., $b_1 > 0$), this would only mean that the data were *consistent* with such a model. It would not mean that a causal interpretation was necessary or incontrovertible. Quite to the contrary, there would be numerous other structural models that involved no lagged effect of job satisfaction on locus of control that were equally consistent with these data. This situation contrasts markedly with the randomized experiment in which alternative third variable models may be rejected *en masse*. But, it is also the case that a demonstration that a synchronous slope does not decay at the same rate as the dependent variable is *more* evidence of a causal effect than would be the synchronous slope alone. In Section 11.6, we will develop some procedures for assessing the plausibility of some alternative explanations of longitudinal relations that are consistent with a linear model that includes a lagged effect.

Temporal Stability of Population Differences

We are now prepared to deal with the sticky wicket that was described earlier as Lord's paradox. The problem arises because variables measuring what are often called *population differences* tend to behave in the same manner as variables that have causal effects. This can be seen by reviewing the means presented in Table 11.1. That table gives the means of the IE scale for 1969 and 1971 and for black and white men separately. From those means, it is easy to see that the racial difference in 1969 is roughly comparable to the racial difference in 1971. The black mean is 2.75 points higher than the white mean in 1969 and 2.63 points higher in 1971. Clearly, there is nothing like the entropic decay seen in our previous examples. The correlation between race and the IE scale is .208 in 1969 and .193 in 1971—whereas the inertia of the IE scale for the combined races is still much less than one.

Suppose that we were to estimate a linear structural equation of the following form:

$$Y_2 = b_0 + b_1X + (b_2 + b_3X)(Y_1 - Y_1') + z \qquad (11.5)$$

where

$$\begin{aligned} X &= 1 \text{ for blacks} \\ & 0 \text{ otherwise} \\ Y_1' &= \text{mean of } Y_1 = 22.748 \end{aligned}$$

and z is independent of Y_1 and X. [Note that this equation allows for an interaction between race (X) and 1969 IE (Y_1) in that the inertia of Y is a function of race ($b_2 + b_3X$). The mean deviation form of Y_1 is used to locate the intercepts at the center of the 1969 distribution of Y.] The OLS estimate of this model is

$$\underset{}{Y_2 = 22.888} + \underset{(.235)}{1.505X} + (\underset{(.020)}{.576} - \underset{(.040)}{.223X})(Y_1 - Y_1') + z$$

Now, the appropriate interpretation of this result is that the intercept of Y_2 regressed on $(Y_1 - Y_1')$ is 1.505 greater for blacks, and the slope of Y_2 regressed on Y_1 is .223

less for blacks than for whites. The larger intercept for blacks indicates that blacks scoring near the center of the 1969 distribution of Y tended to respond to the IE scale more in the "external" direction in 1971 than whites at the center of the 1969 distribution, whereas the greater slope for whites indicates a greater inertia for this characteristic in the white population than in the black population. However, suppose that we were not savy to the fact that X measured a population difference? Or, more subtly, suppose that our X variable was another variable that is closely correlated with race? Would we not be in error to conclude from the significant coefficient of X that X had a causal effect on Y? After all, if we compare the change in IE for blacks with that for whites, we find that they were comparable. Blacks increased by .316 points from 1969 to 1971, whereas whites increased an average of .436. From this result, we might conclude that *whites* had moved in the *external* direction more than blacks—a pattern that seems opposite to our conclusion if b_1 in equation 11.5 were given a causal interpretation.

It may be helpful at this point to recapitulate what has been found up to this point. First, we found that the synchronous relationship between the IE scale and job satisfaction decayed at the same rate as the IE scale when job satisfaction was lagged behind IE (IE for 1971 and job satisfaction for 1969). Second, this result was found to be consistent with a linear structural model in which job satisfaction had no lagged effect on IE. Third, it was found that the relationship between the IE scale and race did not decay between 1969 and 1971. The finding that variables measuring population differences do not behave in the same manner as a variable with a spurious time 1 correlation is a troublesome finding for those who might hope to use the longitudinal behavior of scores to make causal inferences. This is a sobering finding that should inject a bit more realism into such attempts at causal inference.

Before leaving this topic, we should explore the details of these findings a bit more carefully. For example, the meaning of the coefficient b_1 in equation 11.5 is intriguing. The reader should be accustomed to interpreting such numbers in the following manner: when participants are equated on Y_1 (the statistical corollary of "holding Y_1 constant"), then participants with a score one point higher on X have an average Y_2 score that is

$$
\begin{aligned}
[\text{predicted } Y_2 | Y_1 = a, X = 1] - [\text{predicted } Y_2 | Y_1 = a, X = 0] \\
= 1.505 - .223(a - Y_1') \\
= 1.505 - .223(a - 22.748)
\end{aligned}
$$

points higher (or lower) than the average respondent with the lower score on X. Thus, we might be attracted to the conclusion that (where $a = Y_1'$) blacks increased in IE more from 1969 to 1971 than did whites—when whites and blacks *started* at the same IE level in 1969 (i.e., *conditional* change). In fact, this conclusion is correct. Blacks at the mean of 22.748 in 1969 were *below* their population mean of 24.802, whereas whites at the mean of 22.748 in 1969 were *above* their population mean of 22.049 (see Table 11.1). Thus, the fact that whites at the overall 1969 mean of 22.748 tended to decline over the two-year span, while blacks tended to increase (giving rise to the 1.505 difference in 1971), is evidence of a causal effect that is correlated with race. Thus, a somewhat hidden assumption in the causal interpretation of lagged regression coefficients is that all cases are regressing to the same mean. And this is the case only when all population differences are incorporated into the structural model. And the "problem" of population differences failing to decay is simply another case of missing variables in a model.

One approach to incorporating population differences in a structural model is to use dummy variables as in equation 11.5. Another approach is to specify different equations for different populations and use constraints between equations to test various null hypotheses. The latter tactic is particularly appropriate when there is reason to believe that the pattern of disturbance covariances may not be the same within different populations. (Joreskog's LISREL program allows such specifications, so that assessment of interactions involving population differences is possible.) An example of such an analysis is given in Section 11.5 of this chapter.

But the real problem, of course, is when lack of (or failure of) randomization procedures leaves the possibility that *unmeasured* variables with lagged effects are correlated with variables whose lagged effects are of interest (see Figure 11.4). This problem is ultimately unresolvable, since no study can measure all possibly relevant variables; however, in Section 11.6 of this chapter a procedure for testing "restricted" spurious models is presented. This procedure is of some assistance in dealing with the problem of unmeasured population differences as a plausible alternative causal model.

The final problem to be addressed here is that of answering the question: "Was the race of respondents in the NLS sample related to the amount of change in IE over the two-year period from 1969 to 1971 when change was considered relative to the starting point of each *group* rather than the starting point of individuals?" In other words, Did the difference between the populations change over time? This is a very different question from that addressed by the magnitude of b_1 in model 11.5. The question now being asked is whether or not a state of *equilibrium* has been upset between time 1

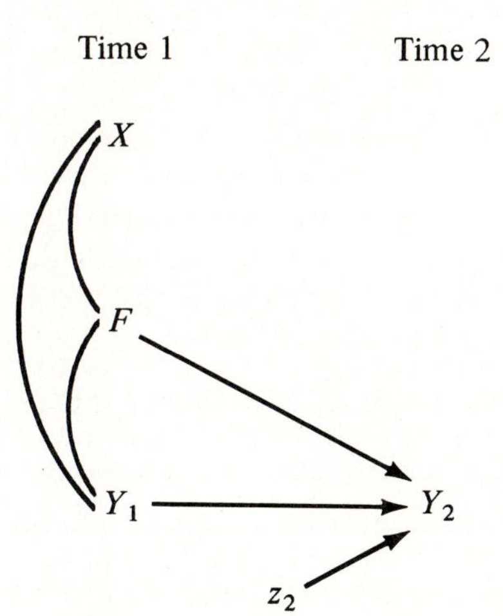

Figure 11.4 A longitudinal model incorporating an unmeasured background variable. If F is unobserved and Y_2 is regressed on Y_1 and X, then a lagged effect will be inappropriately attributed to X. The variable X might be a dummy variable distinguishing between blacks and whites.

and time 2. Equilibrium might be defined here in terms of a temporally stable difference between the means of the two races on the IE scale, a temporally stable slope of IE regressed on race, or temporal stability in the correlation between race and IE.

If equilibrium is defined in terms of the slope (or covariance), then the null hypothesis that

$$\text{Cov}(Y_1, X) = \text{Cov}(Y_2, X) \quad \text{or} \quad c_{Y_2X} = c_{Y_1X} \text{ (slopes)}$$

can be tested with a regression equation of the form:

$$Y_2 = c_0 + c_1X + c_2Y_1 + e \tag{11.6}$$

where the term c_2 is constrained to be one. In this instance, the residual e will be correlated with Y_1, and equation 11.6 may be rewritten for OLS estimation as

$$Y_2 - Y_1 = c_0 + c_1X + e$$

The least squares estimate of equation 11.6 for the 2670 NLS respondents is

$$\begin{aligned} Y_2 = 23.184 - \underset{(.254)}{.121X} + 1.000Y_1 + e \end{aligned}$$

where

$$\begin{aligned} X &= \text{race} \\ &\quad 1 \text{ if black} \\ &\quad 0 \text{ if white} \\ Y_2 &= 1971 \text{ locus of control} \\ Y_1 &= 1969 \text{ locus of control} \end{aligned}$$

Note that the coefficient for race is negative, but not even as large as its standard error in absolute value. This means that the black mean increased .121 points less than the white mean increased between 1969 and 1971 (see the means in Table 11.1).

It is important to reiterate that the regression equation in 11.6 is not a stochastic structural model in which coefficients may be given causal interpretations. Note, for example, that the causal relationship between Y_1 and Y_2 is not incorporated into the equation. The problematic result of treating such an equation as a causal model is easily demonstrated with the NLS data. Suppose that we add job satisfaction (X_2) to the regression equation, we then have

$$Y_2 = c_0 + c_1X_1 + c_2X_2 + 1.00Y_1 + e$$

The least squares estimate of the equation from the NLS data is

$$Y_2 = 24.299 - \underset{(.254)}{.131X_1} - \underset{(.168)}{.725X_2} + 1.000Y_1 + e$$

Note that the coefficient for job satisfaction is now negative. This occurs because, as outlined above, the synchronous relationship decays as X_2 is lagged behind Y. And the constraint that the coefficient of Y_1 is one implies that such decay will occur only if

X_2 is exerting a negative influence on Y (a negative influence that erodes the positive synchronous slope). This can be seen algebraically by solving for the coefficient c_2:

$$\text{Cov}(Y_2,X_2) - \text{Cov}(Y_1,X_2) = c_1 \text{ Cov}(X_2,X_1) + c_2 \text{ Var}(X_2)$$

since $\text{Cov}(X_1,e) = \text{Cov}(X_2,e) = 0$. Similarly, if $c_1 = 0$, then

$$\begin{aligned} c_2 &= [\text{Cov}(Y_2,X_2) - \text{Cov}(Y_1,X_2)]/\text{Var}(X_2) \\ &= (\text{lagged slope}) - (\text{synchronous slope}) \end{aligned}$$

Thus, although equations with unconditional change scores as the dependent variable may be useful for analyses concerning the stability of certain relationships, they should not be confused with causal models for longitudinal processes. The circumstances in which concern with the stability of relationships is appropriate will be addressed in Section 11.5, which deals with analysis of nonequivalent control group designs.

Consider, once again, the coefficient for race in equation 11.5. This coefficient indicates that blacks tend to score a bit higher (in the "external" direction) than whites—on the average, and especially where blacks are compared to whites with similar 1969 scores at the center or "internal" side of the 1969 distribution of IE scores. Clearly, this finding is the result of causal processes that shaped whites to respond in the more internal direction, or, conversely, processes that shaped blacks to respond in the more external direction. However, these processes have come to equilibrium such that population differences are stable, and the causal processes that gave rise to these differences *continue to maintain those differences.* And this latter argument reveals a fundamental assumption in the model building presented in this chapter: that entropic forces will decay any set of differences if there is not some force operating to maintain them (or change them). This means that any slope that does not decay as the time of measurement for the dependent variable advances is indicative of a causal effect. We speak of "population differences" when the differences do not decay, and yet the causal process that maintains those differences is not specified. The variable measuring population membership is thus a proxy for a number of social and biological factors that may not be nearly so fixed as the population distinction. The distinction between males and females, for example, will outlast any of the various social factors that produce or exaggerate various sex differences. In the final analysis, then, investigators must incorporate as many population differences as are relevant to the causal process being studied, and unmeasured population differences take on the same unsettling status as unmeasured third variables that have a more clarified causal effect on the dependent variable of interest.

[It may be of interest to some readers that the null hypothesis of a stable population difference can be tested by placing a constraint on an actual equation model like that in 11.5. Suppose the regression model

$$Y_2 = b_0 + b_1 X + b_2 Y_1 + z_2$$

where z_2 is well behaved. Then the constraint that

$$b_1^2/(1 - b_2) + kb_2 = \text{Cov}(Y_1,Y_2)/\text{Var}(X)$$

where $k = \text{Var}(Y_1)/\text{Var}(X)$ tests the null hypothesis that

$$\text{Cov}(Y_1,X) = \text{Cov}(Y_2,X) \quad \text{or} \quad c_{Y_2X} = c_{Y_1X}$$

This constraint will be of particular interest in Section 11.3, since it defines a state of covariational equilibrium between the variable X and the variable Y where the variable X is presumed to be unchanging.]

The purpose of this section has been to give the reader a sense of how longitudinal patterns of change can be used to identify causal relations. Inferences from such data always involve some assumptions about what patterns will occur in the case of the null hypothesis. The reader is urged to investigate such patterns in a particular substantive area before presuming that patterns found with these attitude data are appropriate. The next section of this chapter comes at the problem of causal models for longitudinal designs from the opposite perspective; whereas we started with real data in this section, and then proposed causal models that would produce the observed patterns, the next section begins with abstract models and proceeds to work out their *logical* implications. Thus, whereas Section 11.2 is more empirical, Section 11.3 is more in the tradition of mathematical modeling, and is thus more rationalistic. In Section 11.6, the rational models and real data will be rejoined.

11.3 A STRUCTURAL MODEL FOR LONGITUDINAL DESIGNS

The goal of this section is the specification of a causal model that will be applicable to many, but not all, research situations involving causal inference from longitudinal data. The model is applicable to true experiments, quasi-experiments, and passive observational designs. The crucial application, however, is to designs other than true experiments.

A concept that becomes central in the development of longitudinal designs is that of equilibrium. Readers who have been primarily concerned with true experiments will probably find this concept rather novel. All experimental designs seek to disturb equilibrium by creating differences that did not previously exist. In "real world" social psychological systems, however, mutual causal influences often come to equilibrium—that is, synchronous slopes, covariances, or correlations are fairly stable over time. Thus, it is extremely useful to concoct plausible causal models that have the property of "coming to equilibrium" after some period of time. Certain developmental or growth processes in humans may, for example, involve an increasing correlation between their characteristics and the characteristics of their varying social environments. At certain points in their life cycle, however, there may be periods of relative stability in these relations. A simple linear model will be used to demonstrate how a model with these characteristics can be specified—and then estimated with longitudinal data. The exercise is instructive not only in that a model with desirable characteristics is found, but it is also instructive of the many sources of confounds and misleading causal inferences that can result from an uncritical application of these models.

An Equilibrating Simulation

The task here is to concoct a two-variable causal model that comes to equilibrium over time. A simple simulation technique will be used to see what happens as lagged causal relations "equilibrate"—i.e., achieve a stable synchronous variance-covariance matrix for the variables in the causal system.

We begin with two variables: X and Y. X and Y can be observed at discrete points in time. The subscript i is used to indicate the time point at which an observation takes place, and i ranges from 1 to m. Thus, X_3 is a variable consisting of observations of X at time point 3, whereas the values of Y_4 are derived from observations at time point 4. The *autocorrelation* of X between time points i and $i + 1$ is

$$\text{Corr}(X_i,X_{i+1}) = \text{Cov}(X_i,X_{i+1})/\text{SD}(X_i)\text{SD}(X_{i+1})$$

A *synchronous* correlation or covariance between X and Y involves X_i and Y_i. A *cross-lagged* correlation or covariance involves X and Y at different time points. For example, if X is lagged behind Y, then $\text{Cov}(Y_6,X_4)$ is a cross-lagged covariance with a lag of two time intervals.

In most research situations, observations are available for only two to six points in time. However, in this model building exercise, we will consider many more time points. Also, in many nonexperimental research situations it is not possible to start at "the beginning" of a causal system; once again, however, for conceptual clarity we will start at the beginning.

The construction of this model will be carried out in a rather abstract manner. However, it may be useful for the reader to apply conceptual labels to the variables X and Y. A good candidate for such labels is the following: let X be a characteristic of a work environment that differs across jobs within a large organization; let Y be a personality characteristic that varies across individuals. Now suppose that a large number of people are hired into different positions in this organization. At time point 1 (starting day), there may be little or no relationship between X and Y across these cases. However, if differences in X produce time-lagged changes in Y, and differences in Y produce time-lagged changes in X, then X and Y will become increasingly correlated. At some point, the system may come to equilibrium, and the correlation between these two variables would stabilize. In this example, variability in people's organizational environment shapes a personality characteristic while people's personalities shape the organizational environment. We would probably expect that the inertia of the organization was greater than that of the individual, but this remains an empirical question in any particular instance.

Now, our task is to construct a causal model using linear equations that may be applicable to systems such as the one described above. We begin with the following:

$$Y_{i+1} = a_0 + a_1X_i + a_2Y_i + z_{i+1} \tag{11.7a}$$

$$X_{i+1} = b_0 + b_1X_i + b_2Y_i + t_{i+1} \tag{11.7b}$$

This model specifies that change in Y is determined by a *lagged effect* of X (with a lag of one time period), the inertia of Y (over a single time period), and a disturbance

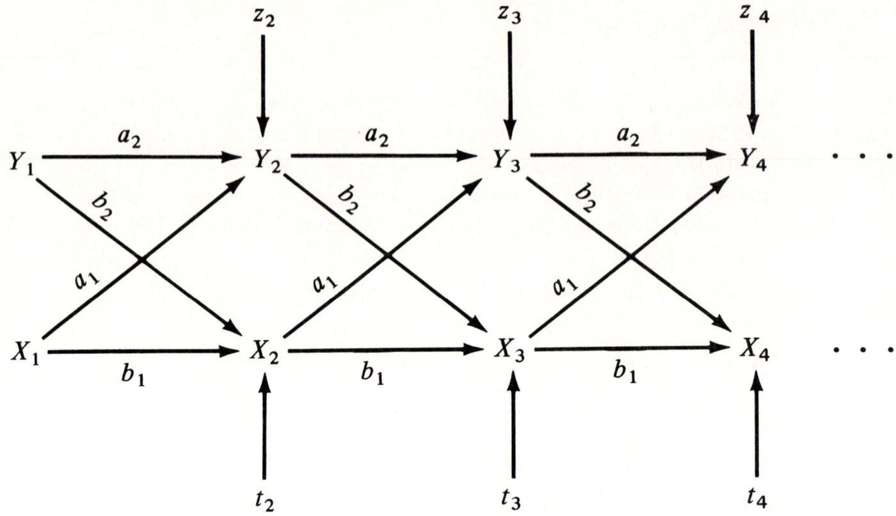

Figure 11.5 A path diagram of the longitudinal model specified in equation 11.7.

z. Similarly, change in X is a lagged function of Y, the inertia of X, and a disturbance t. The structural coefficient a_1 is the lagged effect of X on Y; a_2 is the inertia of Y; b_2 is the lagged effect of Y on X; b_1 is the inertia of X (see Figure 11.5).

The conspicuous limitations that characterize this model include the following: (1) there are no synchronous effects in the model; (2) cross-lagged effects occur only with a lag of one time period (i.e., there are no distributed lags); (3) inertial effects do not extend beyond one time period (this constraint is described as "first-order autoregressive" in the econometric literature); (4) the magnitudes of effects are independent of time; and (5) all relations are linear and additive (no interactions). The procedures used in the following simulation can be extended to assess the behavior of systems involving any or all of these additional complexities. However, the difficulties estimating such models with actual data can be enormous, and this is particularly true when variables are measured with error. There is a need for further simulation research addressed to the following question: When data are generated by more "realistic" longitudinal models (e.g., incorporating continuous and cumulative lagged effects), does the testing of more simplified models (e.g., those constrained to measurements at a few discrete points in time) lead to misleading conclusions? The work of Coleman (1968; cf. Rogosa, 1979) points in this direction; it can be shown that the "cross-lagged effects" model developed below tends not to be misleading about the *direction* of causal influence when the model generating the data is actually a dynamic model (i.e., specified in terms of differential equations for continuous time series).

We start at the beginning by assuming that the correlation between X and Y is zero when $i = 1$. For simplicity, we also presume that both X and Y have unit variances and means of zero when $i = 1$. The first simulation will also assume the following:

$$\text{Cov}(z_i, t_i) = 0 \qquad \text{disturbances are uncorrelated between variables}$$

$$\text{Cov}(z_{i+1}, Y_i) = \text{Cov}(z_{i+1}, X_i) = 0$$

$$\text{Cov}(t_{i+1},Y_i) = \text{Cov}(t_{i+1},X_i) = 0$$
$$a_0 = b_0 = 0$$

The strategy for this first simulation is to specify values of the structural coefficients and then allow the causal system to run through enough time periods that it comes to equilibrium. In order to do this, it is necessary to derive expressions for covariances at subsequent time points—based on the starting conditions (zero correlation between X and Y, unit variance, etc.). We begin by using the method of moments on equations 11.7:

$$\text{Cov}(Y_{i+1},X_i) = a_1 \text{Var}(X_i) + a_2 \text{Cov}(Y_i,X_i) \quad (11.8a)$$
$$\text{Cov}(Y_{i+1},Y_i) = a_1 \text{Cov}(Y_i,X_i) + a_2 \text{Var}(Y_i) \quad (11.8b)$$
$$\text{Var}(Y_{i+1}) = a_1 \text{Cov}(Y_{i+1},X_i) + a_2 \text{Cov}(Y_{i+1},Y_i) + \text{Var}(z_{i+1}) \quad (11.8c)$$
$$\text{Cov}(X_{i+1},X_i) = b_1 \text{Var}(X_i) + b_2 \text{Cov}(Y_i,X_i) \quad (11.8d)$$
$$\text{Cov}(X_{i+1},Y_i) = b_1 \text{Cov}(Y_i,X_i) + b_2 \text{Var}(Y_i) \quad (11.8e)$$
$$\text{Var}(X_{i+1}) = b_1 \text{Cov}(Y_{i+1},X_i) + b_2 \text{Cov}(Y_{i+1},Y_i) + \text{Var}(t_{i+1}) \quad (11.8f)$$
$$\text{Cov}(Y_{i+1},X_{i+1}) = a_1b_1 \text{Var}(X_i) + a_2b_2 \text{Var}(Y_i) + (a_1b_2 + a_2b_1) \text{Cov}(Y_i,X_i) \quad (11.8g)$$

The above seven equations plus the arbitrary (but convenient) specification that

$$\text{Var}(z_i) = 1 - a_2^2 \qquad \text{Var}(t_i) = 1 - b_1^2$$

are sufficient to identify temporally subsequent covariances, variances, and correlations from previous steps. All that remains is the assignment of values for the structural coefficients.

It is important at this point to make a distinction between *system equilibrium* and *structural invariance* of a longitudinal causal system. The former may be defined in terms of temporal stability (or constancy) of patterns of covariance and variance; the latter refers to temporal constancy of the structural model (i.e., the magnitude of causal effects) that describes the causal system. In the following simulations, we will always assume structural invariance. A structurally invariant causal system may come to equilibrium (an equilibrating system), it may expand without bound, or it may collapse. In what follows, we are primarily concerned with structurally invariant causal systems that come to equilibrium. Consideration of structural change, catastrophe theory, and so on is beyond the scope of this chapter.

The first set of structural coefficients to be tried will be a simplified example in that it will be assumed that lagged causation is unidirectional from X to Y. That is, b_2 will be set to zero:

$$a_1 = .2 \qquad a_2 = .4 \qquad b_1 = .7 \qquad b_2 = .0$$

Table 11.5 shows the results of the iterative simulation of this causal system.

The results in Table 11.5 indicate that the causal system came to equilibrium after seven iterations. The synchronous correlation starts at zero and climbs to .187—where it would remain, regardless of how many additional time points were added. Similarly,

Table 11.5 Results of an iterative simulation of the linear causal system specified in equations 11.7 with $a_1 = .2$, $a_2 = .4$, $b_1 = .7$, and $b_2 = .0$.

Time point (i)	$V(Y_i)$	$V(X_i)$	$r(Y_i,X_i)$	$r(Y_{i+1},X_i)$	$r(Y_i,X_{i+1})$	$r(Y_{i+1},Y_i)$
1	1.000	1.000	.000	.196	.000	.392
2	1.040	1.000	.137	.248	.096	.421
3	1.069	1.000	.173	.261	.121	.431
4	1.080	1.000	.183	.265	.128	.435
5	1.083	1.000	.186	.267	.130	.436
6	1.085	1.000	.186	.267	.130	.436
7	1.085	1.000	.187	.267	.131	.436
8	1.085	1.000	.187	...	...	...

Note. Iterations were discontinued when the variance of Y stabilized at the fourth decimal place. The letter V indicates a variance, while the letter r indicates a correlation.

the two cross-lagged correlations increase steadily, with the lagged correlation corresponding to the direction of causality becoming larger than the other cross-lagged correlation (see Figure 11.6). Note, in fact, that the cross-lagged correlation of .267 even exceeds the synchronous correlation in magnitude. If models such as these were the only ones of concern, comparison of the cross-lagged correlations would suffice for causal inference. However, as we shall see, the addition of a lagged effect in the other direction (from Y to X) renders the cross-lagged correlations unreliable indicators of causal direction.

A close look at Table 11.5 reveals some other interesting characteristics of the causal system. First, the variance of Y increases and stabilizes as the system seeks equilibrium while the variance of X remains equal to one. The reason for this pattern is the fact that X is causing changes in Y that increase the variability of Y—while the reverse is not occurring. Second, the autocorrelation of Y increases over time. The reason for this pattern is again the effect of X: X is shaping Y at the same time that X has considerable inertia (.7). The overall effect of shaping by a variable with high inertia is an increase in the stability of the shaped variable. This stability due to shaping by another stable variable is in addition to inertia. Third, though not shown in Table 11.5 explicitly, the autocorrelation of X is invariant with time and is equal to its inertia. The autocorrelation of X with a lag of one time period is constant because X is not being influenced by any variable other than the disturbance, and the disturbance has a constant variance over time periods.

The next concern that arises with such a causal system is its estimation from sample observations. That is, suppose that the causal model generating the relations in Table 11.5 holds for some population, and a researcher samples that population at time points 2 and 3. Are the structural parameters identified by such data? And can good estimates of the identified parameters be obtained from such data? The answer to the first question is clearly yes. In fact, this model is overidentified by such observations. The answer to the second question is that the quality of estimation is dependent on sample size and, if maximum likelihood estimates are used, the normality of the distribution of the variables. (In this particular instance, however, an ML solution is not

Path diagram

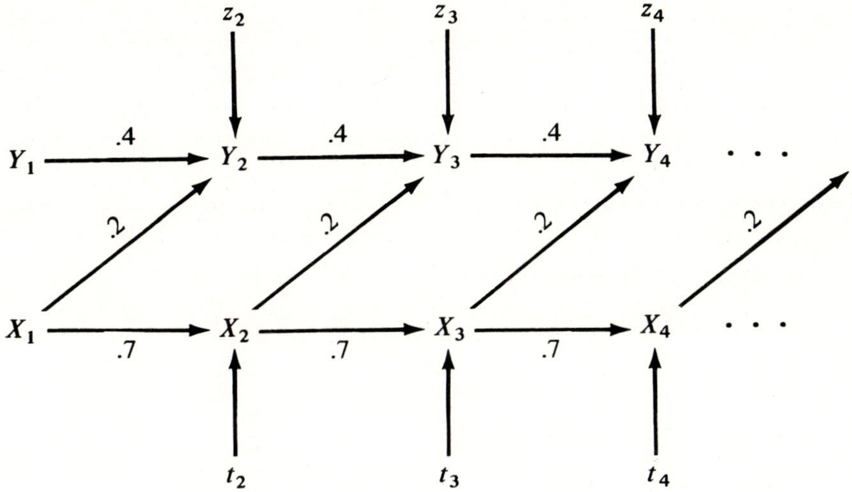

Pattern of correlations

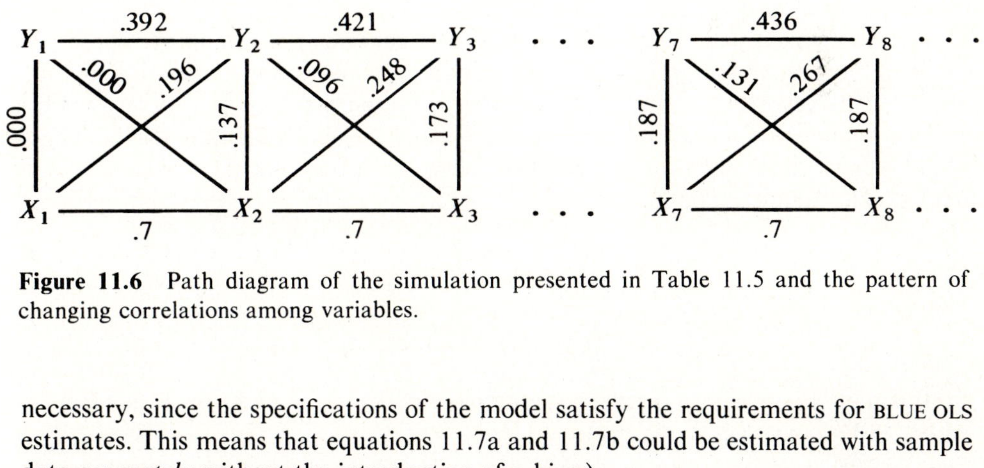

Figure 11.6 Path diagram of the simulation presented in Table 11.5 and the pattern of changing correlations among variables.

necessary, since the specifications of the model satisfy the requirements for BLUE OLS estimates. This means that equations 11.7a and 11.7b could be estimated with sample data *separately* without the introduction of a bias.)

The overidentification of the model specified by equations 11.8 arises because synchronous effects are eliminated and because of the lack of correlation between the disturbances (z and t). These specifications are likely to be dubious in many research situations. Nevertheless, it can be seen that equations 11.8a and 11.8b identify the coefficients a_1 and a_2, equation 11.8c identifies the variance of z, equations 11.8d and 11.8e identify b_1 and b_2, and equation 11.8f identifies the variance of the disturbance t. The remaining equation, 11.8g, thus provides the overidentification of the model. If we allowed z and t to correlate, then the covariance of these disturbances would be identified by equation 11.8g, and the entire model would be just identified.

Problem Use the data in Table 11.5 for time points 2 and 3 to solve for a_1, a_2, b_1, and b_2. Use the usual OLS procedures to obtain the metric and standardized versions of these coefficients. Notice that the results for the metric coefficients are precisely those used at the first stage of the iterative simulation. This little exercise demonstrates how the partial slopes disentangle the synchronous correlation and attribute appropriate effects to each variable.

The case in which two variables are causally linked only because of a lagged effect in one direction is instructive in at least one additional way. Consider the result of multiplying equation 11.7a by X_{i+2} and taking the expectation to obtain:

$$\mathrm{Cov}(Y_{i+1},X_{i+2}) = a_1 \,\mathrm{Cov}(X_i,X_{i+2}) + a_2 \,\mathrm{Cov}(Y_i,X_{i+2})$$

since X must be uncorrelated with all the disturbances z (z is uncorrelated with *all* the determinants of X as this first longitudinal model was specified). The last covariance in this equation can then be expanded to find that

$$\mathrm{Cov}(Y_i,X_{i+2}) = a_1 \,\mathrm{Cov}(X_{i-1},X_{i+2}) + a_2 \,\mathrm{Cov}(Y_{i-1},X_{i+2})$$

It should be clear that this expansion could be substituted in the previous equation, and then $\mathrm{Cov}(Y_{i-1},X_{i+2})$ could be expanded and so on. Now, since X is first-order autoregressive, we have that

$$\mathrm{Cov}(X_i,X_{i+m}) = b_1^m \,\mathrm{Var}(X_i)$$

(It should be noted that the assumption that z_i is uncorrelated with all X_j where $j < i$ is not a necessary assumption for identification of model parameters with longitudinal data. However, it is not an unreasonable addition, and it is of interest for the current argument.) Thus, if we continued to expand the above equation for the lagged covariance (with X leading), and recalled the specification that the variance of X is one, then we would obtain

$$\mathrm{Cov}(Y_i,X_{i+1}) = a_1b_1^2 + a_2(a_1b_1^3 + a_2(a_1b_1^4 + a_2(\ldots$$

which may be rewritten as

$$(a_1b_1^2)[1 + (a_2b_1)^1 + (a_2b_1)^2 + (a_2b_1)^3 + \ldots$$

Fortunately, this series is a familiar one to mathematicians; it is a form of the binomial series, and it converges to the value

$$(a_1b_1^2)[(a_2b_1)^2/(1 - a_2b_1)] = (a_1a_2^2b_1^4)/(1 - a_2b_1)$$

if (a_2b_1) is not equal to one. The purpose of this little exercise is to demonstrate that a two-variable causal system with complete inertia for both variables (a_2 and b_1 equal to one) can never come to equilibrium. That is, the cross-lagged covariance (with the causal variable leading the affected variable) increases without bound. Similarly, the variance of the affected variable increases without bound. Also notice that if Y has no inertia ($a_2 = 0$ and changes in Y_i are not transmitted to Y_{i+1}) then the above covariance between Y_i and X_{i+1} must be zero. However, the variable X could have complete inertia ($b_1 = 1$), even though the variable Y did not; the variable X would then be analogous to a population difference that would come to an equilibrium relationship with Y. The reader should substitute different values of the four parameters into the above equation to see how different values lead to different states of equilibrium.

We can now return to the causal model in equations 11.7 to see what happens when causal effects occur in both directions. For this iterative simulation, we will leave all parameters the same as before, but the structural coefficient b_2 will be fixed at .2 rather than zero. This change means that X and Y have equivalent lagged effects on one another, but the two variables differ in inertia (X has more inertia than Y). The result of the iterative simulation is given in Table 11.6 and Figure 11.7. The results in Table 11.6 are quite complex. The variable with the greater inertia (X) increases in variance at a greater rate than the other variable (Y). The reason for this is the fact that changes in X due to Y do not decay as quickly, since the inertia of X is greater. The cross-lagged correlations are equal for the first time period, but then the correlation with Y lagged behind X becomes larger than the correlation with X lagged behind Y. The reason for this differential between the cross-lagged correlations is the difference in the inertia of the two variables rather than a difference in the magnitude of the cross-lagged *effects* (a_1 and b_2).

The realization that the cross-lagged correlations can be misleading about the "relative directionality" of causal relations is further compounded if we compute the standardized cross-lagged effects:

$$Y_{14} = .214X_{13} + .400Y_{13} + z_{14}$$

$$X_{14} = .700X_{13} + .186Y_{13} + t_{14}$$

In this case, the effect of X on Y appears to be the larger—the opposite of what we might have concluded from the cross-lagged correlations. In fact, of course, both the

Table 11.6 Results of an iterative simulation of the linear causal system specified in equations 11.7 with $a_1 = .2$, $a_2 = .4$, $b_1 = .7$, and $b_2 = .2$.

Time point (i)	V(Y_i)	V(X_i)	r_1	r_2	r_3	r_4	r_5
1	1.000	1.000	.000	.196	.196	.392	.686
2	1.040	1.040	.212	.335	.279	.433	.714
3	1.083	1.123	.271	.376	.310	.451	.733
4	1.106	1.187	.296	.393	.324	.459	.745
5	1.119	1.231	.310	.403	.333	.463	.750
. . .							
12	1.139	1.309	.330	.417	.346	.471	.761
13	1.139	1.309	.330	.417	.346	.471	.761
. . .							

Note. r_1 = Corr(Y_i,X_i); r_2 = Corr(Y_i,X_{i+1}); r_3 = Corr(Y_{i+1},X_i); r_4 = Corr(Y_{i+1},Y_i); and r_5 = Corr(X_{i+1},X_i). The letter V indicates a variance.

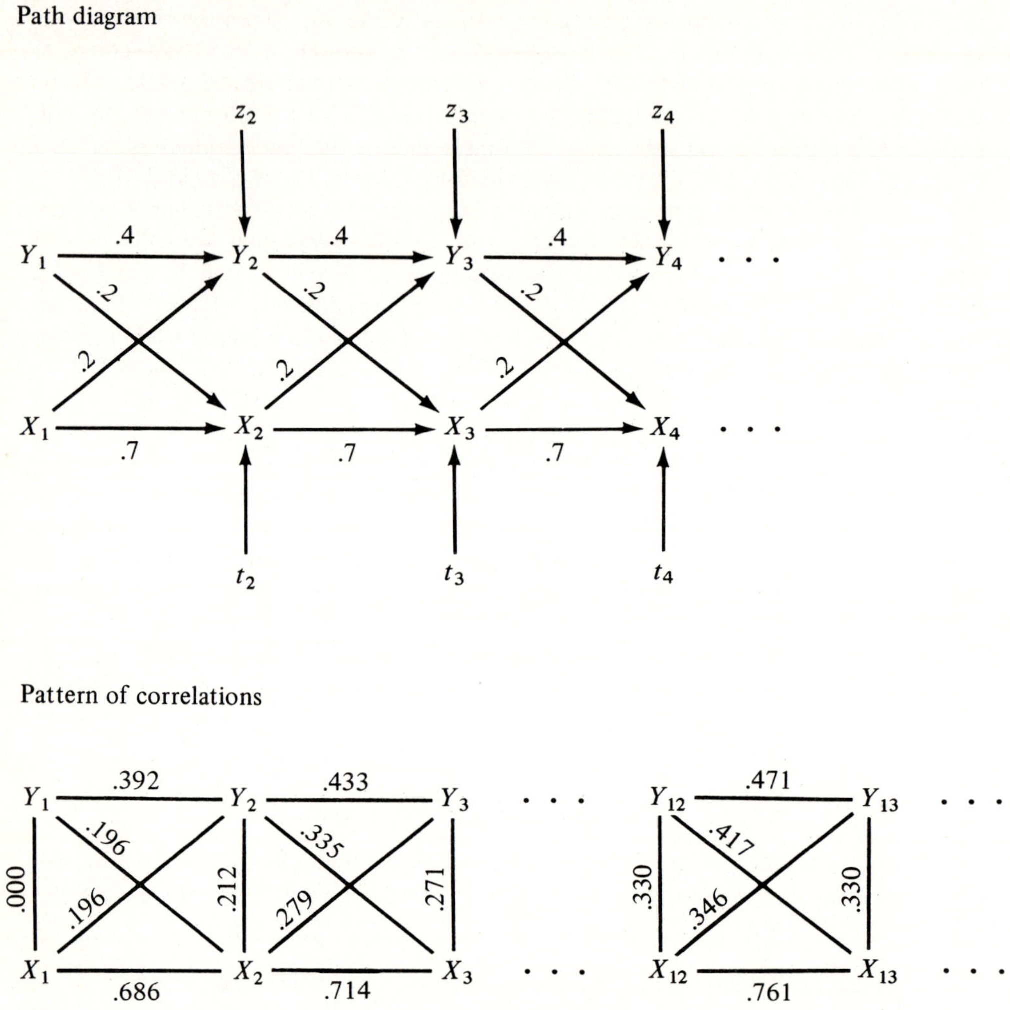

Figure 11.7 Path diagram and pattern of correlations for simulation given in Table 11.6.

differences between the correlations *and* the differences between the standardized effects arise because of the different inertial characteristics of the two variables involved. This can be seen by using these standardized equations as follows:

$$\text{Corr}(Y_{14}, X_{13}) = .214 + .400\ \text{Corr}(Y_{13}, X_{13})$$

$$\text{Corr}(X_{14}, Y_{13}) = .700\ \text{Corr}(Y_{13}, X_{13}) + .186$$

Clearly, the correlation with Y lagged behind X exceeds the correlation with X lagged behind Y because the greater inertia of X "carries" more of the synchronous correlation with it (i.e., the synchronous correlation decays less rapidly when the variable with less inertia is lagged behind the variable with greater inertia than when the lag is in the opposite direction). In one sense, the mutual effects of the two variables on one another are equal; but the way in which differential inertias affect the variances of the two variables renders standardized comparisons as problematic unless the differences are quite large. (The fact that the difference between the standardized effects is less than the difference between the cross-lagged correlations is not accidental. In general, this will be the case with this model.)

Before looking closely at the effects of deviations from the assumptions built into this model on the emergence of relations over time, it is instructive to inspect the consequences of an incorrect choice of lag. For example, suppose that the model used to generate the relations in Table 11.6 operated for some population. However, suppose that a researcher selected sample observations at time points 13 and 15, due to a lack of knowledge concerning the appropriate time span for the lagged effect. How would such a mistake bias estimates of a_1 and b_2? Well, it is straightforward to show that

$$\begin{aligned}
\text{Corr}(Y_{15},X_{13}) &= .214\ \text{Corr}(X_{14},X_{13}) + .400\ \text{Corr}(Y_{14},X_{13}) \\
&= (.214)(.761) + (.400)(.346) \\
&= .301 \\
\text{Corr}(X_{15},Y_{13}) &= .700\ \text{Corr}(X_{14},Y_{13}) + .186\ \text{Corr}(Y_{14},Y_{13}) \\
&= (.700)(.417) + (.186)(.471) \\
&= .380 \\
\text{Corr}(Y_{15},Y_{13}) &= .214\ \text{Corr}(X_{14},Y_{13}) + .400\ \text{Corr}(Y_{14},Y_{13}) \\
&= (.214)(.417) + (.400)(.471) \\
&= .278 \\
\text{Corr}(X_{15},X_{13}) &= .700\ \text{Corr}(X_{14},X_{13}) + .186\ \text{Corr}(Y_{14},X_{13}) \\
&= (.700)(.761) + (.186)(.346) \\
&= .597
\end{aligned}$$

If these values were then used to estimate the structural coefficients from a misspecified model, we would expect to obtain

$$\begin{aligned}
\text{lagged effect of } X \text{ on } Y \text{ (biased estimate of } a_1) &= .219 \\
\text{lagged effect of } Y \text{ on } X \text{ (biased estimate of } b_2) &= .219
\end{aligned}$$

rather than the true values of .20. (The standardized values of these coefficients are .235 and 0.204, respectively, for the lagged effect of X on Y and Y on X.) Thus, the bias in the estimate is not large in this particular instance. In fact, the reader can verify that the use of too lengthy a lag does not lead to disastrously misleading conclusions from this model.

Biases Due to Misspecification

The reader should keep in mind the fact that a pattern of longitudinal correlations can be generated by many processes other than the one developed in this section. Furthermore, even if the causal system is similar to the one specified in equations 11.7, it may not abide by the numerous restrictions that were placed upon the model—especially the assumptions about the lack of correlation between disturbances and other variables. Such cases of misspecification can lead to erroneous estimates of structural coefficients. The following few paragraphs outline the implications of several such deviations.

The first misspecification to be considered is one that does not create much of a problem. Recall that it was assumed that the disturbance z_i was uncorrelated with the disturbance t_i. Suppose now that there existed an "occasion factor" that increased the synchronous covariance of X_i and Y_i, but that this factor decayed completely between time points. Such a factor would spuriously increase the synchronous slope or correlation relating X and Y in the same way that the synchronous correlation between the IE scale and job satisfaction was inflated in Section 11.2 of this chapter. And the reader may recall that the inflated correlation decayed at a rate that rendered an appropriate conclusion concerning the lack of a lagged effect. This previous finding should produce some optimism in the current circumstance.

As can be seen from equations 11.8, we can relax the assumption that

$$\text{Cov}(t_i, z_i) = 0$$

and identify this covariance with equation 11.8g. (We would, however, need to retain the assumption that disturbances at one time point were uncorrelated with disturbances at earlier points in time.) The structural coefficients remain identified by the same pairs of equations in the set of equations 11.8.

The claim that spuriously induced synchronous correlations do not interfere with the estimation of cross-lagged effects is truly extraordinary. It is so extraordinary because the variables that produce that spurious portion of the synchronous correlation need not be measured or explicitly included in the model. Rather, their role is absorbed in the synchronous covariance between the disturbances. Such occasion factors are the Achilles' heel of cross-sectional designs, where they must be measured and controlled explicitly to avoid bias in the estimation of structural coefficients.

Table 11.7 and Figure 11.8 present the results of an iterative simulation based on

Table 11.7 Results of an iterative simulation of the linear causal system specified in equations 11.7 with $a_1 = .2$, $a_2 = .4$, $b_1 = .7$, $b_2 = .2$, and $\text{Cov}(z_{i+1}, t_{i+1}) = .25$.

Time point (i)	$V(Y_i)$	$V(X_i)$	r_1	r_2	r_3	r_4	r_5
1	1.000	1.000	.000	.196	.196	.392	.686
2	1.040	1.040	.452	.482	.366	.472	.738
3	1.123	1.193	.544	.547	.415	.502	.767
4	1.168	1.316	.571	.570	.436	.516	.783
5	1.192	1.340	.584	.582	.447	.523	.792
...							
12	1.231	1.547	.601	.598	.464	.535	.807
13	1.231	1.549	.601	.599	.465	.535	.807
...							

Note. $r_1 = \text{Corr}(Y_i, X_i)$; $r_2 = \text{Corr}(Y_i, X_{i+1})$; $r_3 = \text{Corr}(Y_{i+1}, X_i)$; $r_4 = \text{Corr}(Y_{i+1}, Y_i)$; and $r_5 = \text{Corr}(X_{i+1}, X_i)$. The letter V indicates a variance.

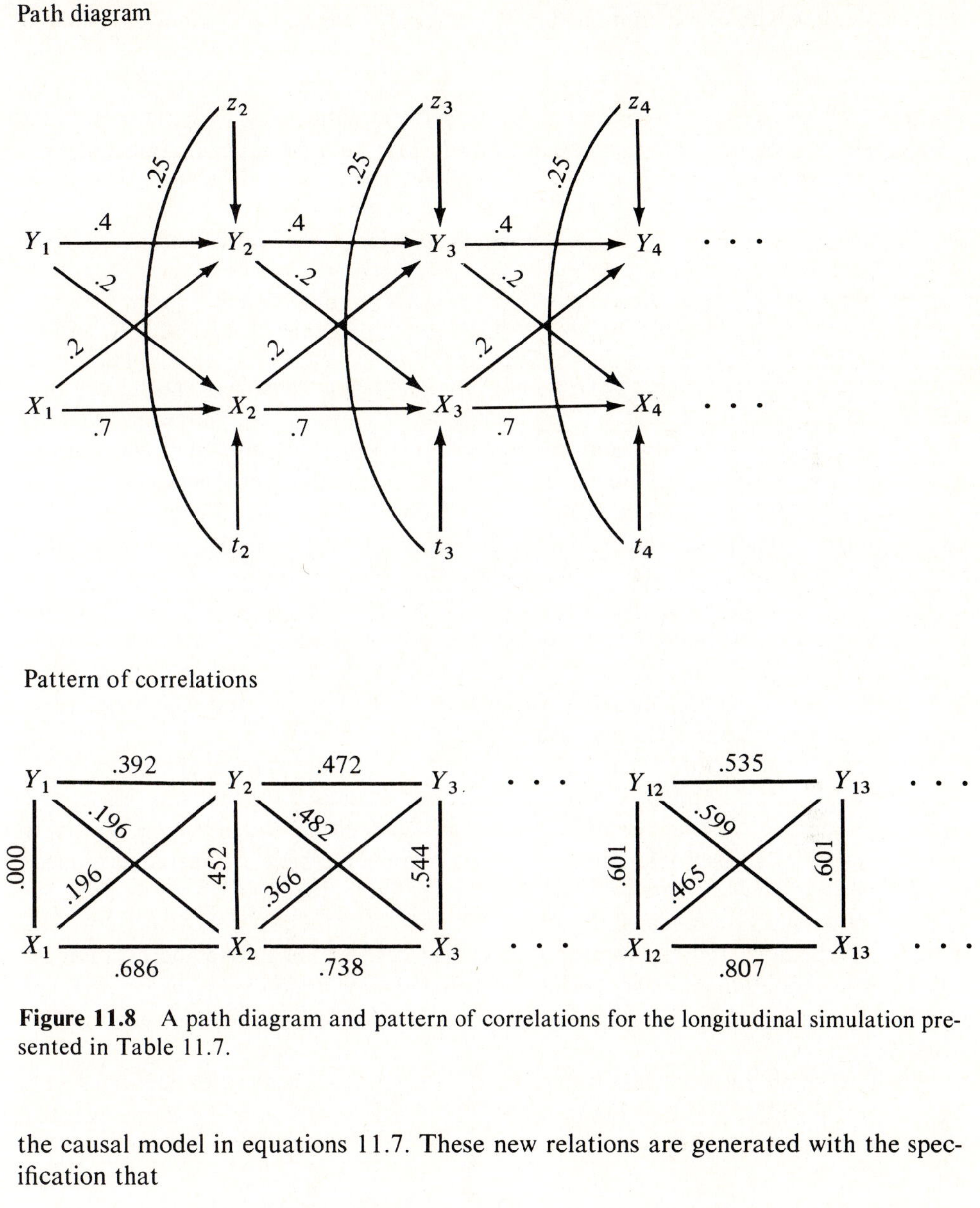

Figure 11.8 A path diagram and pattern of correlations for the longitudinal simulation presented in Table 11.7.

the causal model in equations 11.7. These new relations are generated with the specification that

$$\text{Cov}(t_{i+1}, z_{i+1}) = 0.25 \qquad (i = 1, \ldots, m)$$

If any temporally adjacent relations in Table 11.7 are used to solve for the structural coefficients in equations 11.7, the true values will be obtained. This is the case even though the pattern of correlations in Table 11.7 is quite distinct from that in Table 11.6 (where the synchronous correlation of the disturbances was presumed to be zero). Notice especially that the correlated disturbances raise the synchronous correlation between X and Y to the point that it exceeds both cross-lagged correlations in magnitude. Nevertheless, the actual lagged effects remain identified and easily estimated with large samples.

Problem Use the results in Table 11.7 to solve equations 11.8 for the structural coefficients and the covariance between z and t. Note that equation 11.8g becomes

$$\mathrm{Cov}(Y_{i+1},X_{i+1}) = a_1b_1\,\mathrm{Var}(X_i) + a_2b_2\,\mathrm{Var}(Y_i) + (a_1b_2 + a_2b_1)\,\mathrm{Cov}(Y_i,X_i) + \mathrm{Cov}(z_{i+1},t_{i+1})$$

in this case.

Lest the reader become overly optimistic about the prospects for using longitudinal data to extract spurious associations from lagged causal processes, it is important to remember that this power is reserved only for factors that have zero inertia over the length of one lag—thus, the label "occasion factor." Certain "test situation" factors, mood factors, weather factors, and so on may fit the requirement of complete inertial decay over the causal lag of interest. However, other variables may not decay rapidly enough to be treated as occasion factors. Such factors inflate *cross-lagged* covariances and can introduce a bias in the estimation of lagged effects that is the most damaging: the finding of significant cross-lagged coefficients when no cross-lagged effects are present. We will discuss a partial solution to this problem in Section 11.6 of this chapter, but a complete solution is never possible, since any solution to the third variable problem must assume that the third variable has certain characteristics. And there is always the possibility that some third variable that does not have those characteristics is actually producing the spurious covariation.

The next misspecification that can bias estimates of cross-lagged effects in model 11.7 arises when the stability of X and/or Y is due to factors other than cross-lagged effects and inertia. Another way of stating this is to say that z or t may be positively autocorrelated with itself over time—because misspecification has included unmeasured factors in the estimated disturbance term that influence the inertial variable directly, but not the cross-lagged variable. If such an autocovariance was not estimated in some way, the assumption that it was zero would inflate the estimate of inertia. An overstatement of actual inertia would tend to produce an *under*statement of lagged effects, since the lagged covariance would decay at a faster rate than expected.

The effect of an autocorrelation for z on the longitudinal pattern of equations 11.8 can be seen in Table 11.8 and Figure 11.9. The relations in this table were generated with the presumption that

$$z_{i+1} = (.4/.84)\,z_i + u_{i+1}$$

where u is random. Then we have

$$\mathrm{Cov}(z_{i+1},z_i) = 0.4$$

The problem created by the autocorrelated disturbance is easily seen by using the results in Table 11.8 to regress Y_9 on X_8 and Y_8:

$$Y_9 = .137X_8 + .726Y_8 + e$$

Table 11.8 Results of an iterative simulation of the linear causal system specified in equations 11.7 with $a_1 = .2$, $a_2 = .4$, $b_1 = .7$, $b_2 = .0$, and $\text{Cov}(z_{i+1},z_i) = .40$.

Time point (i)	V(Y_i)	V(X_i)	r_1	r_2	r_3	r_4	r_5
1	1.000	1.000	.000	.000	.200	.400	.700
2	1.040	1.000	.137	.096	.231	.747	.700
3	1.229	1.000	.162	.113	.225	.750	.700
4	1.456	1.000	.158	.110	.223	.744	.700
5	1.530	1.000	.156	.109	.223	.743	.700
. . .							
8	1.555	1.000	.156	.109	.223	.743	.700
9	1.555	1.000	.156	.109	.223	.743	.700
. . .							

Note. $r_1 = \text{Corr}(Y_i,X_i)$; $r_2 = \text{Corr}(Y_i,X_{i+1})$; $r_3 = \text{Corr}(Y_{i+1},X_i)$; $r_4 = \text{Corr}(Y_{i+1},Y_i)$; and $r_5 = \text{Corr}(X_{i+1},X_i)$. The letter V indicates variance.

Clearly, the lagged effect of 0.2 from X to Y is grossly understated by OLS when the disturbance of Y is autocorrelated at this level. The problem here is that z_9 is correlated with Y_8—making OLS estimates biased. Proper specification of the model would either allow for this correlation or distinguish between the systematic and random components of the disturbance.

Fortunately, it is possible to estimate disturbance autocorrelations in some instances. Such estimation requires a flexible linear model such as LISREL and observation at more than two points in time. The details of such an estimation will be demonstrated in Section 11.6 of this chapter. Nevertheless, the disturbance autocorrelation problem is another case of an unmeasured "third" variable (or variables). There is some unmeasured variable that produces stability in Y, and that unmeasured variable is uncorrelated with X. Thus, we find that even variables that are uncorrelated with the cause that is of interest can distort estimates of causal effects from longitudinal designs. Unless that variable is measured and explicitly included in the model, taking account of it necessarily involves assumptions about some characteristics of the variable and its relationship to other variables in the model. (For readers who are used to thinking that variables uncorrelated with the causal variable are of no consequence in causal inference, it is important to recall that estimation of a causal effect from longitudinal data is based upon some model of how the lagged correlation will decay relative to the synchronous correlation. In this context, the inertia of the dependent variable is crucial, and unmeasured variables can distort estimates of that inertia. Once again, the beauty of the true experiment is apparent: successful randomization ensures that the population correlation between the independent and dependent variables is initially zero—which greatly simplifies the problem of modeling the decay process. In fact, when randomization produces a sample synchronous correlation that is close to zero, the problem of modeling decay rate is eliminated altogether.)

The last assumption incorporated in equations 11.8 is the assumption of perfect measurement. When variables are measured with error in longitudinal studies, serious

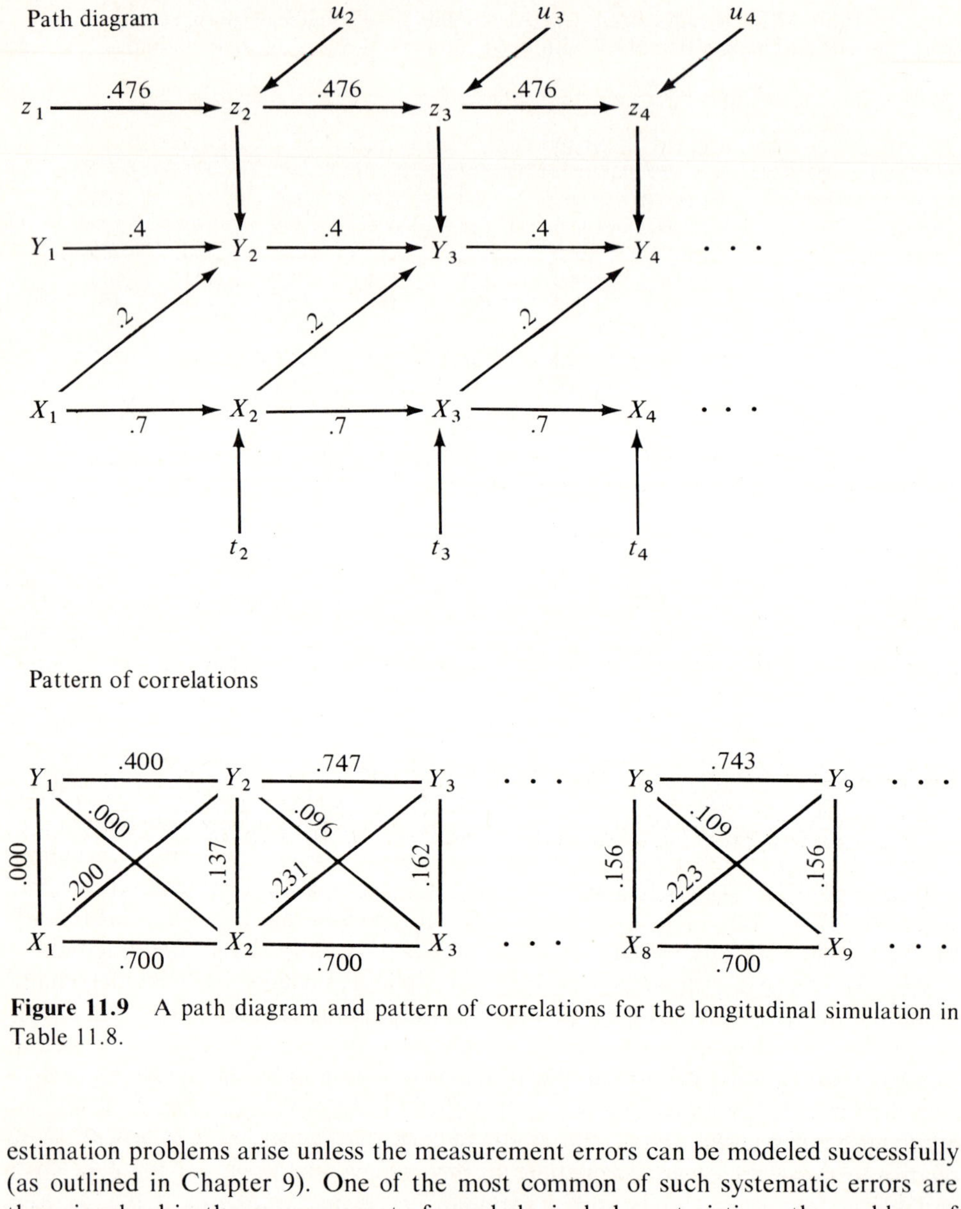

Figure 11.9 A path diagram and pattern of correlations for the longitudinal simulation in Table 11.8.

estimation problems arise unless the measurement errors can be modeled successfully (as outlined in Chapter 9). One of the most common of such systematic errors are those involved in the measurement of psychological characteristics—the problem of "method" factors. When these are present, use of the same instrument at two or more points in time can result in autocorrelated measurement errors that are a special case of the autocorrelated disturbance problem mentioned above. Fortunately, the use of multiple indicators and testable measurement models allows a researcher to estimate these dependencies among measurement errors. Once they are estimated, their role in the longitudinal causal process is not problematic. An explicit example of such model building will be presented in Section 11.6.

11.4 ANALYSIS OF RANDOMIZED (TRUE) LONGITUDINAL EXPERIMENTS

The inferential elegance of the true experiment is nowhere more apparent than in the case of *longitudinal* experimental designs. The elegance derives from two consequences of randomization. First, randomization of the independent variable makes it likely that the correlation between the causal variable and a "pre-manipulation" measure of the dependent variable will be zero. Second, unlike *matched control group designs,* randomization assures that, on the average, members of both the control and experimental groups are subject to the same entropic decay forces when the null hypothesis of no manipulation effect is true.

Furthermore, the use of structural equation models for experimental testing of theories can greatly enhance the ability of social psychological researchers to sort out complex causal relations. The rather unconvincing "manipulation checks" that have characterized the best of experiments in the past can be replaced with precise estimates of causal paths—whether mediated by poorly measured intervening variables, or direct.

To simplify the subsequent discussion, let us return to the recurrent example of the relationship between social threat and authoritarianism. Suppose that the dependent variable, authoritarianism, is the latent variable measured by three observed indicators. Scores on these three indicators are obtained prior to the manipulation of social threat and then again after an appropriate time lag. Three indicators of the latent variable "perceived threat" are also measured at time 1. Immediately after the time 1 observations, participants are randomly sorted into a control group and an experimental group. The experimental group is then subjected to the (supposedly) threatening manipulation. Now, suppose that our theory stipulates that an authoritarian personality style is a time lagged psychological consequence of the belief that one is being threatened. Thus, the manipulation is followed by observation of the three indicators of the latent variable "perceived threat." The time point for this observation will be labeled time 3. Time 4 is then the time point at which the three indicators of authoritarianism are observed for the second time.

The design of this experiment can be outlined as follows:

Time 1: Three indicators of authoritarianism (Y_1, Y_2, Y_3) and three indicators of perceived threat (X_1, X_2, X_3) are observed.

Time 2: Random assignment and the manipulation occur ($X = 0$ for controls, $X = 1$ for experimentals).

Time 3: Three indicators of perceived threat (X_4, X_5, X_6) are observed for a second time.

Time 4: Three indicators of authoritarianism (Y_4, Y_5, Y_6) are observed for a second time.

The goal of the experiment is to demonstrate not only that the manipulation increases authoritarianism, but also to test the hypothesis that feeling threatened *mediates* this effect. Thus, we could construct a linear model that reflects this causal hypothesis and contrast its fit to the data with competing models. Suppose that we adopt the following labels for the latent variables:

$$E_1 = \text{perceived threat at time 1}$$
$$E_2 = \text{perceived threat at time 3}$$
$$E_3 = \text{authoritarianism at time 1}$$
$$E_4 = \text{authoritarianism at time 4}$$
$$F_1 = X \text{ manipulation (no measurement error)}$$

Now, if we presume that relations among variables are linear, the following structural equations could be proposed to model the observed data from the experiment:

$$E_2 = b_1E_1 + b_2E_3 + g_1F_1 + z_2 \tag{11.9a}$$

$$E_4 = b_3E_1 + b_4E_2 + b_5E_3 + g_2F_1 + z_4 \tag{11.9b}$$

(In an actual research situation it would be wise to also include background variables that may explain the time 1 covariance between E_1 and E_3. These variables would be incorporated as F_2, F_3, and so on in both equations 11.9a and 11.9b.) This structural model, together with the yet to be introduced measurement model, is depicted as a path diagram in Figure 11.10. The coefficients of real interest here are g_1 and b_4: the first is the effect of the manipulation on perceived threat, and the second is the effect of perceived threat on authoritarianism. The coefficient g_2, on the other hand, represents the *direct* effect of the manipulation on authoritarianism—i.e., an effect that is not mediated by perceived threat. If g_2 is large relative to g_1, then the mediation hypothesis does not receive much support. On the other hand, if g_1 and b_4 are large relative to g_2, then the mediation hypothesis is supported. If both g_2 and b_4 are small, then there is no indication that the manipulation influenced authoritarianism in any way—whether directly or via perceived threat.

Although the structural model in equations 11.9 is of primary theoretical concern, it can be estimated only in the context of an acceptable measurement model. Thus, before even attempting to estimate the structural coefficients, it would be necessary to assess the fit of the measurement model with the observed data. A plausible measurement model will be specified after writing the structural model in a form congenial to analysis with an ML computer program such as LISREL.

The equations 11.9 may be written in the LISREL format as follows:

$$\underline{B}\,\underline{E} = \underline{G}\,\underline{F} + \underline{z}$$

where

$$\underline{B}\,\underline{E} = \begin{bmatrix} 1 & 0 & 0 & 0 \\ -b_1 & 1 & -b_2 & 0 \\ 0 & 0 & 1 & 0 \\ -b_3 & -b_4 & -b_5 & 1 \end{bmatrix} \begin{bmatrix} E_1 \\ E_2 \\ E_3 \\ E_4 \end{bmatrix}$$

(Note that the coefficients in $\underline{B}$ are opposite in sign—due to the transposition of endogenous variables in the matrix format.)

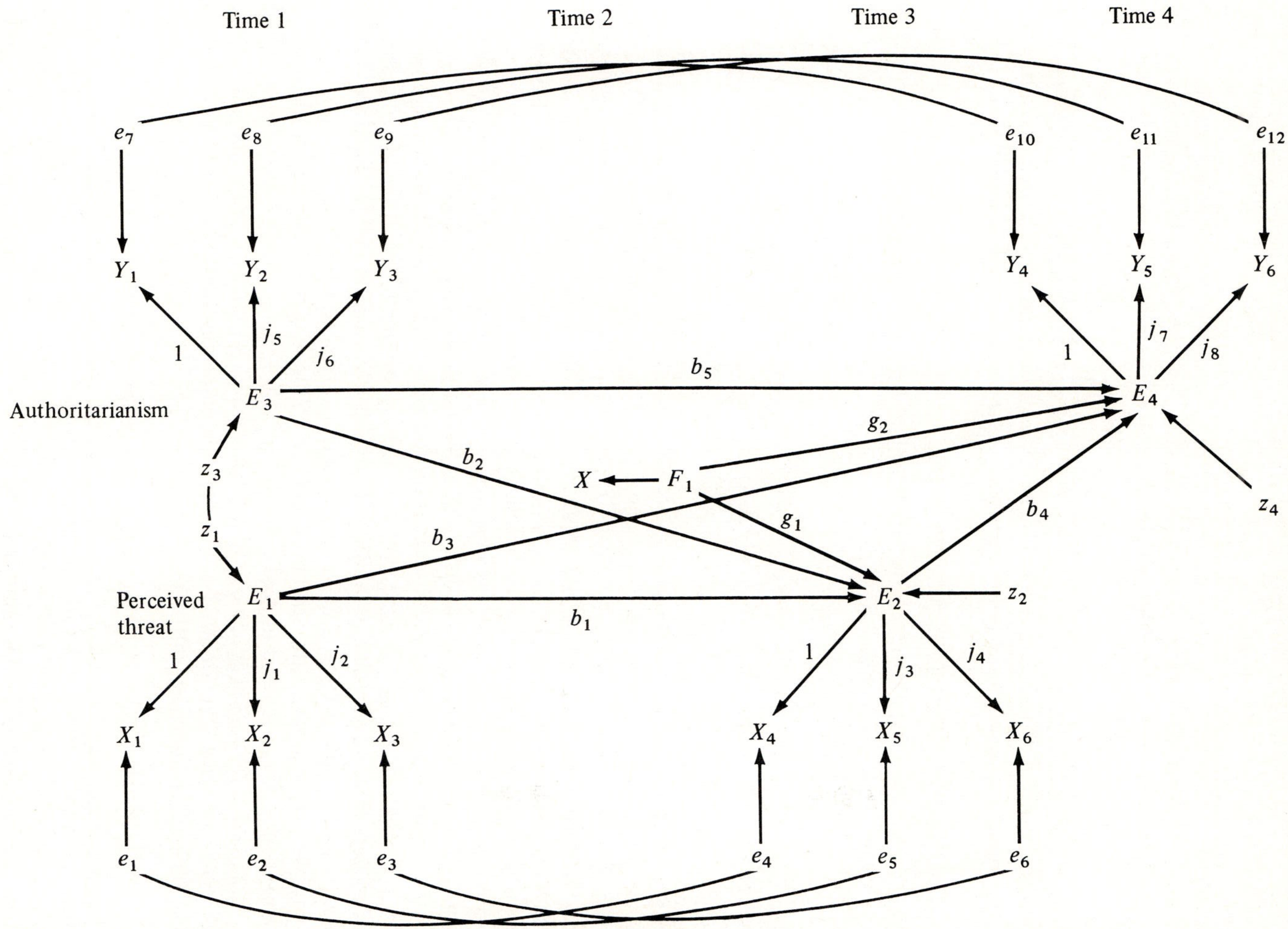

Figure 11.10 A path diagram of a structural model for a hypothetical true (randomized) experiment (equations 11.9).

$$\underline{GF} = \begin{bmatrix} 0 \\ g_1 \\ 0 \\ g_2 \end{bmatrix} [F_1]$$

and

$$\underline{z} = [z_1 \quad z_2 \quad z_3 \quad z_4]'$$

An additional set of specifications that are sufficient to identify the structural portion of the model are given in the following matrices (see Table 10.1 in Chapter 10):

$$\begin{aligned} \underline{PH} &= \text{covariance matrix of } \underline{F} \\ &= [ph_1] \qquad [\text{estimated from sample variance of } X] \\ \underline{PS} &= \text{covariance matrix of } \underline{z} \\ &= \begin{bmatrix} ps_1 & & & \\ 0 & ps_2 & & \\ ps_3 & 0 & ps_4 & \\ 0 & 0 & 0 & ps_5 \end{bmatrix} \end{aligned}$$

Writing the structural portion of the model in this manner reveals several important aspects of specifying this model. First, note that the latent variables E_1 and E_3 are equated with the disturbances z_1 and z_3, respectively. The covariance between these two variables is then treated as exogenous to the model (i.e., unexplained by causal relations) by use of the free parameter ps_3 in $\underline{PS}$. Second, note that most off-diagonal elements of $\underline{B}$ can be fixed at zero because of the longitudinal nature of the experimental design. Third, note that two elements of $\underline{G}$ are fixed at zero because of the randomization procedure that determined variation in F_1. If the randomization procedure failed, then this constraint on $\underline{G}$ will produce a poor fit for the model. Fourth, it is clear that the structural portion of the model is overidentified; there are five free parameters in $\underline{E}$, two in $\underline{G}$, one in $\underline{PH}$, and five in $\underline{PS}$—a total of 13. There are five latent variables with a total of 15 variances and covariances; thus, there are two overidentifying restrictions in the structural portion of the model (deriving from the two elements in $\underline{G}$ that are fixed at zero).

The fit of this model can be compared with numerous other models that include different theoretical hypotheses. For example, the fit of this model could be contrasted with that under the constraint that

$$b_4 = 0 \qquad (b_3 \text{ free})$$

This constraint is consistent with the hypothesis that both authoritarianism and perceived threat are changed by the manipulation directly—there is no mediating role for perceived threat. If this null hypothesis cannot be rejected, then it must be accepted. The difference between the chi-squares for the two models could be used as the basis for deciding whether or not the null hypothesis could be rejected. (Equivalently, if the estimate of b_4 was greater than twice its standard error, then the null hypothesis would be rejected.)

The strength of this design is that actual changes in the dependent and mediating variables would be measured. Furthermore, the use of a randomized manipulation greatly strengthens the plausibility of assuming that z_2 and z_4 are independent of F_1. An extraordinary opportunity to test an hypothesis is provided by this design. However, if perceived threat has too much inertia, then the problem of multicollinearity may occur, and estimates of b_4 would be unreliable. In such an instance, the researcher would have to relinquish the longitudinal observation of perceived threat and delete E_1 (and its indicators) from the model.

It is now necessary to specify the measurement portion of the causal model. The measurement model is crucial in such a design, since it provides some evidence that the observations are at least consistent with the theoretical specifications concerning the variables measured. The measurement model could be specified as follows:

$$X_1 = E_1 + e_1 \tag{11.10a}$$

$$X_2 = j_1 E_1 + e_2 \tag{11.10b}$$

$$X_3 = j_2 E_1 + e_3 \tag{11.10c}$$

$$X_4 = E_2 + e_4 \tag{11.10d}$$

$$X_5 = j_3 E_2 + e_5 \tag{11.10e}$$

$$X_6 = j_4 E_2 + e_6 \tag{11.10f}$$

$$Y_1 = E_3 + e_7 \tag{11.10g}$$

$$Y_2 = j_5 E_3 + e_8 \tag{11.10h}$$

$$Y_3 = j_6 E_3 + e_9 \tag{11.10i}$$

$$Y_4 = E_4 + e_{10} \tag{11.10j}$$

$$Y_5 = j_7 E_4 + e_{11} \tag{11.10k}$$

$$Y_6 = j_8 E_4 + e_{12} \tag{11.10l}$$

and

$$X = F_1 \tag{11.11}$$

Each observed indicator is expressed as a linear function of a latent variable and measurement errors—except for the manipulation variable (X), which is measured without error.

Equations 11.10 and 11.11 can be rewritten in the LISREL format (see Table 10.1) as follows:

$$\underline{Y} = \underline{Ly}\,\underline{E} + \underline{e}$$

$$\underline{X} = \underline{Lx}\,\underline{F} + \underline{d}$$

where

$$\underline{Y} = [X_1 \quad X_2 \quad X_3 \quad X_4 \quad X_5 \quad X_6 \quad Y_1 \quad Y_2 \quad Y_3 \quad Y_4 \quad Y_5 \quad Y_6]'$$

$$\underline{Ly}\,\underline{E} = \begin{bmatrix} 1 & 0 & 0 & 0 \\ j_1 & 0 & 0 & 0 \\ j_2 & 0 & 0 & 0 \\ 0 & 1 & 0 & 0 \\ 0 & j_3 & 0 & 0 \\ 0 & j_4 & 0 & 0 \\ 0 & 0 & 1 & 0 \\ 0 & 0 & j_5 & 0 \\ 0 & 0 & j_6 & 0 \\ 0 & 0 & 0 & 1 \\ 0 & 0 & 0 & j_7 \\ 0 & 0 & 0 & j_8 \end{bmatrix} \begin{bmatrix} E_1 \\ E_2 \\ E_3 \\ E_4 \end{bmatrix}$$

$$\underline{e} = [e_1 \quad e_2 \quad e_3 \quad e_4 \quad e_5 \quad e_6 \quad e_7 \quad e_8 \quad e_9 \quad e_{10} \quad e_{11} \quad e_{12}]'$$

$$\underline{X} = [X]$$

$$\underline{Lx}\,\underline{F} = [1][F_1]$$

and

$$\underline{d} = [0]$$

In order to specify the autocovariances among the errors in the indicators, the following matrix is specified:

$$\underline{Te} = \text{covariance matrix of } e$$

$$= \begin{bmatrix} s_1 \\ 0 & s_2 \\ 0 & 0 & s_3 \\ s_4 & 0 & 0 & s_5 \\ 0 & s_6 & 0 & 0 & s_7 \\ 0 & 0 & s_8 & 0 & 0 & s_9 \\ 0 & 0 & 0 & 0 & 0 & 0 & s_{10} \\ 0 & 0 & 0 & 0 & 0 & 0 & 0 & s_{11} \\ 0 & 0 & 0 & 0 & 0 & 0 & 0 & 0 & s_{12} \\ 0 & 0 & 0 & 0 & 0 & 0 & s_{13} & 0 & 0 & s_{14} \\ 0 & 0 & 0 & 0 & 0 & 0 & 0 & s_{15} & 0 & 0 & s_{16} \\ 0 & 0 & 0 & 0 & 0 & 0 & 0 & 0 & s_{17} & 0 & 0 & s_{18} \end{bmatrix}$$

Note that the six free covariances in $\underline{Te}$ allow for the autocorrelation of measurement errors. This is a crucial specification when the same indicators are used at successive points in time.

The measurement portion of the model is clearly identified. There are eight free parameters in $\underline{Ly}$, 18 in $\underline{Te}$, one in $\underline{PH}$, five in $\underline{PS}$, two in $\underline{G}$, and five in $\underline{B}$. Thus, the model includes 39 free parameters that must be estimated from the observations. The 13 observed variables, on the other hand, generate $(13 \times 13 - 13)/2$ or 78 covariances and 13 variances. Since the model is identified, fitting the model to data generates (91 − 39) = 52 overidentifying restrictions.

The reader may be wondering at this point about the ML assumption that variables are multivariate normal in distribution. The variable distinguishing between the experimental and control groups is a dichotomous variable, and its levels are *fixed* rather than random. Furthermore, if the manipulation is successful, it may alter the variance of the latent variables within groups—thus violating the homoscedasticity or homogeneity of variance assumption. The LISREL program includes a procedure that allows for the specification of a model *for different populations.* That is, a structural model can be specified for the control group and the experimental group. Tests can then be conducted to determine if the same model holds in both populations and whether or not mean differences in latent variables are significantly different between the two populations. Although this procedure will be demonstrated in Section 11.7 with a nonequivalent control group design, the problems created by specification as outlined above will not, in most true experiments with large samples, deserve the added complexities that can be crucial when a control group is nonequivalent at the outset.

11.5 ANALYSIS OF NONEQUIVALENT CONTROL GROUP DESIGNS

As we leave the true experiment, we leave the realm of certitude in causal attribution. And we enter a much more complex and risk-laden realm. Nevertheless, as has been argued previously, many important problems cannot be addressed experimentally. Thus, the task before the social scientist is one of making the best possible inferences based on quasi-experiments or passive observation.

The longitudinal nonequivalent control group design is characterized by manipulation of an independent variable, pre-manipulation and post-manipulation measures of the hypothesized dependent variable, and *lack* of random assignment to groups. Use of this design is not to be encouraged. It incorporates most of the problems found in passive observational designs. But passive observational designs are usually necessary or preferable precisely because manipulation of the causal variables is inappropriate. Thus, the very fact that manipulation is incorporated into a design suggests that randomization should also be included. When lack of randomization is already a fact, or when circumstances make it *extremely* impractical to randomize, then application of structural equation models to the analysis can be very useful.

The problems for causal inference that are created by lack of randomization are essentially the "third variable problem." That is, any causal model that is specified may have excluded one or more variables that introduce a bias into the estimation of causal effects. These problems can be highlighted by a simple example that also serves to restate Lord's paradox.

Suppose that more men than women are included in a control group for a study evaluating a weight-reducing program, while more women than men are included in the program. (The reader might ask why such a silly mistake would be made. The point here is that, unlike sex, many differences between two existing groups are not apparent. The obvious effects of sex differences may be equaled or exceeded by differences that are not measured or even recognized by the investigator.) Now, suppose that people in the control group are matched with people in the program based on preprogram weight. The mean weight for men is 160 pounds, whereas the mean weight for women is 120 pounds. Now, what does this situation mean for men from the control group that are matched with women from the program? Consider persons weighing 140 pounds. If there is no effect of the program, what would we expect over time? Since weight changes closely fit a "linear regression to the mean" pattern, we would expect the men weighing 140 pounds to gain weight (regressing toward their population mean of 160), while women weighing 140 pounds would lose weight (regressing toward their mean of 120). Thus, if we did not know to control for sex differences, we would probably conclude that an ineffective program was effective.

This example also suggests the importance of *random sampling* in quasi-experiments. The internal validity of true experiments is only peripherally related to random sampling from a population of cases; the external validity of true experiments is, of course, almost always dependent upon random sampling. On the other hand, random sampling greatly enhances the internal validity of the longitudinal nonequivalent control group design, since increases or decreases over time are less likely to occur due to entropic decay of an extreme group. But, once again, if random sampling *and* manipulation are possible, most of the blocks to randomization are also not present. Thus, most instances of nonequivalent control groups should not involve random sampling of cases for each group.

Although there is no ultimate solution to the excluded third variable problem, the use of multiple indicator measurement models can enable a researcher to control adequately for measured differences between groups. For example, in the assessment of educational programs in poverty areas, it is clear that comparisons between participant children and children from wealthier areas must be done in the context of controls for the socioeconomic status of the families and/or communities. Without the use of measurement models, it is unlikely that such controls will be adequate. And without adequate controls, we can expect a repeat of the confound described above in the case of a failure to control for sex in evaluation of a weight-reducing program.

Suppose, for example, that a preschool program in a poverty area is made available to families in a particular neighborhood. Now, the best way to evaluate the effect of such a program is to randomly select the participants and control children from the pool of applicants. On the average, then, members of both groups would be from the same population, and the pretreatment tests should have only a small correlation with group assignment. This would mean that any differences emerging over the following years could not be attributed to regression toward different pre-existing population means.

But now suppose that such randomization is not possible, or not chosen for some other reason. We are left with the problem of choosing a control group. One possibility is to select children from another neighborhood. The problem here is that the children attending the program may be from a different population and would therefore regress toward a different population mean. Another possibility would be to obtain the control group from the families in the same neighborhood that did not seek inclusion in the program. In this case, families seeking inclusion may systematically differ from those who do not—and these differences may produce patterns of change that confound assessment of treatment effects. The problem is that without randomization or detailed knowledge of population characteristics, a researcher must make a guess concerning what to expect if the program has no effect.

The question of what to expect in a longitudinal non-equivalent control group design when a treatment has no effect is similar to the question of how to specify the null hypothesis in a statistical model used to analyze results from such a study. Suppose, for example, that the model

$$Y_2 = b_0 + b_1X + b_2F + b_3Y_1 + z \qquad (11.12)$$

where

$$\begin{aligned} Y_2 &= \text{post-treatment measure of dependent variable} \\ Y_1 &= \text{pretreatment measure of dependent variable} \\ X &= \text{treatment variable} \\ F &= \text{other factor(s) influencing } Y_2 \end{aligned}$$

is the true causal model relating reading skills to a preschool program and other factors. Now, if z is independent of the other right-hand variables in equation 11.12, we know that

$$b_1 = [\text{Cov}(Y_2,X) - b_2\,\text{Cov}(F,X) - b_3\,\text{Cov}(Y_1,X)]/\text{Var}(X)$$

However, suppose that F is not measured, then the coefficient a_1 in

$$Y_2 = a_0 + a_1X + a_2Y_1 + e$$

is given by

$$\begin{aligned} a_1 &= [\text{Cov}(Y_2,X) - a_2\,\text{Cov}(Y_1,X)]/\text{Var}(X) \\ &= (c_1 - c_2c_3)/(1 - c_3c_4) \end{aligned}$$

where

$$\begin{aligned} c_1 &= \text{slope of } Y_2 \text{ on } X \\ c_2 &= \text{slope of } Y_2 \text{ on } Y_1 \\ c_3 &= \text{slope of } Y_1 \text{ on } X \\ c_4 &= \text{slope of } X \text{ on } Y_1 \end{aligned}$$

when the residuals e are constrained to a minimum variance. With the model in equation 11.12, it is straightforward to show that

$$a_1 = b_1 + [b_2(c_5 - c_6c_3)/(1 - c_3c_4)]$$

where

$$\begin{aligned} c_5 &= \text{slope of } F \text{ on } X \\ c_6 &= \text{slope of } F \text{ on } Y_1 \end{aligned}$$

Thus, when b_2 is positive, the exclusion of F biases a_1 as an estimate of b_1 in the *positive* direction if

$$c_5 > c_6 c_3$$

and in the *negative* direction if

$$c_5 < c_6 c_3$$

This means that misspecification of the causal model in a program evaluation can give rise to the erroneous conclusion that moderately effective programs are harmful—as has been the case in a number of instances (cf. Campbell and Erlebacher, 1970).

Similar problems of biased estimation arise when F is measured with error. And even more problems arise if Y includes measurement error—especially autocorrelated errors. Fortunately, the measurement models outlined in Chapter 9 can be used to extract these sources of bias from model estimation.

The reader is encouraged to approach the problem of missing third variables through measurement of the theoretically relevant variables and inclusion of additional pretreatment waves of observation. The inclusion of two or more pretreatment measures provides information about the rate of decay of pretreatment differences between the treatment and control groups. This information can be used to test structural models that incorporate unobserved variables. The following discussion outlines one such model.

A Structural Model with an Unmeasured Variable

Some methodologists (e.g., Kenny, 1975) have recommended the analysis of unconditional change scores in the case of a longitudinal nonequivalent control group design. The reason for this is that existing groups (such as neighborhoods or volunteers and non-volunteers) may have stable intergroup differences that do not decay over time. They do not decay because the causal processes maintaining those differences are approaching or have achieved equilibrium. But there are two problems with unconditional change score analyses of two waves of observation. First, a researcher cannot test various hypotheses concerning the equilibrium of intergroup differences; especially in the case of growth studies among children, it is likely that intergroup differences are approaching equilibrium, and differences are expanding or shrinking in the absence of any interventions. Second, the unconditional change score regression equation only tests for deviations from certain hypotheses about group differences (such as the "fan spread hypothesis," cf. Cook and Campbell, 1979) and cannot be used to estimate structural coefficients of a longitudinal model. The inclusion of two or more pretests in a nonequivalent control group design provides an opportunity to partially solve both of these problems. [Another partial solution is achieved through the use of an "instrumental variable" (cf., Linn and Werts, 1977).]

We begin by specifying the following structural model:

$$X = z_4$$

$$Y_1 = z_1$$

$$Y_2 = b_4Y_1 + b_5F + z_2$$
$$Y_3 = b_1Y_2 + b_2F + b_3X + z_3$$

where

$$X = -1 \text{ (control) or } +1 \text{ (treatment)}$$
$$Y_1 = \text{latent pretest score at time 1}$$
$$Y_2 = \text{latent pretest score at time 2 (program onset)}$$
$$Y_3 = \text{latent postest score at time 3 (after program)}$$
$$F = \text{unmeasured variable}$$

This model is depicted in Figure 11.11. Notice that although the measurement model is not given here, it is presumed that the variables (Y_1,Y_2,Y_3) are latent variables in a measurement model. We also assume that the same multiple indicators are used at each wave of observation so that error autocovariances can be estimated and slopes relating observed scores to true scores can be constrained to invariance over time. Adequate fit of such a measurement model assures that the units of the latent variable are comparable over different waves of observation. This means that changes in variance are attributable to factors such as growth rather than changes in the properties of the measuring instrument.

As it stands, the parameters of this model are hopelessly underidentified. However, the hypothesis that the causal system is *structually invariant* identifies all parameters

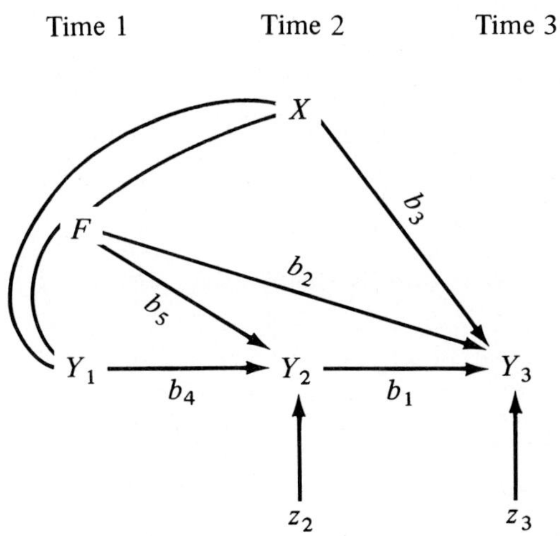

Figure 11.11 Path diagram of structural model incorporating an unmeasured variable for analysis of a longitudinal nonequivalent control group design with three waves of observation. The variable F is unmeasured, the variables Y are presumed to be latent variables in a multiple indicator measurement model.

and achieves one overidentifying restriction that can be used as a test of the hypothesis. The constraints corresponding to the hypothesis of structural invariance are as follows:

$$b_1 = b_4 \text{ (invariant inertia)}$$
$$b_2 = b_5 \text{ (invariant effect of } F)$$
$$\text{Var}(z_2) = \text{Var}(z_3) \text{ (invariant dispersion of disturbances)}$$
$$\text{Var}(F) = 1 \text{ (arbitrary constraint)}$$

If the fit of the model is adequate, then the estimate of b_3 (the effect of the program) can be compared with its estimated standard error. If the fit of the model is poor, then one or more of the assumptions incorporated into the model are probably incorrect. For example, the modeling of Y as first-order autoregressive (no direct effect from Y_1 to Y_3) may be in error. The researcher must then propose alternative models.

The reasoning behind the proposed model is that X may be correlated with Y_1 and Y_2 because of the unmeasured variable F, but X can have a direct effect on Y_3—since Y_3 follows the onset of the intervention program. The magnitude of b_3 is thus the impact of the treatment (e.g., special reading program) on the dependent variable that is of interest (e.g., reading comprehension). The model is quite restrictive, but it is at least testable with three waves of observation. Additional waves of observation prior to and subsequent to the program will provide additional overidentifying restrictions and greater opportunities to identify alternative models.

The following is a listing of statements used by LISREL IV to fit bogus data to the above model.

```
NONEQUIV CONTROL GROUP DESIGN WITH UNOBSERVED VAR-3 WAVES
DATA NI=4 NO=500 MA=CM
LABELS
*
'Y3' 'Y2' 'Y1' 'X'
CM
*
1.256
 .77    1.225
 .48    .68    1.10
-.19   -.25   -.180   1.00
MODEL NY=4 NE=5 LY=FI BE=FI PS=FI,SY TE=ZE
START 1.0 LY(1,1) LY(2,2) LY(3,3) LY(4,4)
FREE BE(1,2) BE(2,3)
FREE BE(1,5) BE(2,5) BE(1,4)
EQ BE(2,5) BE(1,5)
START 1.0 BE(1,1) BE(2,2) BE(3,3) BE(4,4) BE(5,5)
EQUATE BE(1,2) BE(2,3)
START  -.6 BE(1,2) BE(2,3)
START -.2 BE(1,5)  BE(2,5)
FREE PS(1,1) PS(2,2) PS(4,3)
FREE PS(4,4) PS(3,3)
EQ PS(1,1) PS(2,2)
START .6 PS(1,1) PS(2,2)
START 1.0 PS(3,3) PS(4,4) PS(5,5)
START  -.19 PS(4,3)
FREE PS(5,4) PS(5,3)
```

```
START   -.6 PS(5,4)
START .3 PS(5,3)
START .599 PS(1,1) PS(2,2)
START -.630 BE(1,2) BE(2,3)
START -.638 PS(5,4)
START .309  PS(5,3)
START -.08 BE(1,4)
OU MR SE SS TO
```

Problem Suppose that only waves 2 and 3 in the above LISREL IV program were available for analysis. Regress Y_3 on Y_2 and X; show that this result understates the effect of the program. Then regress $(Y_2 - Y_1)$ on X; show that this result also understates the effect of the program.

Now suppose that only waves 1 and 2 were available. What would the fan spread hypothesis lead you to conclude? (The fan spread hypothesis states that intergroup differences expand at the same rate as the standard deviations of characteristics within groups. Thus, in the absence of a treatment effect, we would expect stability in the *correlation* between Y and X over time.)

Due to the generality of the problems arising in the evaluation of Head Start programs, we will now use a structural equation model to analyze data collected by Barnow (1973). Unfortunately, these data are not truly longitudinal in that the dependent measure was only observed at one point in time—after the treatment. Nevertheless, the data lend themselves to a multiple indicator measurement model for background factors, and extension to a truly longitudinal design would be straightforward.

The Case of an Evaluation of a Head Start Program

An argument that has been made many times in this book is that multiple regressions involving variables measured with error can be very misleading. The use of nonequivalent control group designs for the evaluation of Head Start programs has provided researchers with ample opportunity to produce regression equations with what are probably inadequate "statistical controls" for status differences between control and treatment groups (e.g., Cicirelli et al., 1969).

One attempt to apply the structural equation approach to Head Start evaluation data was reported by Magidson (1977). Using data from Barnow (1973), he claimed that taking account of measurement error in socioeconomic status indicators reversed the conclusion drawn about a particular Head Start Program. Whereas the usual regression equation indicated a negative effect of the program, Magidson's LISREL model suggested a *positive* effect. Unfortunately for the advocates of Head Start, the positive coefficient in Magidson's model was not statistically significant. Furthermore, it turned out that Magidson had made an error in computation of the correlation matrix that was discovered only after he and Bentler and Woodward (1978) had published their results (cf. Magidson, 1978).

A reanalysis of Barnow's corrected data for 303 white children will be presented in the following few paragraphs. We will first estimate a model similar to that reported by Magidson and then Bentler and Woodward. This first model assumes that the treatment and control groups are sampled from the same population. As found pre-

Table 11.9 Correlation matrix for evaluation of a Head Start Program (from Magidson, 1978).

	Y_1	Y_2	X_1	X_2	X_3	X_4	X
Y_1	1.000						
Y_2	.652	1.000					
X_1	.263	.275	1.000				
X_2	.246	.215	.516	1.000			
X_3	.217	.255	.258	.285	1.000		
X_4	.116	.190	.337	.209	.407	1.000	
X	−.097	−.094	−.086	−.084	−.220	−.179	1.000

Note. The variables included in this table are described in the text.

viously, the effect of the program is estimated to be positive, but not significant at the 5 percent level.

After outlining this first model, we will move to an approach that will be more generally applicable in such situations. This second approach involves specification of basically the same structural model, but the equality of model parameters between the two groups can be tested rather than simply assumed. This second approach involves specifying the structural model *separately* for each population (treatment and control), and then testing the equality of parameters between populations. The conclusion reached for these data is that the assumption of comparable population characteristics cannot be rejected.

The correlation matrix for Barnow's data is given in Table 11.9. The observed variables in the study included the following:

Y_1 = Illinois Test of Psycholinguistic Abilities

Y_2 = Metropolitan Readiness Test

X_1 = Mother's education (several categories)

X_2 = Father's education (several categories)

X_3 = Father's occupation (status index)

X_4 = Family annual income

X = Treatment (0 = control, 1 = treatment)

Variables X_1 to X_4 are considered background "control" variables, Y_1 and Y_2 are the dependent variables, and X measures the categorical distinction between control and treatment groups.

It is apparent from the correlations between X and the other variables in Table 11.9 that the treatment group was not only lower than the control group on the measures of cognitive ability, it was also lower on all four measures of socioeconomic status. It is also apparent that performance on the "abilities" tests is positively correlated with all four background measures. The question that these data raise is that of what the correlation between X and the abilities measures would be if the children were from comparable status families. One approach to answering this question would be to regress one of the dependent variables on the treatment variable *controlling for* the measured background characteristics. The following results are then obtained

$$Y_1 = \underset{(.067)}{.171X_1} + \underset{(.065)}{.120X_2} + \underset{(.062)}{.142X_3} - \underset{(.063)}{.033X_4} - \underset{(.057)}{.047X} \qquad (11.13a)$$

$$\text{R-square} = .107$$

$$Y_2 = \underset{(.066)}{.183X_1} + \underset{(.065)}{.062X_2} + \underset{(.062)}{.167X_3} + \underset{(.062)}{.042X_4} - \underset{(.056)}{.029X} \qquad (11.13b)$$

$$\text{R-square} = .117$$

where all variables are standardized. Although not large relative to their standard errors, the crucial coefficients of X remain negative.

Upon reflection, it should be clear that demographic reports of parental characteristics are unlikely to be flawless measures of the socioeconomic differences between the control and treatment groups. However, the cluster of background variables may be sufficient to identify a latent variable model that could estimate the measurement error in each indicator. Before drawing any sort of conclusion about the effectiveness of the Head Start Program from equations 11.13, it would be necessary to explore the measurement error question.

One such measurement model may be specified as follows. (For demonstration purposes, and because the subsequent model is specified in metric form, the current model will be specified in terms of the correlation matrix rather than the covariance matrix. In general, this is not a good practice, but since no use is made of the natural metrics in this model, nothing is lost.):

Illinois Test	$Y_1 = E_1 + e_1$
Metro Test	$Y_2 = E_1 + e_2$
Mother's ed	$X_1 = k_1F_1 + k_2F_2 + d_1$
Father's ed	$X_2 = k_3F_1 + k_4F_2 + d_2$
Father's occup	$X_3 = k_5F_1 + k_6F_3 + d_3$
Income	$X_4 = k_7F_1 + k_8F_3$
Treatment	$X = F_4$

Note that this measurement model specifies a general status factor that is measured by all four background indicators. In addition, there is a parental education factor (F_2) and a parental economic factor (F_3). The two ability measures are presumed to have the same correlation with the latent variable E_1, as indicated by the loading of one for both indicators. Finally, it is presumed that the income indicator will be completely explained by F_1 and F_3, such that no measurement error term is needed.

The reader may wonder how such a measurement model is concocted. In fact, it is the result of an interplay between the categories used by the social scientist and the structure of the data. The notion of a single dimension of socioeconomic status, for

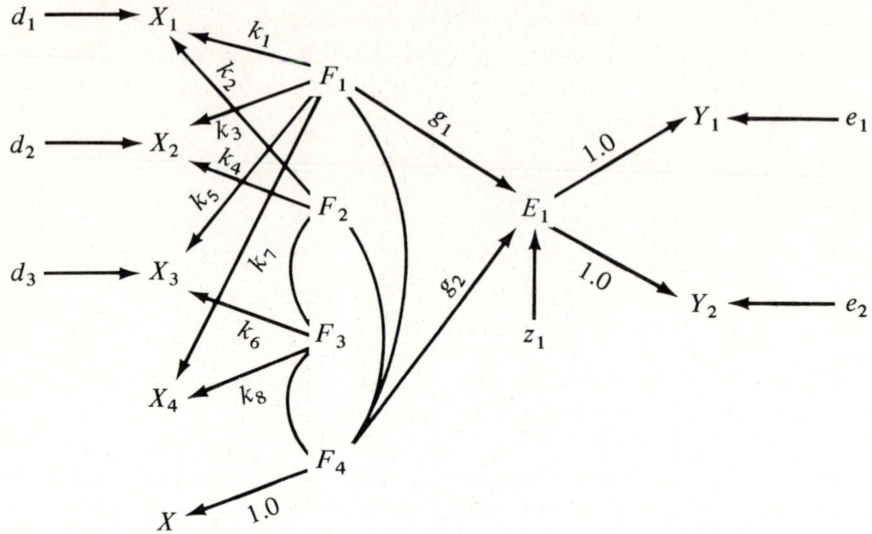

Figure 11.12 A structural equation model for analysis of a nonequivalent control group design with only one wave of observation (see equation 11.14). The model is adapted from Magidson (1977), and the data for testing the model are from Barnow (1973).

example, is a common one among social scientists. Similarly, the idea that income and occupational status are measuring something similar, and the idea that parental educations reflect a common latent variable is from the realm of "common sense" among social scientists. The remaining particulars of the model stem from "playing with the data." That is, several alternative specifications are tested, and some must be discarded.

We are now prepared to specify the remainder of the model. Magidson and Bentler and Woodward proposed the following:

$$E_1 = g_1F_1 + g_2F_4 + z_1 \tag{11.14}$$

as the structural portion of the model. Note that only the general SES factor and the treatment are presumed to affect the cognitive measures. The structural and measurement models are depicted in Figure 11.12.

In LISREL format, the model may be specified as follows:

$$\underline{B}\,\underline{E} = \underline{G}\,\underline{F} + \underline{z}$$

$$\underline{Y} = \underline{Ly}\,\underline{E} + \underline{e}$$

and

$$\underline{X} = \underline{Lx}\,\underline{F} + \underline{d}$$

where

$$\underline{B}\,\underline{E} = [1][E_1]$$

$$\underline{G}\,\underline{F} = [g_1 \quad 0 \quad 0 \quad g_2] \begin{bmatrix} F_1 \\ F_2 \\ F_3 \\ F_4 \end{bmatrix}$$

$$\underline{z} = [z_1]$$

$$\underline{Y} = [Y_1 \quad Y_2]'$$

$$\underline{Ly}\,\underline{E} = \begin{bmatrix} 1 \\ 1 \end{bmatrix} [E_1]$$

$$\underline{e} = [e_1 \quad e_2]'$$

$$\underline{X} = [X_1 \quad X_2 \quad X_3 \quad X_4 \quad X]'$$

$$\underline{Lx}\,\underline{F} = \begin{bmatrix} k_1 & k_2 & 0 & 0 \\ k_3 & k_4 & 0 & 0 \\ k_5 & 0 & k_6 & 0 \\ k_7 & 0 & k_8 & 0 \\ 0 & 0 & 0 & 1 \end{bmatrix} \begin{bmatrix} F_1 \\ F_2 \\ F_3 \\ F_4 \end{bmatrix}$$

$$\underline{d} = [d_1 \quad d_2 \quad d_3 \quad d_4 \quad d_5]'$$

And then, finally, complete specification of the model comes from the following covariance matrices:

$$\underline{PH} = \begin{bmatrix} 1 & & & \\ 0 & 1 & & \\ 0 & ph_1 & 1 & \\ ph_2 & ph_3 & ph_4 & 1 \end{bmatrix}$$

$$\underline{PS} = [ps_1]$$

$$\underline{Te} = \begin{bmatrix} s_1 & \\ 0 & s_1 \end{bmatrix}$$

$$\underline{Td} = \begin{bmatrix} t_1 & & & & \\ 0 & t_2 & & & \\ 0 & 0 & t_3 & & \\ 0 & 0 & 0 & 0 & \\ 0 & 0 & 0 & 0 & 0 \end{bmatrix}$$

Note that the matrix $\underline{PH}$ fixes the variances of the exogenous latent variables at one and constrains the general SES factor (F_1) to be uncorrelated with F_2 and F_3. The matrix $\underline{Te}$ constrains the variances of the endogenous measurement errors to be equal (they are both equal to s_1). And the matrix $\underline{Td}$ is specified such that the measurement errors in the exogenous variables are uncorrelated with one another, and the variance of the errors in F_4 and F_5 is constrained to zero (indicating no measurement error).

The results of the ML estimation of this model with LISREL IV are given in Table 11.10. Note that the estimate of g_2 is positive, but only slightly larger than its standard error. Thus, we might conclude that there was "a trend suggestive of a positive effect."

Table 11.10 Results of the ML estimation of the structural model for data from Barnow's (1973) evaluation of a Head Start Program.

Parameter		Estimate	Std err	Comments
Ly	j_1	1.000	fixed	Standardized value = .807
	j_2	1.000	fixed	Standardized value = .807
Lx	k_1	.478	.094	Mother ed on SES
	k_2	.782	.178	Mother ed on parental ed
	k_3	.448	.086	Father ed on SES
	k_4	.386	.113	Father ed on parental ed
	k_5	.491	.083	Father occ on SES
	k_6	.272	.068	Father occ on parental econ
	k_7	.302	.099	Income on SES
	k_8	.953	.047	Income on parental econ
G	g_1	.548	.108	SES effect on ability
	g_2	.108	.094	Treatment effect, stdized = .13
PH	ph_1	.255	.103	Parental ed w/parental econ
	ph_2	−.370	.119	SES w/treatment
	ph_3	.122	.101	Parental ed w/treatment
	ph_4	−.071	.070	Parental econ w/treatment
PS	ps_1	.383	.095	Standardized value = .588
Te	s_1	.348	.028	Error in ability measures
Td	t_1	.160	.266	Error in mother ed
	t_2	.650	.087	Error in father ed
	t_3	.686	.071	Error in father occ

Note. The chi-square for this model was 7.84 with 8 degrees of freedom; probability = .5505.

[Bentler and Woodward (1978) report a considerably larger standardized effect for a comparable model. However, their estimates were based upon a correlation matrix that was subsequently found to be incorrect (cf. Magidson, 1978), and the ratio of their coefficient to its standard error was comparable to the corrected solution given above.] This result should be somewhat encouraging to supporters of Head Start, since this particular data set was one of the most negative prior to these adjustments for measurement error. But the most important point to be made by this exercise is that the usual procedure of "controlling for social class differences" in multiple regression equations is probably misleading. The inclusion of an index of occupational prestige or parental education as covariates is unlikely to produce adequate statistical adjustments. (An important ancillary finding in this study is the huge effect of SES on reading ability. The standardized form of this effect is 0.679—a value suggesting that almost one-half of the true variance in such test scores is reflective of differences in familial socioeconomic status. This strong association between SES and tests purported to measure cognitive ability will be revisited in a subsequent example.)

The reader should of course keep in mind that the measurement model given above has not been "proven" to be correct with these data. Rather, the hypothesized measurement model has gained some credibility, since it *is* consistent with these data, whereas other alternative measurement models are not. (As an exercise, the reader should test some alternative measurement models to the one given above. This author has tried several alternatives and has been struck with the particular success of the above model. Furthermore, it is the author's expectation that a major theoretical and methodological problem for the social and behavioral sciences will be the construction and testing of competing models of one or more dimensions of SES.)

Separate Population Analysis

The structural model specified in equation 11.14 incorporates the distinction between the control and experimental groups in the Head Start evaluation with the dummy variable F_4. However, when the treatment and control groups are clearly drawn from nonequivalent populations, it may be that some of the parameters in the structural or measurement models do not have the same values in both populations. (This can also arise in randomized experiments when treatments *produce* such change.) For example, it is certainly plausible that reporting errors of parental education might be greater within a treatment group that included more parents with less education—or *vice versa.* The use of a dummy variable in equation 11.14 essentially forces the untested assumption that the structural and measurement parameters are the same for both populations.

Fortunately, LISREL IV allows the simultaneous estimation of structural and measurement models in two or more populations. A sensible approach to such an analysis would be to begin by constraining model parameters to be equal across populations. The likelihood chi-square statistic can then be used to assess the fit of the data to these constraints. If the data are consistent with the assumption of model invariance across populations, then the use of dummy variables (as in equation 11.14) in a single population model is appropriate. However, if there is a poor fit, then the population specific measurement and/or structural models should be explored.

In order to demonstrate the use of LISREL IV for simultaneous estimation in multiple populations, the Barnow data evaluating a Head Start Program will be so analyzed. The goal here is to see if the treatment and control groups differ on the *latent variable* of interest (cognitive ability) when the comparison is based on cases that have been equated on the latent variable SES. The only real difference between the specification of this multi-population model and the model specified with equation 11.14 is that a model will be specified for *each* population (cf. Sorbom, 1978, for a similar analysis of these same data).

The new problem that is added to the comparison of latent variable means between populations is that of identifying differences between these means. This is obviously a problem, since the latent variable scores are not observed. The identification of such means is approached in the same way the identification of other model parameters is approached, except that the expectations of variables are the focus rather than the covariances (expectations of products of variables measured about their respective means). This procedure of identification will be presented only briefly.

Suppose the latent variable F has been measured with three indicators in two populations. Suppose that the following measurement model holds for two populations:

$$Y_1 = j_1 + j_2 F + e_1$$
$$Y_2 = j_3 + j_4 F + e_2$$
$$Y_3 = j_5 + j_6 F + e_3$$

and

$$Y_1' = j_1 + j_2 F' + e_1'$$
$$Y_2' = j_3 + j_4 F' + e_2'$$
$$Y_3' = j_5 + j_6 F' + e_3'$$

where F is a latent variable, the measurement errors are mutually uncorrelated, the measurement errors are uncorrelated with the latent variable, and the prime indicates scores for the second of two populations. These equations indicate that the same model, with the same parameters, holds in both of these populations. An important assumption that was *not* included in this model concerns the means of the latent variable in the two populations. Although we do include the usual assumption that the measurement errors have zero expectations, this may not be the case for F and F'. In fact, the object of our concern at this point is precisely the population means of the latent variable.

Our goal is to see which parameters in this measurement model are identified by observation of Y (and Y'). We know from Chapter 9 that the constraint $j_2 = 1$ identifies j_4 and j_6—the congeneric model. Furthermore, since j_4 and j_6 are just identified in both populations, each is overidentified when the two populations are taken together. The remaining parameters are j_1, j_3, j_5 (the intercepts) and the expected values of F and F'.

In order to sort out the identification of these remaining parameters, consider the expected value of some of the equations:

$$E(Y_1) = j_1 + j_2 E(F)$$

and

$$E(Y_1') = j_1 + j_2 E(F')$$

Now, if j_2 is constrained to one, then we have

$$E(F) - E(F') = E(Y_1) - E(Y_1')$$

At this point we would be stumped except for the fact that our major concern is with the population differences rather than the actual means of the latent variables. This fact allows us to arbitrarily assign a value to one of the population means, say $E(F) = 0$. This constraint then overidentifies the remaining parameters of the model as follows:

$$j_1 = E(Y_1) \qquad j_3 = E(Y_2) \qquad j_5 = E(Y_3)$$

$$E(F') = [E(Y_1') - j_1]/j_2$$

$$j_1 = [E(Y_1') - E(F')]$$
$$j_3 = [E(Y_2') - j_4E(F')]$$
$$j_5 = [E(Y_3') - j_6E(F')]$$

Now that it is clear that differences between population latent variable means can be identified, it is important to see how these differences can be used in a structural model. The goal in this case is to assess what population differences would be if cases were made equivalent on some set of control variables (or covariates). For example, consider a simple structural model in which the latent variable E_2 is influenced by changes in a latent variable E_1:

$$E_2 = b_1E_1 + z_2$$
$$E_1 = z_1$$

where all variables have an expected value of zero. And a similar model holds in a second population:

$$E_2' = g_2 + b_1E_1' + z_2'$$
$$E_1' = g_1 + z_1'$$

except that only the disturbances *necessarily* have a mean of zero in this second population. The differences between the population means of E_1 and E_2 are modeled in the coefficients g_1 and g_2.

It should be clear that g_1 is the difference

$$g_1 = E(E_1') - E(E_1)$$
$$= E(E_1')$$

Thus, g_1 represents the difference between the two populations on the first of our latent variables. Now, the meaning of the intercept g_2 is a bit more subtle. If we take the expectation of the equation for E_2' we have

$$E(E_2') = g_2 + b_1E(E_1')$$
$$= g_2 + b_1g_1$$

where

g_1 = population difference in E_1

b_1 = effect of changes in E_1 on E_2

g_2 = difference between populations in E_2 that is *in addition* to that expected on the basis of population differences on E_1

Thus, estimation of g_2 would be equivalent to the estimation of the effect of some manipulation on the latent variable E_2 in the context of comparing only those cases that were comparable on E_1.

With this additional knowledge of how to identify latent variable means and how to specify multiple population models, we can resume our analysis of the Head Start

Table 11.11 Means and standard deviations of observed variables for white persons in the treatment and control conditions of a Head Start evaluation (from Magidson, 1978).

Variable	Means		Std. dev	
	Treat	Control	Treat	Control
Y_1 = Illinois test	19.67	20.41	3.76	3.90
Y_2 = Metro test	9.56	10.07	2.68	2.72
X_1 = Mother's ed	3.21	3.39	1.03	1.02
X_2 = Father's ed	3.08	3.29	1.28	1.20
X_3 = Father's occ	2.09	2.60	1.07	1.19
X_4 = Income (thousands)	5.36	6.44	2.65	3.24
	(N = 148)	(N = 155)		

evaluation data. Since LISREL IV is only partially geared to the analysis of population differences in latent variable means, the specification of the model will be a bit roundabout (subsequent versions of LISREL and other such programs will probably make the setup of similar problems more straightforward). We begin by attending to the within group means and standard deviations of the variables. These are given in Table 11.11.

Note that the means in Table 11.11 convey the same picture gleaned from the correlations given in Table 11.9. The treatment group is lower on the abilities tests, but it is also lower on the indicators of SES than the control group.

We will first specify the measurement model for each group. The model will initially be invariant across the two groups. Recall that the model used above depicted the four covariates as essentially indicators of a single dimension of SES and some "indicator-specific" factors that can be treated as measurement error. In order to use LISREL IV for this problem, it is necessary to specify all latent variables as endogenous. Then a dummy exogenous variable is equated to an endogenous variable. This little trick allows the specification of a model involving intercepts. The measurement portion of such a model is specified as follows:

$$Y_1 = j_1E_1 + E_2 + e_1 \tag{11.15a}$$

$$Y_2 = j_2E_1 + j_3E_2 + e_2 \tag{11.15b}$$

$$X_1 = j_4E_1 + E_3 + j_5E_4 + e_3 \tag{11.15c}$$

$$X_2 = j_6E_1 + j_7E_3 + E_4 + e_4 \tag{11.15d}$$

$$X_3 = j_8E_1 + j_9E_3 + E_5 + e_5 \tag{11.15e}$$

$$X_4 = j_{10}E_1 + j_{11}E_3 + j_{12}E_5 \tag{11.15f}$$

Note that E_2 is the cognitive ability factor, E_3 is the SES factor, E_4 is a factor specific to parental education indicators, and E_5 is a factor specific to indicators of fathers' economic status. The structural portion of the model is then specified as follows (note that E_1 is equated to F_1 and to one such that the first terms in equations 11.15 are intercepts):

Treatment group:

$$E_1 = F_1 + z_1 = 1 \quad [\mathrm{Var}(z_1) = 0] \tag{11.16a}$$

$$E_2 = b_1 E_3 + z_2 \tag{11.16b}$$

$$E_3 = z_3 \tag{11.16c}$$

$$E_4 = z_4 \tag{11.16d}$$

$$E_5 = z_5 \tag{11.16e}$$

Control group:

$$E_1 = F_1 + z_1 = 1 \quad [\mathrm{Var}(z_1) = 0] \tag{11.17a}$$

$$E_2 = g_1 F_1 + b_1 E_3 + z_2 \tag{11.17b}$$

$$E_3 = g_2 F_1 + z_3 \tag{11.17c}$$

$$E_4 = g_3 F_1 + z_4 \tag{11.17d}$$

$$E_5 = g_4 F_1 + z_5 \tag{11.17e}$$

Note that the only difference between the structural models for the two populations is the inclusion of intercepts for the control group (the usual convention would be to provide the intercepts for the treatment group, but such conventions are arbitrary). Also note that the exogenous variable F_1 is the constant one—as is the endogenous variable E_1. Both these constants serve only the purpose of adding the intercepts to the measurement model for both groups and to the structural model for the control group.

The model specified in equations 11.16 and 11.17 incorporates the constraint that the expected values of the latent variables are all zero in the treatment group (thus the exclusion of intercepts). This means that the parameters g_2, g_3, and g_4 are equal to the means of E_3, E_4, and E_5, respectively. Furthermore, the parameter g_1 is the net difference between the treatment and control groups on E_2 (cognitive ability)—after adjusting for differences between the two groups on E_3 (SES):

$$g_1 = [E(E_2|\text{control}) - E(E_2|\text{treatment})] - b_1 g_2$$

The actual LISREL IV input used to estimate this model is given in Table 11.12.

A thorough understanding of Table 11.12 will require a careful reading of the LISREL IV manual. However, several points can be gleaned even with little knowledge of the actual program. First, the data are input in three forms: correlations (KM), means (MEANS), and standard deviations (SD). Alternative inputs could be the covariance matrix and the means, or the cross products matrix. Second, note that a row of zeros (with a one in the diagonal position) was added to the correlation matrix. This row inputs the correlations between the observed indicators and the constant 1 that is used to specify the intercept terms in the model. It is the fixed X variable. Third, note that a seventh mean of 1.00 is added to the list of means for both populations, whereas a standard deviation of zero is also added for each population. Again, this specification is necessary to incorporate the constant term. Fourth, the LISREL IV program is informed that the X variable is to be treated as "fixed" by the inclusion of the word FIXED on the line beginning with the word MODEL.

Table 11.12 Input to LISREL IV for simultaneous estimation of the model specified in equations 11.16 and 11.17 in two populations

```
REANALYSIS OF BARNOW (1973) — WHITE TREATMENT GROUP
DATA NI=7 NO=148 NG=2 MA=MM
LABELS
*
'LING ABI' 'READINES' 'MOTH ED' 'FATH ED' FATH 'FATH OCC'
  'INCOME' 'TREATMEN'
MEANS
*
19.67 9.56 3.21 3.08 2.09 5.358 1.000
KM
*
1.000
 .6645  1.0000
 .2752   .2564  1.0000
 .2651   .1218   .4659  1.0000
 .2079   .2506   .2532   .2024  1.0000
 .0842   .1981   .3608   .1820   .3773  1.0000
 .0000   .0000   .0000   .0000   .0000   .0000  1.0000
SD
*
3.76 2.68 1.03 1.28 1.07 2.648 0.000
MODEL NY=6 NX=1 NK=1 NE=5 FIXED LY=FU,FI BE=FI GA=FI C
  PS=FI,SY TE=FI
START 1.0 GA(1,1)
PA  LY
*
1  0  0  0  0
1  1  0  0  0
1  0  0  1  0
1  0  1  0  0
1  0  1  0  0
1  0  1  0  1
MA  LY
*
20.05  1.000  0.000  0.000   0.000
 9.82   .600  0.000  0.000   0.000
3.300  0.000  1.000  0.500   0.000
3.190  0.000   .500  1.000   0.000
2.350  0.000   .500  0.000   1.000
5.910  0.000   .500  0.000  13.5
PA  BE
*
0  0  0  0  0
0  0  1  0  0
0  0  0  0  0
0  0  0  0  0
0  0  0  0  0
MA  BE
*
1  0    0   0  0
0  1  -2.0  0  0
0  0    1   0  0
```

```
0 0    0   1 0
0 0    0   0 1
PA  PSI
*
0
0  1
0  0  1
0  0  0  1
0  0  0  1  1
MA  PSI
*
0.0
0.0  1.2
0.0  0.0  0.8
0.0  0.0  0.0  0.5
0.0  0.0  0.0  0.0  0.5
PA  TE
*
1  1  1  1  1  0
MA  TE
*
1.6  1.3  0.5  0.600  0.5  0.0
OU  SE  TO  PM  MR  SS
REANALYSIS OF BARNOW (1973) — WHITE CONTROLS
DATA NI=7 NO=155 MA=MM
LABELS
*
'LING ABI' 'READINES' 'MOTH ED' 'FATH ED' 'FATH OCC'
  'INCOME' 'TREATMEN'
MEANS
*
20.41  10.07  3.39  2.60  6.435  1.000
KM
*
1.000
 .6350  1.0000
 .2392   .2812  1.0000
 .2148   .2967   .5615  1.0000
 .1962   .2338   .2241   .3416  1.0000
 .1146   .1617   .3059   .2154   .3874  1.0000
 .0000   .0000   .0000   .0000   .0000   .0000  1.0000
SD
*
3.90  2.72  1.02  1.20  1.19  3.239  0.000
MODEL NY=6 NX=1 NK=1 NE=5 FIXED LY=IN BE=IN GA=FI PS=IN TE=IN
START 1.0 GA(1,1)
FREE GA(2,1)-GA(5,1)
OUTPUT
```

Note. the data used in this table are from Magidson (1978).

The results of estimating equations 11.16 and 11.17 for the two groups are quite extensive and will only be highlighted here:

For both treatment and control groups:

$$
\begin{aligned}
Y_1 &= \underset{(.30)}{19.68} + E_2 + e_1 \\
Y_2 &= \underset{(.22)}{9.55} + \underset{(.11)}{.74}E_2 + e_2 \\
X_1 &= \underset{(.08)}{3.21} + E_3 + \underset{(.78)}{1.68}E_4 + e_3 \\
X_2 &= \underset{(.09)}{3.04} + \underset{(.20)}{1.09}E_3 + E_4 + e_4 \\
X_3 &= \underset{(.09)}{2.11} + \underset{(.27)}{1.07}E_3 + E_5 + e_5 \\
X_4 &= \underset{(.24)}{5.36} + \underset{(.59)}{1.61}E_3 + \underset{(2.14)}{9.16}E_5
\end{aligned}
$$

For the control group:

$$
\begin{aligned}
E_1 &= F_1 = 1 \\
E_2 &= \underset{(.728)}{-.844} + \underset{(1.35)}{3.99}E_3 + z_2 && [\mathrm{Var}(z_2) = \underset{(1.56)}{5.31}] \\
E_3 &= \underset{(.164)}{.391} + z_3 && [\mathrm{Var}(z_3) = \underset{(.10)}{0.24}] \\
E_4 &= \underset{(.132)}{-.130} + z_4 && [\mathrm{Var}(z_4) = \underset{(.13)}{0.23}] \\
E_5 &= \underset{(.054)}{.049} + z_5 && [\mathrm{Var}(z_5) = \underset{(.05)}{0.10}]
\end{aligned}
$$

$$\text{chi-square} = 31.68,\ df = 27,\ \text{probability} = .244$$

The first important finding here is that the assumption of model invariance between the two populations appears to be a reasonable hypothesis. The likelihood ratio chi-square is close to the degrees of freedom in magnitude, and the probability of .244 is clearly acceptable. (The reader is invited to relax various between population constraints to see if a better fitting result can be achieved.) In view of this good fit, we would expect that the results obtained with a model that used a dummy variable (instead of specifications for separate populations) will give comparable results to these here. In fact, this is the case.

Note that the equation for E_2 above has an intercept term of $-.844$. This value means that the expected value of E_2 in the control group is .844 units *lower* than in the treatment group when differences in SES are partialled. This coefficient may be decomposed as follows:

$$(\text{mean } E_2 | \text{control}) = -.844 + 3.99(.391)$$
$$= -.844 + 1.559$$
$$= .714$$

or

$$.714 - 1.559 = -.844$$

In words, these coefficients mean that the model would have predicted that the control group would be 1.559 points above the treatment group on ability (E_2) because of the higher SES of the control group. However, the treatment group was only .714 points below the control group. Thus, the difference between the 1.559 and the 0.714 may be attributed to the effect of the program. However, as in the previous analysis of these data, the difference of −.844 is only about the magnitude of one standard error, and thus it is quite likely that this result would occur if the program had no effect. This means that a competing explanation of the positive estimate of the effect of the program cannot be rejected—that the program had no effect and the positive estimate was due to chance variation.

11.6 CAUSAL ANALYSIS OF PASSIVE LONGITUDINAL DESIGNS

Introduction This final substantive section of Chapter 11 is not arbitrarily situated at the end of this book; rather, it is the culmination of all that has been developed up to this point. The passive longitudinal design is one that involves no manipulation of variables, but does incorporate observation of change.

It is the author's expectation that passive longitudinal designs will provide the most fruitful soil for the application and further development of structural equation modeling. This expectation stems from two assertions. First, the passive longitudinal design brings otherwise inaccessible causal relations back into the realm of empirical investigation. In abstract terms, this assertion holds that the passive longitudinal design can be the meeting ground between the two methodological parents of the modern human sciences: "grand" social or historical theory and the physical sciences. The offspring might be characterized as a causal analysis of history or a historical investigation of causal relations. In contrast with their offspring, the causal relations of the physical sciences are usually ahistorical, and the evidence of the social theorist or historian has tended to be meager or of dubious generalizability. Second, whereas the results of experiments are often best understood from graphs or charts, formal mathematical models are almost essential to the untangling of complex longitudinal interrelations among variables that are of interest to social and psychological scientists. Linear equations are an obvious starting place for the building of such formal mathematical models.

Some Misconceptions about Passive Longitudinal Designs

If the reader has absorbed what has been said about causal modeling up to this point, then the following few paragraphs will be redundant. However, my experience is that the "what if" status of causal modeling with passive longitudinal data is easily lost in the concreteness of path coefficients.

A number of authors have considerably overstated the case concerning causal inferences from passive longitudinal designs. Perhaps the most extreme is that of Crano and Mellon (1978):

> With the introduction of the cross-lagged panel correlation method . . . , causal inferences based on correlational data obtained in longitudinal panel studies can be made and enjoy the same logical status as those derived in the more standard experimental settings (p. 41, cited by Rogosa, 1980).

As has been argued many times in this book, there is no rival to the randomized experiment where assurance of causal attribution is concerned. Randomization guarantees the long-term validity of the assumption that the disturbances are uncorrelated with the independent variables in the model. Furthermore, Crano and Mellon are referring to the comparison of cross-lagged correlations as the basis for causal inference. And as we saw in section 11.3 and as Rogosa (1980) demonstrates, the cross-lagged correlations can be unequal even though the actual cross-lagged *effects* are zero.

But there is an equivalent and more indirect version of the same fallacy in interpretation of longitudinal data. This other version of the fallacy confuses two conclusions: the first conclusion is that a set of observed relations are consistent with what we would predict from a certain causal model; the second conclusion is that no other causal model except that one estimated could have produced the observed relations. Only with the true experiment can an investigator even approach *both* of these conclusions. With data from a passive longitudinal design, an investigator is constrained to the first conclusion plus the following: the data are *inconsistent* with *certain* other competing models that have been *specifically tested* and *rejected.* It is this powerful capacity of longitudinal data to reject such *restricted* spurious explanations of observed associations that is the primary focus of the remainder of this section.

The above-described fallacy manifests itself most tenaciously in the belief that there is a pattern of longitudinal correlations or path coefficients that could *only* occur if a particular causal effect was occurring among the observed variables. There is no such pattern of covariations. Any observed pattern of covariations can be explained by a plausible model that depicts all observed associations as spurious. If this fact is not obvious to the reader, it should be by the end of Section 11.6.

A Lagged Effects Model

A lagged effects model was specified in Section 11.3 with equations 11.7a and 11.76b (see Figure 11.7):

$$Y_{i+1} = a_0 + a_1X_i + a_2Y_i + z_{i+1}$$
$$X_{i+1} = b_0 + b_1X_i + b_2Y_i + t_{i+1}$$

Indentification of the parameters of this model is straightforward if it is assumed that the disturbances are independent of the lagged variables and that the expected value of the disturbances is zero for each value of the lagged variables.

The most salient exclusions in such a model are the absence of synchronous effects (e.g., from X_3 to Y_3 and/or from Y_3 to X_3) and the absence of lagged effects extending beyond one time period (lag). If the "true" causal process involves any of these excluded causal paths, then estimates of the parameters in equations 11.7 can be misleading about the effect that a manipulation would have—if a manipulation were possible.

A Synchronous Effects Model

There are instances in which the hypothesis of cross-lagged effects is not appropriate. For example, suppose that measurements of mood are obtained for new students at a college. Subsequently, these students become paired as roommates according to some unknown process. After the new roommates interact for a few days, measures of mood are repeated. In such a design, we would not expect a lagged effect of the mood of one roommate on the other; rather, we would only expect synchronous effects.

A causal model for such a longitudinal process could be specified as follows:

$$Y_2 = b_1X_2 + b_2Y_1 + z_1 \tag{11.18a}$$

$$X_2 = b_3Y_2 + b_4X_1 + z_2 \tag{11.18b}$$

where all observed variables are expressed as deviations from their respective means. The identification problem is a bit more complex in this instance, since the usual specification that the disturbances are uncorrelated with the right-hand variables in their equation is clearly untenable. Note, for example, that z_1 is a determinant of Y_2 (equation 11.18a), that Y_2 is, in turn, a determinant of X_2 (equation 11.18b), and that X_2 is a right-hand variable in the equation for which z_1 is the disturbance. The relationship between z_1 and X_2 is depicted in Figure 11.13 as the path leading from z_1 to Y_2 to X_2.

The constraints that allow identification of the parameters of the synchronous effects model turn out to be the same as those that identified the lagged effects model,

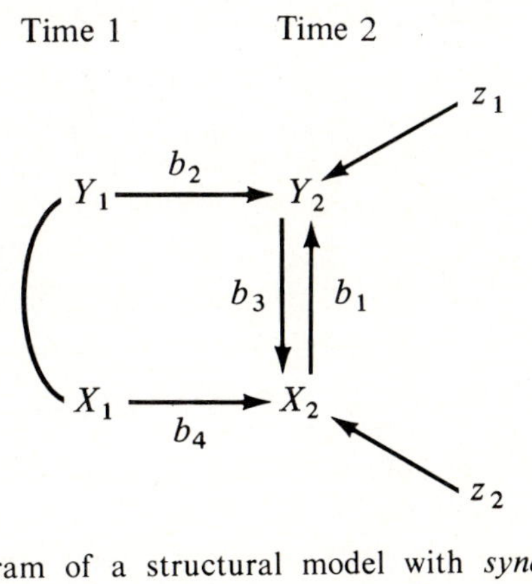

Figure 11.13 Path diagram of a structural model with *synchronous* reciprocal effects between longitudinally observed variables. For a path diagram of a structural model with *lagged* reciprocal effects, see Figure 11.7.

namely, the constraint that the disturbances are uncorrelated with the *lagged* variables (X_1 and Y_1). The reader can demonstrate this identification as an exercise, noting that equation 11.18 implies

$$\begin{aligned}
\mathrm{Cov}(Y_2,X_1) &= b_1\,\mathrm{Cov}(X_2,X_1) + b_2\,\mathrm{Cov}(Y_1,X_1) + \mathrm{Cov}(X_1,z_1)\\
\mathrm{Cov}(Y_2,Y_1) &= b_1\,\mathrm{Cov}(Y_1,X_2) + b_2\,\mathrm{Var}(Y_1) + \mathrm{Cov}(Y_1,z_1)\\
\mathrm{Cov}(X_2,X_1) &= b_3\,\mathrm{Cov}(Y_2,X_1) + b_4\,\mathrm{Var}(X_1) + \mathrm{Cov}(X_1,z_2)\\
\mathrm{Cov}(Y_1,X_2) &= b_3\,\mathrm{Cov}(Y_2,Y_1) + b_4\,\mathrm{Cov}(Y_1,X_1) + \mathrm{Cov}(Y_1,z_2)
\end{aligned}$$

If we extend the notion of the synchronous effects model beyond two waves of observation, then we are led to an investigation of the equilibration process. A structurally invariant synchronous effects model can come to equilibrium in a manner analogous to the equilibrating process of the lagged effects model. For example, if the model in equations 11.18 holds for some population, and the following initial conditions are given

$$\begin{aligned}
\mathrm{Cov}(Y_1,X_1) &= 0\\
\mathrm{Var}(Y_1) &= \mathrm{Var}\,(X_1) = 1.0\\
\mathrm{Var}(z_1) &= .36\\
\mathrm{Var}(z_2) &= .64\\
b_1 &= .15 \quad b_2 = .8 \quad b_3 = .15 \quad b_4 = .6
\end{aligned}$$

then the following equilibrating process would ensue:

Wave (i)	$V(Y_i)$	$V(X_i)$	Y_iX_i	Y_iX_{i-1}	$Y_{i-1}X_i$	Y_iY_{i-1}	X_iX_{i-1}
1	1.000	1.000	.000	—	—	—	—
2	1.070	1.070	.314	.092	.123	.818	.614
3	1.165	1.145	.486	.356	.324	.905	.695
4	1.255	1.201	.589	.503	.442	.998	.762
5	1.331	1.238	.653	.592	.515	1.081	.809
...							
38	1.588	1.321	.804	.780	.689	1.373	.910
39	1.588	1.321	.804	.780	.689	1.373	.910
...							

where the first two columns are variances and the remaining are covariances. It was assumed in this simulation that the disturbances were entirely random at each wave (i.e., no autocorrelations or cross correlations between disturbances). Note that the cross-lagged covariances (and correlations) are asymmetric even though the synchronous effects are equal in both directions: the variable with the greater inertia (Y, in this case) has the larger cross-lagged association when it is lagged behind the more entropic variable (X).

When this causal system has reached equilibrium, it can also be shown that the following covariances are found for variables separated by two lag periods rather than one:

$$\text{Cov}(Y_i,X_{i-2}) = .722 \quad \text{Cov}(Y_{i-2},X_i) = .591$$
$$\text{Cov}(Y_i,Y_{i-2}) = 1.187 \quad \text{Cov}(X_i,X_{i-2}) = .654$$

Problem. Solve for b_1, b_2, b_3 and b_4 in terms of the covariances among X_1, X_{i-1}, Y_i, and Y_{i-1}. Use the covariances among X and Y at time points 4 and 5 in the above simulation to solve for the four structural coefficients.

Help:

$$b_1 = \frac{\text{Cov}(Y_i,X_{i-1})\ \text{Var}(Y_{i-1}) - \text{Cov}(Y_i,Y_{i-1},)\text{Cov}(Y_{i-1},X_{i-1})}{\text{Cov}(X_i,X_{i-1})\ \text{Var}(Y_{i-1}) - \text{Cov}(X_{i-1},Y_i)\ \text{Cov}(Y_{i-1},X_{i-1})}$$

Thus, for iterations 4 and 5:

$$b_1 = \frac{(.592)(1.255) - (1.187)(.804)}{(.809)(1.255) - (.515)(.589)} = .15$$

Now use the data for time points separated by two lags at equilibrium and compute a biased b_1:

$$\text{biased } b_1 = \frac{(.722)(1.588) - (1.187)(.804)}{(.654)(1.588) - (.591)(.804)} = .34$$

Note that the biased estimate of b_1 is quite different from the true value of .15, even though the lag is only twice the appropriate length. Is the estimate of b_3 based upon two lags similarly biased? Can you describe a social psychological system in which this synchronous effects model would be a dubious assumption?

Now that we have generated this covariance matrix from a model with synchronous effects, we are in a position to see whether or not this covariance matrix is also consistent with other models—such as a model with cross-lagged effects rather than synchronous effects or a model with no reciprocal effects whatsoever. This exercise constitutes a beginning for an analysis of the *statistical power* of longitudinal panel data for the rejection of incorrect structural models. In the context of the above covariances, then, we might ask if a cross-lagged effects model fit to three waves of data generated by the synchronous effects model would lead to the conclusion that the cross-lagged effects model should be rejected? The following LISREL IV program setup answers this question:

```
TEST OF CROSS-LAGGED MODEL WITH DATA FROM SYNCH EFFECTS MODEL
DATA NG=1 NO=125 NI=6 MA=CM
LABELS
*
'Y1' 'Y2' 'Y3' 'X1' 'X2' 'X3'
CM
*
1.5877
1.3734  1.5877
1.1874  1.3734  1.5877
 .8043   .7799   .7220  1.3212
 .6886   .8043   .7799   .9097  1.3212
 .5913   .6886   .8043   .6541   .9097  1.3212
```

```
MODEL NY=6 NE=6 LY=ID BE=FI TE=ZE PS=SY,FI
START 1.0 BE(1,1) BE(2,2) BE(3,3) BE(4,4) BE(5,5) BE(6,6)
FREE BE(2,1) BE(3,2) BE(5,4) BE(6,5)
EQUA BE(2,1) BE(3,2)
EQUA BE(5,4) BE(6,5)
START -.7  BE(5,4) BE(6,5)
START -.8 BE(2,1) BE(3,2)
FREE PS(1,1) PS(4,4) PS(4,1)
START 2.0 PS(1,1)
START 1.7 PS(4,4)
START .7 PS(4,1)
FREE PS(2,2) PS(3,3) PS(5,5) PS(6,6)
EQUA PS(2,2) PS(3,3)
EQUA PS(5,5) PS(6,6)
START .36 PS(2,2) PS(3,3)
START .51 PS(5,5) PS(6,6)
FREE BE(2,4) BE(3,5)
EQUA BE(2,4) BE(3,5)
FREE BE(5,1) BE(6,2)
EQUA BE(5,1) BE(6,2)
OU SS SE TO MR PM
```

The resulting chi-square for this lagged effects model is 24.13 (df = 12; Prob = .0195) presuming that 125 cases were observed. This result is certainly encouraging—especially since the chi-square resulting from the fit of a synchronous effects model is zero. However, this encouragement is short lived. By simply freeing the synchronous covariances of the disturbances, the cross-lagged effects model also fits the data perfectly (chi-square = 0.0). [The freeing of these parameters is accomplished in the above LISREL IV program by adding the statements
FREE PS(5,2) PS(6,3)
EQUA PS(5,2) PS(6,3).]
The estimate of this *incorrect* model is

$$Y_3 = .819Y_2 + .092X_2 + z_3$$
$$Y_2 = .819Y_1 + .092X_1 + z_2$$
$$X_3 = .614X_2 + .123Y_2 + z_6$$
$$X_2 = .614X_1 + .123Y_1 + z_5$$

where

$$\text{Cov}(z_6,z_5) = \text{Cov}(z_3,z_2) = .157$$

Note that this solution gives the impression of a greater lagged effect from Y to X than *vice versa*—whereas the "true model" has equivalent *synchronous* effects.

To recapitulate what has been found in the above paragraphs, we have found that two quite different models (one with synchronous reciprocal effects and random disturbances, the other with cross-lagged reciprocal effects and synchronously correlated disturbances) generate exactly the same pattern of covariances (and correlations) among the observed variables. This example should be sobering for enthusiasts of causal modeling with longitudinal data, and it serves as an introduction to the problems of identification discussed in the next section. This example also serves to emphasize the importance of "constraints by design" that can eliminate either synchronous or cross-lagged effects from consideration.

Before proceeding with our discussion of identification problems in structural models for longitudinal designs, it is appropriate at this point to comment on the crude nature of both the synchronous and the cross-lagged effects models that have been developed up to this point. To begin with, it is unlikely that any social psychological model would involve effects that are truly "synchronous." Rather, the effects that are modeled as synchronous are actually cross-lagged effects for which the appropriate lag is much shorter than the period between waves of observation. Second, a bit of reflection reveals that most realistic models of social psychological processes would involve much more continuity than these models involving discrete points in time. The causal effect of one variable on another is more likely to be cumulative over time than a function of a level at a discrete point in time. Third, realistic causal systems are unlikely to be bivariate in nature. Fortunately, the extension of these bivariate models to more general cases of three or more variables is quite straightforward—although laborious.

Nevertheless, there are at least two reasons why the shortcomings of these simple models should motivate careful and skeptical use rather than avoidance. First, more sophisticated models are usually built upon the backs of cruder models. Second, the history of science is replete with examples of instances in which *very* simplistic mathematical models have been of enormous value. Subsequent research into the inaccuracies of oversimplified models has often revealed the inaccuracies to be minimal. Einstein's simple equation relating rest mass to relativistic mass is an obvious case in point. In other instances, even the large inaccuracies of oversimplified models have not been misleading about the dynamics of the system under study; the ideal gas laws are a case in point.

An Identification Problem In some research situations it may be difficult to specify whether lagged or synchronous effects are appropriate. Thus, we can specify a model including both:

$$Y_2 = b_1X_1 + b_2Y_1 + b_3X_2 + z_1 \tag{11.19a}$$

$$X_2 = b_4X_1 + b_5Y_1 + b_6Y_2 + z_2 \tag{11.19b}$$

Even with the constraint that the disturbances are uncorrelated with the lagged variables in the model, the parameters of this model are not identified when two waves are observed. If the model is extended to three waves, it is still not identified even when structural invariance is assumed.

There are two tacks to take when a choice between synchronous and cross-lagged effects has not been determined "by design." The first is inclusion of additional variables in the model that allow additional constraints. For example, inclusion of a background variable in a three-wave design achieves identification if the disturbances are random, the model is structurally invariant, and the background variable has an effect on only one of the two variables that are of primary concern. The constraints in the matrix $\underline{G}$ achieve identification of the cross-lagged and the synchronous effects in the matrix $\underline{B}$ of the LISREL model.

The other tack to the problem of underidentification is simply to constrain certain reciprocal effects to be equal to one another or to constrain a particular effect to a particular value. This latter tack is clearly less informative, since different decisions about which effect to constrain will result in different conclusions. This latter tack is appropriate only in the context of tentative exploration.

Some Spuriousness Models for Null Hypotheses

The model specified with equations 11.19 is clearly not the only possible model that could give rise to a longitudinal pattern of covariation between two variables. Furthermore, the fact that estimates of parameters in a model are obtained is no assurance of the validity of a model.

In order to demonstrate how misleading the estimation of a misspecified causal model can be, we will estimate the lagged effects model in equations 11.7 with data generated from a model in which the associations among the variables are completely spurious. The spurious model to be specified here represents an important class of model—in David Kenny's terminology, the *synchronous common factor model.* Although the spuriousness models used in this chapter are strictly different from Kenny's model, and although comparisons of cross-lagged correlations are not used to test these models, it is important to note that the approach of testing various spurious alternative models stems from Kenny's pioneering work (Kenny, 1975).

Suppose that observations of two variables (Y and X) are obtained at two points in time (1 and 2). And suppose that each of these variables is actually an indicator of the latent variable E. The synchronous and cross-lagged associations between X and Y are completely spurious—due to the fact that each is an indicator of E. Such a model may be specified as follows (see Figure 11.14):

$$\begin{aligned} Y_1 &= j_1E_1 + e_1 \\ X_1 &= j_2E_1 + e_2 \\ Y_2 &= j_3E_2 + e_3 \\ X_2 &= j_4E_2 + e_4 \end{aligned} \tag{11.20}$$

and

$$E_2 = b_1E_1 + z_2 \tag{11.21}$$

where equation 11.20 is essentially a measurement model and equation 11.21 depicts the stability (or perhaps inertia) of E. Note that neither X nor Y have any inertia; all stability in these variables derives from the effect of other variables. For simplicity, assume that all variables are standardized, that the measurement errors in equation 11.20 are uncorrelated with the latent variables and with one another, and that the causal process is structurally invariant (stationary). The stationarity of the process is reflected in the following equalities:

$$j_1 = j_3 \text{ and } j_2 = j_4$$

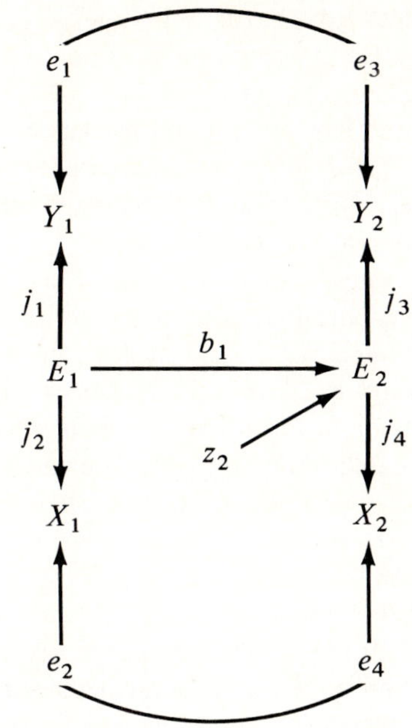

Figure 11.14 A synchronous common factor model for two-wave longitudinal designs.

The covariances among the four observed variables can then be expressed as follows (using the method of moments on equations 11.20 and 11.21):

$$\begin{aligned}
\text{Corr}(Y_1,X_1) &= j_1j_2 \\
\text{Corr}(Y_2,X_2) &= j_3j_4 = j_1j_2 \\
\text{Corr}(Y_2,X_1) &= j_2j_3\,\text{Corr}(E_2,E_1) = j_2j_1b_1 \\
\text{Corr}(Y_1,X_2) &= j_1j_4b_1 = j_1j_2b_1 \\
\text{Corr}(Y_2,Y_1) &= j_1j_3b_1 = j_1^2b_1 \\
\text{Corr}(X_2,X_1) &= j_2j_4b_1 = j_2^2b_1
\end{aligned}$$

Now, suppose that we had misspecified the model generating these data, presuming that it was the lagged model given in equations 11.7. We could then identify the cross-lagged effects of X on Y (a_1 in equation 11.7a) and of Y on X (b_2 in equation 11.7b) as follows:

$$\begin{aligned}
a_1 &= \frac{\text{Corr}(Y_2,X_1) - \text{Corr}(Y_2,Y_1)\,\text{Corr}(Y_1,X_1)}{[1 - \text{Corr}(Y_1,X_1)^2]} \\
&= \frac{(j_2j_1b_1) - (j_1j_1b_1)(j_1j_2)}{[1 - (j_1j_2)^2]} \\
&= \frac{(j_1j_2b_1)(1 - j_1^2)}{[1 - (j_1j_2)^2]}
\end{aligned}$$

which is clearly not zero unless j_1 is $+1$ or -1—the trivial case in which the variable Y is perfectly correlated with the synchronous common factor E. To demonstrate that this value for a_1 could be substantial, let $j_1 = .7$, $j_2 = .6$, and $b_1 = .9$. The resulting value of a_1 is 0.234 with these values assigned to the parameters of the true model. Thus, even though there are *absolutely no cross-lagged effects* in the process that generated these intercorrelations, estimation of the lagged effects model would give the impression of substantial cross-lagged effects.

Thus, before estimating a lagged or synchronous effects model, it is important to be able to demonstrate that a synchronous common factor model can be *rejected* by the observations. It is easiest to accomplish this task if the assumption that the causal system has come to equilibrium is acceptable. However, changes in the validity of an observed variable or changes in structural relations over time may be plausible. Fortunately, the inclusion of a background variable in a model achieves identification of a synchronous common factor model involving two waves of observation. The specification of such a model and its evaluation with actual data will be presented in the context of a subsequent example. The crucial element in this model is the addition of a causal (right-hand) variable to equation 11.21.

The synchronous common factor model is one of two major categories of causal models that can spuriously produce longitudinal covariations that may be misconstrued as arising from cross-lagged or synchronous effects. The other major category is that of the "unmeasured variable model." (The distinction between these two, how-

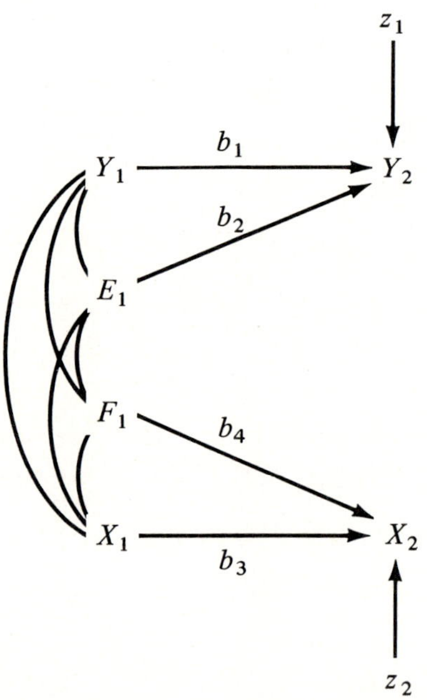

Figure 11.15 Path diagram of a structural model for passive longitudinal designs in which cross-lagged correlations are due to unmeasured variables with lagged effects. The variables E_1 and F_1 are unmeasured in this model.

ever, is more strategic than substantive, since the latter can incorporate the overidentification of the former.) An example of such a model is (see Figure 11.15):

$$Y_2 = b_1 Y_1 + b_2 E_1 + z_1 \qquad (11.22a)$$

$$X_2 = b_3 X_1 + b_4 F_1 + z_2 \qquad (11.22b)$$

where E_1 and F_1 are unmeasured variables that may be related to X_1 and Y_1 in various ways. E_1 has a lagged effect on Y_2 (if b_2 is not zero) and F_1 has a lagged effect on X_2 (if b_4 is not zero). But note that the variables X and Y do not influence one another—either synchronously or with a lag—even though each may have a nonzero inertia. Thus, our interest will be in how a misspecification of the causal model, say assuming the model in equations 11.7, may provide misleading results.

For simplicity, let us again assume that the variables in equations 11.22 are standardized and that the disturbances are uncorrelated with X_1, Y_1, E_1, and F_1. The model in equations 11.22 would then imply that the following pattern of correlations would hold for two waves of observations:

$$\begin{aligned}
r_1 &= \text{Corr}(Y_2,X_1) = b_1 \text{Corr}(Y_1,X_1) + b_2 \text{Corr}(E_1,X_1) \\
r_2 &= \text{Corr}(Y_2,Y_1) = b_1 + b_2 \text{Corr}(E_1,Y_1) \\
r_3 &= \text{Corr}(X_2,X_1) = b_3 + b_4 \text{Corr}(F_1,X_1) \\
r_4 &= \text{Corr}(X_2,Y_1) = b_3 \text{Corr}(Y_1,X_1) + b_4 \text{Corr}(F_1,Y_1) \\
r_5 &= \text{Corr}(Y_2,X_2) = b_1 b_3 \text{Corr}(Y_1,X_1) + b_2 b_4 \text{Corr}(E_1,F_1) \\
&\qquad + b_1 b_4 \text{Corr}(F_1,Y_1) + b_2 b_3 \text{Corr}(E_1,X_1) \\
&\qquad + \text{Cov}(z_1,z_2) \\
r_6 &= \text{Corr}(Y_1,X_1) = \text{given}
\end{aligned}$$

(Note that the time 1 synchronous correlation is not explained by the causal model; thus, it is "exogenous" to the model.) Now suppose that the causal model specified by an investigator who observed these correlations was the lagged effect model in equations 11.7. This model is repeated here with new notation to avoid confusion:

$$Y_2 = a_1 Y_1 + a_2 X_1 + z_3 \qquad (11.23a)$$

$$X_2 = a_3 Y_1 + a_4 X_1 + z_4 \qquad (11.23b)$$

where a_2 is the erroneously specified effect of X_1 on Y_2 and a_3 is the erroneously specified effect of Y_1 on X_2. If we then assume that the variables in equations 11.23 are standardized and that the disturbances are uncorrelated with X_1 and Y_1, the mistaken investigator would identify a_2 and a_3 as follows:

$$a_2 = (r_1 - r_6 r_2)/(1 - r_6 r_6)$$

$$a_3 = (r_4 - r_6 r_3)/(1 - r_6 r_6)$$

But since the correlations entering into this solution for a_2 and a_3 were actually generated by the model in equations 11.22 rather than the model in equations 11.23, estimates of a_2 and a_3 would have the following expectations and meaning:

$$a_2 = \frac{[b_1 r_6 + b_2 \operatorname{Corr}(E_1,X_1)] - r_6[b_1 + b_2 \operatorname{Corr}(E_1,Y_1)]}{(1 - r_6 r_6)} = b_2\,[\operatorname{Corr}(E_1,X_1) - \operatorname{Corr}(E_1,Y_1)]/(1 - r_6 r_6) \quad (11.24a)$$

and

$$a_3 = \frac{[b_3 r_6 + b_4 \operatorname{Corr}(F_1,Y_1)] - r_6[b_3 + b_4 \operatorname{Corr}(F_1,X_1)]}{(1 - r_6 r_6)} = b_4[\operatorname{Corr}(F_1,Y_1) - \operatorname{Corr}(F_1,X_1)]/(1 - r_6 r_6) \quad (11.24b)$$

Equations 11.24a and 11.24b clearly portray the fact that estimation of the erroneous model in equations 11.23 would generally yield nonzero values for a_2 and a_3 even though their true values are zero (the misspecification is in the exclusion of E_1 and F_1, or, alternatively, in the specification that z_3 and z_4 are uncorrelated with X_1 and Y_1). Note in equation 11.24a that a_2 would be zero only if X_1 and Y_1 had the same level of correlation with the unmeasured variable E_1. If all correlations are assumed positive, this would mean that a_2 would be presumed positive if E_1 was more closely associated with X_1 than Y_1, and that a_2 would be presumed *negative* if E_1 was more closely associated with Y_1 than X_1. (This latter case is an instance of what was referred to as the problem of "autocorrelated disturbances" in Section 11.3.) An analogous situation holds for estimates of a_3 based on equation 11.24b.

Equations 11.24 also reveal that the extent of the bias in these estimates of cross-lagged effects increases with the synchronous correlation between X and $Y(r_6)$. Note that as r_6 becomes larger, the bias in estimates of a_2 and a_3 of equations 11.24a and 11.24b will increase if the numerator remains constant. And r_6 becomes large when correlations among the unmeasured and observed variables become large. This can be seen from rewriting the equation for r_5 with the assumption that correlational equilibrium has been reached (i.e., $r_5 = r_6$):

$$r_6 = r_5 = [b_2 b_4 \operatorname{Corr}(E_1,F_1) + b_1 b_4 \operatorname{Corr}(F_1,Y_1) + b_2 b_3 \operatorname{Corr}(E_1,X_1) + \operatorname{Cov}(z_1,z_2)]/(1 - b_1 b_3)$$

Note that when all coefficients are positive the synchronous correlation tends to increase as the correlation between E_1 and F_1 increases. [An interesting special case of the unmeasured lagged factor model is that in which E_1 and F_1 are identical, with a correlation of one. In this instance, the relative bias in estimates of a_2 and a_3 in equations 11.23 are a clear function of which of the two variables, X_1 or Y_1, is more closely correlated with the unmeasured factor. It can be shown that, if b_2 and b_4 in equations 11.22 are both positive, then the misspecified model (11.23) will yield an estimate of a positive cross-lagged effect from the time 1 variable most closely associated with the unmeasured factor and a *negative* cross-lagged effect for the other time 1 variable. Thus, if equation model 11.22 holds in some population, and equation model 11.23 is incorrectly specified with disturbances (z_2, z_3) uncorrelated with X_1 and Y_1, then the time 1 variable most closely associated with the unmeasured factor will give the appearance of having a positive cross-lagged effect (presuming that all correlations and effects are positive).]

There is only one real consolation in the above discussion of the unmeasured variable model for two wave panel designs. That consolation is that when such a model as equation 11.22 holds in some population, and the cross-lagged effects model is erroneously applied, the finding of a cross-lagged effect in one direction and not in the other can be of considerable value. This value derives from a close inspection of equations 11.24. A substantial value of one cross-lagged effect can only occur if the unmeasured variable is more closely correlated with the hypothesized causal variable than with the hypothesized effect variable (or the two correlations are opposite in sign). This means that the bias in estimating equations 11.23 erroneously (i.e., when the model in equations 11.22 is the "true" model) is in the direction of the "domain" that includes both the apparent causal variable and the actual causal variable. Thus, the finding of a substantial cross-lagged effect from X to Y, and none from Y to X, is best interpreted in the following way: if an unmeasured lagged variable model actually holds in the population, then a variable closely associated with X has a substantial cross-lagged effect on Y.

A summary of the argument concerning two-wave longitudinal designs is in order. If a synchronous common factor model holds in a population, and a lagged effects model is incorrectly specified, then estimates of cross-lagged effects can be grossly biased. Fortunately, a fairly general form of the synchronous common factor model can be tested with two waves of observation, if they include appropriate background variables. If an unmeasured variable model holds in some population, and the lagged effects model is incorrectly specified, then biased estimates of cross-lagged effects will also be obtained. The unmeasured variable model is never ultimately rejectable with passive designs; only *specific* candidate variables could be rejected as possibilities through inclusion of them in the model. However, the inclusion of additional waves of observation can achieve overidentification of *certain* unmeasured variable models.

One such model for a three-wave design is specified as follows:

$$\begin{aligned}
Y_3 &= b_1 Y_2 + b_2 E_2 + z_3 \\
Y_2 &= b_1 Y_1 + b_2 E_1 + z_2 \\
Y_1 &= z_1 \\
X_3 &= b_3 X_2 + b_4 E_2 + z_6 \\
X_2 &= b_3 X_1 + b_4 E_1 + z_5 \\
X_1 &= z_4 \\
E_2 &= b_5 E_1 + z_8 \\
E_1 &= z_7
\end{aligned}$$

This model is depicted in Figure 11.16 and is overidentified given the constraints

$$\begin{aligned}
\mathrm{Var}(z_3) &= \mathrm{Var}(z_2) \\
\mathrm{Var}(z_6) &= \mathrm{Var}(z_5) \\
\mathrm{Var}(z_7) &= 1.0
\end{aligned}$$

It is presumed that X and Y are latent variables in a testable multiple indicator measurement model. This particular model is only an example of the numerous

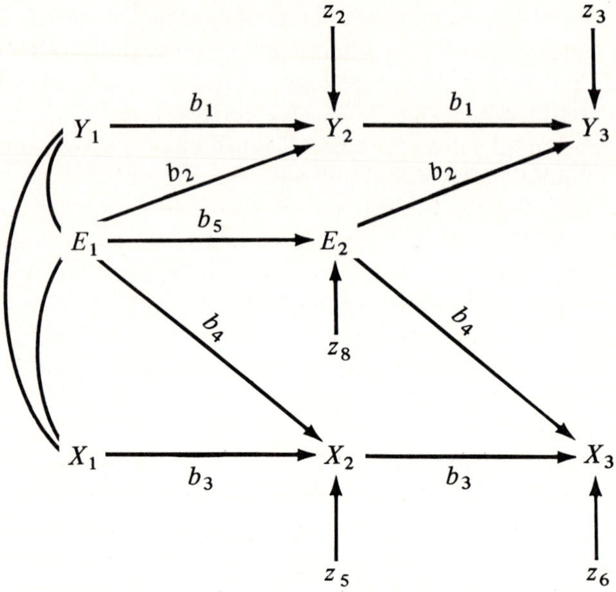

Figure 11.16 Path diagram of a model for longitudinal designs that incorporates an unobserved (or unmeasured) variable. The variables E_1 and E_2 are unmeasured. Also, in practice, the model would include additional measured variables (e.g., background variables). The variables X and Y are directly measured without error or are latent variables in a measurement model.

unmeasured variable models that may be developed for particular circumstances. For example, the effect of the unmeasured variable can be lagged for Y and synchronous for X, or *vice versa*. Also, as is the case in a subsequent example, the unmeasured variable can be given complete inertia and treated as a background variable. Furthermore, some of the structural invariance constraints can be relaxed as more waves of observation are added to the design.

Problem Use a maximum likelihood estimation program such as LISREL to fit the above model to the covariance matrix generated by the synchronous effects model for three waves of observation at equilibrium.

That covariance matrix is repeated here:

```
'Y1' 'Y2' 'Y3' 'X1' 'X2' 'X3'
CM
*
1.5877
1.3734  1.5877
1.1874  1.3734  1.5877
 .8043   .7799   .7220  1.3212
 .6886   .8043   .7799   .9097  1.3212
 .5913   .6886   .8043   .6541   .9097  1.3212
```

The reader will find that the goodness of fit chi-square increases by about 2.8 points for each 100 cases that are presumed to have generated the data. This means that the presumption of several hundred observed cases would be necessary to draw the conclusion that the spuriousness model had a significantly poorer fit with the data than the "true" (synchronous effects) model. (Note that the lagged effects model was "easier" to reject with these data than this spuriousness alternative, which involves unmeasured variables.) Although this exercise provides only a hint concerning the power of data to reject an unmeasured variable model, it should be clear that data sets involving less than several hundred cases may be likely to fail to reject an unmeasured variable model like the one specified above. [A proper assessment of power would require consideration of the effects of sampling error on the estimates of the covariance matrix and consideration of a range of possible values for the parameters (e.g., structural coefficients) of the model.] Thus, in the case of small samples, failure to reject some model depicting longitudinal correlations as spurious must be interpreted with caution. And conversely, in the case of very large samples (several thousand), marginal rejection of a model should be interpreted with caution. This means that rejection with small samples and failure to reject with large samples are more convincing outcomes than failure to reject with small samples and rejection with large samples, even though the true model may involve reciprocal effects.

Longitudinal measurement models are of particular importance in the building of unmeasured variable models. One such model for six waves of observation with two indicators of a latent variable at each wave is given in Figure 11.17. Note that structural invariance is not assumed. For a detailed discussion of identification, see Joreskog and Sorbom (1977).

This process of model testing can be extended indefinitely, with models resistant to rejection emerging over a period of time. The addition of waves of observation will place greater demands on the various models, demands they may not survive. The addition of new variables to a model may reduce what was previously an estimate of a strong lagged effect.

Finally, it is appropriate to point out that the two classes of spurious alternative models outlined above need not be mutually exclusive. A causal system could certainly include some combination of all the patterns of longitudinal causality described above: cross-lagged effects, synchronous effects, associations due to synchronous common factors, and associations due to unmeasured variables. And real social psychological systems are probably this complex, and more so.

The examples given in the following paragraphs illustrate several beginnings in this process.

Example 1: Racial Inequality and Inequality among Whites

For the first example of causal inference from a longitudinal design we will consider data from Dwyer and Taves (1982) on income inequality in U.S. cities. Economists and sociologists have speculated about the causal relationship between black/white economic inequality and economic inequality *within* the white population in the United States for many years. Some have argued that high income whites are the primary beneficiaries of racial discrimination, whereas others have argued that work-

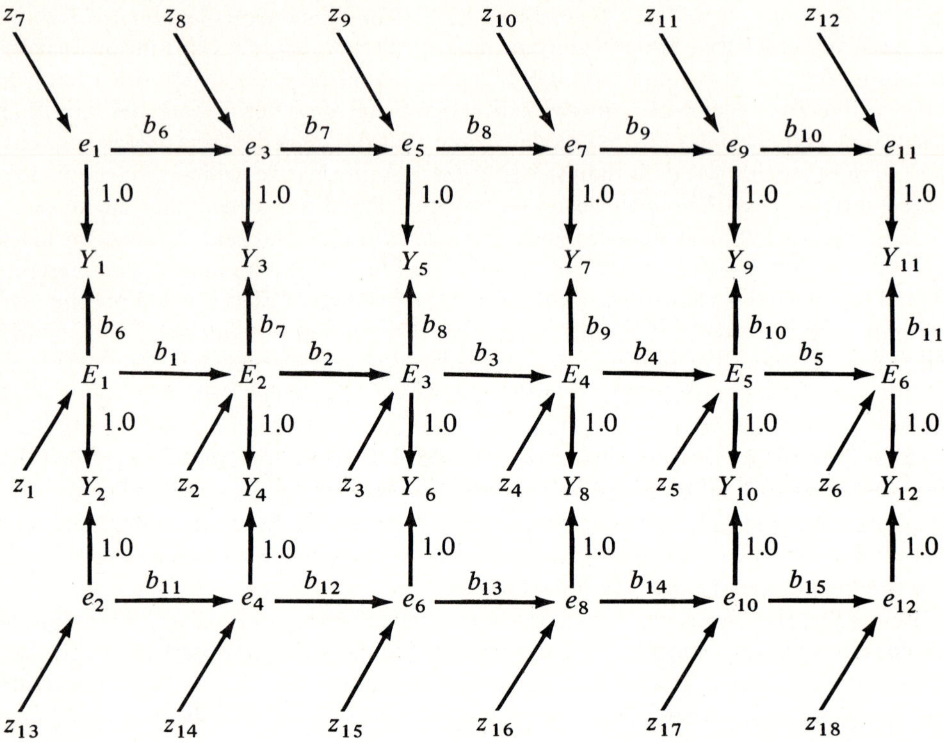

Figure 11.17 Path diagram of a measurement model for longitudinal designs with two indicators of a latent variable at each wave of observation. This model may be estimated with LISREL by specifying the measurement errors (e) as latent variables (E). Identification is achieved by various sets of constraints among the model parameters (e.g., $b_6 = b_{10} = b_{11} = b_{15} = 1$; see Joreskog and Sorbom, 1977).

ing class whites are the primary beneficiaries. Using U.S. cities as the unit of analysis, Michael Reich (1978) found a positive correlation between a black/white ratio of median family incomes and the proportion of white family income received by the middle 50 percent of white families. Similarly, he found a negative correlation between this ratio and the proportion of white family income received by the top 1 percent of white families. This means that in cities where black/white disparities are greatest, high income whites received the larger share of the white income pie.

Our concern then is the causal process that gave rise to this correlation. Does racial inequality improve the position of the wealthy, or do conditions that benefit the wealthy exacerbate racial inequality? Or is this correlation explicable by some spurious model that does not involve a causal relation between these two variables?

In order to explore these questions, data were collected from the 1960 and 1970 Census for 48 U.S. cities (Standard Metropolitan Statistical Areas, SMSAs). The following six variables were among those collected:

Y_1 = black median/white median income, 1960

Y_2 = black median/white median income, 1970

X_1 = income share of middle whites, 1960

X_2 = income share of middle whites, 1970

F_1 = region (1 = south; 0 = not south)

F_2 = percent of families black in SMSA, 1960

The correlations among these six variables are given in Table 11.13.

It is apparent from Table 11.13 that Y and X are stable variables over the ten-year span between the two censuses. Note also that the synchronous correlation increased from 1960 to 1970 (from .486 to .656), suggesting that the system is not in equilibrium.

The first step in our causal analysis of these data is to see if the observed associations are explicable in terms of a synchronous common factor model. Presuming that there is no measurement error in F_1 and F_2, such a model may be specified as follows (see Figure 11.18):

$$Y_1 = j_1E_1 + e_1$$
$$Y_2 = j_2E_2 + e_2$$
$$X_1 = E_1 + e_3$$
$$X_2 = E_2 + e_4$$

and

$$E_1 = g_1F_1 + g_2F_2 + z_1$$
$$E_2 = g_3F_1 + g_4F_2 + b_1E_1 + z_2$$

If we assume that the terms e are uncorrelated with one another, this model is identified with five overidentifying restrictions. If we allow two autocorrelations of the

Table 11.13 Correlations among six variables relevant to income inequality across 48 U.S. cities (from Dwyer and Taves, 1982).

		Y_1	Y_2	X_1	X_2	F_1	F_2
Black/white 1960	Y_1	1.000					
Black/white 1970	Y_2	.845	1.000				
White middle 1960	X_1	.486	.616	1.000			
White middle 1970	X_2	.494	.656	.874	1.000		
Region (south/not)	F_1	−.712	−.592	−.580	−.511	1.000	
Percent black 1960	F_2	−.726	−.617	−.309	−.465	.565	1.000
Standard devs		6.81	6.72	1.27	1.36	4.38	8.42
Means		61.4	63.9	43.6	43.6	.25	12.0

Note. Means are percentages for all variables except region.

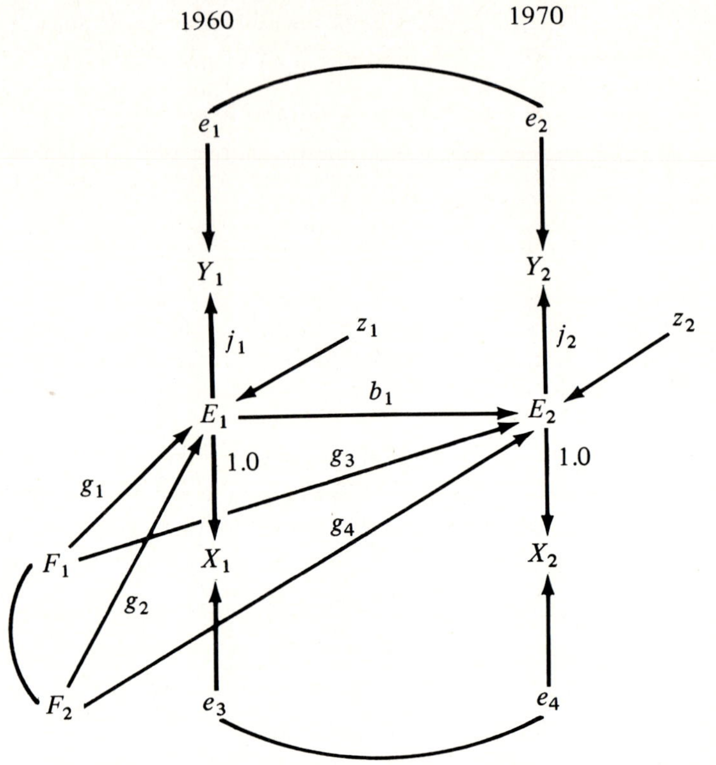

Figure 11.18 Path diagram of a synchronous common factor model for measures of the income share of middle white families (X_1 and X_2) and black/white income disparity (Y_1 and Y_2) in 48 U.S. cities. The fit of this model to the observed data indicated a rejection of the model (chi-square = 20.57; *df* = 6; Prob = .0022).

residuals (e_1 with e_2 and e_3 with e_4), and estimate the model with the data in Table 11.13, we obtain a chi-square of 20.57 with three degrees of freedom (Prob = .0001). Clearly, we can reject this synchronous common factor model, since the observations are extremely unlikely if it held. (Note that the 48 cities used in this example essentially exhaust the population of large U.S. cities. Nevertheless, it is still useful to know how likely certain results would be if the cities were a sample.)

Having successfully rejected one plausible null hypothesis, we will move directly to the fitting of a model that must be specified without measurement errors, since multiple indicators are not available in this data set. A lagged effects model is specified as follows:

$$Y_1 = g_1F_1 + g_2F_2 + z_1$$
$$Y_2 = g_3F_1 + g_4F_2 + b_1Y_1 + b_2X_1 + z_2$$
$$X_1 = g_5F_1 + g_6F_2 + z_3$$
$$X_2 = g_7F_1 + g_8F_2 + b_3Y_1 + b_4X_1 + z_4$$

Table 11.14 Maximum likelihood estimates of two versions of a lagged effects model for data on intercity differences in racial income inequality (see Table 11.13).

	Version 1		Version 2	
Parameter	Estimate	Standardized	Estimate	Standardized
g_1	−.69(.16)	−.44	−.69(.16)	−.44
g_2	−.39(.08)	−.48	−.39(.08)	−.48
g_3	.03(.17)	.02	.28(.16)	.18
g_4	−.01(.09)	−.01	−.05(.08)	−.06
g_5	−.17(.04)	−.60	−.17(.04)	−.60
g_6	.00(.02)	.03	.00(.02)	.03
g_7	−.01(.11)	−.03	.03(.03)	.11
g_8	−.04(.02)	−.27	−.05(.02)	−.31
b_1	.84(.13)	.85	.76(.12)	.77
b_2	.00*	.00*	1.73(.45)	.33
b_3	.00*	.00*	−.01(.02)	−.07
b_4	.61(.08)	.57	.91(.09)	.85
ps_{42}	.31(.62)	.18	.57(.60)	.06
ps_{11}	15.73(3.24)	.34	15.73(3.24)	.34
ps_{22}	12.90(2.66)	.29	9.79(2.02)	.22
ps_{33}	1.07(.22)	.66	1.07(2.02)	.66
ps_{44}	.51(.38)	.28	.42(.09)	.23
ps_{31}	.71(.61)	.08	.71(.61)	.08
chi-square = 16.89 df = 2; Prob = .000				df = 0

Note. The numbers in parentheses are approximate standard errors. Coefficients marked with an asterisk are constrained by hypothesis (fixed).

We will now estimate two versions of this model. In the first, the cross-lagged effects are constrained to zero ($b_2 = b_3 = 0$), while the time 2 disturbances are free to covary. This model would explain the covariances among the variables in terms of the background variables and synchronous correlations between the disturbances. It is another class of null hypothesis for structural models fit to panel data. This model has two overidentifying restrictions and yields a chi-square of 16.89 (Prob = .000). For the second model, we free the cross-lagged effect coefficients. The likelihood ratio chi-square is not of interest in this case, since there are no overidentifying restrictions. The estimates of the parameters for both of these models are given in Table 11.14. The estimates of the lagged effects model (version 2 in Table 11.14) are also presented as a path diagram in Figure 11.19.

Note that the cross-lagged effect of inequality among whites on black/white disparity (path from X_1 to Y_2) is almost four times its standard error (1.73/.45) and has a substantial standardized value (.33). In contrast, the cross-lagged effect from racial disparity to inequality among whites (b_3) is trivial in magnitude. Thus, if this model is the correct one, then black/white inequality is strongly shaped by inequality among whites. If an unmeasured variable model of the form in Figure 11.15 is the true model, then one or more factors closely associated with income inequality among white families in the United States shapes black/white disparities. There is no evidence that the

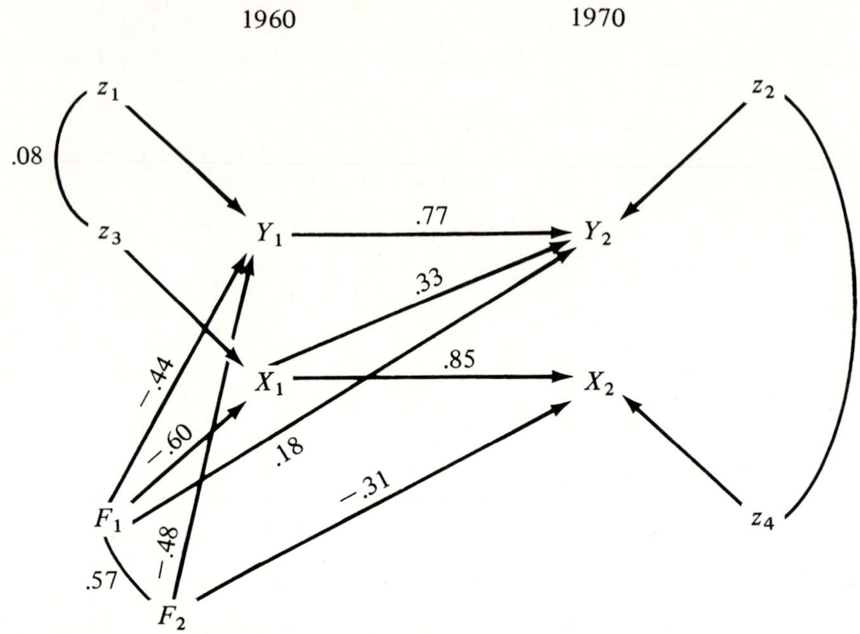

Figure 11.19 Path diagram of the lagged effects model estimated in Table 11.14 (version 2). Path coefficients are standardized. Paths with coefficients not exceeding their standard errors in absolute magnitude are deleted.

black/white disparity has any impact on income distribution among whites (at least for the particular time period and measures involved in the model). This finding directs the attention of those concerned with the determinants of racial inequality to the same processes or structures that produce income inequality among whites.

Further testing of this accepted model could proceed in a number of ways. One would be to add additional waves of observation, say from the 1950 or 1980 U.S. Census. Another would be to measure racial and white income inequality by other methods. The fact that the synchronous correlation increased between 1960 and 1970 suggests that the processes involved are probably not in a state of equilibrium. This fact was advantageous for the current model, since the model begins in 1960 with virtually all the synchronous correlation explained by the exogenous variables. However, the causal process may be structurally invariant such that additional waves of observation would allow identification of lagged and synchronous effects. The availability of only two waves of observation enabled us to test two null hypotheses (synchronous common factor model and the correlated disturbances model), but did not allow testing of an unmeasured variable model.

Example 2: Locus of Control and SES

Example 1 involved a rather small "sample" of cases. With small samples it is "easier" to find an acceptable model than with a large sample. If a model actually holds *exactly* in a population, then the probability of gathering sample observations that will lead to a rejection of the model (type 1 error) is ideally independent of sample size. But since we would not expect any model to be *exactly* true, then the small flaws in a model are more likely to lead to its rejection with larger samples. See Bentler (1980) for references to formalizations of such procedures. For this reason, a researcher may prefer to attend to the magnitude of residuals, rather than the probability of the chi-square, when models are tested with large samples.

The next example comes from a study by Dwyer and Barret-Goldfarb (1982) involving a fairly large sample of white men in the National Longitudinal Survey (NLS). The area of concern here is again an old one in the social and behavioral sciences: the causal relationship between feelings of power or control on the part of individuals and a person's position in a social hierarchy. An individualistic perspective emphasizes the role of personal feelings and attitudes as determinants of success and achievement. The sociological perspective, on the other hand, depicts individual subjectivity as primarily a reflection of social position. Clearly, both these processes operate to varying degrees in different circumstances. But our concern is whether or not one of these "models" holds generally, whereas the other refers to processes that are less prevalent.

The variables used in the model are the following:

Y_1 = locus of control in 1969

Y_2 = locus of control in 1971

Y_3 = annual earnings in 1968

Y_4 = annual earnings in 1970

Y_5 = occupational status 1969

Y_6 = occupational status 1971

X_1 = years of education completed

X_2 = status of first occupation after schooling

Variables Y_1 and Y_2 are scores on Rotter's personality scale that purports to measure the extent to which a person feels that he has control over his destiny—as opposed to the feeling that chance or powerful others are the primary determinants. A higher score on this scale indicates a more "external" orientation.

Variables Y_3 to Y_6 are self-explanatory and are typical indicators of socioeconomic status (SES). Observed variables X_1 and X_2 are also status indicators, but they refer to a period in the respondent's life that precedes real involvement in the world of work. They thus reflect the status of the respondent's family together with whatever educational achievement the individual may have achieved independent of background advantages or disadvantages. The occupational status index used in all cases is Duncan's SEI.

Before testing for lagged or synchronous causal relations between the psychological variable locus of control and SES, it is important to be able to reject the synchronous common factor model. Such a model could be specified as follows:

$$Y_1 = j_1E_1 + e_1$$

$$Y_2 = j_1E_2 + e_2$$

$$Y_3 = j_2E_1 + e_3$$
$$Y_4 = j_2E_2 + e_4$$
$$Y_5 = E_1 + e_5$$
$$Y_6 = E_2 + e_6$$
$$X_1 = F_1 + d_1$$
$$X_2 = k_1F_1 + d_2$$

and

$$E_1 = g_1F_1 + z_1$$
$$E_2 = b_1E_1 + z_2$$

Notice that the locus of control scores, annual earnings, and occupational status are depicted as indicators of a single latent variable in 1969 (E_1) and in 1971 (E_2). No lagged causal relations between the psychological and the SES domains are specified—except for the effect of the background factor F_1. Furthermore, the background latent variable is measured by two methodologically distinct indicators. The covariance matrix that the model must explain is given in Table 11.15.

The fit of the model with the data turns out to be quite good. Allowing for three autocovariances in $\underline{Te}$, the synchronous factor model is identified with 17 overidentifying restrictions. As can be seen in Table 11.16, the chi-square for this model is 23.16 (df = 17, Prob = .14). This means that even with the substantial sample size of 1366, the deviations from the predictions of the model were insufficient to reject this restrictive model. The model is depicted in Figure 11.20.

This result is in sharp contrast to our first example, in which the synchronous common factor model was clearly inconsistent with the observations. Here, with a much larger sample size, the observations would not motivate us to question this model. Other models could be fit to these data, but the knowledge that the associations among the variables are quite consistent with a spuriousness model of this simplicity would certainly detract from the impact of such an exercise.

The reader may treat as an exercise fitting the above model to a comparable covar-

Table 11.15 Covariance matrix for eight variables in a model relating locus of control to SES (N = 1366).

	Y_1	Y_2	Y_3	Y_4	Y_5	Y_6	X_1	X_2
Y_1	32.667							
Y_2	19.102	34.733						
Y_3	−.645	−.580	.280					
Y_4	−.665	−.659	.236	.314				
Y_5	−29.509	−25.413	7.118	7.023	577.739			
Y_6	−28.296	−26.322	7.084	7.256	505.530	580.940		
X_1	−3.665	−3.342	.800	.781	43.689	42.282	10.573	
X_2	−15.279	−11.885	4.080	3.903	249.173	244.653	38.846	415.77

Note. These data are for white middle-aged males surveyed in the National Longitudinal Survey.

Table 11.16 Maximum likelihood estimates of a synchronous factor model for a locus of control measure and indicators of SES.

Parameter	Estimate	Standardized	Parameter	Estimate
j_1	−.075(.008)	−1.485	s_{11}	30.31(1.18)
j_2	.018(.001)	.360	s_{22}	32.70(1.27)
k_1	5.516(.245)	14.639	s_{33}	.152(.007)
b_1	.980(.017)	.978	s_{44}	.183(.009)
g_1	6.220(.288)	.827	s_{55}	179.2(15.7)
ph_{11}	7.041(.456)	1.000	s_{66}	180.1(16.4)
ps_{11}	126.139(16.3)	.316	s_{21}	17.0(01.0)
ps_{22}	17.564(5.84)	.044	s_{43}	.109(.007)
t_{11}	3.532(.285)		s_{65}	114.7(14.6)
t_{22}	201.460(10.8)			

chi-square = 23.1593; df = 17; Prob = .1441

Note. Approximate standard errors are given in parentheses, and standardized coefficients are based on the constraint that the latent variables have unit variance.

iance matrix for 495 *black* respondents in the NLS middle-aged male cohort. You will find that the above model does *not* fit well with the data for black men. This finding lends even more support to the validity of the model for whites, since it demonstrates the "rejectability" of the model in a different population. If specified correctly, these data should yield a chi-square of 74.41 when fit with the same model specified above for white men. This finding is very suggestive of racial differences in the dynamics of beliefs about personal control and power in the United States.

The covariance matrix for 495 black men:

IE69	IE71	AE68	AE70	SEI69	SEI71	EDUC	FRSTSEI
31.725							
11.600	34.269						
−.631	−.327	.232					
−.652	−.288	.194	.249				
−7.086	−7.194	2.108	2.201	262.861			
−1.956	−6.710	2.286	2.462	216.595	273.572		
−4.230	−5.032	.521	.512	23.733	23.624	12.872	
−4.144	−4.821	.372	.552	83.708	76.720	16.918	148.840

Example 3: Feelings of Powerlessness and Perceptions of Control in the Workplace

This next example draws upon a data set described in Chapter 9. The data are from a doctoral dissertation by Wehrle (1982), and the model estimated was developed by Dwyer, Wehrle, and Phelan (1981). A sample of working secretaries was interviewed concerning their work, their leisure, and related feelings. The design was longitudinal with a lag of three months. The data from this study are of particular interest at this point because a multimethod approach was taken to the measurement of hypothetical

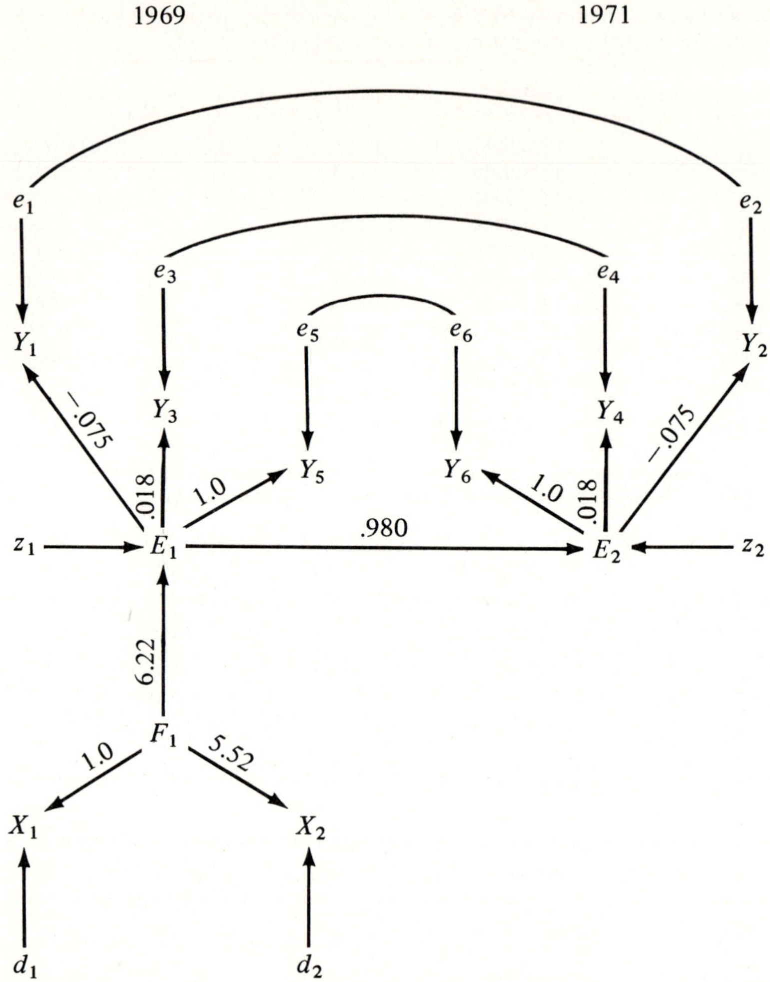

Figure 11.20 Path diagram of a synchronous common factor model fit to data on locus of control and indicators of socioeconomic status from the National Longitudinal Survey (see Table 11.15).

constructs. The two constructs that we will focus on are general feelings of *powerlessness* about the workplace and specific reports of *control over the pace* of work.

Each of these constructs was treated as a latent variable and measured with three different instruments. The three types of measurement were the following:

1. Likert scale
2. Number estimation
3. Rating of narrative by trained rater

Thus, at each point in time, a total of six measures are of interest here—three of powerlessness and three of control over the pace of work. The measurement model for these observed variables may be specified as follows:

Control of pace:

Time 1	Rating	$Y_1 = E_1 + E_5 + e_1$
	Number	$Y_2 = j_1E_1 + E_6 + e_2$
	Likert	$Y_3 = j_2E_1 + E_7 + e_3$
Time 2	Rating	$Y_4 = E_2 + E_5 + e_4$
	Number	$Y_5 = j_1E_2 + E_6 + e_5$
	Likert	$Y_6 = j_2E_2 + E_7 + e_6$

Powerlessness

Time 1	Rating	$Y_7 = j_3E_3 + E_5 + e_7$
	Number	$Y_8 = E_3 + E_6 + e_8$
	Likert	$Y_9 = j_4E_3 + E_7 + e_9$
Time 2	Rating	$Y_{10} = j_3E_4 + E_5 + e_{10}$
	Number	$Y_{11} = E_4 + E_6 + e_{11}$
	Likert	$Y_{12} = j_4E_4 + E_7 + e_{12}$

Note that each substantive latent variable has three observed indicators: powerlessness at time 1 (E_3) and time 2 (E_4) and control of pace at time 1 (E_1) and time 2 (E_2). In addition, three "method factors" have been included: rating (E_5), number (E_6), and Likert (E_7). In order to assure identification, all the loadings on the method factors have been constrained to one. The observed covariances among these 12 indicators are given in Table 11.17.

Allowing the autocovariances of the measurement errors to be free parameters and allowing no covariances among the method factors result in an identified model with 43 overidentifying restrictions. The chi-square for the fit of the model with the data is 56.89 (Prob = .076). The multitrait-multimethod measurement model, then, just

Table 11.17 Convariances among 12 indicators of four latent variables measured at two points in time with three different methods of measurement.

Y_1	Y_2	Y_3	Y_4	Y_5	Y_6	Y_7	Y_8	Y_9	Y_{10}	Y_{11}	Y_{12}
.976											
1.056	3.480										
.339	.699	.517									
.475	.793	.213	1.025								
.949	1.660	.508	1.694	4.625							
.248	.494	.212	.438	.918	.537						
.272	.277	.202	.180	.194	.125	.903					
.756	1.797	.879	.542	1.075	.363	1.522	7.126				
.303	.440	.239	.264	.520	.223	.508	1.577	.933			
.270	.432	.234	.388	.785	.311	.387	1.081	.527	1.117		
.586	1.131	.604	.849	2.087	.794	1.036	3.499	1.363	1.715	6.065	
.340	.603	.238	.315	.690	.341	.441	1.404	.674	.649	1.742	.974

Note. Data are from Wehrle (1982); $N = 119$.

barely avoids rejection at the conventional 5 percent level. (As an exercise, the reader can test some alternative measurement models.)

The null hypothesis that the associations among these indicators are explicable in terms of a synchronous common factor model is especially relevant in this particular example. This is the case because it is not at all obvious that the indicators of "general feelings of powerlessness" are measuring a latent variable that is distinct from the indicators purporting to measure "control over the pace of work." The synchronous model is tested by placing constraints upon the variances and covariances of the latent variables such that E_1 and E_3 are perfectly correlated (time 1) and E_2 and E_3 are perfectly correlated (time 2). Since the LISREL program does not allow constraints placed directly upon covariances among the endogenous latent variables, these constraints must be placed via the covariance matrix for the disturbances $\underline{PS}$:

$$ps_{11} = ps_{33} = ps_{31} \qquad [\text{Corr}(E_3,E_1) = 1]$$

$$ps_{22} = ps_{44} = ps_{42} \qquad [\text{Corr}(E_4,E_2) = 1]$$

$$ps_{21} = ps_{32} = ps_{41} = ps_{43}$$

free loading of Y_1 on E_1

free loading of Y_4 on E_2

Freeing up these two loadings and placing these constraints essentially equates E_1 with E_3 and E_2 with E_4. The constraints add 7 degrees of freedom and estimation of the loadings uses up two; thus the model has an additional 5 degrees of freedom for a total of 48. The goodness of fit chi-square is 193.37; this is clearly a significant deterioration in fit relative to the original model. Thus, we may conclude that the covariances among the indicators are inconsistent with a synchronous common factor model and *not* inconsistent with the hypothesized latent variable model (involving four substantive latent variables and three method factors).

We can now proceed to the structural portion of the model. In order to test the measurement model, the matrix $\underline{B}$ was constrained to be an identity matrix, and elements of $\underline{PS}$ were used to specify the covariance matrix for the latent variables. With the specification of *structural* relations among the latent variables, $\underline{B}$ is used to constrain relations among the latent variables.

For example, the first structural model to be tested is the "no effects" model, which includes neither cross-lagged nor synchronous effects between powerlessness and control over the pace of work. Such a model may be specified as follows:

$$E_1 = z_1$$

$$E_2 = b_1E_1 + b_2E_3 + b_3E_4 + z_2$$

$$E_3 = z_3$$

$$E_4 = b_4E_1 + b_5E_2 + b_6E_3 + z_4$$

$$E_5 = z_5$$

$$E_6 = z_6$$

$$E_7 = z_7$$

where b_2, b_3, b_4, and b_5 are constrained to zero (only the inertial coefficients are left free). These structural constraints on the model increase the degrees of freedom from 43 (for the measurement model) to 48, and the goodness of fit chi-square increases from 56.89 to 70.35 (Prob = .0119). This difference of 13.46 has 5 degrees of freedom and would occur infrequently by chance if the model were true in the population from which the sample was drawn (Prob = .0194). However, this model includes the constraint that z_2 and z_4 are uncorrelated, and this is an unlikely assumption (since "occasion" factors unrelated to the latent variables at time 1 could easily contribute to the synchronous correlation at time 2). And, in fact, freeing this disturbance covariance reduces the degrees of freedom by one (from 46 to 45) and reduces the chi-square from 70.35 to 57.42 (Prob = .1014). The fit of this "no effects" model, then, is quite adequate. It fits the data as well as the measurement model without any structural constraints. The model is depicted in Figure 11.21, and the LISREL IV program for

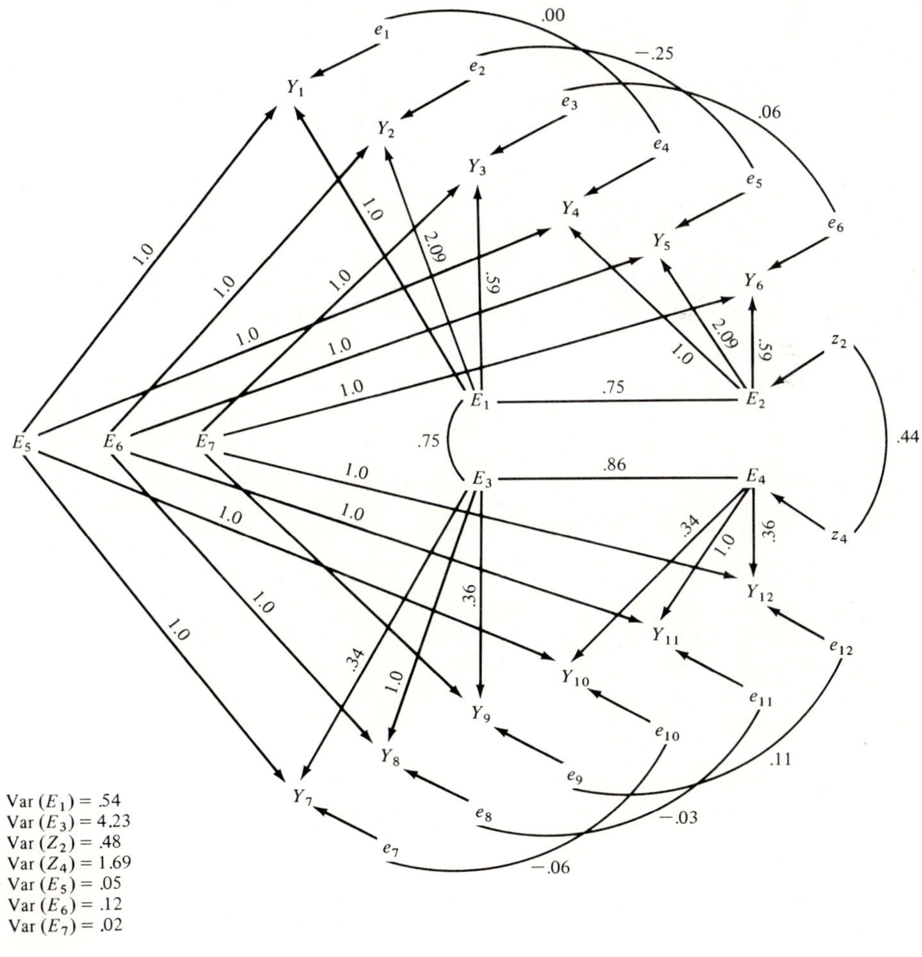

Figure 11.21 Path diagram of a multiple indicator measurement model and longitudinal structural model relating measures of powerlessness and control over the pace of work (data from Wehrle, 1982).

estimation is given below. The reader can verify that the addition of either lagged or synchronous effects fails to improve the fit of the model.

```
NO EFFECTS MODEL FOR POWERLESSNESS AND CONTROL OF PACE--WEHRLE
DATA NI=12 NOBS=119 MA=CM
LABELS
*
'OWPACE1' 'NWPACE1' 'AWPACE1' 'OWPOWER1' 'NWPOWER1' 'AWPOWER1'
'OWPACE2' 'NWPACE2' 'AWPACE2' 'OWPOWER2' 'NWPOWER2' 'AWPOWER2'
KM
*
1.0      .5732   1.0      .4780   .5213   1.0      .2899   .1562
 .2963   1.0      .2865   .3609   .4582   .5999   1.0      .3175
 .2442   .3437   .5536   .6115   1.0      .4745   .4200   .2921
 .1872   .2005   .2704   1.0      .4468   .4137   .3287   .0950
 .1873   .2505   .7780   1.0      .3431   .3614   .4017   .1792
 .1857   .3152   .5901   .5824   1.0      .2588   .2192   .3086
 .3847   .3830   .5164   .3630   .3455   .4014   1.0      .2409
 .2461   .3412   .4425   .5322   .5730   .3404   .3941   .4398
 .6587   1.0      .3483   .3277   .3354   .4698   .5330   .7067
 .3157   .3252   .4709   .6223   .7168   1.0
SD
*
 .9880   1.8654  .7188   .9505   2.6694  .9659   1.0123  2.1505
 .7330   1.0571  2.4628  .9869
SELECTION
1 2 3 7 8 9 4 5 6 10 11 12/
MODEL NY=12 NE=7 LY=FI BE=FI PS=SY TE=SY
PA LY
*
0 0 0 0 0 0 0
1 0 0 0 0 0 0
1 0 0 0 0 0 0
0 0 0 0 0 0 0
0 1 0 0 0 0 0
0 1 0 0 0 0 0
0 0 1 0 0 0 0
0 0 0 0 0 0 0
0 0 1 0 0 0 0
0 0 0 1 0 0 0
0 0 0 0 0 0 0
0 0 0 1 0 0 0
MA LY
*
 1   0   0   0   1   0   0
 1   0   0   0   0   1   0
 1   0   0   0   0   0   1
 0   1   0   0   1   0   0
 0   1   0   0   0   1   0
 0   1   0   0   0   0   1
 0   0   1   0   1   0   0
 0   0   1   0   0   1   0
 0   0   1   0   0   0   1
 0   0   0   1   1   0   0
 0   0   0   1   0   1   0
 0   0   0   1   0   0   1
```

```
EQ LY(2,1) LY(5,2)
EQ LY(3,1) LY(6,2)
EQ LY(7,3) LY(10,4)
EQ LY(9,3) LY(12,4)
PA PS
*
1
0 1
1 0 1
0 0 0 1
0 0 0 0 1
0 0 0 0 0 1
0 0 0 0 0 0 1
FREE PS(4,2)
MA PS
*
 .536
 .000   .453
 .231   .000   .415
 .000   .000   .000   .097
 .0     .0     .0     .0    .1
 .0     .0     .0     .0    .0    .1
 .0     .0     .0     .0    .0    .0    .1
PA BE
*
0 0 0 0 0 0 0
1 0 0 0 0 0 0
0 0 0 0 0 0 0
0 0 1 0 0 0 0
0 0 0 0 0 0 0
0 0 0 0 0 0 0
0 0 0 0 0 0 0
MA BE
*
1.0  0.0  0.0  0.0  0.0  0.0  0.0
-.8  1.0  0    0    0    0    0
  0  0    1    0    0    0    0
  0  0    -.8  1    0    0    0
  0  0    0    0    1    0    0
  0  0    0    0    0    1    0
  0  0    0    0    0    0    1
PA TE
*
1
0 1
0 0 1
1 0 0 1
0 1 0 0 1
0 0 1 0 0 1
0 0 0 0 0 0 1
0 0 0 0 0 0 0 1
0 0 0 0 0 0 0 0 1
0 0 0 0 0 0 1 0 0 1
0 0 0 0 0 0 0 1 0 0 1
0 0 0 0 0 0 0 0 1 0 0 1
ST .5 TE(1,1) TE(2,2) TE(3,3) TE(4,4) TE(5,5) TE(6,6)
ST .5 TE(7,7) TE(8,8) TE(9,9) TE(10,10) TE(11,11) TE(12,12)
OU PM MR SS TO SE
```

With the firm understanding that this spuriousness model cannot be rejected, we may nevertheless explore other models for the sake of demonstration and possible clues for further data collection. The two additional models fit to the data will be the synchronous effects model and a model that includes synchronous effects and one cross-lagged effect (recall that all four reciprocal effects cannot be identified in this simple case of two variables at two points in time).

The synchronous effects model is specified by the following constraints on the coefficients in the structural model:

$$b_2 = b_4 = 0$$

If we constrain z_2 and z_4 to be uncorrelated, we have a model with 44 degrees of freedom and a goodness of fit chi-square of 58.20 (Prob = .0742). The ML estimates of the two important structural equations are as follows:

$$\begin{aligned} E_2 &= \underset{(.144)}{.720E_1} + \underset{(.053)}{.028E_4} + z_2 \\ E_4 &= \underset{(.228)}{.593E_2} + \underset{(.090)}{.774E_3} + z_4 \end{aligned}$$

or in standardized form

$$\begin{aligned} E_2 &= .593E_1 + .068E_4 + z_2 \\ E_4 &= .242E_2 + .723E_3 + z_4 \end{aligned}$$

Thus, if this synchronous effects model holds, the effect of control over the pace of work on feelings of powerlessness is significant, whereas the reverse effect is essentially zero.

In order to emphasize the tenuous nature of the assumption that the synchronous effects model is true, we shall estimate an additional model in which a cross-lagged effect is estimated in addition to the synchronous effects. We might, for argument's sake, reason that although control over the pace of work shapes general feelings of powerlessness in the short run, this more general attitude may shape the choice and development of job situations over time. Thus, we could eliminate the constraint on the path from powerlessness at time 1 to control over pace at time 2 (the coefficient b_2 in the structural model). The ML estimates and their standard errors are as follows:

$$\begin{aligned} E_2 &= \underset{(.134)}{.736E_1} - \underset{(.127)}{.154E_3} + \underset{(.131)}{.183E_4} + z_2 \\ E_4 &= \underset{(.362)}{.277E_2} + \underset{(.101)}{.820E_3} + z_4 \end{aligned}$$

This result essentially returns us to that found earlier with the spuriousness model: none of the effects between the two variables has a significant coefficient. However, if one were to interpret these coefficients, they give an impression opposite to that given by the synchronous effects model. Clearly, the best conclusion to draw from these data is that there is no evidence of a causal relation between powerlessness and control over pace. The finding of a significant coefficient in the synchronous effects model with z_4 and z_2 constrained to be uncorrelated might be considered suggestive for future work, but no more. The addition of more waves of observation would aid in sorting out this relation, as can be seen in the next example.

Example 4: The Case of Three Waves of Observation

The data for this example come from a study by Jessor and Jessor (1977) that was cited by Cook and Campbell (1979, p. 318). The data consist of scores on two variables at three points in time for a sample of 188 male high school students. The two variables of concern are "tolerance of deviance" (X) and reports of "involvement with marijuana" (Y). The correlations among these six variables are given in Table 11.18 along with the other statements necessary to test an unmeasured variable model using LISREL IV.

Note in Table 11.18 that the synchronous correlation is fairly stable over time (.41, .37, and .41). This means that a standardized model involving time invariant parameters may be appropriate. Recall that a longitudinal model may not be at equilibrium (stable synchronous covariance matrix among variables), even though the model is structurally invariant with time. Furthermore, a cross-lagged effects model, for example, can achieve a stable synchronous correlation even though it has not come to equilibrium; in fact, a stable synchronous correlation can result from a model that does not come to equilibrium (a nonequilibrating causal system). In order for equivalence of *standardized* structural coefficients to make sense, it must be assumed that the model has achieved equilibrium (or that a fortuitous relationship exists between the expansion of variances and covariances). In this instance, we shall presume that the stable synchronous correlation is indicative of equilibrium. The unfortunate aspect of this data set is that it does not include multiple indicators; and this means that there is likely to be bias in estimates due to random and autocorrelated measurement errors—since flawed indicators are treated as errorless variables.

A synchronous common factor model was fit to these data intially. Its fit with the data was very poor (chi-square = 179.48; Prob = .0000).

Since three waves of observation are available in this instance, the assumption of both equilibrium and structural invariance allows estimation of an unmeasured variable model in the following standardized form:

$$
\begin{aligned}
X_1 &= z_1 \\
X_2 &= b_1X_1 + b_2E_7 + z_2 \\
X_3 &= b_1X_2 + b_2E_7 + z_3 \\
Y_1 &= z_4 \\
Y_2 &= b_3Y_1 + b_4E_7 + z_5 \\
Y_3 &= b_3Y_2 + b_4E_7 + z_6 \\
E_7 &= z_7
\end{aligned}
$$

$$\text{Constraints: } \mathrm{Var}(z_2) = \mathrm{Var}(z_3);\ \mathrm{Var}(z_5) = \mathrm{Var}(z_6)$$

Note that this model construes the associations between X and Y as due entirely to an unmeasured background variable (E_7). Estimation of this model is achieved using the

Table 11.18 The program setup for an analysis of data reported by Jessor and Jessor (1977) with LISREL IV.

```
SYNCHRONOUS UNMEASURED VARIABLE MODEL-JESSOR JESSOR DATA
DATA NI = 6 NOBS = 188 MA = KM
LABELS
*
'ATT1' 'ATT2' 'ATT3' 'BEH1' 'BEH2' 'BEH3'
KM
*
1.00
 .65  1.00
 .67   .71  1.00
 .41   .26   .21  1.00
 .44   .37   .29   .71  1.00
 .53   .39   .41   .53   .69  1.00
MODEL NY = 6 NE = 9 LY = FI BE = FI,FU PS = FI,SY TE = ZE
START 1.0 LY(1,1) LY(2,2) LY(3,3) LY(4,4) LY(5,5) LY(6,6)
START 1.0 BE(1,1) BE(2,2) BE(3,3) BE(4,4) BE(5,5) BE(6,6) C
          BE(7,7) BE(8,8) BE(9,9)
FREE BE(2,8) BE(3,9) BE(2,1) BE(3,2) BE(5,8) C
      BE(6,9) BE(5,4) BE(6,5)
EQUA BE(2,8) BE(3,9)
EQUA BE(2,1) BE(3,2)
EQUA BE(5,8) BE(6,9)
EQUA BE(5,4) BE(6,5)
FREE BE(8,7) BE(9,8)
START -.9 BE(8,7) BE(9,8)
EQUA BE(8,7) BE(9,8)
FREE PS(4,1) PS(7,1) PS(7,4)
START .41 PS(4,1)
FREE PS(1,1) PS(4,4)
START 1.00 PS(1,1) PS(4,4) PS(7,7)
FREE PS(2,2) PS(3,3) PS(5,5) PS(6,6) PS(8,8) PS(9,9)
EQUA PS(2,2) PS(3,3)
EQUA PS(5,5) PS(6,6)
EQUA PS(8,8) PS(9,9)
ST -.5 BE(2,8) BE(3,9) BE(2,1) BE(3,2) BE(5,8) C
    BE(6,9) BE(5,4) BE(6,5)
ST .4 PS(2,2) PS(3,3) PS(5,5) PS(6,6) PS(8,8) PS(9,9)
FIXED BE(5,8) BE(6,9)
START 0.0 BE(5,8) BE(6,9)
FREE BE(6,8) BE(5,7)
EQUA BE(6,8) BE(5,7)
FIXED BE(8,7) BE(9,8) PS(8,8) PS(9,9)
START -1.0 BE(8,7) BE(9,8)
START 0.0 PS(8,8) PS(9,9)
OU TO MR SS SE
```

Note. This program was initially set up to test an unmeasured variable model with three unmeasured variables. The last few lines of the program specify these three unmeasured variables to be the same variable.

LISREL IV program in Table 11.18. The resulting estimation indicates a good fit (chi-square = 12.58; df = 10; Prob = .25):

$$X_2 = \underset{(.10)}{-.04}X_1 + \underset{(.11)}{.87}E_7 + z_2$$

$$X_3 = \underset{(.10)}{-.04}X_2 + \underset{(.11)}{.87}E_7 + z_3$$

$$Y_2 = \underset{(.04)}{.62}Y_1 + \underset{(.04)}{.24}E_7 + z_5$$

$$Y_3 = \underset{(.04)}{.62}Y_2 + \underset{(.04)}{.24}E_7 + z_6$$

where the numbers in parentheses are approximate standard errors. The only coefficient smaller than its standard error is b_1, the inertia of X. Furthermore, its negative value is rather implausible. The model is depicted in Figure 11.22.

This result is particularly interesting. A single unmeasured variable (E_7) is sufficient to explain all the associations between X and Y when Y is allowed inertia. Furthermore, E_7 explains all the stability of the attitude measure (X) and is strongly correlated with it. Since the correlation between the attitude measure and the unmeasured variable (E_7) is a reasonable value for the validity of such a measure, it is plausible to

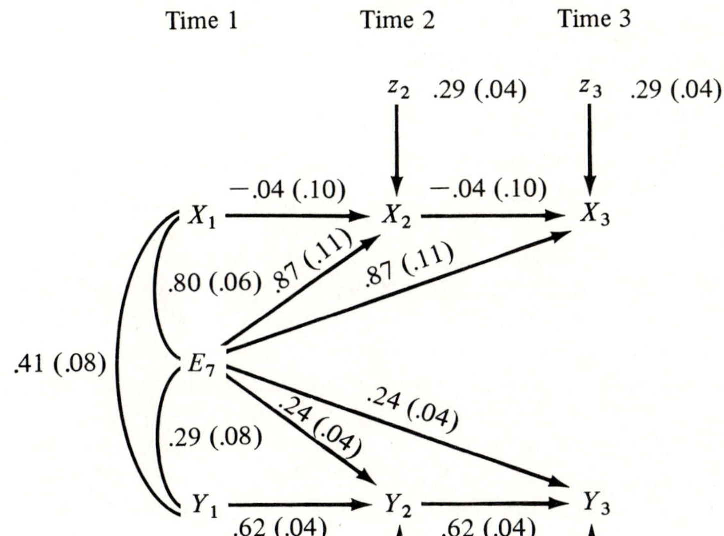

Figure 11.22 Path diagram of an unmeasured variable model fit to longitudinal data on attitude toward deviance (X) and report of marijuana use (Y). The coefficients are standardized, and their approximate standard errors are given in parentheses. Chi-square = 12.58; df = 10; Prob = .24. See Table 11.18 for data.

interpret E_7 as the latent variable measured (with error) by X. If such an interpretation is accepted, then observed change in X over time is due entirely to measurement error. The results would then support a theoretical position emphasizing the influence of stable attitudes on behavior rather than *vice versa*. However, in the absence of method divergent multiple indicators, E_7 has the somewhat ambiguous status of an unmeasured variable that is strongly related to the attitude measure and moderately related to the measure of marijuana use. The unmeasured variable could be a linear combination of background variables such as social class, family organization, and religious training—in which case an induced change in the measured attitude would have no effect on the measured behavior.

We may now proceed to fit additional models to these data. However, the results of such an exercise must be interpreted in light of the fact that a spurious unmeasured variable model is quite consistent with these data.

The next model fit to these data is the "no effects" model where both synchronous and lagged effects in the following structural model are constrained to zero:

$$
\begin{aligned}
X_1 &= z_1 \\
X_2 &= b_1X_1 + b_2Y_1 + b_3Y_2 + z_2 \\
X_3 &= b_1X_2 + b_2Y_2 + b_3Y_3 + z_3 \\
Y_1 &= z_4 \\
Y_2 &= b_4X_1 + b_5X_2 + b_6Y_1 + z_5 \\
Y_3 &= b_4X_2 + b_5X_3 + b_6Y_2 + z_6
\end{aligned}
$$

where

$$
\begin{aligned}
b_2 &= b_4 = 0 \text{ (no cross-lagged effects)} \\
b_3 &= b_5 = 0 \text{ (no synchronous effects)}
\end{aligned}
$$

It is possible in this model to identify the synchronous covariances between disturbances [Cov(z_5,z_2) and Cov(z_6,z_3)]. Freeing these elements of $\underline{PS}$ leaves 13 degrees of freedom for the chi-square, which has a value of 62.90 (Prob = .0000). These data are clearly sufficient to reject the "no effects" model.

A cross-lagged effects model is specified by simply relaxing the constraint that b_2 and b_4 are zero in the structural model. This relaxation uses up two degrees of freedom (reducing the degrees of freedom to 11) and reduces the chi-square to 45.37 (Prob = .0000). This improvement of fit is highly unlikely by chance, but the probability of the data under assumption of the model is still very small. Constraining the lagged effects to be zero once again and freeing the synchronous effects (b_3 and b_5) also yields a model with 10 degrees of freedom. The chi-square for this model is also 45.37. Thus, neither the cross-lagged effects model nor the synchronous effects model achieves a level of fit with the data that is comparable to that of the unmeasured variable model.

Now, the cross-lagged and synchronous effects models fit to these data were quite restrictive, and the reader may be wondering if the relaxation of some constraints would produce an improvement in fit. For example, recalling the unmeasured variable model fit to these data, much of the poor fit of these latter models may arise from the assumption that X is first-order autoregressive (i.e., there is no path from X_1 to X_3). If the element of $\underline{PS}$ corresponding to the covariance between z_1 and z_3 is made a free parameter in the model, then the goodness of fit chi-squares for the cross-lagged and synchronous effects models are significantly reduced. The chi-squàre for the cross-lagged effects model descends from 45.37 to 22.10 (df = 9; Prob = .015); the chi-square for the synchronous effects model descends from 45.37 to 15.15 (df = 9; Prob = .127)—avoiding rejection at the conventional 5 percent level. The maximum likelihood estimates of this synchronous effects model are given in Figure 11.23 and as follows:

$$X_2 = \underset{(.05)}{.56X_1} + \underset{(.06)}{.04Y_2} + z_2$$

$$X_3 = \underset{(.05)}{.56X_2} + \underset{(.06)}{.04Y_3} + z_3$$

$$Y_2 = \underset{(.06)}{.28X_2} + \underset{(.04)}{.62Y_1} + z_4$$

$$Y_3 = \underset{(.06)}{.28X_3} + \underset{(.04)}{.62Y_2} + z_5$$

The rather interesting result in this case is that the attitude (X) has a substantial effect on the behavior (Y), but not *vice versa* (see Figure 11.23). However, interpretation of

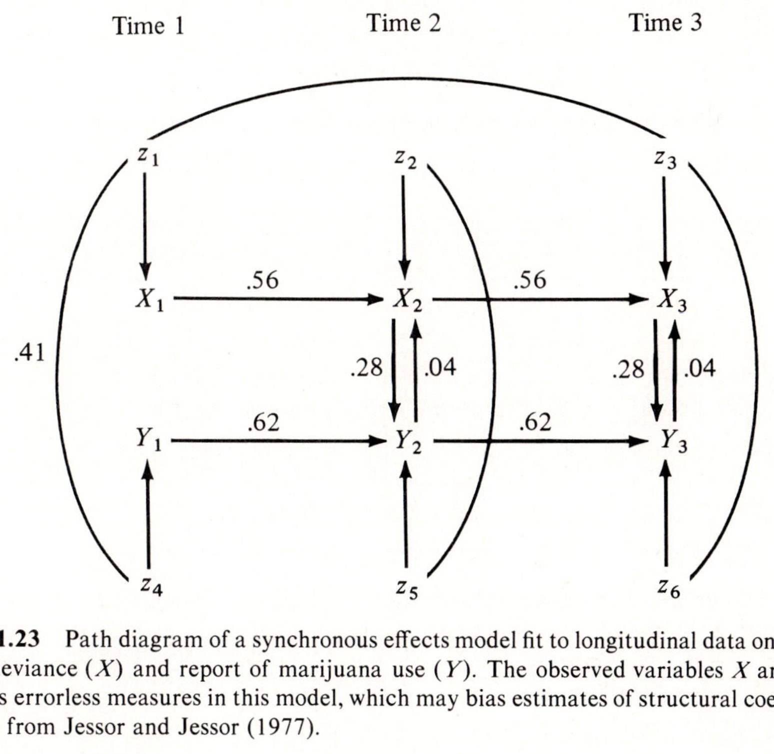

Figure 11.23 Path diagram of a synchronous effects model fit to longitudinal data on attitude toward deviance (X) and report of marijuana use (Y). The observed variables X and Y are treated as errorless measures in this model, which may bias estimates of structural coefficients. Data are from Jessor and Jessor (1977).

these results must be tempered by the fact that freeing the covariance between z_3 and z_1 arose from inspection of the data rather than an *a priori* hypothesis.

The appropriate conclusions to be drawn from these analyses are that (1) the data are insufficient to reject an overidentified model with an unmeasured variable; (2) the data are sufficient to reject two simple models with reciprocal effects between X and Y; and (3) freeing the covariance between two disturbances on the basis of inspecting the data, the data are more consistent with a synchronous effects model than with a cross-lagged effects model. In this latter model, the effect of the attitude on behavior is significant, whereas there is no evidence of an effect in the opposite direction. But any discussion of reciprocal effects must be taken in the context of the finding that an unmeasured variable model cannot be rejected by these data. Thus, the hypothesis that these two variables are spuriously correlated must remain a plausible alternative. There are other models that will also fit the data, and the reader is invited to use the specification in Table 11.18 as a starting point for testing alternative models. (The program given in Table 11.18 specifies the unmeasured variable model.) It is especially important to keep in mind the fact that the models tested here are restricted to those that are testable with three waves of observation on two variables.

Example 5: A Longitudinal Model for School Differences in Reading Achievement

For the last example, we return to the data set reported in Modoono's (1981) doctoral dissertation dealing with reading achievement in New York City junior high schools. Included in Modoono's data set were the following variables:

Y_{75} = median reading score for a junior high school in 1975

Y_{76} = median reading score for a junior high school in 1976

.

.

.

Y_{79}

X_1 = percent of children in the school receiving free lunches under a federal poverty program, 1974

X_2 = median income of families in the police precinct surrounding the school district, 1970 (Census)

X_3 = index for the amount of street litter in the planning district surrounding the school district, 1975

X_4 = percent of children in the school classified as black in 1974

X_5 = percent of children in the school classified as Puerto Rican in 1974

The unit of analysis for these variables is the junior high school, but notice that two of the background variables are measured on geographical districts that do not correspond exactly to the school districts. This means that the median income (X_2) and litter (X_3) measures will include measurement errors in addition to those that arose in their original construction.

One question of interest in educational psychology and the sociology of education is the extent to which differences in the social status of families and communities explain racial and/or cultural differences in children's performance on achievement tests. Conventional wisdom among psychologists holds that measured differences in socioeconomic status between black and white families do not account for the reading achievement differences between black and white children. These residual differences, of course, favor the white population and are used by proponents of genetic explanations of racial differences. However, the reader of this book should react immediately with skepticism to claims resulting from exercises in statistical control of such illusive variables as socioeconomic status (SES). The studies generating this conventional wisdom probably utilized multiple regression techniques involving flawed indicators that were treated as errorless variables. The likely result of this misspecification of the measurement model in these instances would be an insufficient adjustment for differences in SES.

Modoono's data provide an unusual opportunity to approach the question of racial differences and SES with more attention to measurement issues. First, the use of school medians for the dependent variables (rather than the scores of individual children) guarantees that *random* measurement errors in the endogenous variables will be inconsequential in magnitude. There may be "theoretical bias" in the school medians of reading achievement scores due to their measurement of theoretically irrelevant latent variables, but random errors should be eliminated in the process of aggregation. (The *dis*advantage of aggregation, of course, is the danger of the ecological fallacy: that the pattern among aggregates is misleading concerning the pattern of relationships among the individuals that have been aggregated. The model derived from this analysis of aggregates should then be tested with data for individuals.) Second, the inclusion of three indicators of the SES of the school district (X_1, X_2, and X_3) enables us to hypothesize measurement models that take account of measurement errors in those indicators.

Fitting models to these data will proceed in two stages. First, the series of Y variables (reading achievement) will be fit to unmeasured variable models. Second, the background indicators will be fit to measurement models. Finally, these two models (an unmeasured variable model for the endogenous variables and a measurement model for the exogenous indicators) will be combined into one model. The ultimate goal will be to estimate the lagged effects of percent black (X_4) and percent Puerto Rican (X_5) on median reading achievement: that is, are the lagged relationships between these background variables and reading achievement explained by the association between, for example, X_4 and other background variables, or are the lagged relationships consistent with a lagged causal effect? Clearly, the behavioral genetic explanation of racial differences in intelligence would predict a negative effect of the proportion black on reading achievement. (Unfortunately, data for the percent black in each school are not available for the later years in this series. If such were available, then the estimation of synchronous effects would be most sensible. However, the percent black is quite stable over time; for the two years available, 1974 and 1975, the

Table 11.19 Correlations, means, and standard deviations for ten variables from Modoono's (1981) study of reading achievement in New York City intermediate schools (eighth grade).

	Y_{75}	Y_{76}	Y_{77}	Y_{78}	Y_{79}	X_1	X_2	X_3	X_4	X_5
Y_{75}	1.000									
Y_{76}	.947	1.000								
Y_{77}	.933	.934	1.000							
Y_{78}	.940	.936	.947	1.000						
Y_{79}	.914	.906	.905	.938	1.000					
X_1	−.737	−.725	−.675	−.708	−.690	1.000				
X_2	.817	.816	.809	.791	.785	−.581	1.000			
X_3	−.794	−.782	−.779	−.763	−.747	.561	−.869	1.000		
X_4	−.524	−.492	−.459	−.509	−.528	.449	−.382	.362	1.000	
X_5	−.649	−.653	−.674	−.633	−.576	.466	−.668	.674	−.091	1.000
Mean	8.036	7.487	8.038	8.332	8.465	54.60	9298	1.457	37.29	22.80
SD	1.544	1.347	1.251	1.115	1.009	23.46	2533	.1716*	27.26	22.98

Note. This matrix is based upon the 133 schools for which complete data were available.
*The mean and variance of X_3 are arbitrary, since X_3 is a composite of ratings; the variance of X_3 is equated to that of X_2 in subsequent analyses.

autocorrelation of percent black is .995; this result suggests that treating percent black as a stable school characteristic may be appropriate.)

The covariance matrix among these variables and their respective means are given in Table 11.19. The reader should note that X_4 and X_5 are positively skewed—as is often the case with percentaged variables. This means that the assumption of multivariate normality is probably violated in this data set. (A good approach to the solution of this problem would be to transform the three percentage variables according to a logit model, as outlined in Appendix B. However, for simplicity of presentation, this additional step will not be taken in this example.)

We begin by hypothesizing that the covariances among the Y variables are explicable in terms of the inertia of each school and the structurally invariant effect of an unmeasured variable (F_1). The model (model 1, henceforth) is specified as follows:

$$\begin{aligned} E_1 &= Y_{75} = g_1F_1 + z_1 \\ E_2 &= Y_{76} = g_2F_1 + b_1Y_{75} + z_2 \\ E_3 &= Y_{77} = g_2F_1 + b_1Y_{76} + z_3 \\ E_4 &= Y_{78} = g_2F_1 + b_1Y_{77} + z_4 \\ E_5 &= Y_{79} = g_2F_1 + b_1Y_{78} + z_5 \\ E_{10} &= F_1 \end{aligned}$$

where b_1 is the inertia of Y and g_2 is the effect of the unmeasured variable F_1 on median reading achievement. No assumptions concerning whether or not this causal system has come to equilibrium are made. The variance of F_1 is arbitrarily fixed at 1.00, and the covariance between F_1 and z_1 is a free parameter. (In the actual LISREL program, z_1 and F_1 are allowed to correlate via a regression of Y_{75} on F_1; however, the resulting slope g_1 should not be treated as a structural coefficient, since the inertia of Y_{74} is not included in the model.)

In the first version of this model to be tested, we assume that the variance of the disturbances is invariant over time:

$$\text{Var}(z_2) = \text{Var}(z_3) = \text{Var}(z_4) = \text{Var}(z_5)$$

This constraint, together with the assumptions that the structural effects (lagged and inertial) are invariant over time, produces a simple model in which a stable background variable is shaping interschool differences over time. Since the reading achievement variables are scaled in terms of "reading age" for all years, these assumptions of structural invariance are at least plausible. If the units of measurement were not the same for each year, then there would be no basis for assuming structural invariance over time.

The fit of model 1 with the data is not acceptable: chi-square = 27.59, df = 10, Prob = .0021. With only 133 cases, this rejection is convincing. The next step is to relax the constraint that the variance of the disturbances is invariant over time. Estimating these additional parameters takes up 3 degrees of freedom, and the resulting model (model 2) evidences an acceptable fit with the observations: chi-square = 11.11; df = 7; Prob = .134. The difference of 3 degrees of freedom between these two models is associated with an increase in the chi-square of 16.48 (27.59 − 11.11). This difference is clearly unlikely by chance (Prob $<$.01), and the model with constant disturbance variance may be rejected in favor of the latter model. The maximum likelihood estimates of this model are as follows:

$$\begin{aligned}
\hat{g}_1 &= 1.505(.099)\\
\hat{\text{Var}}(z_1) &= .119(.019)\\
\hat{g}_2 &= .593(.093)\\
\hat{b}_1 &= .425(.066)\\
\hat{\text{Var}}(z_2) &= .167(.023)\\
\hat{\text{Var}}(z_3) &= .158(.023)\\
\hat{\text{Var}}(z_4) &= .081(.013)\\
\hat{\text{Var}}(z_5) &= .120(.017)
\end{aligned}$$

where the estimated standard errors are given in parentheses. This unmeasured variable model is also depicted in standardized form in Figure 11.24.

Although the fit of model 2 is adequate to avoid rejection at the usual 5 percent level, there may be an alternative model that more accurately represents the observations. For example, the presumption in model 2 that the determinants of systematic change in reading achievement are unchanging is clearly only an approximation for this time period. That is, the specification of F_1 as a perfectly stable variable is inconsistent with common sense concerning changes in the neighborhoods of New York. However, the changes in such conditions may occur so slowly that model 2 is a reasonable approximation for the five-year period under study.

Standardized

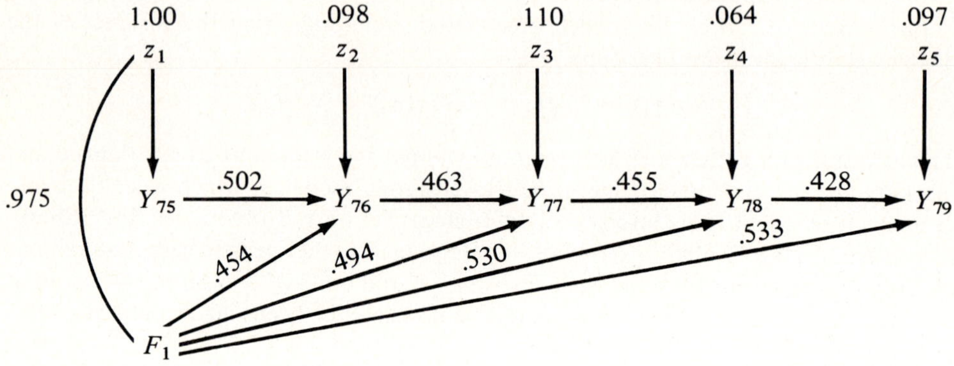

Metric

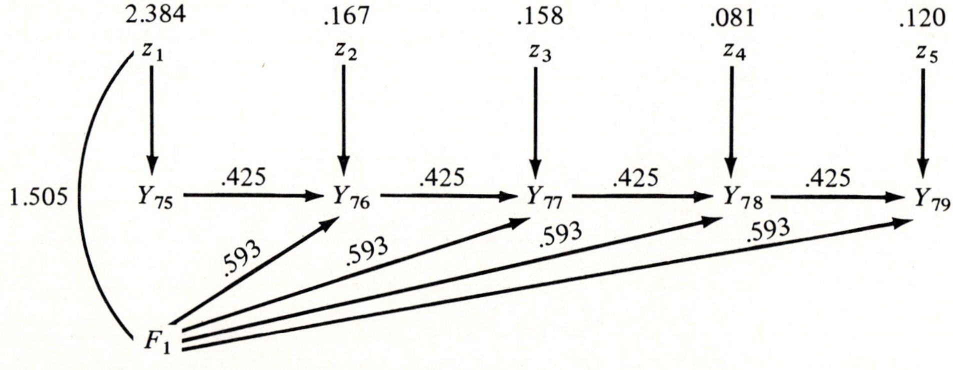

Figure 11.24 Path diagram of an unmeasured variable model (model 2) fit to data on reading achievement in New York City schools (data from Modoono, 1981). The variable F_1 is unmeasured in this model. Chi-square = 11.106; df = 7; Prob = .134.

An alternative to model 2 that takes account of changes in the determinants of school differences in reading achievement may be specified as follows (model 3):

$$E_1 = Y_{75} = g_1 F_1 + z_1$$

$$E_2 = Y_{76} = b_1 E_7 + b_2 Y_{75} + z_2$$

$$E_3 = Y_{77} = b_1 E_8 + b_2 Y_{76} + z_3$$

$$E_4 = Y_{78} = b_1 E_9 + b_2 Y_{77} + z_4$$

$$E_5 = Y_{79} = b_1E_{10} + b_2Y_{78} + z_5$$
$$E_6 = F_1$$
$$E_7 = b_3F_1 + z_7$$
$$E_8 = b_3E_7 + z_8$$
$$E_9 = b_3E_8 + z_9$$
$$E_{10} = b_3E_9 + z_{10}$$

where the disturbances (z) are presumed to be uncorrelated with the observed variables (Y), the unmeasured variables (F and E), and one another. Once again, the variance of F_1 is fixed at 1.0 and the coefficient g_1 is not to be given a causal interpretation. Also, notice that structural invariance has been assumed for structural effects.

Model 3 involves two more parameters than model 2, but the resulting reduction in the magnitude of the chi-square is substantial: chi-square = 1.480; df = 5; Prob = .915. This reduction in the chi-square (from 11.106 for model 2 to 1.480 for model 3) has 2 degrees of freedom and would occur for fewer than 1 percent of random samples if model 2 held in the population. Thus, model 2 may be rejected in favor of model 3. The maximum likelihood estimates of the parameters of model 3 are as follows:

$$\hat{ps}_{11} = \hat{\text{Var}}(z_1) = .103(.040)$$
$$\hat{ps}_{22} = \hat{\text{Var}}(z_2) = .125(.106)$$
$$\hat{ps}_{33} = \hat{\text{Var}}(z_3) = .121(.025)$$
$$\hat{ps}_{44} = \hat{\text{Var}}(z_4) = .049(.017)$$
$$\hat{ps}_{55} = \hat{\text{Var}}(z_5) = .099(.020)$$
$$\hat{ps}_{77} = \hat{\text{Var}}(z_7) = \ldots = \hat{\text{Var}}(z_{10}) = .010(.007)$$
$$\hat{g}_1 = 1.510(.098)$$
$$\hat{b}_1 = 1.145(.201)$$
$$\hat{b}_2 = .154(.106)$$
$$\hat{b}_3 = .924(.016)$$

Model 3 is depicted as a metric and standardized path diagram in Figure 11.25.

We now turn to the assessment of measurement models for the five background indicators that are available. Our first attempt to find such a latent variable structure is to hypothesize that the percent of children on the free lunch program (X_1), the median income in the community (X_2), and the litter index (X_3) are flawed multiple indicators of a single latent dimension: neighborhood socioeconomic status (SES). Since the mean and variance of X_3 (litter index) are arbitrary, the variance of X_3 is set equal to that of X_2 (median income), and the two measures are presumed to be parallel indicators of SES. This presumption arises from an approach to the study of communities in which the physical appearance of a community is presumed to reflect the economic resources of that community.

This measurement model (model 4) is specified as follows:

$$X_1 = k_1F_1 + F_2$$
$$X_2 = F_1 + F_3$$
$$X_3 = -F_1 - F_4$$
$$X_4 = F_5$$
$$X_5 = F_6$$

where F_1, F_5, and F_6 may be intercorrelated, but F_2, F_3, and F_4 are uncorrelated with all other variables in F. (The negatives used in the equation for X_3 are simply for the

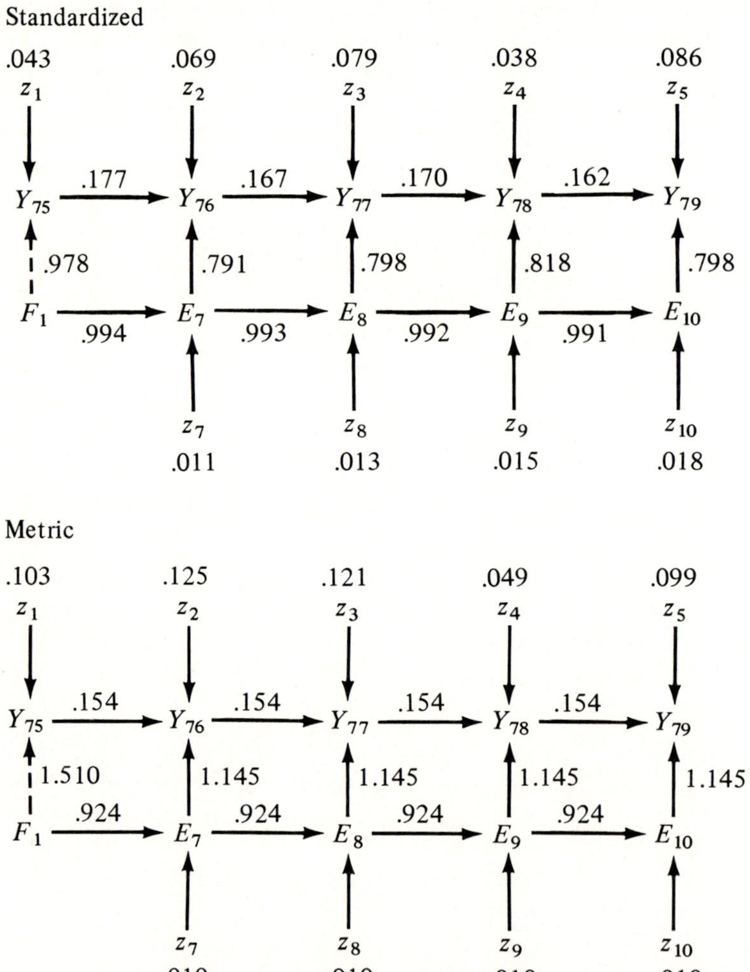

Figure 11.25 Path diagram of an alternative model (model 3) fit to median reading achievement scores in New York City intermediate schools (data from Modoono, 1981).

purpose of reflecting $\overline{X}_3$ so that its components correlate positively with components of X_2.) Note that each of the three SES indicators is partitioned into two components: the SES latent variable F_1 and a factor specific to the indicator. The specific factors are not specified in terms of the usual measurement error terms (d), since it seems likely that some of these measurement errors may be correlated with X_4 and X_5, and it is easier in the LISREL format to specify these correlations when the equations are in this form.

The fit of model 4 with the data is not acceptable given the moderate sample size ($N = 133$): chi-square $= 18.723$, $df = 6$, Prob $= .005$. An alternative model is immediately suggested by the fact that X_2 and X_3 are measured on similar geographic units (police precincts and planning districts, respectively) that are distinct from the school districts (recall that X_1, X_4, and X_5 are based upon school districts). Thus, model 5 is a measurement model analogous to model 4 except that F_3 and F_4 are allowed to covary. The fit of model 5 is adequate: chi-square $= 5.454$; $df = 5$; Prob $= .363$. The maximum likelihood estimates of model parameters are as follows:

$$\hat{ph}_{11} = \hat{\text{V}}\text{ar}(F_1) = 4.380(.084)$$
$$\hat{ph}_{22} = \hat{\text{V}}\text{ar}(F_2) = 2.873(.389)$$
$$\hat{ph}_{33} = \hat{ph}_{44} = \hat{\text{V}}\text{ar}(F_3) = \hat{\text{V}}\text{ar}(F_4) = 2.036(.341)$$
$$\hat{ph}_{55} = \hat{\text{V}}\text{ar}(F_5) = 7.433(.915)$$
$$\hat{ph}_{66} = \hat{\text{V}}\text{ar}(F_6) = 5.281(.650)$$
$$\hat{ph}_{43} = \hat{\text{C}}\text{ov}(F_4,F_3) = 1.197(.341)$$
$$\hat{ph}_{51} = \hat{\text{C}}\text{ov}(F_5,F_1) = -2.857(.619)$$
$$\hat{ph}_{61} = \hat{\text{C}}\text{ov}(F_6,F_1) = -3.738(.583)$$
$$\hat{ph}_{65} = \hat{\text{C}}\text{ov}(F_6,F_5) = -.569(.548)$$
$$\hat{k}_1 = -.775(.084)$$

where the approximate standard errors are given in parentheses. Model 5 is depicted as a path diagram in Figure 11.26. (Note that for ease of computation standard deviations of observed variables have been multiplied by a power of ten such that each standard deviation is greater than one and less than ten. Thus, the percentage variables should be interpreted in terms of 10 percent units rather than 1 percent units.)

Now, we have successfully fit a structural model to the endogenous variables (reading achievement) and a measurement model to the exogenous indicators. Our primary concern will be to combine these two models in a way that gives some information concerning the causal relationship between percent black (and/or percent Puerto Rican) and reading achievement after controlling for SES (F_1).

(As a prelude to the estimation of a combined model, the reader should use the correlations and standard deviations in Table 11.24 to regress Y_{79} on the battery of variables Y_{75}, X_2, X_4, X_5. It will be found that X_4 (percent black) evidences the usual negative effect on reading achievement—even controlling for community income. It is important to remember that treating this regression coefficient as a causal effect is

Metric

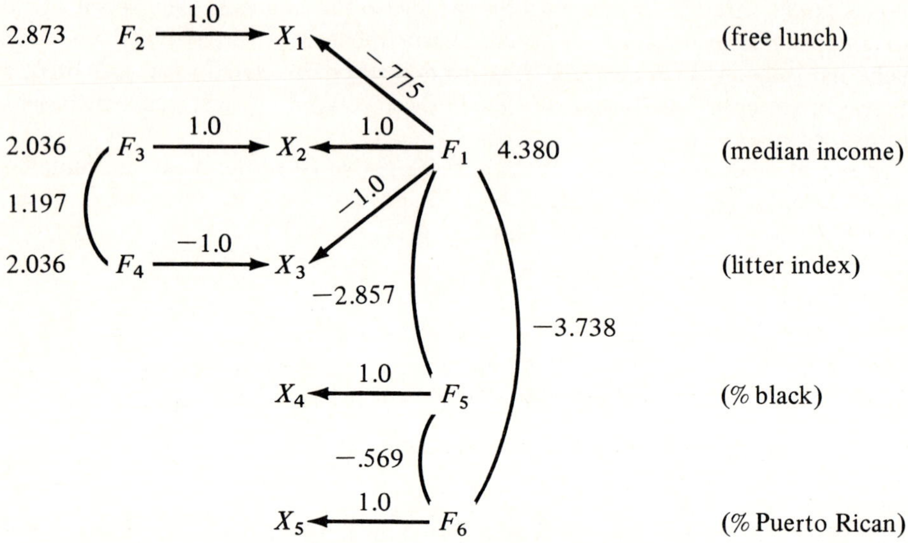

Standardized

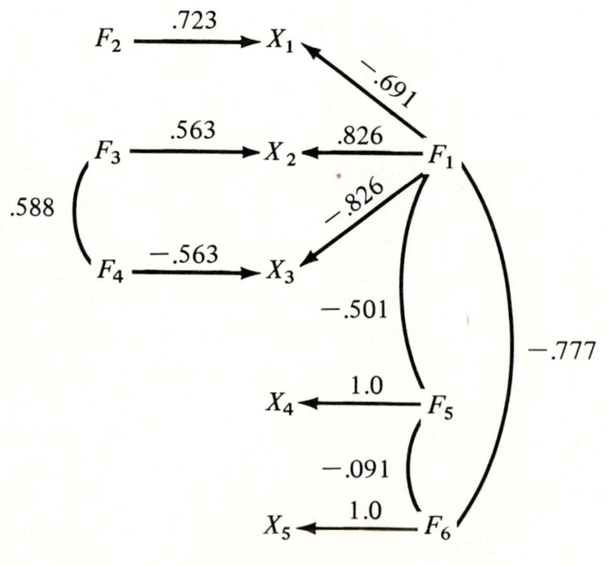

Figure 11.26 A measurement model (model 5) fit to five background indicators for New York City school districts (data from Modoono, 1981).

dependent upon several assumptions—including the one that X_2 is measured without error.)

There are several ways in which the exogenous and endogenous models may be merged. One approach is to hypothesize that the unmeasured variable in model 3 is a linear combination of SES (F_1), percent black (F_5), and percent Puerto Rican (F_6). This is accomplished in LISREL by combining the equations for model 3 and model 5 plus the following equation

$$E_6 = F_1 + g_2F_5 + g_3F_6$$

The resulting model is labeled model 6. Note that in this unifying equation there is no disturbance and the coefficient of F_1 is one. The lack of a disturbance constitutes a hypothesis that the unmeasured variable in model 3 (E_6 or F_1 in model 3) is a surrogate for the three exogenous variables specified in model 5 (F_1, F_5, and F_6 in model 5). Fixing the path from F_1 to E_6 as one in model 6 simply serves to define a metric for the unmeasured variables E_7, E_8, E_9, and E_{10}. An increase of one thousand dollars in 1970 "true" median family income (the metric of the latent variable F_1) would, if the model holds, produce the changes indicated by the structural coefficients leading from E_6. Furthermore, E_7 to E_{10} are similarly scaled.

The equation of E_6 with a linear combination of F_1, F_5, and F_6 implies that the effects of these three variables on reading achievement operate according to the pattern of causal paths defined in model 6. For example, there are no cross-lagged effects from E_7 to Y_{77} and the inertia of F_1, F_5, and F_6 are all equal to b_3. The effect of a 10 percent increase in the number of black students in a school on median reading achievement would then be

$$\text{Effect of 10\% increase in black students} = g_2b_1$$

Since we already know that b_1 is substantial (from estimation of model 5), our primary concern is with the magnitude of g_2. First, however, we must determine whether or not model 6 can be rejected by the data.

The estimation of model 6 may be summarized as follows:

Model 3: 10 parameters, 15 covariances, $df = 5$
Model 5: 10 parameters, 15 covariances, $df = 5$
Model 6: 2 additional parameters (g_2,g_3), 25 additional covariances, 23 additional degrees of freedom, total $df = 5 + 5 + 23 = 33$

Thus, model six adds 23 degrees of freedom to models 3 and 5. The likelihood ratio chi-square for model 6 is 31.710 ($df = 33$, Prob = .531). The overall fit of the model is quite good. Furthermore, the increase in the chi-square relative to the sum of the chi-squares for models 3 and 5 is comparable to the increase of 23 degrees of freedom:

$$31.710 - (1.980 + 5.454) = 24.276$$

The hypothesis that model 6 holds for these data, therefore, cannot be rejected, and the fit of the model with the data is at a level that would be expected if the model were true. Model 6 is depicted as a path diagram in Figure 11.27.

The maximum likelihood estimates of the model parameters are as follows:

parameter	estimate
$= g_1$	$= .957(.133)$
g_2	$.095(.058)$
g_3	$.250(.079)$
b_1	$.767(.156)$
b_2	$.115(.097)$
b_3	$.921(.014)$
$ph_{11} = \text{Var}(F_1)$	$4.519(.717)$
$ph_{22} = \text{Var}(F_2)$	$2.553(.318)$
$ph_{33} = \text{Var}(F_3)$	$1.897(.195)$
$ph_{44} = ph_{33}$	
$ph_{55} = \text{Var}(F_5)$	$7.434(.915)$
$ph_{66} = \text{Var}(F_6)$	$5.281(.650)$
$ph_{43} = \text{Cov}(F_4,F_3)$	$1.058(.195)$
$ph_{51} = \text{Cov}(F_5,F_1)$	$-2.841(.618)$
$ph_{61} = \text{Cov}(F_6,F_1)$	$-3.688(.580)$
k_1	$-.808(.077)$
$ps_{11} = \text{Var}(z_1)$	$.087(.025)$
$ps_{22} = \text{Var}(z_2)$	$.117(.022)$
$ps_{33} = \text{Var}(z_3)$	$.114(.023)$
$ps_{44} = \text{Var}(z_4)$	$.044(.015)$
$ps_{55} = \text{Var}(z_5)$	$.093(.019)$
$ps_{66} = \text{Var}(z_6)$	$.000$(fixed)
$ps_{77} = ps_{88} = ps_{99} = ps_{10,10}$	$.032(.014)$

where approximate standard errors are given in parentheses.

The noteworthy findings in the estimates of model 6 are the *positive* estimates of the effects of percent black (g_2b_1) and percent Puerto Rican (g_3b_1) on median reading achievement scores in New York City intermediate schools. If we accept model 6, then the best estimate of the effect of a ten-point increase in the percent of a school's pupils who are black is that the median reading achievement score will increase by (.095 × .767) years or .073 years. However, the approximate standard error of g_2 is large relative to the estimate of g_2, so we cannot reject the null hypothesis that the product g_2b_1 is zero. In an analogous manner, the best estimate of the effect of a ten-point increase in the Puerto Rican percentage in an intermediate school is .192 (.250 × .767 = .192) year equivalents of reading achievement. In this instance, the estimates of both components of the product are more than three times their standard errors; thus the null hypothesis concerning the effect of an increase in Puerto Rican students

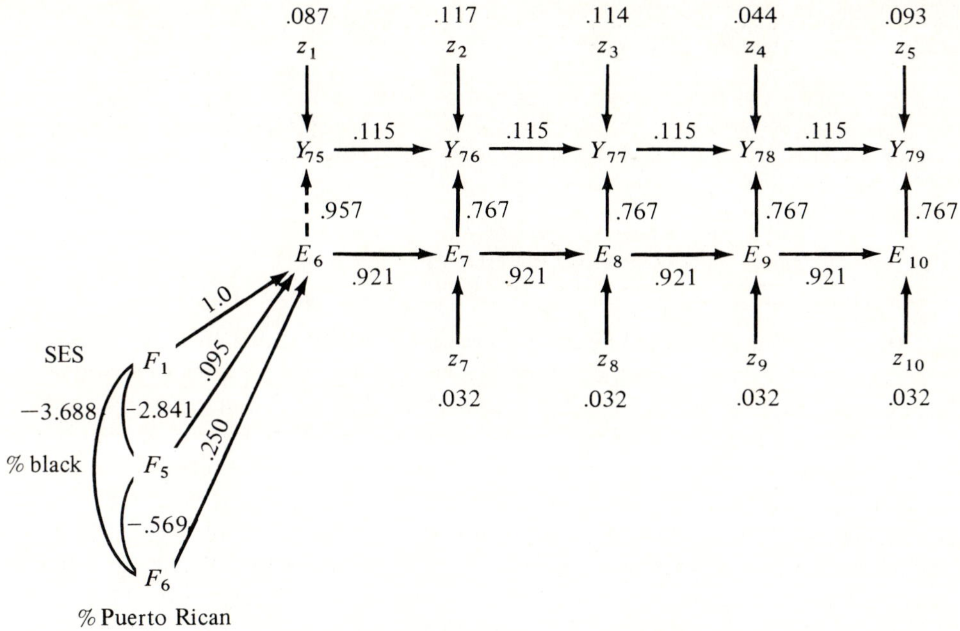

Figure 11.27 Path diagram of combined exogenous and endogenous model (model 6) for median reading achievement in New York City intermediate schools (data from Modoono, 1981). The variance of E_6 is identified by fixing the path from F_1 to E_6 at the value 1.00.

can be rejected with confidence if model 6 holds. This estimate means that a 10 percent increase in the Puerto Rican constituency of a school would boost the median reading achievement by almost one-fifth of a school year. The reader should bear in mind the fact that this estimate of such an "effect" presumes that neighborhood SES does not change as the percent Puerto Rican in the school increases. Clearly, such a pattern of change would be an unlikely event in the real world; thus, it is not surprising that the results of estimating model 6 produce a picture that does not jibe with popular beliefs. Furthermore, the finding of a positive effect of additional minority children in a school is counter to what we would expect from a genetic or "culture of poverty" hypothesis concerning the simple *negative* correlation between median reading achievement and minority student numbers.

Model 6, as always, is not the only model that is consistent with these data. For example, the combined model 6 enables us to identify covariances between errors in the indicators of SES (F_3 and F_4) and percent minority students (F_5 and F_6). The resulting model 7 has 31 degrees of freedom and produces a goodness of fit chi-square of 24.520 (Prob = .789). This decrease in the chi-square relative to model 6 (31.710 − 24.520 = 7.190) is sufficiently unlikely by chance (Prob = .027), but, in truth, it was motivated by inspection of the residuals from model 6. Thus, the probability associated with the decrease in the chi-square is misleading. Nevertheless, it is instructive to see how this "overfitting" modification affects the estimates of the crucial parameters of the model:

Parameter	Maximum likelihood estimate
g_1	.803(.145)
g_2	.092(.101)
g_3	.022(.137)
b_1	.643(.150)
b_2	.115(.097)
b_3	.920(.014)
$ph_{53} = ph_{54} = \text{Cov}(F_5,F_3) = \text{Cov}(F_5,F_4)$	.809(.404)
$ph_{63} = ph_{64} = \text{Cov}(F_6,F_3) = \text{Cov}(F_6,F_4)$	−.955(.414)

Note that in this instance the effects of increased minority students is still estimated as positive, but neither the coefficient for blacks (g_2) or Puerto Ricans (g_3) is greater than its standard error in absolute magnitude. Thus, if model 7 is accepted, increasing the proportion of minority students in an intermediate school appears to have little effect on median reading achievement scores. This conclusion, of course, concerns the effects of increased minority students that are linearly *independent* of median SES in the neighborhood from which students are drawn. The fact that this conclusion contradicts most of previous research in this area is of particular significance. Model 7 is clearly not *the* true model; rather, it is a model that produces 31 overidentifying restrictions on the interrelations of the observed variables, and it is a model whose restrictions are consistent with the observed data. Furthermore, the specification of model 7 points toward the additional observations that would be necessary to further test the model: indicators of minority student constituency and neighborhood SES for successive years. In addition, a more comprehensive analysis of these data would involve assessment of nonlinearities, the use of normalizing transformations, and the testing of additional models.

11.7 CONCLUDING REMARKS

The goal of this chapter was to produce a balanced presentation of causal inference from passive longitudinal data. It is balanced in the sense that both the weaknesses and the strengths of such models are presented. In this concluding section I want to emphasize the fact that successful use of longitudinal models for causal inference requires the development of extensive knowledge within a particular substantive area. This is the case since the assumptions necessary for causal inference in such instances do not have the universality that characterizes randomization.

One way in which to develop the needed knowledge in a specific research area is to begin by developing structural equation models to describe *known* causal relations. For example, a developmental psychologist might begin with the longitudinal relationship between the height of a child (or the length of his legs) and the length of his pants. There is little doubt that changes in the former lead to changes in the latter. However, would it be possible to deduce the direction of causal influence from longitudinal observation of a sample of children at discrete points in time? If a researcher could not succeed in drawing such an inference from sample data, then one should probably hesitate to apply similar modeling techniques to less trivial relationships. On the other hand, if one is successful in developing such models, then a foundation of knowledge concerning patterns of structural invariance can pave the way for research into more interesting problems.

It is my hope that this chapter will contribute to the ability of researchers to find ways of studying that which is of importance rather than that which is amenable to previously available research methodologies.

REFERENCES AND SUGGESTED READING

Barnow, B. S. The effects of head start and socioeconomic status on cognitive development of disadvantaged children. Doctoral dissertation, University of Wisconsin, Madison, 1973.

Bentler, P. M. Multivariate analysis with latent variables: Causal modeling. *Annual review of Psychology,* 1980, *31,* 419–456.

Bentler, P. M., and Speckart, G. An evaluation of models of attitude-behavior relationships. *Psychological Review,* 1979, *86,* 452–464.

Bentler, P. M., and Woodward, J. A. Head start re-evaluation: Positive effects are not yet demonstrable. *Evaluation Quarterly,* 1978, *2,* 493–510.

Bryk, A. S., and Weisberg, H. I. Use of the nonequivalent control group design when subjects are growing. *Psychological Bulletin,* 1977, *84,* 950–962.

Campbell, D. T., and Boruch, R. F. Making the case for randomized assignment to treatments by considering the alternatives: Six ways in which quasi-experimental evaluations in compensatory education tend to underestimate effects. In A. Lumsdaine and C. Bennett (eds.), *Experiment and evaluation,* New York: Academic Press, 1975.

Campbell, D. T., and Erlebacher, A. E. How regression artifacts in quasi-experimental evaluations can mistakenly make compensatory education look harmful. In J. Hellmuth (ed.), *Compensatory education: A national debate* (Vol. 3). *Disadvantaged child.* New York: Brunner/Mazel, 1970.

Cicirelli, V. G., *et al. The impact of Head Start: An evaluation of the effects of Head Start on children's cognitive and affective development.* Athens, Ohio: Ohio University and Westinghouse Learning Corporation, 1969.

Coleman, J. S. The mathematical study of change. In H. M. Blalock and A. B. Blalock (eds.), *Methodology in social research.* New York: McGraw-Hill, 1968.

Cook, T. D., and Campbell, D. T. *Quasi-experimentation: Design and analysis issues for field settings.* Chicago: Rand McNally, 1979.

Cronbach, L. J., and Furby, L. How should we measure 'change'—or should we? *Psychological Bulletin,* 1970, *74,* 68–80.

Cronbach, L. J., Rogosa, D. R., Floden, F. E., and Price, G. Analysis of covariance in nonrandomized experiments: Parameters affecting bias. Occasional paper, Stanford Evaluation Consortium, Stanford University, 1977.

Duncan, O. D. Some linear models for two-wave, two-variable panel analysis. *Psychological Bulletin,* 1969, *72,* 177–182.

Duncan, O. D. Some linear models for two-wave, two-variable panel analysis with one-way causation and measurement error. In H. M. Blalock, A. Agenbegian, F. M. Borodkin, R. Boudon, and V. Capecchi (eds.), *Quantitative sociology.* New York: Academic Press, 1975.

Dwyer, J. H., and Barret-Goldfarb, M. Locus of control and socioeconomic achievement: Some longitudinal models. State University of New York, Stony Brook, 1982.

Dwyer, J. H., and Taves, M. Racial income disparities and the income inequality among white families: A longitudinal analysis. State University of New York, Stony Brook, 1982.

Dwyer, J. H., Wehrle, G., and Phelan, T. A longitudinal structural model fit to multiple method-multiple trait data. State University of New York, Stony Brook, 1981.

Goldberger, A. S. Pitfalls in the resolution of IQ inheritance. In N. E. Morton and C. S. Chung (eds.), *Genetic epidemiology.* New York: Academic Press, 1978.

Heise, D. R. Causal inference from panel data. In E. F. Borgatta and G. W. Bohrnstedt (eds.), *Sociological methodology 1970.* San Francisco: Jossey-Bass, 1969.

Jessor, R., and Jessor, S. L. *Problem behavior and psychosocial development: A longitudinal study of youth.* New York: Academic Press, 1977.

Joreskog, K. G., and Sorbom, D. Statistical models and methods for the analysis of longitudinal data. In D. J. Aigner and A. S. Goldberger (eds.), *Latent variables in socioeconomic models.* Amsterdam: North-Holland, 1977.

Joreskog, K. G., and Wold, H. (eds.), *Systems under indirect observation.*

Kenny, D. A. Cross-lagged panel correlations: A test for spuriousness. *Psychological Bulletin,* 1975, *82,* 887–903.

Linn, R. L., and Werts, C. E. Analysis implications of the choice of a structural model in the nonequivalent control group design. *Psychological Bulletin,* 1977, *84,* 229–234.

Lord, F. M. A paradox in the interpretation of group comparisons. *Psychological Bulletin,* 1967, *68,* 304–305.

Magidson, J. Causal model approach for adjusting for preexisting differences in nonequivalent control-group situations. A general alternative to ANCOVA. *Evaluation Quarterly,* 1977, *1,* 399–420.

Magidson, J. The .05 significance level is not all powerful: Reply to Bentler and Woodward. *Evaluation Quarterly,* 1978, *2,* 511–520.

Markus, G. B. *Analyzing panel data.* Beverly Hills, Calif.: Sage University Paper Series.

McCleary, R., and Hay, R. A. *Applied time series analysis for the social sciences.* Beverly Hills, Calif.: Sage, 1980.

Mednick, S. A., and Harway, M. (eds.) *Longitudinal research in the United States.*

Modoono, S. An investigation into the macro-determinants of scholastic achievement. State University of New York, Stony Brook, doctoral dissertation, 1981.

Nesselroade, P. B., and Baltes, P. B. (eds.) *Longitudinal methodology in the study of behavior and human development.* New York: Academic Press, 1981.

Pierce, D. A., and Haugh, L. D. Causality in temporal systems: Characterizations and a survey. *Journal of Econometrics,* 1977, *5,* 265–293.

Reich, M. *The economics of racism.* Princeton, N.J.: Princeton University Press, 1978.

Rogosa, D. Causal models in longitudinal research: Rationale, formulation and interpretation. In J. R. Nesselroade and P. B. Baltes (Eds.), *Longitudinal research in the study of behavior and development.* New York: Academic Press, 1979.

Rogosa, D. A critique of cross-lagged correlation. *Psychological Bulletin,* 1980, *88,* 245–258.

Rogosa, D. Comparison of some procedures for analyzing longitudinal panel data. *Journal of Economics and Business,* 1980, *32,* 136–151.

Rozelle, R. M., and Campbell, D. T. More plausible rival hypotheses in the cross-lagged panel correlation technique. *Psychological Bulletin,* 1969, *71,* 74–80.

Sorbom, D. An alternative to the methodology for analysis of covariance. *Psychometrika,* 1978, *43,* 381–396.

Wehrle, G. A study of alienation in work and leisure. State University of New York, Stony Brook, 1982.

Werts, C. E., and Linn, R. L. A general linear model for studying growth. *Psychological Bulletin,* 1970, *73,* 17–22.

Appendix A
Rules of Matrix Algebra

This brief appendix introduces the elementary rules of matrix notation and algebra. An understanding of this material is necessary for the reader who wishes to actually be a user of maximum likelihood procedures for the estimation of structural equation models. Casting a series of equations or covariances in terms of matrices is simply a convention for organization that greatly *simplifies* the symbolic representation of a large number of elements. However, like any summarizing abstraction, the abstraction only mystifies if the relation between the concrete and the abstract is not clear from experience. Thus, a matrix equation may reduce hundreds of elements and their interrelations to a few symbols. But the reader must have worked through several examples with the elements of the matrices before the matrix summary makes sense.

We begin by providing a series of definitions and labeling conventions. The static concepts defined are then used as building blocks of an algebra in which the relations between elements are the focus.

A *matrix* is defined as a rectangular array of elements. It is said to be rectangular, since the array of elements is organized into rows and columns. The term *element* is undefined. An element of a matrix may be a building, a person, a number, a variable, etc.

A *data matrix* is a matrix of scores (numbers) for N cases on p variables. A data matrix then has N rows and p columns, and the number in the ith row and jth column of the matrix is the score of the ith case on the jth variable. As an example, suppose that scores on 10 personality dimensions (variables) are gathered on 100 persons (cases). The resulting data matrix would be a rectangular presentation of these 1000 scores where each row contained the 10 scores of a person on the 10 personality measures; similarly, the columns of the data matrix would contain the 100 scores of 100 persons on a particular personality measure.

A matrix is generally denoted by an underlined letter. Thus, Y_1 might be a variable, whereas $\underline{Y}$ might be a data matrix. A small data matrix $\underline{Y}$ might be the following if observations on three variables (Y_1, Y_2, and Y_3) were taken on a sample of four cases:

$$
\begin{array}{c c}
 & \text{variables} \\
 & \begin{array}{ccc} Y_1 & Y_2 & Y_3 \\ \text{---} & \text{---} & \text{---} \end{array} \\
\text{cases}\ \begin{array}{c} 1 \\ 2 \\ 3 \\ 4 \end{array} & \begin{bmatrix} 10 & 12 & 8 \\ 7 & 9 & 11 \\ 15 & 12 & 13 \\ 8 & 10 & 9 \end{bmatrix}
\end{array}
= \underset{(4 \times 3)}{\underline{Y}}
$$

where the expression (4 × 3) under the matrix label $\underline{Y}$ indicates that the matrix has four rows and three columns.

A more general data matrix $\underline{Y}$ may be written as follows:

$$
\begin{array}{c c}
 & \begin{array}{ccccc} Y_1 & Y_2 & \dots Y_j & \dots & Y_p \\ \text{---} & \text{---} & \text{---} & & \text{---} \end{array} \\
\begin{array}{c} 1 \\ 2 \\ \dots \\ i \\ \dots \\ N \end{array} & \begin{bmatrix} Y_{11} & Y_{12} & Y_{1j} & Y_{1p} \\ Y_{21} & Y_{22} \dots & Y_{2j} \dots & Y_{2p} \\ & & & \\ Y_{i1} & Y_{i2} \dots & Y_{ij} \dots & Y_{ip} \\ & & & \\ Y_{N1} & Y_{N2} \dots & Y_{Nj} \dots & Y_{Np} \end{bmatrix}
\end{array}
= \underset{(N \times p)}{\underline{Y}}
$$

where Y_{ij} is a score and Y_j is a variable.

A data matrix can be construed in terms of a p-dimensional Euclidean space. Each variable can be thought of as a dimension (axis), and each row of the matrix then defines a point in this p-dimensional space. In the case of a two-column matrix, the reader is familiar with this representation of a matrix as a scatterplot with vertical and horizontal axes. A $(N \times p)$ data matrix then defines N points in a p-dimensional space.

A matrix with only one column $(N \times 1)$ is called a *column vector* or simply a vector. A matrix with only one row $(1 \times p)$ is called a *row vector.* A vector may be a data vector, or it may be a vector of variables. For example, the row vector

$$[Y_1 Y_2 \dots Y_p]$$

is a random vector of p random variables. This vector of variables may be defined for some population in the same manner that an individual variable is defined.

A *square* matrix is one that has the same number rows as columns, and the *order* of a square matrix is its number of rows (or columns). A *diagonal* matrix is a square matrix in which only the main diagonal elements (from the upper left corner to the lower right corner) are differnt from zero. A *symmetric* matrix is one in which the elements in the upper-right section of the matrix are a mirror image of the elements in the lower-left triangle. That is, for the square matrix $\underline{Y}$, the element Y_{ij} is equal to the element Y_{pq} if $i = q$ and $j = p$. (Covariance and correlation matrices are symmetric.) A *zero* or *null* matrix has all zero elements. A matrix in which all elements are equal to one is known as a *unit* matrix. Finally, an *identity* matrix is a square matrix with ones in the main diagonal and zeros elsewhere.

We now turn to the topic of matrix *operations.* Matrix operations are somewhat analogous to the binary operations of real number algebra (addition, subtraction, multiplication, etc.). The simplest matrix operation is that of *transposition.* The transpose of the the matrix $\underline{Y}$ is obtained by turning the rows of $\underline{Y}$ into the columns of a new matrix denoted $\underline{Y}'$. For example, if

$$\begin{bmatrix} a & b & c \\ d & e & f \end{bmatrix} = \underset{(2 \times 3)}{\underline{Y}}$$

then the transpose of $\underline{Y}$ is given by

$$\begin{bmatrix} a & d \\ b & e \\ c & f \end{bmatrix} = \underset{(3 \times 2)}{\underline{Y}'}$$

The sum of two matrices, say $\underline{Y}$ and $\underline{X}$, is defined only if each matrix has the same order (number of rows and columns). And the sum is defined as a new matrix in which each element is a sum of the two elements in the corresponding position of the two matrices being added. Thus, if

$$\begin{bmatrix} a & b \\ c & d \end{bmatrix} = \underline{Y}$$

and

$$\begin{bmatrix} e & f \\ g & h \end{bmatrix} = \underline{X}$$

then

$$\begin{bmatrix} (a + e) & (b + f) \\ (c + g) & (d + h) \end{bmatrix} = \underline{Y} + \underline{X}$$

defines the sum. Matrix subtraction is defined in an analogous manner.

The *product* of a *scalar* (e.g., a real number) and a matrix is simply defined as a new matrix in which each element of the original matrix is multiplied by the scalar. Thus, if

$$\begin{bmatrix} Y_{11} & Y_{12} & Y_{13} \\ Y_{21} & Y_{22} & Y_{23} \end{bmatrix} = \underline{Y}$$

then the product of the scalar b and the matrix $\underline{Y}$ is

$$\begin{bmatrix} b\,Y_{11} & b\,Y_{12} & b\,Y_{13} \\ b\,Y_{21} & b\,Y_{22} & b\,Y_{23} \end{bmatrix} = b\,\underline{Y}$$

In contrast to matrix addition, *matrix multiplication* is a bit tricky. The product of a matrix $\underline{Y}$ of order $(n \times p)$ and a matrix $\underline{X}$ of order $(m \times q)$ is obtained by mutliplying rows of $\underline{Y}$ by colums of $\underline{X}$ if p is equal to m. For example, if $\underline{Y}$ is of order (2×3) and $\underline{X}$ is of order (3×1), then the product of $\underline{Y}$ and $\underline{X}$ is a new matrix of order (2×1):

$$\underset{(2 \times 3)}{\begin{bmatrix} Y_{11} & Y_{12} & Y_{13} \\ Y_{21} & Y_{22} & Y_{23} \end{bmatrix}} \underset{(3 \times 1)}{\begin{bmatrix} X_{11} \\ X_{21} \\ X_{31} \end{bmatrix}} = \underset{(2 \times 1)}{\begin{bmatrix} Y_{11}\,X_{11} + Y_{12}\,X_{21} + Y_{13}\,X_{31} \\ Y_{21}\,X_{11} + Y_{22}\,X_{21} + Y_{23}\,X_{31} \end{bmatrix}}$$

where elements of the product matrix are defined by summing the products of elements that are in the same row of $\underline{Y}$ as column of $\underline{X}$. Note that matrix multiplication is not necessarily commutative. The product $\underline{X}\,\underline{Y}$, for example, is not even defined, since the number of columns of the matrix multiplying from the left ($\underline{X}$) is not equal to the number of rows in the matrix multiplying from the right ($\underline{Y}$).

The *identity matrix* is so named because it serves as a multiplicative identity for matrices. Multiplying a $(n \times p)$ matrix on the right by a $(p \times p)$ identity matrix produces the original $(n \times p)$ matrix. Similarly, multiplying a $(n \times p)$ matrix on the left by a $(n \times n)$ identity matrix produces the original $(n \times p)$ matrix as well.

The *inverse* of a $(p \times p)$ square matrix $\underline{Y}$ is

$$\underline{Y}^{-1} = \text{inverse of } \underline{Y}$$

such that the matrix product

$$\underline{Y}\,\underline{Y}^{-1} = (p \times p) \text{ identity matrix}$$

If $\underline{Y}$ is a symmetric matrix, then its left- and right-hand inverses are equal.

The *determinant* of a (2×2) matrix is defined as follows:

if

$$\begin{bmatrix} a & b \\ c & d \end{bmatrix} = \underline{Y}$$

then

$$\det(\underline{Y}) = ad - bc$$

defines the determinant of $\underline{Y}$. For a (3×3) matrix, the determinant is defined as follows:

if

$$\begin{bmatrix} a & b & c \\ d & e & f \\ g & h & i \end{bmatrix} = \underline{Y}$$

then

$$\det(\underline{Y}) = a(ei - fh) - b(di - fg) + c(dh - eg)$$

The rule for evaluation of the determinant of a square matrix with more than two rows is the following: take each element in the first row (a, b, and c in the above example), and multiply each by the determinant of the matrix defined by deletion of the row and column containing that element. Then multiply every other element of the first row by negative one (-1) and sum. When this procedure is followed in a stepwise manner, the determinant of a square matrix of any order can be computed. Fortunately, computers handle this task quite easily. A matrix with a determinant of zero has no inverse and is said to be a *singular* matrix. Singular covariance matrices are of particular concern in regression analysis, since solution for the regression coefficients depends upon inversion of the covariance matrix of the right-hand variables.

Some common statistics are defined quite simply with the use of matrices. For example, if $\underline{Y}$ is a $(N \times 1)$ data vector and $\underline{1}$ is a $(N \times 1)$ vector of ones, then the vector product

$$(\underline{Y}'\underline{1})(1/N) = \text{sample mean of } Y$$

where $\underline{Y}'$ is the transpose of $\underline{Y}$. Similarly, if $\underline{Y}$ and $\underline{X}$ are two $(N \times 1)$ data matrices, then the vector product

$$\underline{Y}'\underline{Y} = \text{sum of squares of } Y$$

$$\underline{X}'\underline{X} = \text{sum of squares of } X$$

and

$$\underline{Y}'\underline{X} = \underline{X}'\underline{Y} = \text{cross product of } Y \text{ and } X$$

If we subtract the sample mean of X from each element of $\underline{X}$ and the sample mean of Y from each element of $\underline{Y}$ to obtain the deviation score vectors $\underline{x}$ and $\underline{y}$, then we have

$$(\underline{x}'\underline{y})(1/N) = (\underline{y}'\underline{x})(1/N) = \text{covariance of } Y \text{ and } X$$

$$(\underline{x}'\underline{x})(1/N) = \text{sample variance of } X$$

$$(\underline{y}'\underline{y})(1/N) = \text{sample variance of } Y$$

Similarly, if $\underline{x}$ is a $(N \times p)$ data matrix of mean deviation scores, and $\underline{y}$ is a $(N \times q)$ data matrix of mean deviation scores, then

$$(\underline{x}'\underline{x})(1/N) = \text{variance-covariance matrix of } \underline{X}$$

$$(\underline{y}'\underline{y})(1/N) = \text{variance-covariance matrix of } \underline{Y}$$

and

$$(\underline{x}'\underline{y})(1/N) = \text{covariance matrix of } \underline{X} \text{ with } \underline{Y}$$

These definitions and operations are used on many occasions throughout this book. However, in most instances their meaning is concretely demonstrated so that the reader may learn how to use matrices by following the developments from Chapter 5 to Chapter 10.

ADDITIONAL READING

Bock, R. D. *Multivariate statistical methods in behavioral research.* New York: McGraw-Hill, 1975. (Chapter 2 provides a good introduction to the matrix operations and definitions of most use in the understanding of multivariate statistical problems.)

Appendix B
Logistic Regression of Dichotomous Dependent Variables

As was pointed out in the main body of this book, there are several problems with the use of dichotomous dependent variables in linear models. The first of these problems concerns the linearity of the relationship between a conditional probability and an independent variable. It will be demonstrated here that the *logit transformation* of conditional probabilities (or odds) can linearize such relationships. The second problem raised by linear models with dichotomous dependent variables is that the criterion for additivity is often inappropriate. It will then be shown that a multiplicative model, rather than a linear additive model, is a more appropriate basis for the assessment of interactions in models tested with dichotomous variables. The multiplicative model is linearized by the logit transformation as well, and solution for the parameters of the model is obtained in a manner that is analogous to the application of the method of moments used throughout the main text.

The Linearity Problem

So that we may begin with the concrete, we begin with data from the National Longitudinal Survey (NLS) of older men. Two variables are of concern: the educational attainment of the respondent or son (Y) and the educational attainment of his father (X). If the relationship between these two variables is explored with scatterplots and regression analyses, it is found to be quite well represented by a linear regression equation of the form

$$Y = c_0 + c_1X + u$$

However, suppose that we did not have the actual scores for the variable Y available for analysis. Instead, we only had dichotomous information on whether or not the respondent had graduated from college. Thus, we might be tempted to form a new variable Y_2 such that

$$\begin{aligned} Y_2 = {} & 0 \text{ if } Y < 16 \text{ (son did not graduate from college)} \\ & 1 \text{ if } Y \geq 16 \text{ (son did graduate from college)} \end{aligned}$$

We could then proceed to fit the equation

$$Y_2 = d_0 + d_1X + u_2$$

to the available data. The coefficient d_1 would then be interpreted as the amount of increase in the probability of a son completing college that is associated with an additional year of father's education.

The question that arises at this point is one of the linearity of the relationship between Y_2 and X. Is it reasonable to expect this relationship to be linear when the relationship between Y and X is linear? The answer to this question can be gleaned from the actual NLS data.

Table B.1 gives the number of sons that did and did not graduate from college for each group of fathers with the same educational attainment. The first column in Table B.1 gives the educational attainment of fathers. Thus, each row of the table applies to a group of respondents whose fathers all had the same number of years of education. The second column gives the number of sons who did not graduate from college, whereas column three gives the number of sons who did graduate from college. (The fact that sons rather than fathers were sampled here creates some problems, but these are of no interest at this point.) These frequencies can then be used to compute the probability of a son completing college at each level of fathers' education. Notice that this probability increases from .026 for sons of fathers with no education to .346 for sons of fathers with 16 years of education.

The probabilities given in the third column of Table B.1 are simply the expected values of Y_2 given a particular value of X:

Table B.1 The number of sons graduating from college as a function of their fathers' educational attainment (from the NLS of older men).

(1)	(2)	(3)	(4)	(5)	(6)
	Son				
Fathers' educ (yrs)	Less than college grad	College grad	Prob($Y > 15$)	Odds($Y > 15$)	Log(odds)
0	343	9	.026	.026	−3.641
1	27	1	.036	.037	−3.296
2	53	2	.036	.038	−3.277
3	168	12	.067	.071	−2.639
4	230	5	.021	.022	−3.829
5	170	8	.045	.047	−3.056
6	290	15	.049	.052	−2.962
7	128	9	.066	.070	−2.655
8	687	68	.090	.099	−2.313
9	49	6	.109	.122	−2.100
10	77	11	.125	.143	−1.946
11	22	8	.267	.364	−1.012
12	245	64	.207	.261	−1.342
13	23	11	.324	.478	−.738
14	44	24	.353	.545	−.606
15	14	4	.222	.286	−1.253
16	70	37	.346	.529	−.638

$$E(Y_2|X) = \text{Prob}(Y > 15|\text{father ed} = X)$$

Now, the appropriateness of representing this relationship between Y_2 and X with a linear equation depends upon how well a straight line fits a scatterplot of X against the conditional expectations of Y_2 (column 4 of Table B.1). This relationship between X and $E(Y_2|X)$ is depicted in Figure B.1, where X is plotted on the horizontal axis and $E(Y_2|X)$ is plotted on the vertical axis. Note that the relationship is clearly *not* linear over the full range of X. Thus, if we fit a straight line to these points with OLS procedures, we would obtain a solution in which the residuals would not be independent of X (although they would be uncorrelated with X). In fact, the unweighted OLS line of best fit is

$$\hat{Y}_2 = -.029 + .021X$$

and substitution of values of X into this equation reveals that the probability of college completion tends to be *under*estimated when X is low and *over*estimated when X is in the mid-range.

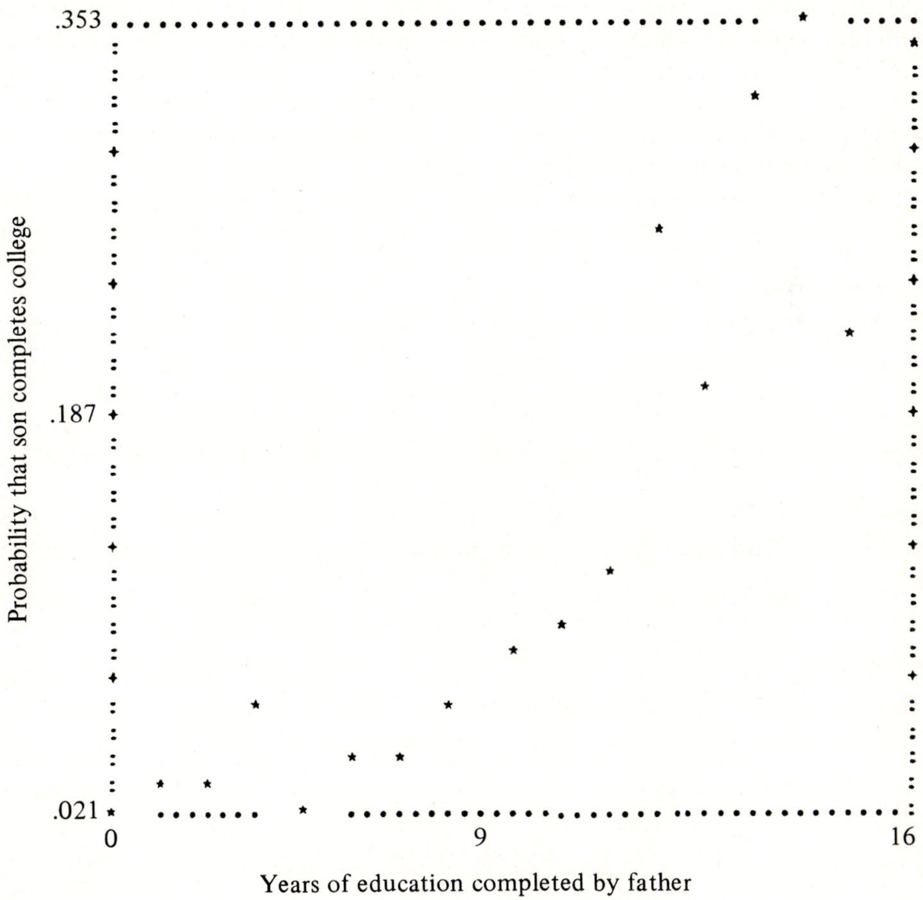

Figure B.1 Scatterplot relating the probability that a son will graduate from college (y-axis) to the number of years of schooling that his father completed (x-axis).

This nonlinear pattern is typical when approximately normally distributed variables are dichotomized. An intuitive sense of why this is the case can be gleaned from some reflection about the normal curve. The crucial insight comes from the realization that the distance of a score from the mean in a normal distribution is *not* a linear function of the probability of its occurrence. If Y is roughly normal in distribution and linearly related to X, then the means of the conditional distributions of Y (conditional, that is, on values of X) increase (or decrease) linearly with increases in X. It is as if the bell-shaped distribution of the Y scores moves one notch higher above the dichotomizing point for each unit increase in X. But the one-notch movement does not produce a constant increase in the probability that Y is above the cut-off point. In fact, as the lower tail of the curve reaches the dichotomizing point, a unit increase in X produces almost no increase in the probability that Y is above the dichotomizing point.

In order to compensate for this nonlinearity, it would be intuitively appealing to transform the conditional probabilities obtained into values of a normal variate (z-scores) corresponding to the points on the normal curve at which such probabilities would occur. For example, the probability of college completion is 0.207 for respondents with fathers completing 12 years of schooling. Consulting a table for the normal distribution reveals that a normal z-score of .82 or greater has a probability of approximately .206. Thus, the probability of .207 could be transformed into a normal variate of .82, and other probabilities could be similarly transformed. This transformation would retrieve the linearity of the original relationship between the variables before dichotomization. Such a procedure is termed *probit analysis*.

However, an approximation to the probit transformation is much simpler to use. This alternative is derived from assuming that the variables (before dichotomization) are distributed according to the *logistic distribution* rather than the normal distribution. For all practical purposes, the logistic distribution is equivalent to the normal distribution, but there is a dramatic computational and parsimony advantage in favor of the logistic. The logistic density is defined as follows:

$$f(X) = 1/(1 + e^{-bX})$$

where X is the variable of interest, b is a "scaling" parameter, e is the base of the natural logarithm (2.71828 . . .), and $f(X)$ is the height of the density function. The beauty of this density function is that it can be solved for bX—the usually unobserved continuous dimension that has been dichotomized. It turns out that the logarithm of the *odds* is equal to bX:

$$\log\{f(X)/[1 - f(X)]\} = bX$$

Now, in the discrete case we approximate $f(X)$ with the conditional probability that Y_2 is one at each value of X and obtain:

$$bX = \log\{P(Y_2 = 1 | X)/[1 - P(Y_2 = 1 | X)]\}$$

or

$$bX = \log[\text{odds}(Y_2 = 1 | X)]$$

where "odds" are defined in the usual sense associated with wagering. Referral to Table B.1 will reveal that the odds of Y_2 being equal to one, given a particular level of X, are simply the ratio of the number of people for whom Y_2 is one (column 3 of Table B.1) divided by the number of people for whom Y_2 is equal to zero (column 2 in Table B.1). And these odds are equal to the ratio of the probability that Y_2 is equal to one divided by one minus that probability. [This relationship can be seen algebraically if we define c as the number of persons for whom $Y_2 = 0$ and d as the number of persons for whom $Y_2 = 1$. Then the probability that Y_2 is one is given by

$$\mathrm{P}(Y_2 = 1) = d/(c + d)$$

and the *odds* that Y_2 is equal to one are

$$\mathrm{odds}(Y_2 = 1) = d/c$$

We then have

$$\begin{aligned}\{\mathrm{P}(Y_2 = 1)/[1 - \mathrm{P}(Y_2 = 1)]\} &= [d/(c + d)]/[1 - d/(c + d)] \\ &= d/[(c + d) - d] \\ &= d/c \\ &= \mathrm{odds}(Y_2 = 1)\end{aligned}$$

which demonstrates the equality.]

This bit of algebra demonstrates that a simple logarithmic transformation of odds will linearize the relationship between Y_2 and X if the relationship between Y and X is linear *and* Y is approximately distributed as the logistic or the normal distribution. The fact that a transformation based on the logistic model is successful in linearizing the relationship between fathers' education and sons' probability of completing college—for the NLS sample, at least—can be seen quite clearly from Figure B.2. In this scatterplot, the natural logarithm of the odds of a son graduating from college (column 6 in Table B.1) are plotted against the number of years of education completed by the father (column 1 of Table B.1). Note that the scatterplot is quite linear and that the scatter of the points is much more uniform (homoscedastic) than the plot of $\mathrm{P}(Y_2 = 1)$ against X in Figure B.1. The unweighted line of best fit to the points in Figure B.2 is as follows:

$$\log[\mathrm{odds}(Y_2 = 1)] = -3.812 + .202X + u_3$$

which may be rewritten as

$$\begin{aligned}\mathrm{odds}(Y_2 = 1) &= e^{-3.812}e^{.202X}e^{u_3} \\ &= (.0221)e^{.202X}e^{u_3} = (.0221)(1.224)^{X}e^{u_3}\end{aligned}$$

by using both sides of the equation as powers of e. Thus, the odds of a son completing college are described as a nonlinear function of the fathers' educational attainment. Note that as X increases, the odds of college completion for the son increase more rapidly. The relationship between $\mathrm{odds}(Y_2 = 1)$ are thus made linear by the logarithmic transformation of the odds and the log of e to the power $(.202X)$. This result suggests that odds obtained with dichotomous variables should be logged before being related to other variables.

The above regression of logged odds on father's education resulted from an OLS

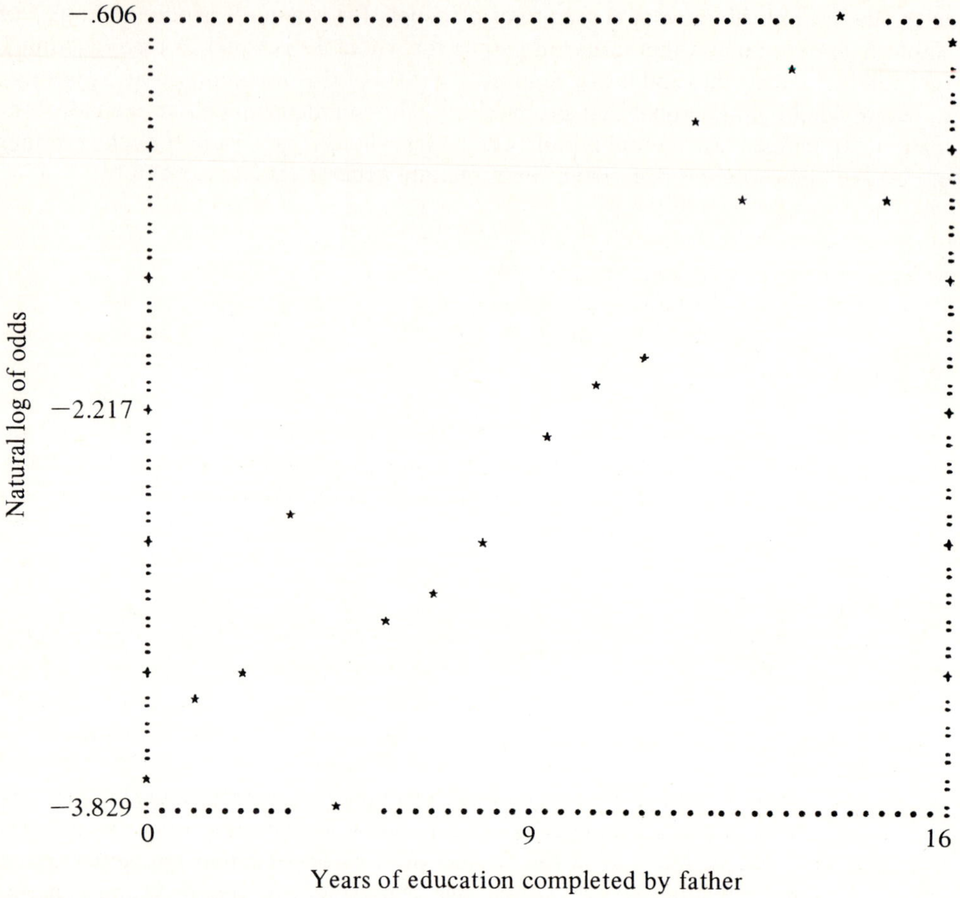

Figure B.2 Scatterplot relating the natural log of the odds that a son completes college (y-axis) to the number of years of education completed by the son's father (x-axis).

regression of column (6) in Table B.1 on column (1) in that same table. Such a regression does not take account of the fact that there are different numbers of persons at the different levels of the right-hand variable (levels of father's education). Not taking account of these differences might bias our estimates of population coefficients or our estimates of standard errors. Both of these shortcomings of applying OLS to these grouped data are solved by the use of a maximum likelihood estimation procedure in which the dichotomous dependent variable is presumed to dichotomize a latent variable that is distributed as the logistic distribution. This procedure is known as *logistic regression,* and the computations are easily performed with the computer program PLR (logistic regression) in the BMD package (Dixon, 1981):

$$\log[\text{odds}\ (Y_2 = 1)] = \underset{(.170)}{-3.994} + \underset{(.016)}{.217}X + u_3$$

where the estimated standard errors are given in parentheses. (Note that the result is essentially the same as the easily understood OLS result given above.) Furthermore, the likelihood ratio chi-square for goodness of fit can be used to test the null hypothesis of linearity against an alternative model in which years of father's education are replaced by 16 dummy variables. The goodness of fit chi-square for the linear model is 17.25 with 15 degrees of freedom (17 levels of X minus one degree of freedom for the intercept and one for the slope). The alternative non-linear model has no degrees of freedom, and thus the null hypothesis of linearity is tested by the chi-square (17.25 − 0.0) with (15 − 0) degrees of freedom. The probability of a chi-square this large or larger is .304, so the null hypothesis of linearity cannot be rejected.

Now that the capacity of the logit model to retrieve linearity from dichotomized variables has been demonstrated, we can specify correlational or structural models involving dichotomous variables. However, before doing so, it should be pointed out that an important shift has taken place in the above discussion. That shift has been away from models specified in terms of individual cases and toward specification in terms of aggregates of cases (based upon a common characteristic, such as fathers' education). The residual u_3, for example, is not the usual deviation from the line of best fit for a particular case, it is the deviation of logged odds for a *group* of cases. From a modeling point of view, this means that a *random* component at the *case* level is not explicitly incorporated into the model. This "sticky wicket" for logistic models does not have a generally accepted solution.

The Problem of Assessing Interaction

As has often been the case in the procedures developed in this book, the more complex techniques are of primary interest in the multivariate case. The lack of linearity for untransformed proportions described above, for example, is of little concern if one is only interested in the direction and approximate magnitude of a bivariate relationship. However, when questions involving several variables are considered, the linearity problem can become crucial. For example, conclusions concerning interactive effects can be very different for the linear and the logit models. The linear model can indicate an interaction when the logit model does not, and *vice versa.* In fact, the two models could even disagree on the direction of an interaction. This fact will be demonstrated with the following example, and the example will serve as an introduction to the specification of logit models.

Suppose that a study of the long-term effects of an educational program is conducted. Two hundred black children and 200 white children are randomly assigned to either participate in the program or to serve as controls. The only dependent variable available is a dichotomous measure of whether or not a participant graduated from college. Suppose that the results of the study were those given in Table B.2. Note in Table B.2 that race is measured by the variable X_1 (black = 0, white = 1), participation in the program is measured by X_2 (control = 0, experimental = 1), and graduation from college is indicated by the variable Y (not a college graduate = 0, a college graduate = 1).

From an inspection of the frequencies in the first two columns of Table B.2, it is

Table B.2 Contrived results of an experiment assessing the long-term effects of a special education program.

	Frequencies				
	Not coll grad ($Y = 0$)	Coll grad ($Y = 1$)	$P(Y = 1)$	Odds ($Y = 1$)	Log(odds)
Blacks ($X_1 = 0$)					
control($X_2 = 0$)	90(c)	10(d)	.10	.1111	−2.198
exper ($X_2 = 1$)	80(e)	20(f)	.20	.2500	−1.386
Whites ($X_1 = 1$)					
control ($X_2 = 0$)	80(g)	20(h)	.20	.2500	−1.386
exper ($X_2 = 1$)	64(i)	36(j)	.36	.5625	−0.575

Note. The letters next to the frequencies in the first two columns are labels for those cell frequencies.

clear that the program had an effect on the likelihood of graduating from college. For both blacks and whites, a greater proportion of experimentals than controls graduated. However, there is another question that may be of theoretical or policy interest: Was the effect of the program greater for whites than for blacks? In order to answer this question, we must decide upon a measure of the magnitude of the *effect* of the program. Once this has been settled, then statistical procedures can be used to test the null hypothesis that the effect of the program was the same within each racial group (the null hypothesis of no interaction between race and the effect of the program).

An obvious approach to measuring the magnitude of the effect of the program on college completion is to compare the probability (or percentage) completing college in the control and experimental conditions. Thus, the "effect" of the program for blacks would be

$$P(Y = 1 | X_1 = 0, X_2 = 1) - P(Y = 1 | X_1 = 0, X_2 = 0) = .20 - .10$$
$$= .10$$

or an increase in the probability of college completion of 10 percent. Similarly, the "effect" of the program for whites would be

$$P(Y = 1 | X_1 = 1, X_2 = 1) - P(Y = 1 | X_1 = 1, X_2 = 0) = .36 - .20$$
$$= .16$$

or an increase in the probability of college completion of 16 percent. Putting aside concerns with hypothesis testing for the moment, this result seems to suggest a greater effect of the program for whites than for blacks. However, is an increase from 10 percent to 20 percent necessarily less than an increase from 20 to 36 percent? Clearly, we can only answer this question if we make some assumptions about the metrics that underlie the observed dichotomous variables, and our earlier discussion of the logit and probit distributions suggests a direction for making these assumptions.

We might, for example, assume that Y is a dichotomization of an underlying variable that is approximately normal in distribution. Under this assumption, the logarithmic transformation of the odds of college graduation is a more appropriate metric for questions of the magnitude of effects than was the probability of college completion. The odds of college completion and the natural logarithm of those odds are given in the fourth and fifth columns of Table B.2. The resulting differences between the log of odds is of special interest:

For blacks:

$$\begin{aligned}\log[\text{odds}(Y = 1 \mid X_1 = 0, X_2 = 1)] &- \log[\text{odds}(Y = 1 \mid X_1 = 0, X_2 = 0)] \\ &= \log(20/80) - \log(10/90) \\ &= \log(.2500) - \log(.1111) \\ &= -1.386 - (-2.198) \\ &= .812\end{aligned}$$

For whites:

$$\begin{aligned}\log[\text{odds}(Y = 1 \mid X_1 = 1, X_2 = 1)] &- \log[\text{odds}(Y = 1 \mid X_1 = 1, X_2 = 0)] \\ &= \log(36/64) - \log(20/80) \\ &= -.575 - (-1.386) \\ &= .811\end{aligned}$$

Thus, in terms of the latent dimension of educational achievement that has presumably been dichotomized by Y, the "effect" of the program appears to have been almost identical for the two racial groups. The program boosted the logit for blacks from -2.198 to -1.386, whereas the program boosted the logit for whites from -1.386 to $-.575$. In each case, the effect of the program was an increase of approximately .81 units along the latent dimension.

Why is there such a discrepancy between the conclusions drawn from the probabilities and the conclusions drawn from the logits (logs of odds)? If the logit model assumptions are correct, the reason for this discrepancy is the fact that more metric change is needed in the tails of the normal (or logistic) distribution than in the center to achieve the same change in probabilities. Thus, the same metric change in the latent variable that produced a 10 percent increase for blacks (who begin farther out in the tail of the distribution than whites) produces a 16 percent difference for whites.

If it is accepted that the logit will generally be the more appropriate basis for estimates of the magnitude of effects, then we need a procedure for specifying causal or correlational models in terms of logits. Such models are usually referred to as *logit* models, and their specification will be outlined in the following section.

Specification of Logit Models for Dichotomous Variables

Before specifying a logit model for the data in Table B.2, the more familiar linear model will be fit to those data. We begin with the usual linear model:

$$Y = b_0 + b_1X_1 + b_2X_2 + b_3(X_1X_2) + u \tag{B.1}$$

where u is a disturbance for individual cases. If we take the conditional expectation of equation B.1 for various values of X_1 and X_2, it becomes apparent how the model represents the data in Table B.2:

$$\begin{aligned}
E(Y|X_1 = 0, X_2 = 0) &= P(Y = 1|X_1 = 0, X_2 = 0) = b_0 = d/(c + d)\\
E(Y|X_1 = 0, X_2 = 1) &= P(Y = 1|X_1 = 0, X_2 = 1) = b_0 + b_2 = f/(e + f)\\
E(Y|X_1 = 1, X_2 = 0) &= P(Y = 1|X_1 = 1, X_2 = 0) = b_0 + b_1 = h/(g + h)\\
E(Y|X_1 = 1, X_2 = 1) &= P(Y = 1|X_1 = 1, X_2 = 1) = b_0 + b_1 + b_2 + b_3\\
&= j/(i + j)
\end{aligned} \tag{B.2}$$

Now, equation B.2 can be used as outlined in Chapter 4 to solve for the parameters of the model (b_0, b_1, . . .) in terms of the frequencies (c, d, . . .). And estimates of those parameters can then be obtained from the observed frequencies. The parameter estimate that is of particular interest here is the interaction effect b_3:

$$\begin{aligned}
\text{est}(b_3) &= (.36 - .20) - (.20 - .10)\\
&= .16 - .10\\
&= .06
\end{aligned}$$

This result reveals that the linear model in equation B.1 treats the changes in Y as the appropriate metric for assessing interactions. The interaction is estimated as a difference between two differences in probabilities.

An alternative equation that also models the data in Table B.2 is a *multiplicative* equation model:

$$\text{odds}(Y = 1|X_1, X_2) = a_0 a_1^{X_1} a_2^{X_2} a_3^{(X_1 X_2)} \tag{B.3}$$

where a_0, a_1, a_2, and a_3 are the parameters of the model that are raised to powers determined by the values of X_1 and X_2. Inputting values of X_1 and X_2 into this model, we can see how it represents the four odds in Table B.2:

$$\begin{aligned}
\text{odds}(Y = 1|X_1 = 0, X_2 = 0) &= O_1 = a_0(1)(1)(1) = d/c\\
\text{odds}(Y = 1|X_1 = 0, X_2 = 1) &= O_2 = a_0(1)a_2(1) = f/e\\
\text{odds}(Y = 1|X_1 = 1, X_2 = 0) &= O_3 = a_0 a_1(1)(1) = h/g\\
\text{odds}(Y = 1|X_1 = 1, X_2 = 1) &= O_4 = a_0 a_1 a_2 a_3 = j/i
\end{aligned} \tag{B.4}$$

where the parameters are replaced by the number 1 when they are raised to a power of zero. The equations B.4 can now be solved such that the parameters of the model are expressed as functions of the frequencies:

$$\begin{aligned}
a_0 &= O_1 = d/c\\
a_2 &= O_2/O_1 = (f/e)/(d/c) = cf/de\\
a_1 &= O_3/O_1 = (h/g)/(d/c) = ch/dg\\
a_3 &= (O_4/O_3)/(O_2/O_1) = (jg/hi)/(cf/de)
\end{aligned} \tag{B.5}$$

where the solution for a_0 is an odds, a_1 and a_2 are ratios of odds or *odds ratios* (also called *cross product ratios*), and a_3 is a ratio of odds ratios. These parameters can be estimated by substitution of the observed frequencies in Table B.2 to obtain:

$$\begin{aligned}
\text{est}(a_0) &= 10/90 = .111 \\
\text{est}(a_1) &= (20/80)/(10/90) = 2.250 \\
\text{est}(a_2) &= (20/80)/(10/90) = 2.250 \\
\text{est}(a_3) &= [(36/64)/(20/80)]/[(20/80)(10/90)] = 1.0
\end{aligned}$$

Now, the interesting finding is that the estimate of a_3 is 1.0. In a multiplicative equation model such as B.3, a parameter with a value of one contributes nothing to the model. A value of one in a multiplicative model is analogous to a value of zero in a linear model. Thus, the multiplicative model has given us a picture of the interaction between race and the educational program that is consistent with our deductions from the logits—no interaction.

The connection between the logits and the parameters of the multiplicative model can be made clear by taking the logarithm of both sides of the equation B.3 to obtain:

$$\begin{aligned}
\log[\text{odds}(Y = 1 \mid X_1, X_2)] &= \log(a_0) + X_1 \log(a_1) + X_2 \log(a_2) \\
&\quad + X_1 X_2 \log(a_3) \qquad \text{(B.6)} \\
\log(\text{est odds}) &= -2.197 + .811X_1 + .811X_2 + 0X_1X_2
\end{aligned}$$

The reader can verify that equation B.6 models the logits in Table B.2 in the same manner that equation B.3 models the odds and equation B.1 models the probabilities.

As was introduced in Chapter 7 on ANOVA, the use of dichotomous variables coded as 0 or 1 is easy to follow in simple cases. But with more complex models, problems can arise. For this reason, it is probably preferable to estimate log linear models that employ the coding scheme in which variables take on the values -1 and $+1$ (and zero when a variable has more than two levels). In this case, the parameters of the model take on a somewhat different meaning. We find that

$$\begin{aligned}
O_1 &= a_0 a_3 / a_1 a_2 \\
O_2 &= a_0 a_2 / a_1 a_3 \\
O_3 &= a_0 a_1 / a_2 a_3 \\
O_4 &= a_0 a_1 a_2 a_3
\end{aligned}$$

and

$$\begin{aligned}
a_0 &= (O_1\ O_2\ O_3\ O_4)^{.25} \\
a_1 &= (O_3\ O_4 / O_1\ O_2)^{.25} \qquad \text{(B.7)} \\
a_2 &= (O_2\ O_4 / O_1\ O_3)^{.25} \\
a_3 &= (O_1\ O_4 / O_2\ O_3)^{.25}
\end{aligned}$$

The reader can verify that a_3 is again equal to one for the data in Table B.2, whereas estimates of a_0, a_1, and a_2 will change relative to the estimates obtained with the 0,1 coding of X_1 and X_2. The latter parameters are interpreted as effects in the classic ANOVA sense. For example, the parameter a_1 is analogous to the multiplicative *average* of the effect of a change in X_1 on the odds of college completion ($Y = 1$) at the two different levels of X_2. This can be seen by rewriting the second equation in B.7 as

$$a_1 = [(O_3/O_1)(O_4/O_2)]^{.25}$$

where

$$O_3/O_1 = \text{effect of } X_1 \text{ for controls}$$

$$O_4/O_2 = \text{effect of } X_1 \text{ for experimentals}$$

[It should be noted that raising the product of these two odds ratios to the power .25 produces a parameter that is actually the square root of the multiplicative average of the two odds ratios. This is the case because the $-1/+1$ coding scheme divides the effect of X_1 into a_1 and $(1/a_1)$.]

The direct estimates of the parameters of the model B.4 (using the $-1/+1$ coding scheme for X_1 and X_2) are as follows:

$$\text{est } a_0 = .25$$

$$\text{est } a_1 = 1.50$$

$$\text{est } a_2 = 1.50$$

$$\text{est } a_3 = 1.00$$

Note that whereas a_1 and a_2 were equal to conditional odds ratios with the 0/1 coding (est $a_1 = 2.25$), the $-1/+1$ coding scheme produces estimates of these parameters that are square roots of odds ratios ($1.5 =$ square root of 2.25).

Another perspective on the model B.3 with the $-1/+1$ coding scheme is gained by taking the logarithm of B.3 as in equation B.6. Note in this instance that the logarithm of each effect parameter (e.g., a_1) is added or *subtracted* as the independent variables change. Thus, the $\log(a_1)$, for example, is one-half of the change in the log of the odds that is produced by a change in X_1. Similarly, a_1 is the *square root* of the multiplicative change in the odds that is produced by a change in X_1.

Before proceeding to an example with actual data, it is important to reiterate that the multiplicative form of model B.3 should not be construed to mean that the relationships among the latent variables are nonlinear. On the contrary, the logit model assumes that the dichotomous variables (observed) relate multiplicatively precisely because the variables of interest (the latent variables) are distributed logistically and are linearly related.

Hypothesis Testing with Logit Models

The logic of hypothesis testing with logit models is analogous to that which we applied to the testing of hypotheses about linear structural equation models that were estimated with maximum likelihood (ML) procedures. A logit model may be specified, for example, in its *saturated* form. This means that all possible interactions are included and the parameters of the model are just identified by the available cell frequencies. In order to test a null hypothesis that one (or more) of the parameters of the model is one (or, equivalently, that the log of the parameter is zero), that parameter is simply fixed at one and the effects on the likelihood function can be used to compute a chi-square statistic. The degrees of freedom for the chi-square are equal to the number of constrained parameters, and the relative fit of various models can be compared in a manner analogous to procedures outlined in Chapters 9 and 11.

In order to apply ML procedures to the estimation of a model such as that in equation B.3, it is necessary to make some assumption about the distribution of the *observed* variables. Fortunately, the ML estimates of logit model parameters and estimates of their standard errors are the same under the assumption of several different distributions: the Poisson distribution, the multinomial distribution, or the product-multinomial distribution (cf. Fienberg, 1977, Appendix II). This means that logit models, such as the one in equation B.3, can be applied to a wide range of sampling situations without concern that the ML procedure will provide biased estimates or standard errors. The model itself may be erroneous for various reasons, but the concern about multivariate normality in the linear equation case is not necessary in the case of log linear models. In other words, accurate statements about the population of observed dichotomous (or polychotomous) variables can be made with ML estimation even though the assumptions about the relationships among the the dichotomous variables and latent variables may not hold. In fact, there are instances in which the observed variable is assumed to be measured without error and without reference to any latent variable. This point is rather subtle and deserves further clarification.

Recall that the assumption of a logistically distributed latent variable was central to the development of the multiplicative model in equation B.3. However, this multiplicative model may be of use in situations where no such assumptions about latent variables are involved. Estimation of a multiplicative relation among observed variables is a matter independent of the validity of the causal interpretation of parameters or assumptions about relations between those observed variables and latent variables.

The basic building block of hypothesis testing with a logit model is a link between the magnitude of parameter estimates and a probability distribution. That link is provided by an estimate of the standard error of the odds ratio (or cross product ratio). In the case of the 2 × 2 table, where c, d, e, and f are the frequencies

c	d
e	f

and

$$\text{odds ratio} = (cf/de) = OR_1$$

the estimated standard error of OR_1 is given in large samples by

$$\text{est std err } [\log(OR_1)] = [(1/c)(1/d)(1/e)(1/f)]^{.5} \qquad \text{(B.8)}$$

And the real advantage of equation B.8 is that it holds under various assumptions about the sampling distributions of the observed variables. Also, it can be shown that OR_1 is the maximum likelihood estimate of the population odds ratio under any of several sampling distribution assumptions. A significance test can then be performed for an odds ratio by dividing its natural logarithm by the above standard error. The resulting ratio is distributed as a normal deviate in large samples. Similarly, an approximate 95 percent confidence interval of the log of the odds ratio can be obtained by adding and subtracting 1.96 times the standard error to and from the log of the observed odds ratio. The 95 percent confidence interval about the odds ratio itself can then be obtained by taking the anti-logarithm of the upper and lower bound of the logarithmic interval.

If one or more parameters in a logit model have been constrained to one (or the log constrained to zero), then the model is no longer *saturated.* Furthermore, a likelihood ratio chi-square can be computed that serves as a test of the constraint(s) as a null hypothesis. This likelihood ratio is analogous to that developed for linear models except that the right-hand variables are presumed to be distributed as the multinomial distribution if dichotomous, and as the multivariate normal distribution if continuous. In the case of a single dichotomous right-hand variable predicting a dichotomous dependent variable, the likelihood ratio chi-square is simply

$$\text{chi-square} = [\log(\text{odds ratio})/\text{std err}(\log \text{odds ratio})]^2$$

which is distributed as the chi-square distribution with one degree of freedom. In more complex cases, the degrees of freedom are equal to the number of parameters constrained to equal one, and the computation of the likelihood ratio requires a computer. Nevertheless, the logic of the likelihood ratio chi-square is straightforward: it becomes larger as the differences between observed odds (or the frequencies defining those odds) and the odds predicted by the model become greater. The predicted cell frequencies are analogous to the predicted covariance matrix in the case of an overidentified linear equation model.

We now turn to an empirical example for clarification and demonstration of the use of a widely available computer program for fitting logit models to categorical data.

A Computational Example

Epidemiology, the statistical study of disease determinants, is a branch of the human sciences that is particularly suited to the application of logit models. The reason for this is that the dependent variable in epidemiological research is usually a dichotomous variable, and one of the two categories of this variable is often much less likely than the other. For example, in a survey of chemical workers, the dependent variable may be a dichotomy defined by whether or not a respondent has suffered from cancer. Clearly, only a few percent of any sample of workers will fall in the disease category of the dichotomous dependent variable. This means that a linear model fit to the probability of cancer in different subgroups of the sample could lead to erroneous conclusions about interactions and standard errors. A log linear model, and its derivative, the logit model, are therefore more useful in the epidemiological context.

Our data for this example are drawn from a study conducted by Lennart Hardell and his colleagues at the University of Umea, Sweden (Hardell, Eriksson, Lenner, and Lundgren, 1981). The study concerns the effect of exposure to various chemicals on the probability of developing a type of cancer known as lymphoma. The particular chemicals studied are of particular interest because they are related to the herbicide used during the Viet Nam War. The herbicide is popularly known as "Agent Orange"

Table B.3 Results of a case control study relating lymphoma to phenoxy acid and/or chlorophenol exposure (from Hardell et al., 1981).

Chlorophenol exposure	Phenoxy acid exposure	$Y = 0$ Controls	$Y = +1$ Cases	Log[odds($Y = 1$)]
None ($X_2 = X_3 = -1$)	None ($X_1 = -1$)	284	94	−1.106
	Yes ($X_1 = +1$)	16	25	.446
Low ($X_2 = 1, X_3 = 0$)	None ($X_1 = -1$)	19	14	−.305
	Yes ($X_1 = +1$)	7	11	.452
High ($X_2 = 0, X_3 = +1$)	None ($X_1 = -1$)	8	20	.917
	Yes ($X_1 = +1$)	1	5	1.609

and is classified chemically as a combination of *chlorinated phenoxyacetic acids* (or just *phenoxy acids*). There has been much concern among U.S. veterans of that war that they are suffering from an excess of various health problems due to exposure to phenoxy acid herbicides.

The design of Hardell's study is known as a *case-control design* or *case referent design*. The essence of the study is that persons suffering from a particular disease, such as lymphoma, are located from hospital records. Each such person is then referred to as a *case*. For each case that is found, one or more *control* respondents are identified. The controls are selected very carefully so that they are matched with the case on age, sex, neighborhood of residence, etc.

The goal of the design is then to see if the cases are more likely to have been exposed to phenoxy acids than the controls. Note here that the cases and controls are selected on the basis of the dependent variable (whether or not they have suffered from malignant lymphoma). Queries concerning exposure to various chemicals are then conducted. In Hardell's study, interviews focused upon exposure to two types of chemicals: phenoxy acid and a related cluster of compounds known as *chlorophenols*.

The results of Hardell's data collection are presented in Table B.3. Note that the odds of being a case (or the probability of being a case) increase as we move down Table B.3 from non-exposed categories to exposed categories.

These data present us with several questions that a logit model would be of assistance in addressing. First, To what extent does exposure to phenoxy acid increase the odds of lymphoma in this sample, and how likely are these results if there is no such relationship in the population from which these cases and controls were drawn? Second, What is the effect of exposure to chlorophenols on the odds of lymphoma and is this effect interactive with that of exposure to phenoxy acids? In order to address these questions, we specify a logit model of the following form:

$$\log[\text{odds}(Y = 1 \mid X_1, X_2, X_3)] = \log(a_0) + \log(a_1)X_1 + \log(a_2)X_2 + \log(a_3)X_3 + \log(a_4)X_1X_2 + \log(a_5)X_1X_3 \quad \text{(B.9)}$$

where

$$Y = \begin{cases} 0 \text{ for controls} \\ +1 \text{ for cases} \end{cases}$$

$$X_1 = \begin{cases} -1 \text{ if no exposure to phenoxy acids} \\ +1 \text{ if exposed to phenoxy acids} \end{cases}$$

$$X_2 = \begin{cases} -1 \text{ if no exposure to chlorophenols} \\ +1 \text{ if exposure to low-grade chlorophenols} \\ 0 \text{ otherwise} \end{cases}$$

$$X_3 = \begin{cases} -1 \text{ if no exposure to chlorophenols} \\ +1 \text{ if exposure to high-grade chlorophenols} \\ 0 \text{ otherwise} \end{cases}$$

Note that the coding scheme for the dummy variables is precisely that used for the analysis of variance in Chapter 7. [Also note that this model does not take account of the matching process. In some instances, but not with these particular data, failure to adjust for matching can produce a biased estimate; see Breslow and Day (1980).]

Equation B.9 can be transformed into a multiplicative equation model of the following form by simply raising the Naperian constant e to the exponent given on each side of equation B.9:

$$\text{odds}(Y = 1 | X_1, X_2, X_3) = a_0 a_1^{X_1} a_2^{X_2} a_3^{X_3} a_4^{X_1 X_2} a_5^{X_1 X_3} \tag{B.10}$$

We could now solve for estimates of each of the six parameters of this logit model (a_0, a_1, a_2, a_3, a_4, and a_5) by expressing the odds (or log of odds) of being a case in each row of Table B.3. For example,

$$94/284 \text{ estimates } a_0(1/a_1)(1/a_2)(1/a_3)(a_4)(a_5)$$

$$25/16 \text{ estimates } a_0(a_1)(1/a_2)(1/a_3)(1/a_4)(1/a_5)$$

and so on, or

$$-1.106 \text{ estimates } \log(a_0) - \log(a_1) - \log(a_2) - \log(a_3) + \log(a_4) + \log(a_5)$$

$$.446 \text{ estimates } \log(a_0) + \log(a_1) - \log(a_2) - \log(a_3) - \log(a_4) - \log(a_5)$$

and so on. This model is just identified (saturated) by the observed data: note that six odds are observed and six model parameters are estimated. This means that the likelihood ratio for the fit of the model will be zero. However, the standard error of some of the estimates may be large relative to the magnitude of the estimate—suggesting that a constraint may be placed upon the model.

The first question to be addressed with these data is whether or not exposure to phenoxy acid interacts with exposure to chlorophenols in its effect on the odds of lymphoma. The null hypothesis of no interaction is tested by placing the constraint $a_4 = a_5 = 1$ on the model in equation B.9. With this constraint, the model is overidentified, and the likelihood ratio can be used to test the fit of the model with the observed data.

In order to fit the just identified (saturated) model in equation B.9 to the observations and then fit the "no interaction" constrained model, it is convenient to use a packaged program such as that provided in the widely available BMDP series (Dixon, 1981). The appropriate programs in the BMDP series are labeled P4F (entitled "Two-way and multiway frequency tables—Measures of association and log-linear models) and PLR ("logistic regression")."

The BMDP statements needed for the analysis are as follows:

```
/ PROBLEM       TITLE IS 'ANALYSIS OF HARDELL DATA WITH BMDP4F'.
/ INPUT         VARIABLES ARE 3.
                TABLE IS 2, 2, 3.
                FORMAT IS '(4F5.0)'.
/ VARIABLE      NAMES ARE OUTCOME, PHENOXY, CHLORO.
/ TABLE         INDICES ARE OUTCOME, PHENOXY, CHLORO.
/ TABLE         ROW IS PHENOXY.
                COLUMN IS OUTCOME.
                CONDITION IS CHLORO.
/ CATEGORY      CODES(1) ARE 1, 2.
                NAMES(1) ARE 'CONTROL', 'CASES'.
                CODES(2) ARE 1, 2.
                NAMES(2) ARE 'NO', 'YES'.
                CODES(3) ARE 1, 2, 3.
                NAMES(3) ARE NONE, LOW, HIGH.
/ TABLE         INDICES ARE OUTCOME, PHENOXY, CHLORO.
/ FIT           MODEL IS OPC.
                MODEL IS OP, OC, PC.
                MODEL IS OC, PC.
                MODEL IS OP, PC.
                MODEL IS O, P, PC.
/ PRINT         LAMBDA. BETA. LRCHI. VAR.
/ END
  284    94    16    25
   19    14     7    11
    8    20     1     5
```

```
/ PROBLEM        TITLE IS 'ANALYSIS OF HARDELL DATA WITH BMDPLR'.
                 TIME=30.
/ INPUT          VARIABLES=4.
                 CASES=12.
                 FORMAT IS '(3F1.0,F4.0)'.
/ VARIABLE        NAMES=OUTCOME,PHENOXY,CHLORO,TALLY.
/ GROUP           CODES(1)=0,1.
                 NAMES(1)=CONTROL,CASE.
                 CODES(2)=0,1.
                 NAMES(2)=NOT,EXPOSED.
                 CODES(3)=0,1,2.
                 NAMES(3)=NOT,LOW,HIGH.
/ REGRESSION      DEPEND=OUTCOME.
                 COUNT=TALLY.
                 CATEG=PHENOXY,CHLORO.
                 MODEL=PHENOXY*CHLORO.
                 METHOD=MLR.
```

```
/  END
100  94
110  25
101  14
111  11
102  20
112   5
000 284
010  16
001  19
011   7
002   8
012   1
```

The output from this program will produce the following results:

	Likelihood ratio	*df*
Just identified (saturated) model	0.00	0
No interaction ($a_4 = a_5 = 1$)	$1.58(p = .45)$	2
No effect of phenoxy ($a_1 = a_4 = a_5 = 1$)	$23.19(p < .01)$	3
No effect of chloroph ($a_2 = a_3 = a_4 = a_5 = 1$)	$28.71(p < .01)$	4
No effects ($a_1 = a_2 = a_3 = a_4 = a_5 = 1$)	$56.16(p < .01)$	5

These results indicate that any model that excludes effects due to exposure to phenoxy acids or chlorophenols can be rejected by the data. However, the "no interaction" model cannot be rejected as a plausible hypothesis. This means that the effects of phenoxy acid and chlorophenol exposure on the logged odds of lymphoma are additive. There is no evidence that exposure to both produces some sort of synergistic effect, and there is no evidence that exposure to one of the chemicals diminishes the effect of exposure to the other.

The ML estimates of the additive model are as follows:

$$\log[\text{odds}(Y = 1)] = \underset{(.192)}{.422} + \underset{(.143)}{.666X_1} - \underset{(.243)}{.257X_2} + \underset{(.284)}{1.092X_3} + \log(u)$$

And in the multiplicative form,

$$\text{odds}(Y = 1) = (1.525)(1.946^{X_1})(.773^{X_2})(2.980^{X_3})(u)$$

where u is a multiplicative "error" term or residual. The reader can verify that this model reproduces the observed odds quite accurately.

The meaning of the parameter estimates may not be immediately obvious. For example, the estimate 1.946 of a_1 may be construed as the square root of an odds ratio or cross product ratio that reflects the increased odds of lymphoma attributed to phenoxy acid exposure. In this instance, the square of 1.946 is approximately 3.8. Thus, the odds of lymphoma are 3.8 times greater for persons exposed to phenoxy acids (relative to those not exposed). In other words, the approximate *relative risk* of lymphoma is 3.8 for those exposed to phenoxy acids. A 95 percent confidence interval for this odds ratio (or relative risk) can be obtained by noting the increase in the likelihood ratio under the constraint $a_1 = 1$. We found that the likelihood ratio increased from 1.58 ($df = 2$) for the no interaction model to 23.19 ($df = 3$) under the additional constraint that $a_1 = 1$. The approximate standard error of the estimate of $\log(a_1)$ is then given by

$$\begin{aligned} \text{approx std err } \{\text{est}[\log(a_1)]\} &= .666/(23.19 - 1.58)^{.5} \\ &= .145 \end{aligned}$$

and therefore

$$\text{approx std err } 2\{\text{est}[\log(a_1)]\} = .290$$

This then means that the 95 percent confidence interval about the log of the odds ratio is

$$2(.666) - 1.96(.290) \text{ to } 2(.666) + 1.96(.290)$$

or

$$.764 \text{ to } 1.900$$

Converting these logs to odds ratios gives the 95 percent confidence interval about the point estimated odds ratio:

$$\begin{aligned} \text{point estimate} &= 3.8 \\ 95\% \text{ confidence interval} &= 2.1 \text{ to } 6.7 \end{aligned}$$

Problem According to the estimated model with no interactive terms, what is the predicted relative risk of lymphoma with exposure to *both* phenoxy acid and high-grade chlorophenol? [Hint:

$$\begin{aligned} \text{odds}(Y = 1 \mid X_1 = 1, X_3 = 1) &= (1.525)(1.946)(1.0)(2.980) \\ \text{odds}(Y = 1 \mid X_1 = -1, X_3 = -1) &= (1.525)(1/1.946)(1/.773)(1/2.980) \\ \text{odds ratio} &= (1.946)(1.946)(.773)(2.980)(2.980) = 26.0] \end{aligned}$$

Concluding Remarks

In many, if not most, practical instances, the differences between conclusions drawn with the linear model and a logit model are minimal. This is the case because differences between the two are substantial only when the extremes of distributions are of central concern. Thus, an epidemiologist concerned with rare diseases would do well to pay attention to developments in logit models, whereas most researchers will probably settle for linear models in their day-to-day work. A particularly surprising or important empirical finding with a linear model should certainly be checked against the conclusions that would be drawn from the logit model.

REFERENCES AND SUGGESTED READING

Dixon, W. J. (Ed.) *BMDP statistical software, 1981*. Berkeley, Calif.: University of California Press, 1981.

Breslow, N. E. and Day, N. E. *Statistical methods in cancer research. Vol. 1—The analysis of case-control studies.* Lyon: International Agency for Research on Cancer, 1980.

Fienberg, Stephen E. *The analysis of cross-classified categorical data.* Cambridge, Mass.: MIT Press, 1977.

Hardell, L., Eriksson, M., Lenner, P., and Lundgren, E. Malignant lymphoma and exposure to chemicals, especially organic solvents, chlorophenols and phenoxy acids: A case control study. *British Journal of Cancer,* 1981, *43,* 169–176.

Appendix C

χ^2 Table

df	$P = 0.99$	0.98	0.95	0.90	0.80	0.70	0.50	0.30	0.20	0.10	0.05	0.02	0.01
1	0.000157	0.000628	0.00393	0.0158	0.0642	0.148	0.455	1.074	1.642	2.706	3.841	5.412	6.635
2	0.0201	0.0404	0.103	0.211	0.446	0.713	1.386	2.408	3.219	4.605	5.991	7.824	9.210
3	0.115	0.185	0.352	0.584	1.005	1.424	2.366	3.665	4.642	6.251	7.815	9.837	11.341
4	0.297	0.429	0.711	1.064	1.649	2.195	3.357	4.878	5.989	7.779	9.488	11.668	13.277
5	0.554	0.752	1.145	1.610	2.343	3.000	4.351	6.064	7.289	9.236	11.070	13.388	15.086
6	0.872	1.134	1.635	2.204	3.070	3.828	5.348	7.231	8.558	10.645	12.592	15.033	16.812
7	1.239	1.564	2.167	2.833	3.822	4.671	6.346	8.383	9.803	12.017	14.067	16.622	18.475
8	1.646	2.032	2.733	3.490	4.594	5.527	7.344	9.524	11.030	13.362	15.507	18.168	20.090
9	2.088	2.532	3.325	4.168	5.380	6.393	8.343	10.656	12.242	14.684	16.919	19.679	21.666
10	2.558	3.059	3.940	4.865	6.179	7.267	9.342	11.781	13.442	15.987	18.307	21.161	23.209
11	3.053	3.609	4.575	5.578	6.989	8.148	10.341	12.899	14.631	17.275	19.675	22.618	24.725
12	3.571	4.178	5.226	6.304	7.807	9.034	11.340	14.011	15.812	18.549	21.026	24.054	26.217
13	4.107	4.765	5.892	7.042	8.634	9.926	12.340	15.119	16.985	19.812	22.362	25.472	27.688
14	4.660	5.368	6.571	7.790	9.467	10.821	13.339	16.222	18.151	21.064	23.685	26.873	29.141
15	5.229	5.985	7.261	8.547	10.307	11.721	14.339	17.322	19.311	22.307	24.996	28.259	30.578
16	5.812	6.614	7.962	9.312	11.152	12.624	15.338	18.418	20.465	23.542	26.296	29.633	32.000
17	6.408	7.255	8.672	10.085	12.002	13.531	16.338	19.511	21.615	24.769	27.587	30.995	33.409
18	7.015	7.906	9.390	10.865	12.857	14.440	17.338	20.601	22.760	25.989	28.869	32.346	34.805
19	7.633	8.567	10.117	11.651	13.716	15.352	18.338	21.689	23.900	27.204	30.144	33.687	36.191
20	8.260	9.237	10.851	12.443	14.578	16.266	19.337	22.775	25.038	28.412	31.410	35.020	37.566
21	8.897	9.915	11.591	13.240	15.445	17.182	20.337	23.858	26.171	29.615	32.671	36.343	38.932
22	9.542	10.600	12.338	14.041	16.314	18.101	21.337	24.939	27.301	30.813	33.924	37.659	40.289
23	10.196	11.293	13.091	14.848	17.187	19.021	22.337	26.018	28.429	32.007	35.172	38.968	41.638
24	10.856	11.992	13.848	15.659	18.062	19.943	23.337	27.096	29.553	33.196	36.415	40.270	42.980
25	11.524	12.697	14.611	16.473	18.940	20.867	24.337	28.172	30.675	34.382	37.652	41.566	44.314
26	12.198	13.409	15.379	17.292	19.820	21.792	25.336	29.246	31.795	35.563	38.885	42.856	45.642
27	12.879	14.125	16.151	18.114	20.703	22.719	26.336	30.319	32.912	36.741	40.113	44.140	46.963
28	13.565	14.847	16.928	18.939	21.588	23.647	27.336	31.391	34.027	37.916	41.337	45.419	48.278
29	14.256	15.574	17.708	19.768	22.475	24.577	28.336	32.461	35.139	39.087	42.557	46.693	49.588
30	14.953	16.306	18.493	20.599	23.364	25.508	29.336	33.530	36.250	40.256	43.773	47.962	50.892

For degrees of freedom (df) greater than 30, the expression $\sqrt{2x^2} - \sqrt{2n' - 1}$ may be used as a normal deviate with unit variance, where n' is the number of degrees of freedom.

Reproduced from *Statistical Methods for Research Workers*, 6th ed., with the permission of the author, R. A. Fisher, and his publisher, Oliver and Boyd, Edinburgh.

Critical Values of F

The first F ratio is the critical value for $\alpha = .05$, the second for $\alpha = .01$, and the third for $\alpha = .001$.

df for denominator	α	*df* for numerator 1	2	3	4	5	6	7	8	9	10	12	15	20	24	30	40	60	∞	*df* for denominator
3	.05	10.1	9.55	9.28	9.12	9.01	8.94	8.89	8.85	8.81	8.79	8.74	8.70	8.66	8.64	8.62	8.59	8.57	8.53	3
	.01	34.1	30.8	29.5	28.7	28.2	27.9	27.7	27.5	27.3	27.2	27.1	26.9	26.7	26.6	26.5	26.4	26.3	26.1	
	.001	167	148	141	137	135	133	132	131	130	129	128	127	126	126	125	125	124	123	
4	.05	7.71	6.94	6.59	6.39	6.26	6.16	6.09	6.04	6.00	5.96	5.91	5.86	5.80	5.77	5.75	5.72	5.69	5.63	4
	.01	21.2	18.0	16.7	16.0	15.5	15.2	15.0	14.8	14.7	14.5	14.4	14.2	14.0	13.9	13.8	13.7	13.7	13.5	
	.001	74.1	61.2	56.2	53.4	51.7	50.5	49.7	49.0	48.5	48.1	47.4	46.8	46.1	45.8	45.4	45.1	44.7	44.1	
5	.05	6.61	5.79	5.41	5.19	5.05	4.95	4.88	4.82	4.77	4.74	4.68	4.62	4.56	4.53	4.50	4.46	4.43	4.36	5
	.01	16.3	13.3	12.1	11.4	11.0	10.7	10.5	10.3	10.2	10.1	9.89	9.72	9.55	9.47	9.38	9.29	9.20	9.02	
	.001	47.2	37.1	33.2	31.1	29.8	28.8	28.2	27.6	27.2	26.9	26.4	25.9	25.4	25.1	24.9	24.6	24.3	23.8	
6	.05	5.99	5.14	4.76	4.53	4.39	4.28	4.21	4.15	4.10	4.06	4.00	3.94	3.87	3.84	3.81	3.77	3.74	3.67	6
	.01	13.7	10.9	9.78	9.15	8.75	8.47	8.26	8.10	7.98	7.87	7.72	7.56	7.40	7.31	7.23	7.14	7.06	6.88	
	.001	35.5	27.0	23.7	21.9	20.8	20.0	19.5	19.0	18.7	18.4	18.0	17.6	17.1	16.9	16.7	16.4	16.2	15.7	
7	.05	5.59	4.74	4.35	4.12	3.97	3.87	3.79	3.73	3.68	3.64	3.57	3.51	3.44	3.41	3.38	3.34	3.30	3.23	7
	.01	12.2	9.55	8.45	7.85	7.46	7.19	6.99	6.84	6.72	6.62	6.47	6.31	6.16	6.07	5.99	5.91	5.82	5.65	
	.001	29.2	21.7	18.8	17.2	16.2	15.5	15.0	14.6	14.3	14.1	13.7	13.3	12.9	12.7	12.5	12.3	12.1	11.7	
8	.05	5.32	4.46	4.07	3.84	3.69	3.58	3.50	3.44	3.39	3.35	3.28	3.22	3.15	3.12	3.08	3.04	3.01	2.93	8
	.01	11.3	8.65	7.59	7.01	6.63	6.37	6.18	6.03	5.91	5.81	5.67	5.52	5.36	5.28	5.20	5.12	5.03	4.86	
	.001	25.4	18.5	15.8	14.4	13.5	12.9	12.4	12.0	11.8	11.5	11.2	10.8	10.5	10.3	10.1	9.92	9.73	9.33	
9	.05	5.12	4.26	3.86	3.63	3.48	3.37	3.29	3.23	3.18	3.14	3.07	3.01	2.94	2.90	2.86	2.83	2.79	2.71	9
	.01	10.6	8.02	6.99	6.42	6.06	5.80	5.61	5.47	5.35	5.26	5.11	4.96	4.81	4.73	4.65	4.57	4.48	4.31	
	.001	22.9	16.4	13.9	12.6	11.7	11.1	10.7	10.4	10.1	9.80	9.57	9.24	8.90	8.72	8.55	8.37	8.19	7.81	
10	.05	4.96	4.10	3.71	3.48	3.33	3.22	3.14	3.07	3.02	2.98	2.91	2.85	2.77	2.74	2.70	2.66	2.62	2.54	10
	.01	10.0	7.56	6.55	5.99	5.64	5.39	5.20	5.06	4.94	4.85	4.71	4.56	4.41	4.33	4.25	4.17	4.08	3.91	
	.001	21.0	14.9	12.6	11.3	10.5	9.93	9.52	9.20	8.96	8.75	8.45	8.13	7.80	7.64	7.47	7.30	7.12	6.76	

df for denominator	α	*df* for numerator 1	2	3	4	5	6	7	8	9	10	12	15	20	24	30	40	60	∞	*df* for denominator
11	.05	4.84	3.98	3.59	3.36	3.20	3.09	3.01	2.95	2.90	2.85	2.79	2.72	2.65	2.61	2.57	2.53	2.49	2.40	11
	.01	9.65	7.21	6.22	5.67	5.32	5.07	4.89	4.74	4.63	4.54	4.40	4.25	4.10	4.02	3.94	3.86	3.78	3.60	
	.001	19.7	13.8	11.6	10.3	9.58	9.05	8.66	8.35	8.12	7.92	7.63	7.32	7.01	6.85	6.68	6.52	6.35	6.00	
12	.05	4.75	3.89	3.49	3.26	3.11	3.00	2.91	2.85	2.80	2.75	2.69	2.62	2.54	2.51	2.47	2.43	2.38	2.30	12
	.01	9.33	6.93	5.95	5.41	5.06	4.82	4.64	4.50	4.39	4.30	4.16	4.01	3.86	3.78	3.70	3.62	3.54	3.36	
	.001	18.6	13.0	10.8	9.63	8.89	8.38	8.00	7.71	7.48	7.29	7.00	6.71	6.40	6.25	6.09	5.93	5.76	5.42	
13	.05	4.67	3.81	3.41	3.18	3.03	2.92	2.83	2.77	2.71	2.67	2.60	2.53	2.46	2.42	2.38	2.34	2.30	2.21	13
	.01	9.07	6.70	5.74	5.21	4.86	4.62	4.44	4.30	4.19	4.10	3.96	3.82	3.66	3.59	3.51	3.43	3.34	3.17	
	.001	17.8	12.3	10.2	9.07	8.35	7.86	7.49	7.21	6.98	6.80	6.52	6.23	5.93	5.78	5.63	5.47	5.30	4.97	
14	.05	4.60	3.74	3.34	3.11	2.96	2.85	2.76	2.70	2.65	2.60	2.53	2.46	2.39	2.35	2.31	2.27	2.22	2.13	14
	.01	8.86	6.51	5.56	5.04	4.70	4.46	4.28	4.14	4.03	3.94	3.80	3.66	3.51	3.43	3.35	3.27	3.18	3.00	
	.001	17.1	11.8	9.73	8.62	7.92	7.44	7.08	6.80	6.58	6.40	6.13	5.85	5.56	5.41	5.25	5.10	4.94	4.60	
15	.05	4.54	3.68	3.29	3.06	2.90	2.79	2.71	2.64	2.59	2.54	2.48	2.40	2.33	2.29	2.25	2.20	2.16	2.07	15
	.01	8.68	6.36	5.42	4.89	4.56	4.32	4.14	4.00	3.89	3.80	3.67	3.52	3.37	3.29	3.21	3.13	3.05	2.87	
	.001	16.6	11.3	9.34	8.25	7.57	7.09	6.74	6.47	6.26	6.08	5.81	5.54	5.25	5.10	4.95	4.80	4.64	4.31	
16	.05	4.49	3.63	3.24	3.01	2.85	2.74	2.66	2.59	2.54	2.49	2.42	2.35	2.28	2.24	2.19	2.15	2.11	2.01	16
	.01	8.53	6.23	5.29	4.77	4.44	4.20	4.03	3.89	3.78	3.69	3.55	3.41	3.26	3.18	3.10	3.02	2.93	2.75	
	.001	16.1	11.0	9.01	7.94	7.27	6.80	6.46	62.0	5.98	5.81	5.55	5.27	4.99	4.85	4.70	4.54	4.39	4.06	
17	.05	4.45	3.59	3.20	2.96	2.81	2.70	2.61	2.55	2.49	2.45	2.38	2.31	2.23	2.19	2.15	2.10	2.06	1.96	17
	.01	8.40	6.11	5.18	4.67	4.34	4.10	3.93	3.79	3.68	3.59	3.46	3.31	3.16	3.08	3.00	2.92	2.83	2.65	
	.001	15.7	10.7	8.73	7.68	7.02	6.56	6.22	5.96	5.75	5.58	5.32	5.05	4.78	4.63	4.48	4.33	4.18	3.85	
18	.05	4.41	3.55	3.16	2.93	2.77	2.66	2.58	2.51	2.46	2.41	2.34	2.27	2.19	2.15	2.11	2.06	2.02	1.92	18
	.01	8.29	6.01	5.09	4.58	4.25	4.01	3.84	3.71	3.60	3.51	3.37	3.23	3.08	3.00	2.92	2.84	2.75	2.57	
	.001	15.4	10.4	8.49	7.46	6.81	6.35	6.02	5.76	5.56	5.39	5.13	4.87	4.59	4.45	4.30	4.15	4.00	3.67	

Critical Values of F (cont.)

The first F ratio is the critical value for $\alpha = .05$, the second for $\alpha = .01$, and the third for $\alpha = .001$.

df for denominator	α	*df* for numerator 1	2	3	4	5	6	7	8	9	10	12	15	20	24	30	40	60	∞	*df* for denominator
19	.05	4.38	3.52	3.13	2.90	2.74	2.63	2.54	2.48	2.42	2.38	2.31	2.23	2.16	2.11	2.07	2.03	1.98	1.88	19
	.01	8.18	5.93	5.01	4.50	4.17	3.94	3.77	3.63	3.52	3.43	3.30	3.15	3.00	2.92	2.84	2.76	2.67	2.49	
	.001	15.1	10.2	8.28	7.27	6.62	6.18	5.85	5.59	5.39	5.22	4.97	4.70	4.43	4.29	4.14	3.99	3.84	3.51	
20	.05	4.35	3.49	3.10	2.87	2.71	2.60	2.51	2.45	2.39	2.35	2.28	2.20	2.12	2.08	2.04	1.99	1.95	1.84	20
	.01	8.10	5.85	4.94	4.43	4.10	3.87	3.70	3.56	3.46	3.37	3.23	3.09	2.94	2.86	2.78	2.69	2.61	2.42	
	.001	14.8	9.95	8.10	7.10	6.46	6.02	5.69	5.44	5.24	5.08	4.82	4.56	4.29	4.15	4.00	3.86	3.70	3.38	
21	.05	4.32	3.47	3.07	2.84	2.68	2.57	2.49	2.42	2.37	2.32	2.25	2.18	2.10	2.05	2.01	1.96	1.92	1.81	21
	.01	8.02	5.78	4.87	4.37	4.04	3.81	3.64	3.51	3.40	3.31	3.17	3.03	2.88	2.80	2.72	2.64	2.55	2.36	
	.001	14.6	9.77	7.94	6.95	6.32	5.88	5.56	5.31	5.11	4.95	4.70	4.44	4.17	4.03	3.88	3.74	3.58	3.26	
22	.05	4.30	3.44	3.05	2.82	2.66	2.55	2.46	2.40	2.34	2.30	2.23	2.15	2.07	2.03	1.98	1.94	1.89	1.78	22
	.01	7.95	5.72	4.82	4.31	3.99	3.76	3.59	3.45	3.35	3.26	3.12	2.98	2.83	2.75	2.67	2.58	2.50	2.31	
	.001	14.4	9.61	7.80	6.81	6.19	5.76	5.44	5.19	4.99	4.83	4.58	4.33	4.06	3.92	3.78	3.63	3.48	3.15	
23	.05	4.28	3.42	3.03	2.80	2.64	2.53	2.44	2.37	2.32	2.27	2.20	2.13	2.05	2.01	1.96	1.91	1.86	1.76	23
	.01	7.88	5.66	4.76	4.26	3.94	3.71	3.54	3.41	3.30	3.21	3.07	2.93	2.78	2.70	2.62	2.54	2.45	2.26	
	.001	14.2	9.47	7.67	6.70	6.08	5.65	5.33	5.09	4.89	4.73	4.48	4.23	3.96	3.82	3.68	3.53	3.38	3.05	
24	.05	4.26	3.40	3.01	2.78	2.62	2.51	2.42	2.36	2.30	2.25	2.18	2.11	2.03	1.98	1.94	1.89	1.84	1.73	24
	.01	7.82	5.61	4.72	4.22	3.90	3.67	3.50	3.36	3.26	3.17	3.03	2.89	2.74	2.66	2.58	2.49	2.40	2.21	
	.001	14.0	9.34	7.55	6.59	5.98	5.55	5.23	4.99	4.80	4.64	4.39	4.14	3.87	3.74	3.59	3.45	3.29	2.97	
25	.05	4.24	3.39	2.99	2.76	2.60	2.49	2.40	2.34	2.28	2.24	2.16	2.09	2.01	1.96	1.92	1.87	1.82	1.71	25
	.01	7.77	5.57	4.68	4.18	3.86	3.63	3.46	3.32	3.22	3.13	2.99	2.85	2.70	2.62	2.54	2.45	2.36	2.17	
	.001	13.9	9.22	7.45	6.49	5.89	5.46	5.15	4.91	4.71	4.56	4.31	4.06	3.79	3.66	3.52	3.37	3.22	2.89	
26	.05	4.23	3.37	2.98	2.74	2.59	2.47	2.39	2.32	2.27	2.22	2.15	2.07	1.99	1.95	1.90	1.85	1.80	1.69	26
	.01	7.72	5.53	4.64	4.14	3.82	3.59	3.42	3.29	3.18	3.09	2.96	2.82	2.66	2.58	2.50	2.42	2.33	2.13	
	.001	13.7	9.12	7.36	6.41	5.80	5.38	5.07	4.83	4.64	4.48	4.24	3.99	3.72	3.59	3.44	3.30	3.15	2.82	

df for denominator	α	*df* for numerator: 1	2	3	4	5	6	7	8	9	10	12	15	20	24	30	40	60	∞	*df* for denominator
27	.05	4.21	3.35	2.96	2.73	2.57	2.46	2.37	2.31	2.25	2.20	2.13	2.06	1.97	1.93	1.88	1.84	1.79	1.67	27
	.01	7.68	5.49	4.60	4.11	3.78	3.56	3.39	3.26	3.15	3.06	2.93	2.78	2.63	2.55	2.47	2.38	2.29	2.10	
	.001	13.6	9.02	7.27	6.33	5.73	5.31	5.00	4.76	4.57	4.41	4.17	3.92	3.66	3.52	3.38	3.23	3.08	2.75	
28	.05	4.20	3.34	2.95	2.71	2.56	2.45	2.36	2.29	2.24	2.19	2.12	2.04	1.96	1.91	1.87	1.82	1.77	1.65	28
	.01	7.64	5.45	4.57	4.07	3.75	3.53	3.36	3.23	3.12	3.03	2.90	2.75	2.60	2.52	2.44	2.35	2.26	2.06	
	.001	13.5	8.93	7.19	6.25	5.66	5.24	4.93	4.69	4.50	4.35	4.11	3.86	3.60	3.46	3.32	3.18	3.02	2.69	
29	.05	4.18	3.33	2.93	2.70	2.55	2.43	2.35	2.28	2.22	2.18	2.10	2.03	1.94	1.90	1.85	1.81	1.75	1.64	29
	.01	7.60	5.42	4.54	4.04	3.73	3.50	3.33	3.20	3.09	3.00	2.87	2.73	2.57	2.49	2.41	2.33	2.23	2.03	
	.001	13.4	8.85	7.12	6.19	5.59	5.18	4.87	4.64	4.45	4.29	4.05	3.80	3.54	3.41	3.27	3.12	2.97	2.64	
30	.05	4.17	3.32	2.92	2.69	2.53	2.42	2.33	2.27	2.21	2.16	2.09	2.01	1.93	1.89	1.84	1.79	1.74	1.62	30
	.01	7.56	5.39	4.51	4.02	3.70	3.47	3.30	3.17	3.07	2.98	2.84	2.70	2.55	2.47	2.39	2.30	2.21	2.01	
	.001	13.3	8.77	7.05	6.12	5.53	5.12	4.82	4.58	4.39	4.24	4.00	3.75	3.49	3.36	3.22	3.07	2.92	2.59	
40	.05	4.08	3.23	2.84	2.61	2.45	2.34	2.25	2.18	2.12	2.08	2.00	1.92	1.84	1.79	1.74	1.69	1.64	1.51	40
	.01	7.31	5.18	4.31	3.83	3.51	3.29	3.12	2.99	2.89	2.80	2.66	2.52	2.37	2.29	2.20	2.11	2.02	1.80	
	.001	12.6	8.25	6.59	5.70	5.13	4.74	4.44	4.21	4.02	3.87	3.64	3.40	3.15	3.01	2.87	2.73	2.57	2.23	
60	.05	4.00	3.15	2.76	2.53	2.37	2.25	2.18	2.10	2.04	1.99	1.92	1.84	1.75	1.70	1.65	1.59	1.53	1.39	60
	.01	7.08	4.98	4.13	3.65	3.34	3.12	2.95	2.82	2.72	2.63	2.50	2.35	2.20	2.12	2.03	1.94	1.84	1.60	
	.001	12.0	7.76	6.17	5.31	4.76	4.37	4.09	3.86	3.69	3.54	3.32	3.08	2.83	2.69	2.55	2.41	2.25	1.89	
120	.05	3.92	3.07	2.68	2.45	2.29	2.18	2.09	2.02	1.96	1.91	1.83	1.75	1.66	1.61	1.55	1.50	1.43	1.25	120
	.01	6.85	4.79	3.95	3.48	3.17	2.96	2.79	2.66	2.56	2.47	2.34	2.19	2.03	1.95	1.86	1.76	1.66	1.38	
	.001	11.4	7.32	5.78	4.95	4.42	4.04	3.77	3.55	3.38	3.24	3.02	2.78	2.53	2.40	2.26	2.11	1.95	1.54	
∞	.05	3.84	3.00	2.60	2.37	2.21	2.10	2.01	1.94	1.88	1.83	1.75	1.67	1.57	1.52	1.46	1.39	1.32	1.00	∞
	.01	6.63	4.61	3.78	3.32	3.02	2.80	2.64	2.51	2.41	2.32	2.18	2.04	1.88	1.79	1.70	1.59	1.47	1.00	
	.001	10.8	6.91	5.42	4.62	4.10	3.74	3.47	3.27	3.10	2.96	2.74	2.51	2.27	2.13	1.99	1.84	1.66	1.00	

Index